VOLUME EIGHTY SEVEN

Advances in
FOOD AND NUTRITION
RESEARCH

VOLUME EIGHTY SEVEN

ADVANCES IN
FOOD AND NUTRITION
RESEARCH

Edited by

FIDEL TOLDRÁ

Department of Food Science
Instituto de Agroquimica y Tecnología de Alimentos (CSIC)
Valencia, Spain

ACADEMIC PRESS

An imprint of Elsevier

ELSEVIER

Academic Press is an imprint of Elsevier
50 Hampshire Street, 5th Floor, Cambridge, MA 02139, United States
525 B Street, Suite 1650, San Diego, CA 92101, United States
The Boulevard, Langford Lane, Kidlington, Oxford OX5 1GB, United Kingdom
125 London Wall, London, EC2Y 5AS, United Kingdom

First edition 2019

Notices

Knowledge and best practice in this field are constantly changing. As new research and
experience broaden our understanding, changes in research methods, professional practices,
or medical treatment may become necessary.

Practitioners and researchers must always rely on their own experience and knowledge in
evaluating and using any information, methods, compounds, or experiments described
herein. In using such information or methods they should be mindful of their own safety and
the safety of others, including parties for whom they have a professional responsibility.

To the fullest extent of the law, neither the Publisher nor the authors, contributors, or editors,
assume any liability for any injury and/or damage to persons or property as a matter of
products liability, negligence or otherwise, or from any use or operation of any methods,
products, instructions, or ideas contained in the material herein.

ISBN: 978-0-12-816049-7
ISSN: 1043-4526

For information on all Academic Press publications
visit our website at https://www.elsevier.com/books-and-journals

Working together
to grow libraries in
developing countries

www.elsevier.com • www.bookaid.org

Publisher: Zoe Kruze
Acquisition Editor: Sam Mahfoudh
Editorial Project Manager: Shellie Bryant
Production Project Manager: Denny Mansingh
Cover Designer: Victoria Pearson

Typeset by SPi Global, India

CONTENTS

CONTRIBUTORS

José Miguel Aguilera
Department of Chemical and Bioprocess Engineering, Pontificia Universidad Católica de Chile, Santiago, Chile

Alaa El-din A. Bekhit
Department of Food Science, University of Otago, Dunedin, New Zealand

Franck Carbonero
Department of Food Science and Center for Human Nutrition, University of Arkansas, Fayetteville, AR, United States

Alan Carne
Department of Biochemistry, University of Otago, Dunedin, New Zealand

Ruizeng Gu
Beijing Engineering Research Center of Protein & Functional Peptides, China National Research Institute of Food and Fermentation Industries, Beijing, PR China

Yakun Hou
Department of Food Science, University of Otago, Dunedin, New Zealand

Fuhuai Jia
Ningbo Yu Fang Tang Biological Science and Technology Co., Ltd., Ningbo, Zhejiang, PR China

Bum-Keun Kim
Division of Strategic Food Research, Korea Food Research Institute, Seoul, South Korea

Laura Lavefve
Department of Food Science and Center for Human Nutrition, University of Arkansas, Fayetteville, AR, United States; Direction des Etudes Et Prestations (DEEP), Institut Polytechnique UniLaSalle, Beauvais, France

Frédéric Leroy
Research Group of Industrial Microbiology and Food Biotechnology (IMDO), Faculty of Sciences and Bioengineering Sciences, Vrije Universiteit Brussel, Brussels, Belgium

Guoming Li
Beijing Engineering Research Center of Protein & Functional Peptides, China National Research Institute of Food and Fermentation Industries, Beijing, PR China

Wenying Liu
Beijing Engineering Research Center of Protein & Functional Peptides, China National Research Institute of Food and Fermentation Industries, Beijing, PR China

Jun Lu
Beijing Engineering Research Center of Protein & Functional Peptides, China National Research Institute of Food and Fermentation Industries, Beijing, PR China

Yong Ma
Beijing Engineering Research Center of Protein & Functional Peptides, China National
Research Institute of Food and Fermentation Industries, Beijing, PR China

Daya Marasini
Department of Food Science and Center for Human Nutrition, University of Arkansas,
Fayetteville, AR, United States

Michelle McConnell
Department of Microbiology, University of Otago, Dunedin, New Zealand

Sarah O'Connor
CHU de Québec Research Center; Department of Kinesiology, Faculty of Medicine,
Université Laval, Québec, QC, Canada

Dong June Park
Division of Strategic Food Research, Korea Food Research Institute, Seoul, South Korea

Sandra Rodrigues
Mountain Research Centre (CIMO), Escola Superior Agrária/Instituto Politécnico de
Bragança, Bragança, Portugal

Iwona Rudkowska
CHU de Québec Research Center; Department of Kinesiology, Faculty of Medicine,
Université Laval, Québec, QC, Canada

Amin Shavandi
Department of Food Science, University of Otago, Dunedin, New Zealand

Severiano Silva
Veterinary and Animal Research Centre (CECAV), Universidade Trás-os-Montes e Alto
Douro, Vila Real, Portugal

Wen-Hao Su
Food Refrigeration and Computerised Food Technology (FRCFT), School of Biosystems
and Food Engineering, Agriculture & Food Science Centre, University College Dublin
(UCD), National University of Ireland, Dublin, Ireland

Da-Wen Sun
Food Refrigeration and Computerised Food Technology (FRCFT), School of Biosystems
and Food Engineering, Agriculture & Food Science Centre, University College Dublin
(UCD), National University of Ireland, Dublin, Ireland

Alfredo Teixeira
Mountain Research Centre (CIMO), Escola Superior Agrária/Instituto Politécnico de
Bragança, Bragança, Portugal

Yuchen Wang
Beijing Engineering Research Center of Protein & Functional Peptides, China National
Research Institute of Food and Fermentation Industries, Beijing, PR China

Yuqing Wang
Beijing Engineering Research Center of Protein & Functional Peptides, China National
Research Institute of Food and Fermentation Industries, Beijing, PR China

PREFACE

This Volume 87 of *Advances in Food and Nutrition Research* is compiling eight chapters reporting the latest developments in a wide variety of relevant topics such as bioactive peptides from corn gluten meal and their health effects, dietary fatty acids and the metabolic syndrome risks, plant-based fermented foods and their impact on human health, production of functional and health compounds from marine wastes, analysis of roots and tubers by hyperspectral techniques, latest advances in sheep and goat meat products, the scenario for particular alimentations, and biosocial complexities of meat consumption.

The first chapter describes the in vitro and in vivo functions of peptides from corn gluten meal like antihypertensive, hepatoprotective, antiobesity, antimicrobial, antioxidative, mineral-binding, and accelerating alcohol metabolism. Such identified bioactive peptides may constitute relevant active ingredients for the prevention or treatment of such diseases and, based on their health benefit effects, they could be used in the formulation of functional foods, nutraceuticals, and natural drugs. The second chapter is bringing new insights on the contribution of dietary fatty acids to the evolution or prevention of the metabolic syndrome, but also on the identification of many gene variants showing relevant gene–diet interactions with dietary fatty acids intake that affect the metabolic response. The challenges for the implementation of personalized nutrition and its integration in public policies are also discussed. The third chapter deals with less-studied plant-based foods like fermented vegetables (kimchi and sauerkraut) and plant-based fermented drinks (kombucha). The potential mechanisms involved in health benefits are discussed especially their impact on the gut microbiota. The fourth chapter is revising the use of marine wastes like fishbone, skin, oil, internal organs, and collagen, as sources of functional and bioactive ingredients for foods or bioactive health substances for pharmaceutical applications. It describes how marine wastes may be processed for the extraction of added-value substances like lipids, bioactive peptides, enzymes, calcium supplement, gelatin, pigments, chitin, and chitosan. The fifth chapter is focusing on the use of hyperspectral techniques, majorly spectroscopy and hyperspectral imaging, as analytical tools for the rapid information about external or internal defects including sprout, bruise, and hollow heart, and identifying different grades of food quality attributes of roots and tubers (such as potato, sweet potato, cassava, yam, taro, and sugar beet), also for the rapid determination of physical properties and chemical constituents. The sixth chapter reviews the recent advances in sheep and goat meat products research, especially focusing on the use of imaging and spectroscopic methods (real-time ultrasonography, computed tomography, computer vision, near-infrared spectroscopy, hyperspectral imaging, Raman spectroscopy) for predicting body composition, carcass, and meat quality. Future trends in sheep and goat meat products research are also presented and discussed. The seventh chapter deals with the particular alimentations of the gourmet and the frail elderly. So, it focuses on the gastronomy and pleasure of eating and the role of modern chefs but also on the growing food and nutrition requirements of the aging population, discussing also the availability of nutritious and enjoyable products to attend such demands. The eighth and last chapter deals with the biosocial complexities of meat consumption. It outlines the historical and biosocial need for meat and analyzes how its transformative effects have contributed to a polarized discourse on diet and well-being in academia and society at large, explaining the reasons behind the endless debates between the advocates and adversaries of meat eating.

This volume is presenting the combined efforts of 27 professionals from 10 countries (China, Canada, United States, France, New Zealand, Ireland, Portugal, Chile, South Korea, and Belgium) with diverse expertise and background. The Editor wishes to thank the production staff and all the contributors for sharing their experience and for making this book possible.

Fidel Toldrá

Editor

CHAPTER ONE

Functions and Applications of Bioactive Peptides From Corn Gluten Meal

Guoming Li*, Wenying Liu*, Yuqing Wang*, Fuhuai Jia[†],
Yuchen Wang*, Yong Ma*, Ruizeng Gu*, Jun Lu*,[1]

*Beijing Engineering Research Center of Protein & Functional Peptides, China National Research Institute of Food and Fermentation Industries, Beijing, PR China
[†]Ningbo Yu Fang Tang Biological Science and Technology Co., Ltd., Ningbo, Zhejiang, PR China
[1]Corresponding author: e-mail address: johnljsmith@163.com

Contents

Abstract

Corn protein has been identified as an important source of bioactive peptides. Such peptides can be released during hydrolysis induced by proteolytic enzymes or microbial fermentation. Corn peptides have been found to exhibit different functions *in vitro* and *in vivo* such as antihypertensive, hepatoprotective, anti-obesity,

1

antimicrobial, antioxidative, mineral-binding and accelerating alcohol metabolism. To date, 22 sequences of bioactive corn peptides have already been identified. There is an increasing commercial interest in the production of corn peptides with the purpose of using them as active ingredients, which may find use in the treatment of liver injury, hypertension, dental carries, oxidative stress, mineral malabsorption and obesity. These bioactive peptides may be used in formulation of functional foods, nutraceuticals, and natural drugs because of their health benefit effects.

1. INTRODUCTION

1.1 Corn Gluten Meal and Corn Protein

Proteins are an essential component of the human diet and serve as a key source of the amino acids necessary to synthesize functional and structural proteins. Unfortunately, despite estimates that the world has a shortage of millions of tons of protein every year, numerous protein resources remain underutilized. A typical example of this is corn gluten meal (CGM). Corn is the only cereal crop indigenous to the Americas and one of the most important food and industrial crops in the United States. It is a warm-season crop that requires higher growing temperatures than other grains. The global annual production of corn is approximately 560 million metric tons, of which the United States alone accounts for about half. Yellow dent corn is the major variety grown for animal feed, food ingredients, and industrial products (Shukla & Cheryan, 2001). CGM is a major byproduct of the wet milling of corn and contains at least 60% total protein. >840,000 tons of CGM are produced annually in China, and 9% of the total output of corn is processed by wet milling (Wu et al., 2014). Traditionally, CGM has been predominantly used as animal feed, despite its very low bioavailability *in vivo*, or otherwise discarded, owing to its low water solubility and imbalanced amino acid profile (Li, Han, & Chen, 2008). On a dry basis, CGM contains 67–71% (w/w) of proteins, 21–26% (w/w) of carbohydrates including 12–15% (w/w) starch, 3–7% (w/w) of fat, 1–2% (w/w) of fiber, and 1–2% (w/w) of ash (He et al., 2018).

Corn protein has a low aqueous solubility and is deficient in certain essential amino acids such as Lys and Trp, which greatly limits its application in the food industry, and consequently, it is mostly utilized as animal feed. Modification of the proteins in CGM to permit its utilization in the food industry would increase its market value and range of applications (Wang et al., 2014). The major protein fractions of CGM are zein and glutelin, which account for 68% and 28%, respectively, of the total protein weight. The various fractions have been referred to as glutelin-1, alcohol-soluble reduced glutelin (ASG), zein-2, zein-like, g-zein, C-zein, D-zein, and reduced soluble protein (Esen, 1987; Paulis, James, & Wall, 1969; Paulis & Wall, 1977; Wilson, 1985). These fractions contain varying amounts of sulfur amino acids, such as cystine and Met (Sodek & Wilson, 1971). Zein is the major storage protein in corn, and it is rich in Glu/Gln, Ala, Leu, and Pro and deficient in Lys and Trp (Lin et al., 2011). Zein of biochemical purity used in typical experimental studies is mainly composed of two distinct bands when analyzed by SDS-PAGE, with molecular weights of 23 and 21 kDa, alongside minor bands at 13 and 9.6 kDa. Zein contains a high proportion of non-polar and hydrophobic amino acid residues buried inside the protein structure, which is responsible for its poor aqueous solubility. Its molecular structure consists of a helical wheel conformation in which nine homologous repeating units are arranged in an antiparallel manner stabilized by hydrogen bonds (Liu, Sun, Wang, Zhang, & Wang, 2005).

1.2 Development of Corn Peptides

In recent studies, various proteins from low-value commercial sources such as rotifer cultures (Lee, Hong, Jeon, Kim, & Byun, 2009), fish byproducts (Bougatef et al., 2008), algae waste (Sheih, Wu, & Fang, 2009), and poultry industry residue (Rossi, Flores, Heck, & Ayub, 2009) were subjected to enzymatic treatments to obtain protein hydrolysates. These hydrolysates can be effective for improving the functional properties of proteins. Small peptides possess some advantages compared to proteins, such as low osmotic pressure, fast absorption, good taste, and low antigenicity. Moreover, besides having preserved and usually enhanced biological value, these hydrolysates can be characterized by functional properties superior to the original proteins. Furthermore, the hydrolysates are a source of bioactive peptides (Lin et al., 2011).

Bioactive peptides are small fragments of proteins that can be beneficial for health (Moller, Scholz-Ahrens, Roos, & Schrezenmeir, 2008). In the past few years, researchers have discovered that certain small peptides derived from the hydrolysis of food proteins can play important roles in regulating the autonomic nervous system, activating cellular immunity, ameliorating cardiovascular dysfunction, and reducing oxidative stress and the effects of aging. These peptides may exhibit various biological effects, such as antioxidative (Cumby, Zhong, Naczk, & Shahidi, 2008), antihypertensive (Wu, Aluko, & Muir, 2009), immunomodulatory (Yang et al., 2009), antimicrobial (Liu et al., 2008), and anticancer (Kannan, Hettiarachchy, Johnson, & Nannapaneni, 2008) activities. Furthermore, it has been reported that dipeptides and tripeptides are more efficiently absorbed into the body from the intestinal tract than free amino acids (Arai, 1978). Functional peptides can be used in foods as additives or texture enhancers and as active pharmaceutical ingredients.

It has been reported that the bioavailability of CGM can be improved significantly by enzymatic hydrolysis, because the hydrolysates contain many small peptides, especially dipeptides and tripeptides, which can be absorbed more efficiently than either the intact proteins or the free amino acids. Moreover, the ingestion of amino acid residues from the partial enzymatic hydrolysis of whole protein has beneficial effects on protein synthesis. The development of methods for hydrolyzing CGM to generate peptides that could be used as foods, food additives, or active pharmaceutical ingredients would increase the market value and range of applications of this industrial byproduct. Many reports on the preparation of bioactive peptides from CGM, such as antioxidant peptides, hepatoprotective peptides, and antihypertensive peptides (Li, Wen, Li, Zhang, & Lin, 2011). Owing to its particular amino acid composition, which is rich in hydrophobic amino acids such as Leu, Ala, and Phe, CGM proteins are considered a good source of bioactive peptides (Li et al., 2008). Over recent years, corn peptides, a novel food prepared from CGM by enzymatic hydrolysis or microbial fermentation, have attracted considerable interest owing to their various bioactive properties, including antioxidant activity (Guo, Sun, He, Yu, & Du, 2009; Li, Guo, Hu, Xu, & Zhang, 2007; Li et al., 2008; Miao et al., 2010; Yu, Lv, He, Huang, & Han, 2012; Zhang, Zhang, & Li, 2012; Zhuang, Tang, Dong, Sun, & Liu, 2013), improvements in lipid profiles, and ability to accelerate alcohol metabolism and protect against alcohol-induced liver injury (Guo et al., 2009; Li et al., 2007; Ma, Zhang, Yu, He, & Zhang, 2012; Yu, Li, He, Huang, & Zhang, 2013). Further research and development of corn peptides might effectively increase the market value of CGM (Wu et al., 2014).

There have been numerous reports on the production of corn peptides. Miyoshi et al. (1991) used thermolysin to prepare corn peptides from α-zein, a major component of maize endosperm protein, and isolated several peptides with inhibitory activity toward angiotensin-converting enzyme (ACE). Kim, Whang, Kim, Koh, and Suh (2004) reported the preparation of corn peptides with inhibitory activity toward ACE along with the solubility and moisture sorption of the hydrolysate and the influence of prior heat treatment on the hydrolysis. Kim, Whang, and Suh (2004) also described a method for improving the ACE inhibitory activity and emulsifying and foaming properties using ultrafiltration membranes. Lin et al. (2011) developed a pilot-scale production process to enhance the value of proteins obtained from CGM. Their results demonstrated that the pilot-scale production of corn peptides represents a practical way to utilize CGM. Based on the antihypertensive activity of orally administered corn peptides in spontaneously hypertensive rats, corn peptides may be promising as antihypertensive drugs and functional foods.

2. PREPARATION OF CORN PEPTIDES

2.1 Preparation Using Proteolytic Enzymes

Enzymes are specific biocatalysts produced by living cells. Both single-enzyme and more complex systems involving multiple enzymes have been applied to the production of corn peptides. The single-enzyme hydrolysis methods utilize individual enzymes to hydrolyze zein into bioactive corn peptides. The commonly used enzymes can be classified into three types, namely, acidic proteases, alkaline proteases, and neutral proteases. The multiple-enzyme methods rely on the combined action of several proteases, and the synergistic effects and optimum ratio of the various enzymes should be considered, along with the optimal reaction conditions for each enzyme. The multiple-enzyme approach is faster, more convenient, and affords a higher yield than the single-enzyme method.

The potential applications of corn protein are limited owing to its poor aqueous solubility, which can be improved by enzymatic hydrolysis. The solubility of CGM can be increased by acid hydrolysis, alkaline hydrolysis, or enzymatic hydrolysis. Among these methods, enzymatic hydrolysis is regarded as the most appropriate method because of its controllable and mild reaction conditions, large-scale commercial availability, and high product quality. Nevertheless, the traditional enzymatic hydrolysis of proteins has many disadvantages, such as the low utilization rate of the enzyme, low conversion rate of the substrate, long reaction time, and high energy consumption. These drawbacks are largely attributable to the sterically hindered conformation of the substrate protein, which renders the peptide bonds inaccessible to hydrolysis by the protease. Therefore, the development of more efficient enzymatic hydrolysis methods would be of great value.

Several studies concerning the preparation of corn peptides using proteolytic enzymes have been reported. For example, in the study of Lin et al. (2011), wet corn protein isolate obtained from the heat treatment of CGM was re-suspended in distilled water to a protein concentration of 6% (w/w) in the same reactor. In the first step, enzymatic hydrolysis was performed using crude alkaline proteases from *Bacillus licheniformis* (enzyme:substrate ratio = 1:100) at 55 °C and pH 8.5 for 3 h. The resulting mixture was subjected to a second enzymatic hydrolysis step using crude neutral proteases from *Bacillus subtilis* (enzyme:substrate ratio = 1:100) at 45 °C and pH 7.0 for 2 h. During the hydrolysis, the optimal pH and temperature were maintained using NaOH and a circulating water bath, respectively. The hydrolysis was terminated by increasing the temperature to 90 °C for 15 min to inactivate the proteases. The obtained corn peptides displayed low molecular weights, with 96.77% of the peptides below 1000 Da. Compared with the one-step enzymatic hydrolysis methods applied in other studies (Lee et al., 2009; Wu et al., 2009), the two-step hydrolysis promoted the release of shorter oligopeptides from the corn proteins and afforded a higher yield.

Lu, Chen, and Tang (2000) used trypsin to obtain corn peptides with different degrees of hydrolysis and examined the influence of hydrolysis on the functional properties. Suh, Whang, Kim, Bae, and Noh (2003) prepared corn peptides exhibiting powerful ACE inhibitory activity via the hydrolysis of corn gluten using six commercial proteases, and the peptides obtained using Flavourzyme displayed the highest ACE inhibitory activity using ultrafiltration membranes Li et al. (2011) described the preparation of corn peptides by the proteolysis of CGM using alkaline protease. Yamaguchi, Takada, Nozaki, Ito, and Furukawa (1996) also reported the production of corn peptides by proteolysis using alkaline protease from alkaliphilic *Bacillus* A-7. Wu et al. (2014) prepared corn peptides from CGM by proteolysis using the alkaline protease Alcalase.

Jin, Liu, Zheng, Wang, and He (2016) reported the hydrolysis of CGM using Alcalase, Flavourzyme, Alcalase followed by Flavourzyme, and vice versa. The results revealed that the sequential action of the two proteases led to more effective proteolysis. At a substrate concentration of 10%, Alcalase afforded a degree of hydrolysis of 17.83%, which was higher than that obtained using Flavourzyme (3.65%). The hydrolysate obtained by sequential treatment with Alcalase followed by Flavourzyme exhibited improved antioxidant activities and was further purified.

The preparation of corn peptides by enzymatic hydrolysis leads to the product having a bitter taste, because the pH adjustments during the enzymatic hydrolysis introduce salts and the proteolysis causes the exposure of hydrophobic groups, which are responsible for bitterness. The degree of bitterness is related to the choice of enzyme, temperature, pH, and other factors. This makes the use of enzymatic hydrolysis of corn peptides subject to certain restrictions. However, the enzymatic hydrolysis of CGM also possesses certain advantages. In particular, the enzymatic hydrolysis method can improve the solubility and stability of proteins by improving the protein composition and functional properties without affecting its nutritional value.

The China National Research Institute of Food and Fermentation Industries Co., Ltd. has used enzyme mixtures to produce food-derived peptides such as corn peptides, wheat peptides, soy peptides, fish collagen peptides, and black chicken peptides, and optimized the amounts and hydrolysis conditions for two enzyme preparations. Industrial-scale production has been achieved, in which over 85% of the final peptide products possessed a molecular weight of < 1000 Da.

2.2 Preparation by Microbial Fermentation

Microbial fermentation is a new method for producing peptides and has various advantages. The typical microorganisms used include mold, bacteria, and yeast. In this approach, the active microbial cells produce a rich protease system that can act directly on high-molecular-weight plant proteins and degrade them into smaller proteins and peptides. The advantages of this strategy lie in the diversity of microbial proteases, high levels of protease activity, short production cycles, and low production costs. LeBlanc, Matar, Valdéz, LeBlanc, and Perdigon (2002) isolated peptides from milk fermented with *Lactobacillus helveticus*. Owing to its high proteolytic activity, *L. helveticus* was able to release peptidic compounds during milk fermentation, as shown by the degree of proteolysis and size-exclusion HPLC elution profiles, and the release of bioactive peptides by lactic acid bacteria can have important implications for the modulation of the cellular immune response. Qian et al. (2011) prepared peptides from skim milk fermented with *Lactobacillus delbrueckii* ssp. *bulgaricus* LB340, and investigated their antioxidant, antihypertensive, and immunomodulatory activities. The results showed the potential of the milk fermented with *L. delbrueckii* ssp. *bulgaricus* LB340 as a functional food. Microbial fermentation can also lead to the production of flavor substances such as alcohol and lactic acid, which can directly mask bitterness. The prepared product has good organoleptic properties such as flavor and color, and this method is promising for further development.

The preparation of corn peptides by microbial fermentation relies on the enzymatic system of the microorganism to hydrolyze the CGM. As the molecular weight of the corn peptides prepared by microbial fermentation is relatively low, the digested mixture can be first centrifuged to remove high-molecular-weight, non-hydrolyzed proteins, and the ultra-molecular-weight proteins and peptides can be removed by ultrafiltration. Finally, the filtrate can be subjected to Sephadex G–25 gel-filtration chromatography to afford only low-molecular-weight corn peptides. The preparation of corn peptides by microbial fermentation reduces the production of bitterness, leading to a simpler and cheaper production process compared with enzymatic hydrolysis. The disadvantage is the low conversion rate.

In recent years, there have been some significant achievements in the preparation of corn peptides using microbial fermentation. The enzymes used to hydrolyze the CGM by microbial fermentation are converted into simple small peptides. This method can not only reduce the production cost but also shorten the corn peptides through the microbial metabolism itself. The peptide bonds between the peptides are transferred and recombined, so that the R groups of the hydrophobic amino acids at the two ends of the corn peptide are re-modified, the amino acid polarity is changed, the hydrophilicity of the corn peptide is increased, and the corn peptide becomes more water soluble. This process can also reduce the bitter taste of corn peptides. Therefore, screening for strains capable of secreting proteases into the environment would be very promising for the production of corn peptides by the fermentative hydrolysis of CGM.

Zhang, Huang, Zhu, and Song (2009) used B. subtilis Is–45 to ferment CGM, using 12% of inoculum, 40 mesh CGM, a substrate concentration of 5%, an initial pH of 8.0, a liquid volume of 100 mL per 500 mL flask, a culture temperature of 41 °C, and a shaking speed of 180 rpm, and the yield of corn peptides was reported to reach 82.7% after 63 h of fermentation. The relative molecular weights of the corn peptides obtained after isolation and purification were approximately 5128, 3715, and 1513 Da.

2.3 Chemical Composition and Physiochemical Properties of Corn Peptides

The chemical compositions and physiochemical properties of corn peptides are distinct from those of CGM, and the properties of proteins can be changed and improved through controlled hydrolysis. Lin et al. (2011) applied two–step enzymatic hydrolysis and multistage separation to generate corn peptides with low molecular weights, 96.77% of which were

<1000 Da, and analyzed the chemical compositions of CGM and the corn peptides. The CGM had a protein content of 63.83%, along with a high sugar content of 18.40%, a lipid content of 8.98%, and an ash content of 2.78% (on a dry basis). The protein composition of the corn peptides was as high as 91.61%, while the contents of ash, lipid, and sugar were 5.56%, 0.52%, and 0.82%, respectively. Amino acid composition analysis (Table 1) revealed that CGM and the corn peptides contained similar relative abundances of all amino acids. Therefore, the enzymatic hydrolysis did not significantly alter the amino acid composition of the corn proteins. This finding is consistent with the results of Li et al. (2008). The corn peptides

Table 1 Amino Acid Composition (Dry Basis) of Corn Peptides and CGM (%)[a]

Amino Acid	Corn Peptides	CGM
Alanine	9.68	9.22
Arginine	1.79	2.77
Aspartic acid[b]	5.28	5.55
Cysteine	0.36	1.09
Glutamic acid[c]	24.21	22.37
Glycine	1.61	2.26
Histidine	1.36	1.73
Isoleucine	3.49	3.35
Leucine	18.27	17.12
Lysine	0.25	1.42
Methionine	2.51	2.63
Phenylalanine	5.82	6.16
Proline	8.29	8.18
Serine	4.83	4.74
Threonine	2.84	2.93
Tryptophan	0.26	0.34
Tyrosine	5.51	4.22
Valine	3.63	3.93
Total	100	100

[a]Data correspond to the average and standard deviation of three independent experiments.
[b]Aspartic acid and asparagine.
[c]Glutamic acid and glutamine.

were rich in Glu/Gln, Leu, Ala, Pro, Phe, and Tyr, which accounted for 24.21%, 18.27%, 9.68%, 8.29%, 5.82%, and 5.51%, respectively.

Li et al. (2011) described the preparation of corn peptides from CGM by proteolysis with alkaline protease. The corn peptides, whose major fraction is an oligopeptide, did not contain any free amino acids. The amino acid composition of the corn peptides was similar to that of CGM, i.e., rich in Ala, branched-chain amino acids (Leu, Ile, and Val), and Pro, but poor in basic amino acids.

Corn peptides have a higher aqueous solubility than CGM and many desirable physicochemical properties. The structural properties of corn peptides are responsible for the improved solubility, which is mainly reflected in high solubility across a wide pH range, the formation of homogenous solutions with no precipitation, and floe phenomenon. In addition, corn peptides have good solubility even under extreme conditions such as low pH. This feature makes corn peptides widely used in the acidic beverage industry. Furthermore, corn peptide drinks have the advantages of low viscosity, good taste, high glutamate content, and strong brain function, making corn peptide beverages popular.

Corn peptides have good thermal stability. Previous studies have shown that after subjecting corn peptides to temperatures of 80–100 °C in a water bath for about 2 h, the components remained essentially unchanged and the ratio of the various molecular weight intervals did not exceed 2%. Previous experiments demonstrated that corn peptides are stable to heat and have little change in function.

Aqueous solutions of corn peptides exhibit a low viscosity, as reflected by the fact that the viscosity of these solutions is not substantially affected by the concentration. Under normal temperature conditions, even at a high concentration of 50%, the viscosity is only 9 mPa·s, and the fluidity is also good. The viscosity of corn peptides is also not substantially influenced by temperature.

Corn peptides have low molecular weight with interesting structures and functions and have a high nutritional value. The proportions of eight of the essential amino acids required by the human body are very high. Corn peptides can be obtained by enzymatic hydrolysis using *in vitro* digestion. The index increased several times. In addition, corn peptides are actively absorbed by the body, which allows them to be easily digested and may reduce the burden on the gastrointestinal tract, thereby improving gastrointestinal dysfunction and similar conditions.

3. PHYSIOLOGICAL FUNCTIONS OF CORN PEPTIDES

3.1 Antihypertensive Function

In recent years, hypertension, which is associated with a high risk of cerebrovascular, cardiac, and renal complications, has become the most common yet serious chronic health problem worldwide (Huang, Sun, He, Dong, & Li, 2011). Lifestyle modifications and diet therapy are two of the most important tools in the prevention and treatment of hypertension (Hermansen, 2000; Huang et al., 2011). Although synthetic ACE inhibitors such as captopril, enalapril, and lisinopril have been developed and are effective for decreasing blood pressure, some undesirable side effects have been reported, such as coughing, dizziness, headache, abnormal (metallic or salty) taste, and kidney and liver problems (Huang et al., 2011). Therefore, natural ACE inhibitory peptides are a promising alternative for the treatment and prevention of hypertension (Yang, Tao, Liu, & Liu, 2007). An increasing number of studies have suggested that some food proteins have functions other than energetic and nutritional roles (Huang et al., 2011). In particular, biologically active peptides with antihypertensive effects have been studied extensively (Suh, Whang, & Lee, 1999).

ACE (dipeptidyl carboxypeptidase, EC 3.4.15.1) is a key enzyme in the renin–angiotensin system that plays a crucial role in regulating blood pressure and fluid and electrolyte balance (Martínez-Maqueda, Miralles, Recio, & Hernández-Ledesma, 2012; Meisel & Fitzgerald, 2000; Suh et al., 1999). ACE inhibition is the most common mechanism underlying blood-pressure-lowering effects, and ACE inhibitors are widely used in clinics to reduce the morbidity and mortality of patients with hypertension and related diseases (Huang et al., 2011). Research has indicated that CGM, with its high protein content, may be a good source for preparing ACE inhibitory peptides owing to its high proportions of hydrophobic amino acids and Pro (Huang et al., 2011; Kim, Whang, Kim, et al., 2004).

Based on the nature of their interaction with ACE, ACE inhibitory peptides can generally be classified into three categories: (1) true inhibitors, which have high affinity for ACE but are resistant to ACE-mediated cleavage and prevent the reaction between ACE and its endogenous substrates such as angiotensin I; (2) substrates for ACE, which are hydrolyzed into inactive or markedly less active fragments; and (3) "prodrugs," which are converted by ACE with the release of highly active fragments. Previous studies have demonstrated that the actual substrates, which showed apparent ACE inhibitory activities in a screening assay, are inactive after oral administration because they are hydrolyzed by ACE to inactive fragments (Fujita, Yokoyama, & Yoshikawa, 2000; Fujita & Yoshikawa, 1999; Yang et al., 2007). Several peptides with potent ACE inhibitory activity have been isolated from CGM (Huang et al., 2011; Miguel & Aleixandre, 2006). Yang et al. (2007) isolated an ACE inhibitory dipeptide, Ala–Tyr, which can be considered a true ACE inhibitor because its IC_{50} value was unchanged by preincubation with ACE. Ala–Tyr was found to exhibit *in vivo* antihypertensive activity after oral administration to spontaneously hypertensive rats. A maximum reduction of systolic blood pressure of 9.5 mmHg was observed 2 h after oral administration of 50 mg/kg of Ala–Tyr.

The stability of ACE inhibitory corn peptides to thermal treatment and a simulated gastrointestinal environment has also been investigated (Wang, Chen, et al., 2015; Wang, Ding, Wang, Zhang, & Liu, 2015). The results revealed that the corn peptides were relatively stable toward the gastrointestinal proteases pepsin and pancreatin. Huang et al. (2011) isolated fractions of corn peptides based on molecular weight ($M_w < 1$ kDa, $M_w < 3$ kDa, and $M_w < 5$ kDa). The $M_w < 3$ kDa peptide fraction exhibited the highest ACE inhibitory activity, followed by the $M_w < 1$ kDa fraction and then the $M_w < 5$ kDa fraction. In another study, the ACE inhibitory activity of corn peptides increased with decreasing molecular weight when three peptide fractions ($M_w < 4$ kDa, 4 kDa $< M_w < 6$ kDa, $M_w > 6$ kDa) were assayed at the same concentration of 1.0 mg mL^{-1} (Wang, Chen, et al., 2015; Wang, Ding, et al., 2015).

Furthermore, several studies of the primary structures of corn peptides with antihypertensive activity have been conducted recently. The common features of the antihypertensive peptides were found to include short sequences (2–12 amino acid residues), a high proportion of hydrophobic residues, and the presence of Pro at the C-terminus (Puchalska, Luisa Marina, & Concepción García, 2013).

An ACE inhibitory hexapeptide (Pro-Ser-Gly-Gln-Tyr-Tyr), which contains a hydrophobic amino acid at the N-terminus, a basic amino acid at the center, and a Tyr at the C-terminus, was also isolated from corn gluten hydrolysate (Huang et al., 2011; Suh et al., 1999; Yang et al., 2007). A dipeptide (Ala–Tyr) was isolated from the CGM hydrolysate and found to exhibit a high ACE inhibitory activity. Miyoshi et al. (1991) also separated a tripeptide (Leu-Arg-Pro) from α-zein hydrolysate, which displayed potent ACE inhibitory activity in spontaneously hypertensive rats (Huang et al., 2011).

The ACE inhibitory activities of a series of dipeptides demonstrated the presence of Trp, Tyr, Pro, or Phe at the C-terminus and a branched-chain aliphatic amino acid at the N-terminus afforded peptides capable of binding to ACE as a competitive inhibitor (Cheung, Wang, Ondetti, Sabo, & Cushman, 1980; Suh et al., 1999).

3.2 Antioxidative Function

Reactive oxygen species (ROS), such as superoxide and hydroxyl radicals and non-radical species, are mainly generated through normal physiological reactions in the human body (Jin et al., 2016; Wang, Ding, et al., 2015). However, when the production of free radicals exceeds the ability of the body to clear them, oxidative stress will occur, which can result in deleterious effects on cell membranes, lipid, proteins, and DNA in humans (Jin et al., 2016; Wang, Chen, et al., 2015; Wang, Ding, et al., 2015; Yongvanit, Pinlaor, & Bartsch, 2012). This can promote aging and initiate various diseases, such as cardiovascular disease, diabetes, cancer, rheumatoid arthritis, and neurological disorders (Lee, Koo, & Min, 2004; Wang, Chen, et al., 2015; Wang, Ding, et al., 2015). However, the antioxidant–prooxidant balance in the human body can vary with age and other factors, such as exposure to environmental pollutants, fatigue, excessive caloric intake, and high-fat diets. With increasing age, the plasma and cellular antioxidant potentials as well as the absorption of nutrients, including antioxidants, gradually diminish (Samaranayaka & Li-Chan, 2011). Although living organisms contain several intracellular biological antioxidant defense systems, these are often insufficient to counteract the excessive ROS present in the organism, especially under conditions of oxidative stress. Therefore, the dietary intake of exogenous antioxidants is also critical for maintaining the oxidation–reduction balance (Johansen et al., 2005; Wang, Chen, et al., 2015; Wang, Ding, et al., 2015).

Owing to the potential risks and toxicity associated with synthetic antioxidants such as butylated hydroxyanisole (BHA), butylated hydroxytoluene (BHT), and tert-butylhydroquinone (TBHQ), the use of these substances in food has been subject to strict limitations (Delgado-Zamarreño, González-Maza, Sánchez-Pérez, & Carabias Martínez, 2007; Jin et al., 2016; Li et al., 2010). Therefore, the antioxidants present in natural foods have attracted increasing interest as alternatives to synthetic antioxidants (Li et al., 2010; Wang, Chen, et al., 2015; Wang, Ding, et al., 2015). Peptide antioxidants are more attractive candidates than non-peptidic antioxidants as dietary ingredients for promoting human health because of their multiple functions, such as antihypertensive, opioid, and cholesterol-lowering activities (Dávalos, Miguel, Bartolomé, & López-Fandiño, 2004; Hernández-Ledesma, Miralles, Amigo, Ramos, & Recio, 2005; Samaranayaka & Li-Chan, 2011). In recent years, corn peptides obtained from CGM by enzymatic hydrolysis have attracted considerable interest as a novel food in view of their antioxidant activity and low cost (Li et al., 2010; Wu et al., 2014; Zhuang et al., 2013).

Antioxidant compounds can inhibit oxidation by various mechanisms, such as scavenging of free radicals or chelation of metal ions. Thus, there are numerous ways to evaluate the antioxidant activity of a particular compound and the use of multiple assays is preferable to obtain a comprehensive understanding (Huang, Majumder, & Wu, 2010; Jin et al., 2016). As the antioxidative mechanism measured and reaction conditions used may differ between assays, test compounds may show different results depending on the assay system. Specific assays to measure the antioxidative capacities of peptides or peptide mixtures have not yet been developed or standardized. Therefore, in the literature, combinations of assays that are commonly used to measure the antioxidative capacities of non-peptidic antioxidants have been applied to peptides as well (Samaranayaka & Li-Chan, 2011).

Owing to the complexity of the oxidative processes that occur in foods and biological systems as well as the distinct mechanisms through which different antioxidants may act, the selection of a single method suitable for evaluating the overall antioxidative potential of a food is not an easy task. There are several comprehensive reviews covering antioxidative mechanisms (Samaranayaka & Li-Chan, 2011). Because of the presence of multiple hydrogen atoms and electron-donating amino acid residues in a single peptide sequence, there is a high probability of some peptides exerting their activities through more than one reaction mechanism. Furthermore, *in vitro* studies using various chemical assays have indicated the potential of food-derived peptides to act as antioxidative agents and modulate a variety of oxidative processes in the human body (Samaranayaka & Li-Chan, 2011).

The antioxidant capacity of corn peptides has been studied through various chemical assays. A free-radical scavenging assay using 2,2-diphenyl-1-picrylhydrazyl (DPPH), a superoxide anion radical eliminating assay, a hydroxyl radical eliminating assay using rhodamine B, a linoleic acid autoxidation assay, and a hydrogen-donating activity assay using potassium ferricyanide were used (Li et al., 2010). The results revealed that the corn peptides included peptide components that (1) donated protons and reacted with free radicals to convert them into more stable products and terminate the radical chain reaction; (2) removed superoxide anions generated from pyrogallol autoxidation and exhibited dose-dependent scavenging of the superoxide anion; (3) served as a reducing agent to eliminate hydroxyl radicals before they attacked rhodamine B; (4) scavenged radicals; and (5) possessed reducing power (Li et al., 2010). The antioxidant efficacies of corn peptide fractions were also examined with respect to their scavenging activities toward 2,2'-azino-bis (3-ethylbenzothiazoline-6-sulfonic acid) (ABTS·), DPPH·, and superoxide anion (O_2·$^-$) radicals (Tang et al., 2010). The results demonstrated that the free-radical scavenging activity of the zein hydrolysate was related to both the molecular weight and the hydrophobicity of the peptides. In the study of Zhou et al., corn protein hydrolysates obtained using three microbial proteases were separated into several fractions by sequential ultrafiltration, and the 1–3 kDa fraction was found to exhibit the highest activity in scavenging peroxyl radicals (Wang, Chen, et al., 2015; Wang, Ding, et al., 2015; Zhou, Sun, & Canning, 2012). CGM was hydrolyzed using Protamex and it was found that the antioxidant activities of the corn peptides were highly correlated with smaller peptide length and higher content of antioxidative amino acids (Wang, Chen, et al., 2015; Wang, Ding, et al., 2015; Zhou et al., 2015). Corn peptide fractions were also investigated by Wang et al., and the results revealed that two particular corn peptide fractions ($M_w < 1$ kDa and 1 kDa $< M_w < 3$ kDa) exhibited good hydroxyl radical, superoxide anion radical, and ABTS· radical scavenging activity and oxygen radical absorbance capacity (Wang, Chen, et al., 2015; Wang, Ding, et al., 2015). An investigation into the cytoprotective effects of corn peptides on Caco-2 cells exposed to H_2O_2 indicated that they can not only scavenge the intracellular ROS directly but also activate signaling pathways related to the cellular defense system against oxidative stress (Hancock, Desikan, & Neill, 2001; Wang, Chen, et al., 2015; Wang, Ding, et al., 2015).

A myriad of interpretations have been put forth to explain the antioxidant properties of peptides. For example, the antioxidative potency of peptides containing Leu has been attributed to its long aliphatic side–chain group that is conceivably capable of interaction with acyl chains of susceptible fatty acids (Mendis, Rajapakse, & Kim, 2005). His exhibits strong radical scavenging activity due to the decomposition of its imidazole ring (Wade & Tucker, 1998). The radical scavenging activities of Tyr and Trp have been explained by the reactivities of their phenolic and indolyl moieties, respectively, which function as hydrogen donors (Pihlanto, 2006). Although it is not possible to establish precise structure–activity relationships for zein peptides on the basis of the present data, it is reasonable to speculate that hydrophobic amino acids present in the peptide sequences contributed greatly to the DPPH· radical scavenging activity, whereas hydrophilic amino acid residues were largely responsible for the ABTS· scavenging activity (Tang et al., 2010). Mendis et al. purified a radical scavenging peptide and suggested that the presence of His, Leu, Gly, and Pro played important roles in its activity (Mendis et al., 2005; Tang et al., 2010). Chen et al. studied the antioxidant activity of designed peptides and reported that His and Pro played important roles in the antioxidative behavior (Chen, Muramoto, Yamauchi, & Nokihara, 1996). The presence of a terminal Leu was also shown to be associated with the radical scavenging activity of buckwheat peptides (Ma, Xiong, Zhai, Zhu, & Dziubla, 2010; Tang et al., 2010).

Several amino acids such as His, Leu, Lys, Met, Trp, and Tyr are generally accepted to be antioxidants that contribute to the scavenging of free radicals (Li, Li, He, Qian, 2011a, 2011b; Wang, Chen, et al., 2015; Wang, Ding, et al., 2015). The antioxidant properties of peptides depend on their amino acid composition, structure, hydrophobicity, and other factors. For example, the antioxidant peptide Tyr-Phe-Cys-Leu-Thr contains two hydrophobic amino acid residues (Phe and Leu), which are usually present in antioxidant peptides, as well as two aromatic residues (Tyr and Phe), which may play critical roles as free-radical scavengers. This is because aromatic amino acids can donate protons to electron-deficient radicals (Rajapakse, Mendis, Jung, Je, & Kim, 2005). Furthermore, the presence of the Cys residue in the center of the antioxidant peptide can also facilitate its free-radical scavenging properties, because the thiol group in Cys can interact with radicals directly (Harman, Mottley, & Mason, 1984). Thus, all of these specific amino acid residues might contribute to the high free-radical scavenging activity of the peptide Tyr-Phe-Cys-Leu-Thr (Wang, Chen, et al., 2015; Wang, Ding, et al., 2015).

Jin et al. (2016) reported that the sequence and structure contribute to the bioactivity of corn peptides. It was also reported that repeated di- or tripeptide sequences within a peptide might play a role in antioxidant activity (Jin et al., 2016; Siow & Gan, 2013). Based on these concepts, Jin et al. (2016) predicted that a repetitive sequence of LL in a peptide might result in higher antioxidant activity. Hydrophobic interactions between hydrophobic amino acid residues might also enhance the antioxidant activity of peptides. The APLA motif in the peptide CSQAPLA was speculated to play a crucial role in its good antioxidant activity. A similar sequence, PALA, is present in the peptide DPALATEPDPMPF, which was isolated from Nile tilapia scale in a previous study and reported to exhibit strong DPPH radical-scavenging ability (Jin et al., 2016; Ngo, Qian, Ryu, Park, & Kim, 2010).

The presence of certain amino acids in a peptide is equally critical for antioxidant activity. Aromatic amino acids (Tyr, Trp, Phe, and His) can convert radicals to stable molecules by donating electrons (Jin et al., 2016). For example, some peptides contain a Tyr at the N-terminus, which plays a role in antioxidant activity. Hydrophobic amino acids can improve the solubility of peptides in lipids, which facilitates the interaction between peptides and radical species. The abundance of hydrophobic amino acids in the three peptides might be correlated with their antioxidant activities; a similar conclusion was also reached in previous reports (Jin et al., 2016; Qian, Jung, & Kim, 2008). Furthermore, the sulfhydryl group of Cys may act as a radical scavenger, protecting tissues from oxidative stress (Hernández-Ledesma et al., 2005; Jin et al., 2016). The Cys residue in Cys-Ser-Gln-Ala-Pro-Leu-Ala might contribute to the antioxidant activity of this peptide (Jin et al., 2016). Cys, His, Lys, Met, Trp, and Tyr are examples of amino acids that have been reported to possess antioxidant activity. Aromatic amino acids could donate protons to electron-deficient radicals. This property could improve the radical-scavenging properties of the amino acid residues (Ren, Zhao, Cui, You, & Wang, 2009; Wang, Chen, et al., 2015; Wang, Ding, et al., 2015). Even though these chemical assays give an insight to the potential biological activity of these food-derived antioxidants, further analysis are necessary such as investigating the fate of peptides during GI digestion as described previously, and their permeability through cellular membranes (Samaranayaka & Li-Chan, 2011). Peptides with antioxidative properties have been recognized to have great potential for use as nutraceuticals, pharmaceuticals, and substitutes for synthetic antioxidants (Freitas et al., 2013).

3.3 Hepatoprotective Function

It has been estimated that almost 4% of all deaths worldwide (6.2% for men and 1.1% for women) are attributable to alcohol (Wu et al., 2014). The primary target organ of alcohol is the liver, despite its enormous capacity to remove toxic substances and synthesize useful compounds (Yu et al., 2012). Chronic exposure to high doses of alcohol results in many pathophysiological changes in hepatic cellular function due to either the alcohol itself or its metabolism, such as increased activities of hepatic enzymes, inhibition of the β-oxidation of fatty acids in mitochondria, peroxidation of lipids

(Arteel, 2003; Yu, Wang, Wang, Lin, & Liu, 2017), generation of acetaldehyde, formation of free radicals, oxidative stress (Gutiérrez-Ruiz et al., 1999; Kumar et al., 2011; Yu et al., 2017), and an increased risk of liver injury leading to alcoholic liver disease (ALD) (Szabo, Petrasek, & Bala, 2012; Yu et al., 2017). Hepatic steatosis is the initial stage of alcoholic liver injury, and it can progress to steatohepatitis, fibrosis, and cirrhosis, which is the third leading (at 16.6%) cause of alcohol-attributable deaths worldwide (Wu et al., 2014).

Hepatic fibrosis, characterized by the deposition of excess extracellular matrix, is a wound-healing response to chronic or repeated liver injury due to biological factors or chemical drugs such as carbon tetrachloride and thioacetamide (Lv et al., 2013; Williams, 2006).

The hepatoprotective effect of corn oligopeptides has been confirmed by recent studies. Zhang et al. (2012) investigated the influence of corn oligopeptides on alcohol-induced morphological changes of the liver in rats. Histopathological examination revealed no changes in the livers of the normal control group. Alcohol administration induced the assembly of lipid droplets in the hepatocytes, and the corn oligopeptide treatment effectively prevented these changes. Yu et al. (2012) studied the influence of corn germ meal albumin peptides on liver damage induced by carbon tetrachloride and found that low-molecular-weight corn peptides exerted a hepatoprotective effect. Corn peptides were also shown to exhibit a hepatoprotective effect against thioacetamide-induced liver fibrosis (Lv et al., 2013). The antifibrotic effect of the corn peptides was demonstrated by decreased inflammation and fibril connective tissue formation in the rats treated with the corn peptides, based on histological and morphological analysis (Lv et al., 2013). Several bioactive peptides have been reported to exhibit protective effects against the liver injury induced by D-galactosamine or acetaminophen (Zhang et al., 2012). In addition, Zhang et al. (2012) reported that corn oligopeptides derived from CGM by enzymatic hydrolysis have a protective effect on early alcoholic liver injury in rats (Yu et al., 2017). Wu et al. (2014) further presented that corn peptides prepared from CGM have a protective effect on alcoholic liver injury via modulation of lipid metabolism and oxidative stress in men who chronically consume alcohol (Yu et al., 2017). It was also reported that corn peptides can protect against acute liver injury induced by alcohol (Li et al., 2010, 2007). Further studies in mice have demonstrated that corn peptides can provide significant protection against liver injuries induced by alcohol, D-galactosamine, Bacillus Calmette–Guérin/lipopolysaccharide, and carbon tetrachloride (Ma et al., 2012; Sui, 2009; Yu et al., 2012).

An increased aspartate transaminase/alanine transaminase (AST/ALT) ratio indicates an inflammatory state in liver cells that might lead to hepatic fibrosis (Li & Friedman, 1999; Lv et al., 2013). Yu et al. (2017) used a mouse model to evaluate the effects of corn germ meal albumin peptide fractions on blood alcohol metabolism and reported that alcohol administration significantly increased the activities and levels of AST, ALT, malondialdehyde, and triglycerides ($P < 0.01$) and significantly reduced the activities of superoxide dismutase and catalase and levels of glutathione ($P < 0.01$) compared with the control group. These changes were significantly prevented by the application of the $M_w < 1$ kDa peptide fraction with a high Fischer's ratio of $F > 3$ (indicating a high molar ratio of branched amino acids to aromatic amino acids) at 800 mg/kg bw. These results demonstrate that the peptides exerted a significant protective effect against chronic alcohol-induced liver injury, possibly by enhancing the *in vivo* antioxidant ability (Yu et al., 2017). Moreover, the hepatoprotective effects of corn peptides (200 mg/kg bw) were also demonstrated by significant decreases in the serum aminotransferase (AST, ALT) activities and liver malondialdehyde levels and significant increases in the superoxide dismutase activity and glutathione level in the liver ($P < 0.01$), compared with the control group (Yu et al., 2012). The ALT and AST activities of the animals treated with 200 mg/kg bw of corn peptides were similar to those of animals treated with 50 mg/kg bw silymarin, a milk thistle extract with antioxidant activity traditionally used to treat liver disorders, indicating that the two treatments exerted a similar hepatoprotective effect (Yu et al., 2012).

Hepatic injury causes liver malfunction, which results in increased protein catabolism and a deficit of branched-chain amino acids, while the blood concentration of aromatic amino acids increases. These amino acids can then reach the brain in high amounts, where they can mimic the effects of certain neurotransmitters and/or compete with others (Yu et al., 2012). In addition, oligopeptides with high Fischer's ratio have been used in certain medical diets (Fischer, 1990) for the treatment of patients with liver diseases (Yu et al., 2012). Owing to its high proportion of branched-chain amino acids, corn protein can be hydrolyzed to afford oligopeptides with a high Fischer's ratio ($F > 20$, with a Tyr/Phe content of <2% of the total amino acid residues). In particular, the hexapeptide PVVLID was identified in the corn peptides, which may contribute to the hepatoprotective function (Lv et al., 2013).

Ethanol inhibits gluconeogenesis and lowers the plasma alanine concentration by inhibiting alanine release from muscle rather than by increasing alanine conversion to lactate (Ma et al., 2012; Yamaguchi, Nishikiori,

Ito, & Furukawa, 1997). The depletion of nicotinamide adenine dinucleotide (NAD^+) leads to a transient pause of tricarboxylic acid (TCA) cycle activity. Yamaguchi et al. suggested that the effect of corn peptides on facilitating alcohol metabolism may be due to the increased supply of NAD^+ as a result of the increased levels of Ala and Leu, especially Leu, from the corn peptides. Leu helps in maintaining a normal TCA cycle by promoting alcohol metabolism. Furthermore, Leu is one of the branched–chain amino acids, which are associated with amino acid metabolism in contracting skeletal muscle as well as the generation of Ala by serving as a donor of amino groups to pyruvic acid in muscle (Ma et al., 2012; Yamaguchi, Nishikiori, et al., 1997; Yamaguchi, Nozaki, Ito, & Furukawa, 1997).

Additionally, there is increasing evidence to support the notion that, when considering the many factors that contribute to the pathogenesis of ALD, oxidative stress plays an important role (Bjelakovic et al., 2011; Gao et al., 2011; Zhang et al., 2012). Previous studies have suggested that the generation of free radicals and subsequent oxidative stress are the key factors responsible for ALD (Caro & Cederbaum, 2004; Xu, Leo, & Lieber, 2003; Yu et al., 2017). There is also increasing evidence that free radicals and ROS play crucial roles in the various steps that initiate and regulate the progression of liver disease, independently of the original causative agent (Loguercio & Federico, 2003; Yu et al., 2012). Further studies are necessary to elucidate the underlying mechanism of the hepatoprotective function of corn peptides.

3.4 Anti-Obesity Function

Obesity has become increasingly prevalent over recent years as a consequence of the global increase in living standards. Obesity is harmful to health and a primary factor in the development of many chronic physical and psychological diseases. Therefore, strategies to promote weight loss have attracted a great deal of attention. A number of studies have investigated the potential anti-obesity effects of corn protein hydrolysates and enriched peptides. Sun, Tian, and Shi (2017) examined the influence of dietary supplementation with corn peptides on body composition and factors related to hepatic lipid metabolism in obese rats fed a high-fat diet. The rats that were fed a diet supplemented with the corn peptides exhibited significantly decreased body weight and perirenal fat mass index compared with the control rats. Furthermore, hepatic adipose triglyceride lipase (ATGL) levels were significantly higher upon supplementation with corn peptides, whereas the expression of tumor necrosis factor-α (TNF-α) was clearly suppressed compared with the control group. These results demonstrate that corn peptides can lead to weight reduction in animal models by decreasing the amount of adipose tissue, which may be related to the expression of ATGL and TNF-α in the liver and subsequent promotion of lipid metabolism. Shi et al. (2013) investigated the effects of corn peptides on the plasma levels of metabolic hormones such as insulin, leptin, glucagon, monocyte chemo-attractant protein 1 (MCP-1), pancreatic polypeptide (PP), and peptide YY (PYY) in obese rats to elucidate the mechanistic role of these hormones in the body weight reduction. A high-fat diet increased the levels of leptin, insulin, glucagon, and MCP-1 in obesity-susceptible rats, and also increased the level of glucagon and reduced the level of PYY in obesity-resistant rats. The administration of corn peptides was found to reduce the levels of glucagon and MCP-1 and increase the level of PYY, which may be the mechanism responsible for energy expenditure and lipid oxidation leading to weight loss. Lu, Liu, Wang, and Lou (2016) also evaluated the effects of corn peptide supplementation on fat loss and blood lipid profile in obese rats fed a high-fat diet and investigated the role of ATGL and lipoprotein lipase (LPL). Compared with the control group, the animals treated with the corn peptides showed a clear decrease in body weight and perirenal and epididymal fats. Furthermore, the protein levels of ATGL in the liver and LPL in adipose tissues were significantly increased in the rats treated with corn peptides. These results indicated that the decreases in body weight and perirenal and epididymal fats induced by the corn peptides may be related to the increased protein levels of ATGL in the liver and LPL in adipose tissues. However, more research is necessary to determine the mechanism of the anti-obesity effects of corn peptides.

3.5 Effect on Alcohol Metabolism

Alcohol abuse and alcoholism are serious public health and socioeconomic problems in many countries. It has been estimated that almost 4% of all deaths worldwide (6.2% for men and 1.1% for women) are attributable to alcohol, and the harmful use of alcohol is the leading risk factor for death in men aged 15–59 (Wu et al., 2014). Excessive alcohol consumption can result in malnutrition owing to the displacement of other nutrients from the diet, possibly leading to conditions such as various vitamin and mineral deficiencies, liver diseases (e.g., fatty liver, hepatocirrhosis), breast cancer, hyperlipidemia, hyperuricemia, ketoacidosis, and neuronal injury associated with neuronal degeneration (Li et al., 2011; Yamaguchi, Nozaki, et al., 1997). It is well established that the liver is the principal organ responsible for the elimination of ingested alcohol and contains the enzymatic systems responsible for ethanol oxidation. The main pathway for ethanol metabolism is mediated by alcohol dehydrogenase (ADH). Under normal circumstances, ethanol is first oxidized by ADH to form acetaldehyde, which is then further oxidized by acetaldehyde dehydrogenase (ALDH) in the presence of the cofactor NAD to produce acetate, which is subsequently converted into carbon dioxide via the TCA cycle. Therefore, the ADH activity plays a key role in the elimination of ethanol. As the initial product of ethanol metabolism, acetaldehyde is considered a major candidate toxin involved in alcohol-induced injury to tissues and cells and has also been shown to form protein adducts leading to organ and tissue damage. The accumulation of acetaldehyde further leads to symptoms such as cardiac arrhythmias, nausea, anxiety, and facial flushing (Li et al., 2011).

Several investigators have demonstrated that alcohol metabolism is accelerated in the presence of certain amino acids, such as Gly, Lys, Met, Pro, and Ala. For example, Yang, Ito, Morimatsu, Furukawa, and Kimura (1993) reported that the alcohol intake by stroke-prone spontaneously hypertensive (SHR-SP) rats increased upon the addition of Pro, Lys, and Thr to the ethanol solution, and found that Pro and Lys could regulate the alcohol metabolism.

It was also reported that corn peptides can accelerate the metabolism of alcohol and reduce its plasma concentration. Corn peptides were found to be capable of attenuating acute hepatic injury induced by ethanol. Li et al. (2011) demonstrated that corn peptides could enhance ADH activity both *in vitro* and *in vivo* and consequently accelerate alcohol metabolism. Furthermore, they tested the safety of corn peptides as a health food by feeding them to Wistar rats for 30 days. The *in vitro* and *in vivo* effects of corn peptides prepared from CGM by proteolysis using an alkaline protease and fractions of the corn peptides from Sephadex G-15 and G-10 columns on ADH activity were studied. The results showed that the corn peptides and fraction 3 of the corn peptides from the Sephadex G-10 column enhanced the *in vitro* ADH activity. Furthermore, such activity was superior to that of glutathione, which was also found even in the presence of the ADH inhibitor pyrazole. In the *in vivo* experiments, the animals were fed with different dosages of corn peptides and with a dose of Chinese distilled spirit orally, and sacrificed for the measurement of ADH activity. The *in vivo* results indicated that the corn peptides enhanced the hepatic ADH activity. To test the safety of corn peptides as a health food, a 30-day feeding test was performed. No obvious toxic effects were detected in the treated Wistar rats.

Yamaguchi, Nishikiori, Ito, and Furukawa (1996) and Yamaguchi, Takada, et al. (1996) prepared corn peptides by proteolysis of CGM using alkaline protease. The amino acid composition of the resulting peptides was rich in Ala, branched-chain amino acids (Leu, Ile, and Val), and Pro, but poor in basic amino acids. The intragastric administration of the corn peptides prior to ethanol intake significantly mitigated the subsequent increase in blood ethanol concentration and led to elevated plasma concentrations of Ala, branched-chain amino acids, and Pro in SHR-SP rats. Yamaguchi, Nishikiori, et al. (1997) and Yamaguchi, Nozaki, et al. (1997) investigated the influence of corn peptides on plasma amino acid concentrations and alcohol metabolism in SHR-SP rats with respect to the interaction between corn peptides and ethanol in the gastrointestinal tract. They administered corn peptides in ethanol or corn peptides alone intragastrically. In the case of the intragastric administration of corn peptides alone, ethanol was given intraperitoneally in due consideration of the influence of individual amino acid levels in the plasma by ethanol loading. SHR-SP rats were chosen as the experimental animals in this study because they showed a high ethanol preference and did not show any genetic change in ADH and ALDH activities. This study showed that the effective absorption of corn peptides coexistent with ethanol through the gastrointestinal tract results in the elevation of plasma Ala, Leu, and Pro concentrations, which may enhance alcohol metabolism due to the potential supply of NAD^+.

Yamaguchi, Nishikiori, et al. (1996) demonstrated previously that corn peptides administration before alcohol intake results in the significant reduction of increase in blood ethanol level and the elevation in plasma Ala, Leu, and Pro levels in SHR-SP and humans. Concerning alcohol metabolism in animals, the ability to metabolize alcohol is affected by nutritional conditions and the rats fed with a low protein diet had poorer ability for alcohol metabolism due to the deficiency of amino acids, especially essential amino acids. Acute alcohol loading causes a decrease in plasma alanine in humans as well as rabbits, but branched-chain amino acid levels and other amino acid levels are not affected. The decrease in plasma alanine concentration is correlated with saving in alanine release from the skeletal muscle. Alanine, which serves to convey amino groups and carbon substrates from muscle to liver, is known to be the principal amino acid extracted from man's splanchnic circulation in both postabsorptive and prolonged starvation states.

The increase in the blood is likely to reflect a sufficient supply of NAD^+. Leucine is one of the branched–chain amino acids that are associated with amino acid metabolism in contracting skeletal muscle as well as generating alanine as a donor of amino groups to pyruvic acid in muscle. They reported that administration of Pro and Lys to SHR–SP rats increased in preference for alcohol and accelerated alcohol metabolism. Corn peptides were effective in the reduction of the increase in blood ethanol level and the mechanism of action might be due to the potential supply of NAD^+ by the elevation of alanine and leucine resulting from corn peptides, but neither by the delay of ethanol release from stomach nor malabsorption of ethanol in the gastrointestinal tracts (Yamaguchi, Nishikiori, et al., 1997).

3.6 Other Functions

In addition to the wide range of functions discussed above, some corn peptides also exhibit antimicrobial or metal-binding activities. The antimicrobial spectra of corn peptides have been found to cover a broad range of pathogenic organisms, such as *Escherichia*, *Helicobacter*, *Listeria*, *Salmonella*, and *Staphylococcus* species, yeasts, and filamentous fungi (García-Olmedo, Molina, Alamillo, & Rodríguez-Palenzuéla, 1998). Corn peptides are recognized to be an important component of innate immunity, particularly at mucosal surfaces such as the lungs and small intestine that are constantly exposed to a range of potential pathogens. An amphiphilic sequence and a positive charge are recognized as major structural motifs determining the interaction with bacterial membranes, which have been accepted as a common target in their mechanism of action. It has been demonstrated that some corn-derived antibacterial peptides can reach intracellular targets, which involves interaction with the microbial cellular lipid bilayer and eventual disintegration of the cell membrane. Although the bioactivity of corn peptides after parenteral administration has been demonstrated, no studies have yet been reported on the possible antimicrobial effects of these peptides after oral administration. Antimicrobial peptides may find interesting applications in the field of food safety. For example, in laboratory tests they have been proven to protect fresh meat by inhibiting bacterial growth and preventing bacteria from attaching to meat surfaces (Epand & Vogel, 1999; Zasloff, 2002).

The metal-binding ability of bioactive peptides has been widely acknowledged for many years. Some corn peptides possess physicochemical properties that enable the chelation of various ions such as Ca, Mg, and Fe, along with trace elements that include Zn, Ba, Cr, Ni, Co, and Se (Kitts, 1994). Most minerals from food will be dissociated at the low pH found in the stomach, and will subsequently be transferred to the duodenum. They may gradually become insoluble as the pH increases. Interestingly, corn peptides could form soluble nanometer-sized particles with mineral ions to increase mineral solubility at intestinal pH, and these could easily be absorbed in the distal small intestine. In general, the chelating power of corn peptides is highly correlated with the difference of corn peptides fractions, as well as the conditions of pH, ionic strength, and temperature. Although only a limited number of short-term studies in humans have been conducted, several studies performed in rats have demonstrated that dietary corn peptides not only increase mineral absorption but also stimulate mineral incorporation. Further research is necessary to elucidate the effects of corn peptides on mineral absorption. As corn peptides can bind and solubilize minerals, they have been considered physiologically beneficial in the prevention of osteoporosis, dental caries, hypertension, and anemia (Kitts, 1994; Miyoshi, Kaneko, Ishikawa, Tanaka, & Maruyama, 1995; Scholz-Ahrens & Schrezenmeir, 2002).

4. IDENTIFICATION OF BIOACTIVE CORN PEPTIDES

4.1 Instrumental Techniques

Bioactive peptides are specific fragments of proteins that are inactive within the parent protein but once released, they impart positive functions or benefits to human health (Freitas et al., 2013; Hettiarachchy, Sato, Marshall, & Kannan, 2012; Sarmadi & Ismail, 2010).

The most popular technique for identifying bioactive peptides involves their separation by gradient high-performance liquid chromatography (HPLC) on a reversed-phase (RP) column and subsequent analysis by mass spectrometry (MS). Peptide sequencing can be performed by processing the MS/MS spectra using an MS database or manual calculations (Chen, Kwon, Kim, & Zhao, 2005; Wang, Chen, et al., 2015; Wang, Ding, et al., 2015). The identified peptides can then be synthesized for quantification and evaluation of their bioactivity by chemical assays or *in vitro* and *in vivo* experiments.

HPLC-MS/MS using electrospray ionization (ESI) is commonly applied, especially for short peptides (Li et al., 2004; Ma et al., 2012; Tang et al., 2010; Wang, Chen, et al., 2015; Wang, Ding, et al., 2015; Yang et al., 2007). Some researchers have also used a quadrupole time-of-flight mass spectrometer (Q-TOF) with ESI to identify bioactive peptides

(Jin et al., 2016; Puchalska et al., 2013; Tang & Zhuang, 2014; Wang et al., 2014; Zheng et al., 2006). MALDI-TOF/TOF mass spectroscopy has also been applied to identify amino acid sequences (Wang, Chen, et al., 2015; Wang, Ding, et al., 2015). Moreover, Suh et al. (1999) determined the amino acid sequence directly by the Edman degradation method using a JEOL JAS-47K peptide sequencer. The common MS databases include LC/MSD Trap Data Analysis (Lv et al., 2013) and SCIEX Triple Quad software (Lin et al., 2011).

4.2 Structural Characterization of Corn Peptides

Corn peptides are small peptides that typically contain 2–20 amino acids and have molecular masses in the range from 300 to 1000 Da (Li et al., 2010; Wang, Chen, et al., 2015; Wang, Ding, et al., 2015; Zhang et al., 2012). In fact, Lin et al. reported that the majority (96.77%) of their corn oligopeptides had molecular masses of <1000 Da and were rich in Glu/Gln, Leu, Ala, Pro, Phe, and Tyr, which accounted for 24.21%, 18.27%, 9.68%, 8.29%, 5.82%, and 5.51%, respectively (Lin et al., 2011). The average molecular weight of the corn oligopeptides was 363 Da, the average molecular weight of the constituent amino acids was 137 Da, and the mean peptide length was approximately 2.7 (Lin et al., 2011).

Furthermore, corn peptides are rich in branched-chain amino acids, with an F-value of >20, which was reported to be useful for the clinical treatment of liver cirrhosis (Lv et al., 2013). The high abundance of Leu, Ala, Pro, Phe, and Tyr in corn oligopeptides was also considered important for their ACE inhibitory and antihypertensive activities (Fang et al., 2008).

The sequences of some bioactive corn peptides are listed in Table 2 alongside their biological functions.

4.3 Absorption and Transportation of Corn Peptides

Orally administered proteins and peptides are digested and degraded by numerous enzymes in the gastrointestinal tract before being absorbed through the intestinal epithelium (Shimizu, 2004). Studies in Caco-2 cell monolayer systems, a good intestinal model for drug absorption first described systematically by Hidalgo, Raub, and Borchardt (1989), demonstrated that the size and lipophilicity are two critical parameters that determine the permeability of peptides through the intestinal epithelium (Artursson, Palm, & Luthman, 1996; Ding et al., 2014). Low-molecular-weight and hydrophobic peptides are absorbed more easily in an intact form.

Table 2 Amino Acid Sequences of Some Bioactive Corn Peptides and Their Biological Functions

Sequence	Function	References
Ala-Tyr	Antihypertensive	Lin et al. (2011)
Leu-Arg-Pro	Antihypertensive	Puchalska et al. (2013)
Leu-Ser-Pro	Antihypertensive	Puchalska et al. (2013)
Leu-Gln-Pro	Antihypertensive	Puchalska et al. (2013)
Pro-Ser-Gly-Gln-Tyr-Tyr	Antihypertensive	Suh et al. (1999)
Met-Ile/Leu-Pro-Pro	Antioxidant and antihypertensive	Wang, Chen, et al. (2015) and Wang, Ding, et al. (2015)
Phe-Pro-Leu-Glu-Met-Met-Pro-Phe	Antioxidative	Zheng et al. (2006)
Gly-His-Lys-Pro-Ser	Antioxidative	Zhuang et al. (2013)
Gln-Gln-Pro-Gln-Pro-Trp	Antioxidative	Wang et al. (2014)
Ala-Tyr	Antioxidative	Zhou et al. (2015)
Leu-Asp-Tyr-Glu	Antioxidative	Li et al. (2004)
Pro Phe and Leu Pro Phe	Antioxidative	Tang and Zhuang (2014)
Tyr-Ala and Leu-Met-Cys-His	Antioxidative	Tang et al. (2010)
Tyr-Phe-Cys-Leu-Thr	Antioxidative	Wang, Chen, et al. (2015) and Wang, Ding, et al. (2015)
Tyr-Ala	Antioxidative	Tang et al. (2010)
Leu-Met-Cys-His	Antioxidative	Tang et al. (2010)
Cys-Ser-Gln-Ala-Pro-Leu-Ala	Antioxidative	Jin et al. (2016)
Gln-Leu-Leu-Pro-Phe	Facilitates alcohol metabolism	Ma et al. (2012)
Phe-Leu-Leu-Gln	Hepatoprotective	Freitas et al. (2013)
Pro-Val-Val-Leu-Ile-Asp	Hepatoprotective	Freitas et al. (2013)
Phe-Leu-Pro-Gln	Hepatoprotective	Freitas et al. (2013)
Leu-Leu-Pro-Phe	Hepatoprotective	Freitas et al. (2013)
Gln-Leu-Leu-Pro-Phe	Hepatoprotective	Freitas et al. (2013)
Leu-Met-Leu-Phe/Met-Leu-Leu-Phe	Hepatoprotective	Freitas et al. (2013)

However, there have been few studies on low-molecular-weight antioxidant peptides derived from CGM (Wang, Chen, et al., 2015; Wang, Ding, et al., 2015).

Caco-2 cell monolayers are the most commonly reported system in the literature for studying the intestinal permeability of bioactive compounds owing to their similarity to intestinal endothelial cells (Liu & Finley, 2005; Vermeirssen et al., 2005). Upon culturing as a monolayer, Caco-2 cells differentiate to form tight junctions between cells to serve as a model for the paracellular movement of compounds across the monolayer. Furthermore, Caco-2 cells express transporter proteins, efflux proteins, and phase II conjugation enzymes to permit the modeling of a variety of transcellular pathways as well as the metabolic transformation of test substances (Samaranayaka & Li-Chan, 2011). Small di- and tripeptides may be absorbed intact across the brush border membrane via the H^+-coupled PepT1 transporter system (Samaranayaka & Li-Chan, 2011; Vermeirssen et al., 2002). Larger water-soluble peptides can cross the intestinal barrier paracellularly via the tight junction between cells, whereas highly lipid-soluble peptides may diffuse via the transcellular route (Miguel et al., 2008; Samaranayaka & Li-Chan, 2011). Peptides may also enter the enterocytes via endocytosis, which entails membrane binding and vesiculation of the material (Samaranayaka & Li-Chan, 2011; Ziv & Bendayan, 2015). The intestinal basolateral membrane also possesses a peptide transporter that facilitates the exit of hydrolysis-resistant small peptides from the enterocyte into the portal circulation (Gardner, 1984; Samaranayaka & Li-Chan, 2011). Furthermore, the contribution of each route and the ability of individual peptides to cross the membrane depend on their molecular size and other structural characteristics such as hydrophobicity, as well as their resistance to brush border peptidases (Samaranayaka & Li-Chan, 2011; Satake et al., 2002).

5. APPLICATIONS OF CORN PEPTIDES

The concept of functional and clinical nutritional foods was first established in Japan and other developed countries in the early 1990s, bringing great hope to people to improve their health and delay aging. Global sales of functional and clinical nutritional foods were close to 100 billion USD in 2010. The United States is currently the world's largest consumer of functional foods, accounting for 38% of global sales. In recent years, with the intensification of environmental pollution and the extension of human life, cerebrovascular disease, malignant tumors, and diabetes have become increasingly serious threats to human health, such that there is an urgent demand for developing new drugs and health foods to prevent and treat these diseases. This provides broad prospects for the application of foods or drugs based on bioactive peptides.

Food safety and security concerns restrict the application of antioxidants (e.g., BHA, BHT, TBHQ) to foods. Although some natural antioxidants such as plant extracts are known, their use can influence the flavor and color of foods. Therefore, the development of efficient and non-toxic peptide antioxidants has become a major research topic. Corn peptides can effectively inhibit lipid oxidation and are easy to use with other fat-soluble antioxidants. Moreover, studies have implicated free radicals in the pathogenesis of many diseases and conditions such as oxidative stress. Corn peptides effectively inhibit the production of free radicals and can be used as nutritional regulators and stabilizers by adding them to beverages, dairy products, and grains to serve as antioxidants or functional and clinical nutritional foods (Freitas et al., 2013; Jin et al., 2016; Li et al., 2010; Tang et al., 2010).

Hypertension is a serious condition that is responsible for >12 million deaths per year worldwide, which is higher than any other disease. ACE inhibitory peptides derived from the proteolysis of food proteins under mild conditions or extracted from fermented foods can significantly lower blood pressure (Yang et al., 2007). Corn peptides contain antihypertensive components that only exert their antihypertensive effect on hypertensive patients and do not reduce blood pressure in healthy patients, thereby avoiding excessive blood pressure reduction. In addition to the observation of antihypertensive activity in *in vitro* experiments, animal experiments and human clinical trials have also demonstrated antihypertensive effects. The use of health foods to improve the physiological state is becoming increasingly accepted for the prevention of hypertension and other chronic non-communicable diseases. As hypertensive patients often require lifelong treatment, the development of antihypertensive health foods based on corn peptides would be of great value (Huang et al., 2011; Suh et al., 1999; Yang et al., 2007).

Obesity involves physiological and biochemical changes that result in excessive fat deposition and a series of pathophysiological changes in the body. Adult women and men with a proportion of adipose tissue above 30% or 25%, respectively, are considered obese. Studies have shown that obesity is associated with at least 20 diseases, such as cardiovascular disease, diabetes, tumors, cholecystitis, respiratory dysfunction, osteoarthrosis, kidney disease, and endocrine disorders. Obesity causes the body's organs to be in a state of overload and can lead to metabolic abnormalities, such as gout, gallstones, and pancreatic diseases. Research has demonstrated that dietary protein can promote energy metabolism more than dietary fat and sugar. Therefore, as long as sufficient protein intake can be ensured, the remaining nutrients in the diet can be reduced to a minimum to achieve the goal of scientific weight loss (Shi et al., 2013). Corn peptides could activate the sympathetic nervous system and brown adipose tissue, prevent adipose absorption, promote lipid metabolism, and reduce subcutaneous fat. The intake of corn peptides guarantees the weight loss of the nitrogen balance. As peptides are absorbed from the small intestine more easily than amino acids, dieters obtain energy from stored fat owing to alteration of the metabolism to promote the TCA cycle. Experiments have been conducted to compare the amount of heat released after the ingestion of foods supplemented with various proteins, and the results demonstrated that the food supplemented with corn peptides produced the greatest amount of heat, indicating that corn peptides have a greater effect on promoting energy metabolism than other proteins. Consequently, corn peptides may prove useful as a diet food for obese individuals and a nutritional food supplement for weight loss treatment (Lu et al., 2016; Shi et al., 2013; Sun et al., 2017).

The liver is an essential organ for life-sustaining activities and is closely involved in the metabolism of sugars, lipids, proteins, vitamins, hormones, and many other bioactive substances. Alcoholic liver, fatty liver, hepatitis, and other liver diseases have become increasingly prevalent over recent years and pose a serious threat to human health (Lv et al., 2013; Williams, 2006; Wu et al., 2014). With the increasing incidence of liver disease and a growing emphasis on health, the search for functional foods and drugs that can alleviate liver damage and improve liver function has become a hot topic in research. Currently, most treatments for liver injury involve chemical medicines, although these can be expensive and have deleterious side effects on the metabolism. Furthermore, as most medicines are metabolized in the liver, improper administration can aggravate the liver injury. In addition, patients with impaired liver function can exhibit various metabolic abnormalities, including disorders involving protein synthesis and amino acid metabolism. Because liver disease patients are often anorexic with abdominal distension, nausea, vomiting, and other symptoms that affect normal eating, they are very susceptible to malnutrition owing to insufficient energy and nutrient intake. Liver disease patients can also suffer from dysfunctional intestinal digestion and absorption of sugars, fats, proteins, and other nutrients, and therefore excessive protein intake could lead to a negative nitrogen balance. Consequently, the development of foods that can meet the physiological needs of patients with liver disease is crucial for ensuring the supply of high-quality proteins and maintaining the nitrogen balance in the body. Corn peptides are quickly and completely absorbed and do not increase the functional burden on the gastrointestinal tract, liver, and kidneys. Corn peptides also possess a high nutritional value and to some degree can correct the negative nitrogen balance in patients with liver disease, making them very suitable for protein supplements and nutritional improvement in patients. Focusing on this characteristic, many corn peptide products are developed worldwide, and some are already commercialized (Table 3). In China, the prevalence rate of hepatopathy has been higher than world average level for a long period and has been threatening the development many under-developed areas (WHO, 2017). Some important reasons to

Table 3 Some Products Using Corn Peptides

Company	Product Name	Product Form	Main Functions
Perfect (China) Co., Ltd.	Perfect Corn Peptide and Brown Rice Germ Tablet	Tablet	Liver protection; anti-alcoholic
A&Z Pharmaceutical, Inc.	UFIT Peptide Liver Support	Tablet	Liver protection; anti-alcoholic
Jilin Sky Scenery Food Co., Ltd.	Baogan Maydis Stigma Beverage	Beverage	Anti-alcoholic; sub-health regulatory
Inner Mongolia Xind Beverage Co., Ltd.	Qiaomi Peptide Beverage	Beverage	Anti-alcoholic; sub-health regulatory
Guangxi Endo Food Co., Ltd.	No. of Peptide Functional Beverage	Beverage	Anti-alcoholic; sub-health regulatory
SARAYA Co., Ltd.	Gains Sport Jelly	Jelly	Stamina supplementation
SARAYA Co., Ltd.	Collagen Aid TOP	Powder (solid drink)	Skin complexion improvement; nutritional supplementation

pose this threat is inadequate sanitary and excessive drinking. In these circumstances, foods with healthy claims in liver protection and anti ethanol induced damage are the most concentrated designed and developed. For instance, a tablet made of corn peptides and brown rice germ with main functions of liver protection and acceleration of ethanol metabolism is developed by Perfect (China) Co., Ltd. This product creates >30 million CNY income per year for the company. Another corn peptides anti-alcoholic tablet product with similar healthy claim designed by A&Z Pharmaceutical, Inc. will be available on the US market. Moreover, several corn peptides enhanced beverages are also designed for people who have demands in sub-health regulatory.

Based on the physiological characteristics of patients with liver disease and the problems associated with simply relying on a conventional diet to deliver nutrition, combined with other natural functional factors, the development of new hepatoprotective foods based on corn peptides with higher absorption and utilization rates and promising physiological function is becoming a hot spot in the field of peptide research (Bjelakovic et al., 2011; Ma et al., 2012; Yamaguchi, Nishikiori, et al., 1997; Yamaguchi, Nozaki, et al., 1997; Yu et al., 2017; Zhang et al., 2012). In Japan, some clinical nutrition foods were developed to improve the nutritional statues of post-surgery patients. Nevertheless, due to the high content of branch-chain amino acids in corn peptides, some jelly formal stamina supplements are also welcomed in Japan market.

Along with the continual expansion of market demands of corn peptides, regulations and standards for guiding the production and use of corn peptides were gradually set and improved. In China, corn peptides were authenticated as the first peptide products to be included in the New Resource Food list by State Food & Drug Administration in 2008. After years of using corn peptides as important food material, a professional standard of oligopeptides powder of corn (QB/T 4707-2014) was also issued by the Ministry of Industry and Information Technology of the People's Republic of China in 2014 to standardize the production and use.

6. SUMMARY AND PROSPECTS

CGM is a major byproduct of the corn starch industry and over 840,000 tons are produced annually in China. However, despite its high protein content of approximately 60% (w/w), CGM is mainly used as an animal feed or discarded, owing to its low water solubility and severely imbalanced amino acid profile. In recent years, corn peptides prepared from CGM by enzymatic hydrolysis or microbial fermentation have attracted considerable interest as novel foods because of their various bioactive properties, which include antioxidant, antihypertensive, anti-obesity, antimicrobial, hepatoprotective, and metal-binding activities and the ability to accelerate alcohol metabolism. These corn peptides are typically small sequences of 2–20 amino acids with molecular masses between 300 and 1000 Da. To date, 22 sequences of bioactive corn peptides have been identified. The research and development of corn peptides and their products is expected to effectively increase the market value of CGM.

The chemical synthesis of dipeptides in the early 20th century marked the birth of the field of peptide science. Through the multidisciplinary efforts of scientists all over the world, peptide science has continued to grow and develop. Peptides mediate many important physiological and biochemical functions in organisms. More than 1000 bioactive peptides are known to play roles in the central and peripheral nervous systems, cardiovascular system, immune system, and digestive system. Peptides also interact with receptors to mediate information exchange between cells and participate in biochemical processes such as metabolism, pain, regeneration, and immune response. In recent years, numerous bioactive peptides have been discovered, isolated, identified, and characterized. Peptide science has influenced many fields of science and aspects of human life, including medicine, chemistry, biology, pharmacy, nutrition, the food industry, and animal husbandry.

The development of biotechnology has resulted in major changes in pharmaceutical chemistry research. Modern molecular biology has become the driving force for the development of pharmaceutical chemistry and allowed the accumulation of a great deal of knowledge relating to peptides. Therefore, the development of peptidic products and basic molecular biology research have a natural link. With the completion of the Human Genome Project and the advent of the protein age, an increasing number of bioactive peptides with clinical applications are expected to be discovered and characterized. Their functional role will also be elucidated, making it possible to control diseases that seriously threaten human health at the molecular level. To date, corn peptides have been applied as functional components in a variety of foods, such as low-antigen food, baby food, sports food, calcium-absorption-promoting food, antihypertensive food, and sobering-up food, by researchers in Japan and other developed countries.

Corn peptides are currently produced mainly by enzymatic hydrolysis. Consequently, the obtained products contain a mixture of peptides, although the content of active peptides is uncertain, and the product distribution can vary depending on the raw materials and enzymes used. The application of biotechnology to the production of corn peptides is promising. By repeat superposition of oligopeptide genes, genes encoding different peptide fragments could be fused, allowing the molecular weights of the resulting peptide fragments to be controlled via the introduction of enzymatic cleavage sites, such that desirable peptide fragments could be conveniently obtained by cleavage *in vitro* or after exposure to human digestive enzymes. Transgenic technology has become an important tool for studying bioactive peptides, making it possible to obtain bioactive peptide products more efficiently.

In summary, the presence of bioactive proteins in the human diet has been known for many years; however, it is only recently that the various specialized physiological functions of bioactive peptides have been elucidated in greater detail. Common sources of bioactive peptides include milk, fish, meat, eggs, and cereals, some of which have been investigated in animal models and clinical studies. Corn-derived bioactive peptides play vital roles in human health and nutrition. The possibility of designing new dietary products and drugs based on corn peptides to help prevent or control diet-related chronic diseases appears promising. In the future, further studies will be necessary to establish the *in vivo* efficacy of corn peptides as well as their long-term effects during regular long-term oral administration. Corn peptides represent a highly interesting resource for possible exploitation as active ingredients in health-promoting foods.

REFERENCES

Arai, S. (1978). Nutritional effect of food peptides in special relevance to their specificity for intestinal absorption. *Eiyo To Shokuryo, 31*, 247–253 [in Japanese].

Arteel, G. E. (2003). Oxidants and antioxidants in alcohol-induced liver disease. *Gastroenterology, 124*, 778–790.

Artursson, P., Palm, K., & Luthman, K. (1996). Caco-2 monolayers in experimental and theoretical predictions of drug transport. *Advanced Drug Delivery Reviews, 22*, 67–84.

Bjelakovic, G., Nikolova, D., Gluud, L. L., Bjelakovic, M., Nagorni, A., & Gluud, C. (2011). Antioxidant supplements for liver diseases. *Cochrane Database of Systematic Reviews, 16*, CD007749.

Bougatef, A., Nedjar-Arroume, N., Ravallec-Ple, R., Leroy, Y., Guillochon, D., Barkia, A., et al. (2008). Angiotensin I-converting enzyme (ACE) inhibitory activities of Sardinella (*Sardinella aurita*) by-products protein hydrolysates obtained by treatment with microbial and visceral fish serine proteases. *Food Chemistry, 111*, 350–356.

Caro, A. A., & Cederbaum, A. I. (2004). Oxidative stress, toxicology, and pharmacology of CYP2E1. *Annual Review of Pharmacology and Toxicology, 44*, 27–42.

Chen, Y., Kwon, S. W., Kim, S. C., & Zhao, Y. (2005). Integrated approach for manual evaluation of peptides identified by searching protein sequence databases with tandem mass spectra. *Journal of Proteome Research, 4*, 998–1005.

Chen, H. M., Muramoto, K., Yamauchi, F., & Nokihara, K. (1996). Antioxidant activity of designed peptides based on the antioxidative peptide isolated from digests of a soybean protein. *Journal of Agricultural and Food Chemistry, 44*, 2619–2623.

Cheung, H. S., Wang, F. L., Ondetti, M. A., Sabo, E. F., & Cushman, D. W. (1980). Binding of peptide substrates and inhibitors of angiotensin-converting enzyme. Importance of the COOH-terminal dipeptide sequence. *Journal of Biological Chemistry, 255*, 401–407.

Cumby, N., Zhong, Y., Naczk, M., & Shahidi, F. (2008). Antioxidant activity and water-holding capacity of canola protein hydrolysates. *Food Chemistry, 109*, 144–148.

Dávalos, A., Miguel, M., Bartolomé, B., & López-Fandiño, R. (2004). Antioxidant activity of peptides derived from egg white proteins by enzymatic hydrolysis. *Journal of Food Protection, 67*, 1939–1944.

Delgado-Zamarreño, M. M., González-Maza, I., Sánchez-Pérez, A., & Carabias Martínez, R. (2007). Analysis of synthetic phenolic antioxidants in edible oils by micellar electrokinetic capillary chromatography. *Food Chemistry, 100*, 1722–1727.

Ding, L., Zhang, Y., Jiang, Y., Wang, L., Liu, B., & Liu, J. (2014). Transport of egg white ACE-inhibitory peptide, Gln-Ile-Gly-Leu-Phe, in human intestinal Caco-2 cell monolayers with cytoprotective effect. *Journal of Agricultural and Food Chemistry, 62*, 3177–3182.

Epand, R. M., & Vogel, H. J. (1999). Diversity of antimicrobial peptides and their mechanisms of action. *Biochimica et Biophysica Acta, 1462*, 11–28.

Esen, A. (1987). A proposed nomenclature for the alcohol-soluble proteins (zeins) of maize (*Zea mays* L.). *Journal of Cereal Science, 5*, 117–128.

Fang, H., Luo, M., Sheng, Y., Li, Z. X., Wu, Y. Q., & Liu, C. (2008). The antihypertensive effect of peptides: A novel alternative to drugs. *Peptides, 29*, 1062–1071.

Fischer, J. E. (1990). Branched-chain-enriched amino acid solutions in patients with liver failure: An early example of nutritional pharmacology. *Journal of Parenteral and Enteral Nutrition, 14*, S249–S256.

Freitas, A. C., Andrade, J. C., Silva, F. M., Rocha-Santos, T. A., Duarte, A. C., & Gomes, A. M. (2013). Antioxidative peptides: Trends and perspectives for future research. *Current Medicinal Chemistry, 20*, 4575–4594.

Fujita, H., Yokoyama, K., & Yoshikawa, M. (2000). Classification and antihypertensive activity of angiotensin I-converting enzyme inhibitory peptides derived from food proteins. *Journal of Food Science, 65*, 564–569.

Fujita, H., & Yoshikawa, M. (1999). LKPNM: A prodrug-type ACE-inhibitory peptide derived from fish protein. *International Immunopharmacology, 44*, 123–127.

Gao, B., Seki, E., Brenner, D. A., Friedman, S., Cohen, J. I., Nagy, L., et al. (2011). Innate immunity in alcoholic liver disease. *American Journal of Physiology: Gastrointestinal and Liver Physiology, 300*, 516–525.

García-Olmedo, F., Molina, A., Alamillo, J. M., & Rodríguez-Palenzuéla, P. (1998). Plant defense peptides. *Biopolymers, 47*, 479–491.

Gardner, M. L. (1984). Intestinal assimilation of intact peptides and proteins from the diet—A neglected field? *Biological Reviews of the Cambridge Philosophical Society, 59*, 289–331.

Guo, H., Sun, J., He, H., Yu, G. C., & Du, J. (2009). Antihepatotoxic effect of corn peptides against Bacillus Calmette-Guerin/lipopolysaccharide-induced liver injury in mice. *Food and Chemical Toxicology, 47*, 2431–2435.

Gutiérrez-Ruiz, M. C., Quiroz, S. C., Souza, V., Bucio, L., Hernández, E., Olivares, et al. (1999). Cytokines, growth factors, and oxidative stress in HepG2 cells treated with ethanol, acetaldehyde, and LPS. *Toxicology, 134*, 197–207.

Hancock, J. T., Desikan, R., & Neill, S. J. (2001). Role of reactive oxygen species in cell signalling pathways. *Biochemical Society Transactions, 29*, 345–350.

Harman, L. S., Mottley, C., & Mason, R. P. (1984). Free radical metabolites of L-cysteine oxidation. *Journal of Biological Chemistry, 259*, 5606–5611.

He, Y.-Q., Ma, C.-Y., Pan, Y., Yin, L.-J., Zhou, J., Duan, Y., et al. (2018). Bioavailability of corn gluten meal hydrolysates and their effects on the immune system. *Czech Journal of Food Sciences, 36*, 1–7.

Hermansen, K. (2000). Diet, blood pressure and hypertension. *British Journal of Nutrition, 83*, S113–S119.

Hernández-Ledesma, B., Miralles, B., Amigo, L., Ramos, M., & Recio, I. (2005). Identification of antioxidant and ACE-inhibitory peptides in fermented milk. *Journal of the Science of Food and Agriculture, 85*, 1041–1048.

Hettiarachchy, N. S., Sato, K., Marshall, M. R., & Kannan, A. (Eds.), (2012). *Bioactive food proteins and peptides: Applications in human health.* New York: CRC Press [chapter 1].

Hidalgo, I. J., Raub, T. J., & Borchardt, R. T. (1989). Characterization of the human colon carcinoma cell line (Caco-2) as a model system for intestinal epithelial permeability. *Gastroenterology, 96*, 736–749.

Huang, W. Y., Majumder, K., & Wu, J. P. (2010). Oxygen radical absorbance capacity of peptides from egg white protein ovotransferrin and their interaction with phytochemicals. *Food Chemistry, 123*, 635–641.

Huang, W., Sun, J., He, H., Dong, H., & Li, J. (2011). Antihypertensive effect of corn peptides, produced by a continuous production in enzymatic membrane reactor, in spontaneously hypertensive rats. *Food Chemistry, 128*, 968–973.

Jin, D., Liu, X., Zheng, X., Wang, X., & He, J. (2016). Preparation of antioxidative corn protein hydrolysates, purification and evaluation of three novel corn antioxidant peptides. *Food Chemistry, 204*, 427–436.

Johansen, J. S., Harris, A. K., Rychly, D. J., & Ergul, A. (2005). Oxidative stress and the use of antioxidants in diabetes: Linking basic science to clinical practice. *Cardiovascular Diabetology, 4*, 5.

Kannan, A., Hettiarachchy, N., Johnson, M. G., & Nannapaneni, R. (2008). Human colon and liver cancer cell proliferation inhibition by peptide hydrolysates derived from heat-stabilized defatted rice bran. *Journal of Agricultural and Food Chemistry, 56*, 11643–11647.

Kim, J. M., Whang, J. H., Kim, K. M., Koh, J. H., & Suh, H. J. (2004). Preparation of corn gluten hydrolysate with angiotensin I converting enzyme inhibitory activity and its solubility and moisture sorption. *Process Biochemistry, 39*, 989–994.

Kim, J. M., Whang, J. H., & Suh, H. J. (2004). Enhancement of angiotensin I converting enzyme inhibitory activity and improvement of the emulsifying and foaming properties of corn gluten hydrolysate using ultrafiltration membranes. *European Food Research and Technology, 218*, 133–138.

Kitts, D. D. (1994). Bioactive substances in food: Identification and potential uses. *Canadian Journal of Physiology and Pharmacology, 72*, 423–434.

Kumar, K. J. S., Chu, F. H., Hsieh, H. W., Liao, J. W., Li, W. H., Lin, C. C., et al. (2011). Antroquinonol from ethanolic extract of mycelium of *Antrodia cinnamomea* protects hepatic cells from ethanol-induced oxidative stress through Nrf-2 activation. *Journal of Ethnopharmacology, 136*, 168–177.

LeBlanc, J. G., Matar, C., Valdéz, J. C., LeBlanc, J., & Perdigon, G. (2002). Immunomodulating effects of peptidic fractions issued from milk fermented with *Lactobacillus helveticus*. *Journal of Dairy Science, 85*, 2733–2742.

Lee, J. K., Hong, S., Jeon, J. K., Kim, S. K., & Byun, H. G. (2009). Purification and characterization of angiotensin I converting enzyme inhibitory peptides from the rotifer, *Brachionus rotundiformis*. *Bioresource Technology*, *100*, 5255–5259.

Lee, J., Koo, N., & Min, D. B. (2004). Reactive oxygen species, aging, and antioxidative nutraceuticals. *Comprehensive Reviews in Food Science and Food Safety*, *3*, 21–33.

Li, D., & Friedman, S. L. (1999). Liver fibrogenesis and the role of hepatic stellate cells: New insights and prospects for therapy. *Journal of Gastroenterology and Hepatology*, *14*, 618–633.

Li, H., Guo, P., Hu, X., Fang, P., Li, X., & Zhang, X. (2010). Antioxidant properties and possible mode of action of corn protein peptides and zein peptides. *Journal of Food Biochemistry*, *34*(s1), 44–60.

Li, H., Guo, P., Hu, X., Xu, L., & Zhang, X. (2007). Preparation of corn (*Zea mays*) peptides and their protective effect against alcohol-induced acute hepatic injury in NH mice. *Biotechnology and Applied Biochemistry*, *47*, 169–174.

Li, X., Han, L., & Chen, L. (2008). In vitro antioxidant activity of protein hydrolysates prepared from corn gluten meal. *Journal of the Science of Food and Agriculture*, *88*, 1660–1666.

Li, Y. W., Li, B., He, J., & Qian, P. (2011a). Quantitative structure–activity relationship study of antioxidative peptide by using different sets of amino acids descriptors. *Journal of Molecular Structure*, *998*, 53–61.

Li, Y. W., Li, B., He, J., & Qian, P. (2011b). Structure activity relationship study of antioxidative peptides by QSAR modeling: The amino acid next to C-terminus affects the activity. *Journal of Peptide Science*, *17*, 454–462.

Li, X., Li, X., Wu, X., Hua, W., Hou, R., Huang, Y., et al. (2004). Preparation and structural characterization of a new corn anti-oxidative peptide. *Chemical Research in Chinese Universities*, *25*, 466–469.

Li, H., Wen, L., Li, S., Zhang, D., & Lin, B. (2011). In vitro and in vivo effects and safety assessment of corn peptides on alcohol dehydrogenase activities. *Chemical Research in Chinese Universities*, *27*, 820–826.

Lin, F., Chen, L., Liang, R., Zhang, Z., Wang, J., Cai, M., et al. (2011). Pilot-scale production of low molecular weight peptides from corn wet milling byproducts and the antihypertensive effects in vivo and in vitro. *Food Chemistry*, *124*, 801–807.

Liu, Z. Y., Dong, S. Y., Xu, J., Zeng, M. Y., Song, H. X., & Zhao, Y. H. (2008). Production of cysteine-rich antimicrobial peptide by digestion of oyster (*Crassostrea gigas*) with alcalase and bromelin. *Food Control*, *19*, 231–235.

Liu, R. H., & Finley, J. (2005). Potential cell culture models for antioxidant research. *Journal of Agricultural and Food Chemistry*, *53*, 4311–4314.

Liu, X., Sun, Q., Wang, H., Zhang, L., & Wang, J.-Y. (2005). Microspheres of corn protein, zein, for an ivermectin drug delivery system. *Biomaterials*, *26*, 109–115.

Loguercio, C., & Federico, A. (2003). Oxidative stress in viral and alcoholic hepatitis. *Free Radical Biology and Medicine*, *34*, 1–10.

Lu, X. X., Chen, X. H., & Tang, J. Z. (2000). Studies on the functional property of enzymatic modified corn protein. *Journal of Food Science*, *21*, 13–15.

Lu, L., Liu, G., Wang, X. H., & Lou, S. J. (2016). Effects of supplement of corn peptides combined with aerobic exercise on lipolysis key enzymes: Adipose triglyceride lipase and lipoprotein lipase of obese rats. *Chinese Journal of Applied Physiology*, *32*, 326–331.

Lv, J., Nie, Z. K., Zhang, J. L., Liu, F. Y., Wang, Z. Z., Ma, Z. L., et al. (2013). Corn peptides protect against thioacetamide-induced hepatic fibrosis in rats. *Journal of Medicinal Food*, *16*, 912–919.

Ma, Y., Xiong, Y. L., Zhai, J., Zhu, H., & Dziubla, T. (2010). Fractionation and evaluation of radical-scavenging peptides from in vitro digests of buckwheat protein. *Food Chemistry*, *118*, 582–588.

Ma, Z. L., Zhang, W. J., Yu, G. C., He, H., & Zhang, Y. (2012). The primary structure identification of a corn peptide facilitating alcohol metabolism by HPLC-MS/MS. *Peptides*, *37*, 138–143.

Martínez-Maqueda, D., Miralles, B., Recio, I., & Hernández-Ledesma, B. (2012). Antihypertensive peptides from food proteins: A review. *Food and Function*, *3*, 350–361.

Meisel, H., & Fitzgerald, R. J. (2000). Opioid peptides encrypted in intact milk protein sequences. *British Journal of Nutrition*, *84*, S27–S31.

Mendis, E., Rajapakse, N., & Kim, S. K. (2005). Antioxidant properties of a radical-scavenging peptide purified from enzymatically prepared fish skin gelatin hydrolysate. *Journal of Agricultural and Food Chemistry*, *53*, 581–587.

Miao, F. S., Yu, W. Q., Wang, Y. G., Wang, M. J., Liu, X. Y., & Li, F. L. (2010). Effects of corn peptides on exercise tolerance, free radical metabolism in liver and serum glutamic-pyruvic transaminase activity of mice. *African Journal of Pharmacy and Pharmacology*, *4*, 178–183.

Miguel, M., & Aleixandre, A. (2006). Antihypertensive peptides derived from egg proteins. *Journal of Nutrition*, *136*, 1457–1460.

Miguel, M., Dávalos, A., Manso, M. A., de la Peña, G., Lasunción, M. A., & López-Fandiño, R. (2008). Transepithelial transport across Caco-2 cell monolayers of antihypertensive egg-derived peptides. PepT1-mediated flux of Tyr-Pro-Ile. *Molecular Nutrition and Food Research*, *52*, 1507–1513.

Miyoshi, S., Ishikawa, H., Kaneko, T., Fukui, F., Tanaka, H., & Maruyama, S. (1991). Structures and activity of angiotensin-converting enzyme inhibitors in an α-zein hydrolysate. *Agricultural and Biological Chemistry*, *55*, 1313–1318.

Miyoshi, S., Kaneko, T., Ishikawa, H., Tanaka, H., & Maruyama, S. (1995). Production of bioactive peptides from corn endosperm proteins by some proteases. *Annals of the New York Academy of Sciences*, *750*, 429–431.

Moller, N. P., Scholz-Ahrens, K. E., Roos, N., & Schrezenmeir, J. (2008). Bioactive peptides and proteins from foods: Indication for health effects. *European Journal of Nutrition*, *47*, 171–182.

Ngo, D. H., Qian, Z. J., Ryu, B. M., Park, J. W., & Kim, S. K. (2010). In vitro antioxidant activity of a peptide isolated from Nile tilapia (*Oreochromis niloticus*) scale gelatin in free radical-mediated oxidative systems. *Journal of Functional Foods*, *2*, 107–117.

Paulis, J. W., James, C., & Wall, J. S. (1969). Comparison of glutelin proteins in normal and high-lysine corn endosperms. *Journal of Agricultural and Food Chemistry*, *17*, 1301–1305.

Paulis, J. W., & Wall, J. S. (1977). Fractionation and characterization of alcohol-soluble reduced corn endosperm glutelin proteins. *Cereal Chemistry*, *54*, 1223–1228.

Pihlanto, A. (2006). Antioxidative peptides derived from milk proteins. *International Dairy Journal*, *16*, 1306–1314.

Puchalska, P., Luisa Marina, M., & Concepción García, M. (2013). Development of a high-performance liquid chromatography–electrospray ionization-quadrupole-time-of-flight-mass spectrometry methodology for the determination of three highly antihypertensive peptides in maize crops. *Journal of Chromatography A*, *1285*, 69–77.

Qian, Z. J., Jung, W. K., & Kim, S. K. (2008). Free radical scavenging activity of a novel antioxidative peptide purified from hydrolysate of bullfrog skin, *Rana catesbeiana Shaw*. *Bioresource Technology*, *99*, 1690–1698.

Qian, B., Xing, M., Cui, L., Deng, Y., Xu, Y., Huang, M., et al. (2011). Antioxidant, antihypertensive, and immunomodulatory activities of peptide fractions from fermented skim milk with *Lactobacillus delbrueckii* ssp. *bulgaricus* LB340. *Journal of Dairy Research*, *78*, 72–79.

Rajapakse, N., Mendis, E., Jung, W. K., Je, J. Y., & Kim, S. K. (2005). Purification of a radical scavenging peptide from fermented mussel sauce and its antioxidant properties. *Food Research International*, *38*, 175–182.

Ren, J. Y., Zhao, M. M., Cui, C., You, L. J., & Wang, H. Y. (2009). Isolation and identification of antioxidant peptides from hydrolyzed grass carp protein. *Food Science, 30,*
13–17.

Rossi, D. M., Flores, S. H., Heck, J. X., & Ayub, M. A. Z. (2009). Production of high-
protein hydrolysate from poultry industry residue and their molecular profiles. *Food
Biotechnology, 23,* 229–242.

Samaranayaka, A. G. P., & Li-Chan, E. C. Y. (2011). Food-derived peptidic antioxidants:
A review of their production, assessment, and potential applications. *Journal of Functional
Foods, 3,* 229–254.

Sarmadi, B. H., & Ismail, A. (2010). Antioxidative peptides from food proteins: A review.
Peptides, 31, 1949–1956.

Satake, M., Enjoh, M., Nakamura, Y., Takano, T., Kawamura, Y., Arai, S., et al. (2002).
Transepithelial transport of the bioactive tripeptide, Val-Pro-Pro, in human intestinal
Caco-2 cell monolayers. *Bioscience, Biotechnology, and Biochemistry, 66,* 378–384.

Scholz-Ahrens, K. E., & Schrezenmeir, J. (2002). Inulin, oligofructose and mineral
metabolism—Experimental data and mechanism. *The British Journal of Nutrition, 87,*
S179–S186.

Sheih, I. C., Wu, T. K., & Fang, T. J. (2009). Antioxidant properties of a new antioxidative
peptide from algae protein waste hydrolysate in different oxidation systems. *Bioresource
Technology, 100,* 3419–3425.

Shi, R. F., Wang, S. L., Lou, S. J., Li, X., Wang, R., & Wang, X. H. (2013). In *Corn peptides
ingestion with exercise attenuate body fat mass and enhance lean body mass in obese rats. The 11th
China nutrition science congress & international DRIs summit evolution of DRIs: Nutrition science
and practice based on evidence.*

Shimizu, M. (2004). In *Food-derived peptides and intestinal functions. International conference on
food factors.*

Shukla, R., & Cheryan, M. (2001). Zein: The industrial protein from corn. *Industrial Crops
and Products, 13,* 171–192.

Siow, H. L., & Gan, C. Y. (2013). Extraction of antioxidative and antihypertensive bioactive
peptides from *Parkia speciosa* seeds. *Food Chemistry, 141,* 3435–3442.

Sodek, L., & Wilson, C. M. (1971). Amino acid composition of proteins isolated from normal, opaque-2, and floury-2 corn endosperms by a modified Osborne procedure. *Journal
of Agricultural and Food Chemistry, 19,* 1144–1150.

Suh, H. J., Whang, J. H., Kim, Y. S., Bae, S. H., & Noh, D. O. (2003). Preparation of angiotensin I converting enzyme inhibitor from corn gluten. *Process Biochemistry, 38,*
1239–1244.

Suh, H. J., Whang, J. H., & Lee, H. (1999). A peptide from corn gluten hydrolysate that is
inhibitory toward angiotensin I converting enzyme. *Biotechnology Letters, 21,* 1055–1058.

Sui, Y. (2009). Hepatoprotective effects of corn peptides against D-galactosamine-induced
liver injury in mice. *Journal of the Chinese Cereals and Oils Association, 5,* 36–39.

Sun, P., Tian, X. Y., & Shi, R. F. (2017). Effects of dietary corn peptide and exercise on
hepatic ATGL and TNF-α in obese rats. *Chinese Journal of Physiology, 33,* 117–120.

Szabo, G., Petrasek, J., & Bala, S. (2012). Innate immunity and alcoholic liver disease. *Digestive Diseases and Sciences, 30,* S55–S60.

Tang, X., He, Z., Dai, Y., Xiong, Y. L., Xie, M., & Chen, J. (2010). Peptide fractionation
and free radical scavenging activity of zein hydrolysate. *Journal of Agricultural and Food
Chemistry, 58,* 587–593.

Tang, N., & Zhuang, H. (2014). Evaluation of antioxidant activities of zein protein fractions.
Journal of Food Science, 79, 2174–2184.

Vermeirssen, V., Augustijns, P., Morel, N., Van, C. J., Opsomer, A., & Verstraete, W.
(2005). In vitro intestinal transport and antihypertensive activity of ACE inhibitory
pea and whey digests. *International Journal of Food Sciences and Nutrition, 56,* 415–430.

Vermeirssen, V., Deplancke, B., Tappenden, K. A., Van, C. J., Gaskins, H. R., & Verstraete, W. (2002). Intestinal transport of the lactokinin Ala-Leu-Pro-Met-His-Ile-Arg through a Caco-2 Bbe monolayer. *Journal of Peptide Science*, *8*, 95–100.

Wade, A. M., & Tucker, H. N. (1998). Antioxidant characteristics of L-histidine. *Journal of Nutritional Biochemistry*, *9*, 308–315.

Wang, Y., Chen, H., Wang, X., Li, S., Chen, Z., Wang, J., et al. (2015). Isolation and identification of a novel peptide from zein with antioxidant and antihypertensive activities. *Food & Function*, *6*, 3799–3806.

Wang, L., Ding, L., Wang, Y., Zhang, Y., & Liu, J. (2015). Isolation and characterisation of in vitro and cellular free radical scavenging peptides from corn peptide fractions. *Molecules*, *20*, 3221–3237.

Wang, X., Zheng, X., Kopparapu, N., Cong, W., Deng, Y., Sun, X., et al. (2014). Purification and evaluation of a novel antioxidant peptide from corn protein hydrolysate. *Process Biochemistry*, *49*, 1562–1569.

WHO (2017). Global hepatitis report, 2017 [OL]. http://apps.who.int/iris/bitstream/10665/255016/1/9789241565455-eng.pdf?ua = 1.

Williams, R. (2006). Global challenges in liver disease. *Hepatology*, *44*, 521–526.

Wilson, C. M. (1985). A nomenclature for zein polypeptides based on isoelectric focusing and sodium dodecyl sulfate polyacrylamide gel electrophoresis. *Cereal Chemistry*, *62*, 361–365.

Wu, J. P., Aluko, R. E., & Muir, A. D. (2009). Production of angiotensin I-converting enzyme inhibitory peptides from defatted canola meal. *Bioresource Technology*, *100*, 5283–5287.

Wu, Y., Pan, X., Zhang, S., Wang, W., Cai, M., Li, Y., et al. (2014). Protective effect of corn peptides against alcoholic liver injury in men with chronic alcohol consumption: A randomized double-blind placebo-controlled study. *Lipids in Health and Disease*, *13*, 192–200.

Xu, Y., Leo, M. A., & Lieber, C. S. (2003). Lycopene attenuates alcoholic apoptosis in HepG2 cells expressing CYP2E1. *Biochemical and Biophysical Research Communications*, *308*, 614–618.

Yamaguchi, M., Nishikiori, F., Ito, M., & Furukawa, Y. (1996). Effect of corn peptide on alcohol metabolism and plasma free amino acid concentrations in healthy men. *European Journal of Clinical Nutrition*, *50*, 682–688.

Yamaguchi, M., Nishikiori, F., Ito, M., & Furukawa, Y. (1997). The effects of corn peptide ingestion on facilitating alcohol metabolism in healthy men. *Bioscience, Biotechnology, and Biochemistry*, *61*, 1474–1481.

Yamaguchi, M., Nozaki, O., Ito, M., & Furukawa, Y. (1997). Effect of corn peptide administration on plasma amino acid concentration and alcohol metabolism in stroke-prone spontaneously hypertensive rats. *Journal of Clinical Biochemistry and Nutrition*, *22*, 77–89.

Yamaguchi, M., Takada, M., Nozaki, O., Ito, M., & Furukawa, Y. (1996). Preparation of corn peptide from corn gluten meal and its effect on alcohol metabolism in stroke-prone spontaneously hypertensive rats. *Journal of Nutritional Science and Vitaminology*, *42*, 219–231.

Yang, S. C., Ito, M., Morimatsu, F., Furukawa, Y., & Kimura, S. (1993). Effects of amino acids on alcohol intake in stroke-prone spontaneously hypertensive rats. *Journal of Nutritional Science and Vitaminology*, *39*, 55–61.

Yang, Y., Tao, G., Liu, P., & Liu, J. (2007). Peptide with angiotensin I-converting enzyme inhibitory activity from hydrolyzed corn gluten meal. *Journal of Agricultural and Food Chemistry*, *55*, 7891–7895.

Yang, R. Y., Zhang, Z. F., Pei, X. R., Han, X. L., Wang, J. B., Wang, L. L., et al. (2009). Immunomodulatory effects of marine oligopeptide preparation from Chum Salmon (*Oncorhynchus keta*) in mice. *Food Chemistry*, *113*, 464–470.

Yongvanit, P., Pinlaor, S., & Bartsch, H. (2012). Oxidative and nitrative DNA damage: Key events in opisthorchiasis-induced carcinogenesis. *Parasitology International, 61*, 130–135.

Yu, G. C., Li, J. T., He, H., Huang, W. H., & Zhang, W. J. (2013). Ultrafiltration preparation of potent bioactive corn peptide as alcohol metabolism stimulator in vivo and study on its mechanism of action. *Journal of Food Biochemistry, 37*, 161–167.

Yu, G. C., Lv, J., He, H., Huang, W., & Han, Y. (2012). Hepatoprotective effects of corn peptides against carbon tetrachloride-induced liver injury in mice. *Journal of Food Biochemistry, 36*, 458–464.

Yu, Y., Wang, L., Wang, Y., Lin, D., & Liu, J. (2017). Hepatoprotective effect of albumin peptides from corn germ meal on chronic alcohol-induced liver injury in mice. *Journal of Food Science, 82*, 2997–3004.

Zasloff, M. (2002). Antimicrobial peptides of multicellular organisms. *Nature, 415*, 389–395.

Zhang, Z., Huang, F., Zhu, H., & Song, J. (2009). Production of corn peptide by fermentation with *Bacillus subtilis* ls-45. *Journal of the Chinese Cereals and Oils Association, 24*, 37–41.

Zhang, F., Zhang, J., & Li, Y. (2012). Corn oligopeptides protect against early alcoholic liver injury in rats. *Food and Chemical Toxicology, 50*, 2149–2154.

Zheng, X., Li, L., Liu, X., Wang, X., Lin, J., & Li, D. (2006). Production of hydrolysate with antioxidative activity by enzymatic hydrolysis of extruded corn gluten. *Applied Microbiology and Biotechnology, 73*, 763–770.

Zhou, C., Hu, J., Ma, H., Yagoub, A. E., Yu, X., Owusu, J., et al. (2015). Antioxidant peptides from corn gluten meal: Orthogonal design evaluation. *Food Chemistry, 187*, 270–278.

Zhou, K., Sun, S., & Canning, C. (2012). Production and functional characterisation of antioxidative hydrolysates from corn protein via enzymatic hydrolysis and ultrafiltration. *Food Chemistry, 135*, 1192–1197.

Zhuang, H., Tang, N., Dong, S. T., Sun, B., & Liu, J. B. (2013). Optimisation of antioxidant peptide preparation from corn gluten meal. *Journal of the Science of Food and Agriculture, 93*, 3264–3270.

Ziv, E., & Bendayan, M. (2015). Intestinal absorption of peptides through the enterocytes. *Microscopy Research and Technique, 49*, 346–352.

> CHAPTER TWO

Dietary Fatty Acids and the Metabolic Syndrome: A Personalized Nutrition Approach

Sarah O'Connor*,†, Iwona Rudkowska*,†,1

*CHU de Québec Research Center, Université Laval, Québec, QC, Canada
†Department of Kinesiology, Faculty of Medicine, Université Laval, Québec, QC, Canada
¹Corresponding author: e-mail address: iwona.rudkowska@crchudequebec.ulaval.ca

Contents

Abstract

Dietary fatty acids are present in a wide variety of foods and appear in different forms and lengths. The different fatty acids are known to have various effects on metabolic health. The metabolic syndrome (MetS) is a constellation of risk factors of chronic diseases. The etiology of the MetS is represented by a complex interplay of genetic and environmental factors. Dietary fatty acids can be important contributors of the evolution or in prevention of the MetS; however, great interindividual variability exists in the

Advances in Food and Nutrition Research, Volume 87
ISSN 1043-4526
https://doi.org/10.1016/bs.afnr.2018.07.004

response to fatty acids. The identification of genetic variants interacting with fatty acids might explain this heterogeneity in metabolic responses. This chapter reviews the mechanisms underlying the interactions between the different components of the MetS, dietary fatty acids and genes. Challenges surrounding the implementation of personalized nutrition are also covered.

1. INTRODUCTION

Dietary fatty acids are nutritive macronutrients providing an important amount of energy for the human body but are also important contributors of flavor and palatability in foods. The biological functions of dietary fatty acids are diverse, some have detrimental and others beneficial effects on global health. Recent shifts in dietary patterns in modern societies contributed to an increase in the prevalence of the metabolic syndrome (MetS), a cluster of risk factors including abdominal obesity, inflammation, hyperglycemia, dyslipidemia and high blood pressure. Each component of the MetS has a multifactorial etiology represented by a complex interplay between genetic and environmental factors. Thus, genetic background has a key role in the development of the MetS and the variability of responses to lifestyle modifications. Dietary fatty acids, especially the different types of fats, have major roles in the evolution or in the prevention of the MetS; yet, great disparities in the response to fatty acids have been observed between individuals. In parallel, genomics examines the interaction between diet and genes in order to identify which individuals might greater benefit from changes in the intake of dietary fatty acids in prevention or management of the MetS. The objective of this chapter is to explore the effects of total dietary fats and different types of dietary fatty acids on the MetS and its components by integrating recent knowledge generated by genomic research. First, the nature and sources of dietary fatty acids are described, followed by the comparison between actual recommendations and consumption around the world. Next, the role of genomics in the understanding of gene–diet interactions in health and diseases are covered, followed by a brief description of the MetS etiology. Further, the roles of different fatty acids in the prevention, evolution or management of the components of the MetS are covered, using specific genetic and transcriptomic approaches. Finally, the challenges surrounding personalized nutrition and its integration in public policies are covered.

2. DIETARY FATTY ACIDS

2.1 Nature and Properties of Dietary Fatty Acids

Fatty acids are carbon chains of different lengths and degrees of saturation, composed of a methyl group at one end and a carboxyl group at the other. When all carbons are saturated with hydrogen, the fatty acids are called saturated fatty acids (SFA). The complete hydrogenation of SFA gives the molecules a flat configuration and higher melting point, often resulting in solid fats at room temperature (Rustan & Drevon, 2005). Most dietary SFA are composed of 12–22 carbons atoms, yet shorter-chain and longer-chain fatty acids occur naturally in foods. Dietary medium-chain fatty acids (MCFA) contain 6–10 carbons and are always saturated with carbons. MCFA are more efficiently absorbed and are transported directly to the liver through the portal vein without incorporating chylomicrons unlike their longer-chain counterparts. These major differences give MCFA unique metabolic effects in comparison with long-chain fatty acids (Marten, Pfeuffer, & Schrezenmeir, 2006). In contrast with SFA, unsaturated fatty acids have one or multiple double bonds due to unsaturation of some carbons with hydrogen. Unsaturated fatty acids with a single double bond are called mono-unsaturated fatty acids (MUFA), while fatty acids composed of two or more double bonds are called poly-unsaturated fatty acids (PUFA) (Rustan & Drevon, 2005). The number and location of double bonds on the molecules give fatty acids their unique physiochemical properties and their proper identification. PUFA, which include omega-3 and omega-6 PUFA, are essential to all mammals since they cannot be synthetized from other fatty acids. The omega-6 and omega-3 terminology relates to the position of the first double bond from the methyl end (Merched & Chan, 2013). The vast majority of unsaturated fatty acids in nature have a *cis*-configuration, giving the molecules a kink shape decreasing significantly their melting point. In contrast, a *trans*-configuration of unsaturated fatty acids, known as *trans* fatty acids (TFA), gives the molecules a similar configuration as SFA and higher melting point than their *cis*-configured counterparts (Rustan & Drevon, 2005).

2.2 Actual Consumption and Dietary Requirements of Dietary Fatty Acids

The main sources of total and individual fatty acids along with the actual consumption and the main dietary requirements are presented in Table 1.

Table 1 Sources of Dietary Fatty Acids, Global Intake and Actual Dietary Guidelines

Fatty Acids	Main Dietary Sources	Consumption of Dietary Fatty Acids	Main Dietary Requirements
Total fatty acids	Animal sources: Lard, butter, visible fat on meat products, egg yolk, full-fat milk, cheese, cream, fatty fish, fish oils Vegetal sources: Vegetable oils, avocado, nuts and seeds, margarine Other sources: potato chips and other fatty snack foods, deep fried foods, cakes, pastries, biscuits, pies	Global consumption: 20–35%TE (11.1–46.2%TE) Canada: Between 30% and 33% TE for most age groups United States: 34%TE	RDA (Institute of Medicine, 2006): 20–35%TE for adults WHO (2015): <30%TE from fat, supports boiling instead of frying AHA/ACC (2013): No limit provided, supports the consumption of low-fat dairy products (for reduction of LDL-C and blood pressure) (Eckel et al., 2013) Academy of Nutrition and Dietetics (2014): 20–35%TE (Vannice & Rasmussen, 2014) NLA (2015)[a]: No limit imposed (Jacobson et al., 2015)
Saturated fatty acids	Animal sources: Meat products and dairy products Vegetal sources: Palm oil, palm kernel oil, coconut oil, cocoa butter, fully hydrogenated vegetable oils	Global consumption: 9.4%TE (2.3–23.5%TE) Canada: 10.2–10.9%TE United States: 11%TE Western Europe: 12.6%TE Asia Pacific: 8.8%TE Northern Africa//Middle East: 10.3%TE	RDA (Institute of Medicine, 2006): Non-essential fatty acids; no RDA fixed, suggests to limit dietary intake WHO (2015): <10%TE, replacing with unsaturated fats. Suggests removing fatty part of meat and limits the consumption of foods rich in SFA AHA/ACC (2013): Reduce SFA intake to achieve 5–6%TE (for reduction of LDL-C) (Eckel et al., 2013) Academy of Nutrition and Dietetics (2014): <7%TE, maximum 10% (Vannice & Rasmussen, 2014) NLA (2015)[a]: <7%TE (Jacobson et al., 2015)
Mono-unsaturated fatty acids	Animal sources: Beef, lard Vegetal sources: Olive oil, canola oil, avocados, nuts (cashews, almonds, macadamia nuts)	Global consumption: 12%TE (2.3–23.5%) Canada: 12.5%TE United States: 12.5%TE	RDA (Institute of Medicine, 2006): Non-essential fatty acids; no RDA fixed WHO (2015): Supports to replace SFA with unsaturated fats AHA/ACC (2013): Supports the consumption of non-tropical oils (for reduction of LDL-C and blood pressure) (Eckel et al., 2013) Academy of Nutrition and Dietetics (2014): 15–20%TE

	Sources	Consumption	Recommendations
			(Vannice & Rasmussen, 2014) NLA (2015)[a]: Suggests to adopt a Mediterranean-style diet (rich in olive oil) (Jacobson et al., 2015)
Poly-unsaturated fatty acids		Canada: 5.6%TE United States: 7.0%TE	Academy of Nutrition and Dietetics (2014): 3–10%TE (Vannice & Rasmussen, 2014)
Omega-6 poly-unsaturated fatty acids	Animal sources: Meat, poultry and eggs (small amounts of arachidonic acid) Vegetal sources: safflower oil, soybean oil, corn oil, nuts (brazil nuts)	Global consumption: 5.9%TE (2.5–8.5%). Canada: 4.2%TE United States: 6.7%TE Western Europe: 5.2%TE Asia Pacific: 4.4%TE Northern Africa//Middle East: 5.9%TE	RDA (Institute of Medicine, 2006): 5–10%TE for all ages WHO (2015): Supports to replace SFA with unsaturated fats AHA/ACC (2013): Supports the consumption of non-tropical oils (for reduction of LDL-C and blood pressure) (Eckel et al., 2013) NLA (2015)[a]: No limit imposed (Jacobson et al., 2015)
Omega-3 poly-unsaturated fatty acids	Animal sources: Fatty fish, fish oil, seafood Vegetal sources: Flaxseed, chia, walnuts, canola oil	Global consumption: 163 mg/day, vegetal sources: 1371 mg, marine sources: 163 mg Canada: Vegetal sources: 2085 mg, marine sources: 82 mg United States: Vegetal sources: 127 mg, marine sources: 41 mg Western Europe: Vegetal sources: 1120 mg, marine sources: 351 mg Asia Pacific: Vegetal sources: 1128 mg, marine sources: 701 mg Northern Africa//Middle East: Vegetal sources: 1176 mg marine sources, 112 mg	RDA (Institute of Medicine, 2006): 0.6–1.2%TE for all ages WHO (2015): Supports to replace SFA with unsaturated fats AHA/ACC (2013): Supports the consumption of non-tropical oils (for reduction of LDL-C and blood pressure) (Eckel et al., 2013) NLA (2015)[a]: No limit imposed (Jacobson et al., 2015)

Continued

Table 1 Sources of Dietary Fatty Acids, Global Intake and Actual Dietary Guidelines—cont'd

Fatty Acids	Main Dietary Sources	Consumption of Dietary Fatty Acids	Main Dietary Requirements
Trans fatty acids	Industrial sources: processed foods with partially hydrogenated oils, margarine Natural sources: Ruminant meat and dairy products	Global consumption: 1.4%TE (0.2–6.5%) Canada: 1.9%TE United States: 1.1%TE Western Europe: 1.15%TE Asia Pacific: 1.0%TE Northern Africa/Middle East: 2.4%TE	RDA (Institute of Medicine, 2006): Non-essential fatty acids; no RDA fixed, suggestion to limit dietary intake WHO (2015): <1%TE, replace by unsaturated fats, suggests avoiding processed foods containing trans fatty acids AHA/ACC (2013): Reduce the % of calories from TFA (for reduction of LDL-C and blood pressure) (Eckel et al., 2013) Academy of Nutrition and Dietetics (2014): <1%TE (Vannice & Rasmussen, 2014) NLA (2015)[a]: Limit TFA intake (Jacobson et al., 2015)

[a]NLA guidelines for the management of dyslipidemia.

ACC: American College of Cardiology; AHA: American Heart Association; LDL-C: low-density lipoprotein cholesterol, NLA: National Lipid Association; RDA: recommended dietary allowance; SFA: saturated fatty acids; TFA: *trans* fatty acids; WHO: World Health Organisation; %TE: % total energy intake.

Adapted from Bowen, K. J., Sullivan, V. K., Kris-Etherton, P. M., & Petersen, K. S. (2018). Nutrition and cardiovascular disease—An update. *Current Atherosclerosis Reports*, 20(2), 8. https://doi.org/10.1007/s11883-018-0704-3; Harika, R. K., Eilander, A., Alssema, M., Osendarp, S. J. M., & Zock, P. L. (2013). Intake of fatty acids in general populations worldwide does not meet dietary recommendations to prevent coronary heart disease: A systematic review of data from 40 countries. *Annals of Nutrition and Metabolism*, 63(3), 229–238; Micha, R., Khatibzadeh, S., Shi, P., Fahimi, S., Lim, S., Andrews, K. G., … Mozaffarian, D. (2014). Global, regional, and national consumption levels of dietary fats and oils in 1990 and 2010: A systematic analysis including 266 country-specific nutrition surveys. *British Medical Journal*, 348, g2272. https://doi.org/10.1136/bmj.g2272; Wanders, A. J., Zock, P. L., & Brouwer, I. A. (2017). Trans fat intake and its dietary sources in general populations worldwide: A systematic review. *Nutrients*, 9(8), 840. https://doi.org/10.3390/nu9080840.

The main sources of dietary fatty acids are vegetable oils, meat and dairy products, fish and fish oils, nuts, seeds and fatty processed foods (Table 1) (Australian Government, 2015; Rustan & Drevon, 2005). Most fatty acids are incorporated in triglycerides in food (98%) with small quantities in dietary phospholipids (Institute of Medicine, 2006). Energy intake from lipids varies considerably between geographic locations. Mean global dietary fat intake ranges between 20% and 35% of total energy in data from 25 countries (Harika, Eilander, Alssema, Osendarp, & Zock, 2013). The Institute of Medicine recommends consuming around 20–35% of energy from dietary fats to maintain optimal health in adults (Institute of Medicine, 2006). Similarly, the World Health Organisation (WHO) recommends limiting fatty acid intake up to 30% of energy intake in order to ease weight maintenance (World Health Organisation, 2015). Finally, the American Heart Association/American College of Cardiology (AHA/ACC) Guideline on lifestyle does not specified limits on fatty acids; however, low-fat dairy products are encouraged (Table 1) (Eckel et al., 2013).

Main sources of SFA are animal products such as meat and dairy products, and some vegetable oils like coconut oil and palm oil (Table 1) (Australian Government, 2015; Rustan & Drevon, 2005). Palmitic acid (16:0) is the most common SFA in food, followed by stearic acid (18:0) and myristic acid (14:0). MCFA are common in coconut oil (representing more than 50% of fatty acids) and dairy milk, composing around 4–12% of dairy fat (Marten et al., 2006; Rustan & Drevon, 2005). Mean global SFA intake in adults was estimated at 9.4% of total energy intake in 2010, with 61.8% of the world's population consuming less than 10% of total energy intake (Table 1) (Micha et al., 2014). In the United States, median intake of SFA is estimated at 11% of total energy, coming mostly from processed foods such as burgers, sandwiches, snacks and candies (Harika et al., 2013; Ruiz-Núñez, Dijck-Brouwer, & Muskiet, 2016; Wang & Hu, 2017). No RDA has been fixed for SFA since these fats are not essential to the human body (Institute of Medicine, 2006). Yet, the WHO and other national food guides recommend to limit SFA intake to a maximum of 10% of total energy (Health Canada, 2007; U.S. Department of Health and Human Services and U.S. Department of Agriculture, 2015; World Health Organisation, 2015). Other institutions, such as the AHA/ACC, the National Lipid Association (NLA) and the Academy of Nutrition and Dietetics, recommend to limit SFA intake to 5–7% of total energy in order to prevent cardiovascular diseases (Table 1) (Eckel et al., 2013; Jacobson et al., 2015; Vannice & Rasmussen, 2014).

The main sources of MUFA are vegetable oils, including olive and canola oils, nuts, peanuts, avocados and animal products (Table 1) (Australian Government, 2015; Ros, 2003). Dietary MUFA are provided at approximately 50% by animal sources, mostly oleic acid (18:1, n-9) (Institute of Medicine, 2006; Ros, 2003). *Cis*-palmitoleic acid (16:1, n-7) is another MUFA present in some seed oils, macadamia and blue-green algae (Australian Government, 2015; Rustan & Drevon, 2005). Median global consumption of MUFA was estimated at 12% of total energy and varies around 2.3–23.5% across countries (Table 1) (Harika et al., 2013; Micha et al., 2014). No RDA has been elaborated for MUFA because these fatty acids can be synthetized endogenously and thus, are not essential (Institute of Medicine, 2006). Since the beneficial effects of MUFA on health are still debated, no specific amount of MUFA to consume has been specified by most health institutions, with the exception of the Academy of Nutrition and Dietetics which suggests around 15–20% of total energy intake from MUFA (Vannice & Rasmussen, 2014). The WHO supports the replacement of SFA by unsaturated oils including MUFA-rich oils (World Health Organisation, 2015). The NLA suggests to adopt a Mediterranean diet (MedDiet) (known to be rich in olive oil) in order to prevent diseases (Table 1) (Jacobson et al., 2015).

The omega-6 PUFA family is mainly represented by linoleic acid (LA) (18:2, n-6) in foods, which major sources are sunflower oil, soy oil, corn oil, nuts and seeds (Table 1) (Monteiro et al., 2014; Wang & Hu, 2017). Arachidonic acid (ARA) (20:4, n-6) is another omega-6 PUFA found in small quantities in meat and eggs (Australian Government, 2015). Omega-3 PUFA are found in vegetal and animal sources. Alpha-linolenic acid (ALA) (18:3, n-3) is present in plant-based food sources, principally flaxseed oil, chia seeds and walnuts (Australian Government, 2015; Monteiro et al., 2014). Long-chain omega-3 PUFA, namely eicosapentaenoic acid (EPA) (20:5, n-3) and docosahexaenoic acid (DHA) (22:6, n3), are found in fatty fishes such as salmon, herring and mackerel (Table 1) (Monteiro et al., 2014; Rustan & Drevon, 2005). EPA and DHA have greater biological actions in the body than ALA; however, conversion from dietary ALA to EPA and DHA is around 6% and 3%, respectively (Gerster, 1998). Global consumption of omega-6 PUFA was around 5.9% of total energy in 2010, ranging from 2.5% to 8.5% across countries (Table 1). During the past decades, omega-6 LA consumption has greatly increased in Western countries with the use of vegetable oils in replacement of animal fats for cooking. In 2010, around 52% of the world population was consuming the recommended amount of omega-6 PUFA. For omega-3 PUFA, the global mean intake from plants was 1371 mg/day, while marine omega-3 PUFA was around 160 mg/day, with important variations between countries depending on the availability of marine products (Table 1) (Micha et al., 2014). Given omega-6 and omega-3 PUFA are essential fatty acids, specific RDA have been fixed, representing 5–10% of total energy for omega-6 PUFA and 0.6–1.2% of total energy for ALA (Institute of Medicine, 2006). Knowing that there is limited conversion of ALA to EPA and DHA, most dietary institutions around the world recommend two servings of fish per week (around 500 mg of omega-3 PUFA/day) for the general population and up to 1 g/day of EPA and DHA for coronary patients (Delgado-Lista et al., 2016; Rustan & Drevon, 2005).

Dietary TFA can originated either from industrial sources (iTFA) or natural ruminant sources (rTFA). Most TFA consumed are from industrial sources (Micha & Mozaffarian, 2009). iTFA are by-products of the hydrogenation of vegetable oils, initially with *cis*-configurations. Hydrogenation process uses heating cycles to add hydrogens to double bonds in order to saturate fatty acids with hydrogen. Hydrogenation increases melting point of fatty acids, resulting in solid and stable fats at room temperature (Nishida & Uauy, 2009). Therefore, iTFA are solely present in processed foods containing hydrogenated oils such as fried goods, cookies, commercial cakes, crackers, fried potatoes, potato chips, popcorn, margarine or shortening (Table 1). The main iTFA is elaïdic acid (18:1, *trans*-9) (Dhaka, Gulia, Ahlawat, & Khatkar, 2011). rTFA are present mostly in dairy products and ruminant meat (Table 1) (Tardy, Morio, Chardigny, & Malpuech-Brugère, 2011). *Trans* vaccenic acid (TVA) (18:1, *trans*-11) is a rTFA found in all ruminant food sources, while *trans* palmitoleic acid (TPA) (*trans* 16:1, n-7) is mainly found in dairy products (Dhaka et al., 2011; Tremblay & Rudkowska, 2017). Conjugated fatty acids (CLA) are rTFA represented by different isomers of position and configuration of LA. The isomer 9*cis*-11*trans* and 10*trans*-12*cis* are the most abundant in nature (Kim, Kim, Kim, & Park, 2016). Worldwide, the mean consumption of TFA was estimated at 1.4% of total energy with a 28-fold difference between countries (from 0.2% to 6.5%) (Table 1). 99.4% of the world adult population had higher intakes of TFA than the limit of <0.5% of total energy recommended in 2010 (Micha et al., 2014). Regarding rTFA, human consumption is estimated at <0.5% of total energy intake due to low occurring in natural foods (Micha & Mozaffarian, 2009). CLA intake is estimated at around 100–200 mg/day in the United States (Kim et al., 2016). Overall, roughly 80% of TFA intake comes from industrial sources in comparison with approximately 20% from natural sources; yet, great variations are observed between geographical regions and countries (Dhaka et al., 2011). No RDA has been fixed for iTFA or rTFA; however, the Institute of Medicine recommends to keep TFA intake as low as possible (Institute of Medicine, 2006). Likewise, major dietary institutions suggest to limit and/or avoid TFA (Eckel et al., 2013; Jacobson et al., 2015; Micha et al., 2014; Vannice & Rasmussen, 2014; World Health Organisation, 2015).

3. PERSONALIZED NUTRITION USING OMICs

Actual dietary requirements surrounding dietary fatty acids have evolved according to strong epidemiological evidences on fat intake at a population level. The main dietary requirements are usually built following a *"one size fit all"* model in order to suit the needs of the majority of the population; yet, accumulating evidence shows great disparities in individual responses to dietary fatty acids. These discrepancies could be explained partially by the unique genetic background of individuals, but also by the interactions between genes and environmental factors, such as diet. For example, Schaefer et al. (1997) observed an overall reduction of low-density lipoprotein cholesterol (LDL) levels after the administration of a National Cholesterol Education Program (NCEP) Step 2 diet, which consists of a strict dietary approach of $\leq$30% of total energy from fat, $<$7% of total energy from SFA and $<$200 mg of dietary cholesterol per day for the treatment of hypercholesterolemia. Yet, great variations of individual LDL levels in response to the NCEP Step 2 diet were observed, ranging from +3% to $-$55% and +13% to $-$39% in men and women, respectively (Schaefer et al., 1997). From these important interindividual variabilities emerged the development of personalized nutrition, which has grown in popularity since the beginning of the millennium (Corella, Coltell, Mattingley, Sorlí, & Ordovas, 2017). Personalized nutrition is a novel field of research studying the variability of responses to diet in relation with genotype and environmental factors in order to better prevent or manage diseases at an individual level (de Toro-Martín, Arsenault, Després, & Vohl, 2017).

The increasing interest in personalized nutrition nowadays is mostly contributive of genomics. Genomics is a field of research focusing on the characterization of genome, which represent all of an individual's genes. Genomics is part of "omics" technologies, known as different fields of research aiming at measuring as many biological molecules as possible (genes, RNA, proteins, metabolites, lipids, etc.) in order to achieve a systematic overview of biological phenomena (Benkeblia, 2012). Genomics also studies the interactions of genes with others genes and with environmental factors (National Human Genome Research Institute, 2016). Nutrigenomics is a discipline derived from genomics seeking a global understanding of gene–diet interactions. Nutrigenomics is often used as a general term including nutrigenetics and gene transcription studies, or transcriptomics (Fig. 1). Nutrigenetics is a field of research studying the different phenotype responses to a given diet or nutrient in relation with individual genotypes (Corella et al., 2017). Genetic variations can have numerous manifestations; single nucleotide polymorphisms (SNPs) genetic variations are the most widely studied to predict given phenotypes (Ning & Kaput, 2009). SNPs can be used to identify individuals with predispositions to some diseases according to the presence/absence of some risk alleles. Since many genes are usually involved in multifactorial diseases, genome-wide association studies (GWAS) can be used to analyze millions of SNPs simultaneously and identify the relevant SNPs related to a given disease or risk factor. Following the identification of individual SNPs, the relevant SNPs for a specific phenotype can be clustered in genetic risk scores (GRSs). GRSs represent scores calculated according to the number of risk alleles and can be useful to discriminate individuals presenting a high genetic risk of disease from those at low risk (Corella et al., 2017). The combination of multiple SNPs makes GRSs usually more informative than individual SNPs to understand the link between genetic background and phenotypes (Fitó et al., 2016). Despite the discovery of multiple genetic variants

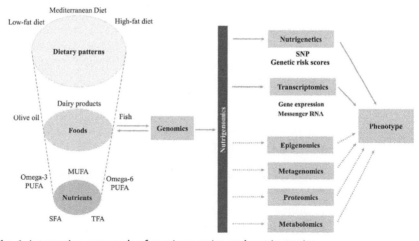

Fig. 1 Integrative approach of nutrigenomics and nutrigenetics.

associated with health and diseases, the genetic loci identified to date only give a limited explanation of total heritability of diseases. This "missing" heritability is likely to be associated with interactions between genes and environment, such as diet (Heianza & Qi, 2017). The identification of gene-diet interactions can discriminate responders (the anticipated outcome) of non-responders (no response to the intervention), and adverse responders (a detrimental and often unexpected outcome) to a dietary intervention (Corella et al., 2017; Fenwick et al., 2018). Overall, the association between SNPs or GRSs and diet allows the identification of individuals which could benefit from adopting or avoiding some dietary patterns in prevention of diseases, and thus, applying personalized nutritional advices to these individuals in particular.

Nutrigenomics studies how diet affects gene expression and metabolism according to specific genetic variants, health status or following the administration of a given diet. OMICs technologies such as transcriptomics are different tools used alone or in combination to understand globally how nutrition affects gene transcription, protein expression and global metabolism. Transcriptomics studies the expression of genes by observing variations in messenger RNA levels. Each type of cells has unique gene expression signatures which vary with different treatments and health status. The utilization of transcriptomics in mechanistic studies has been a huge step toward a more profound understanding of metabolic activities in health and diseases. In nutrition research, transcriptomics can give useful clues about beneficial or detrimental effects of dietary treatments in the progression or management of diseases (Mine, Miyashita, & Shahidi, 2009). Overall, the utilization of nutrigenomics in personalized nutrition can give a comprehensive picture of the dynamic effects of diet on gene expression and metabolism. Further, the identification of subgroups of individuals with similar genetic profile might better predict the influence of foods or dietary patterns on chronic diseases risk factors (de Toro-Martín et al., 2017; Mine et al., 2009).

4. THE METABOLIC SYNDROME

Obesity has reached epidemic proportions in the last 50 years. Obesity is strongly related with multiple metabolic dysfunctions such as low-grade chronic inflammation, insulin resistance, high blood pressure or dyslipidemia. Therefore, the characterization of the MetS, defined as a cluster of major risk factors of chronic diseases, arouses as an efficient diagnosis to identify individuals at high risk of developing chronic diseases (Grundy et al., 2005). During the last decades, many definitions have been proposed for the diagnosis of the MetS; yet, there is still no consensus on the relative importance of each risk factors (de la Iglesia et al., 2016). Most definitions suggest that individuals presenting at least three out of the five following components would be diagnose with the MetS (Alberti et al., 2009):

- Abdominal obesity with elevated waist circumference (≥ 94 cm in males and ≥ 80 cm in females) according to the ethnicity specifications.
- Dyslipidemia, with increased triglycerides (1.7 mmol/L or treatment for elevated triglycerides) and reduced high-density lipoprotein (HDL) levels (1.03 mmol/L in males, 1.29 mmol/L in females or specific treatment for low HDL levels).
- Raised blood pressure, with systolic blood pressure ≥ 130 mmHg, diastolic blood pressure ≥ 85 mmHg or treatment for hypertension.
- Hyperglycemia, with fasting plasma glucose levels ≥ 5.6 mmol/L or type 2 diabetes.

The prevalence of the MetS drastically increased in the last decades along with obesity rates and sedentary lifestyle. Worldwide, the prevalence of the MetS is around 20–25% of the global population with major differences between ethnicities and geographic regions (Alberti et al., 2009; O'Neill & O'Driscoll, 2015). People suffering from the MetS have 2-fold increased risk of mortality, are three times more at risk of stroke or myocardial infarction and five times more at risk of suffering from type 2 diabetes compared with healthy individuals (Grundy et al., 2005; International Diabetes Federation, 2006). Unfortunately, the prevalence of the MetS is unlikely to decrease without major changes in lifestyle.

4.1 The Etiology of the Metabolic Syndrome

The MetS has a complex etiology with strong inter-dependent relationships between major tissues in the body. The etiology of the MetS is summarized in Fig. 2. Many experts consider abdominal obesity as the main trigger of the MetS since the other components are originating from impairment of adipocytes gene expression and functionality. To date, the etiology of the MetS is not fully understood and numerous mechanisms remain unknown. Further, there exists great variability in the manifestation of the MetS between individuals. These discrepancies can be explained in part by environmental, ethnic and genetic variations

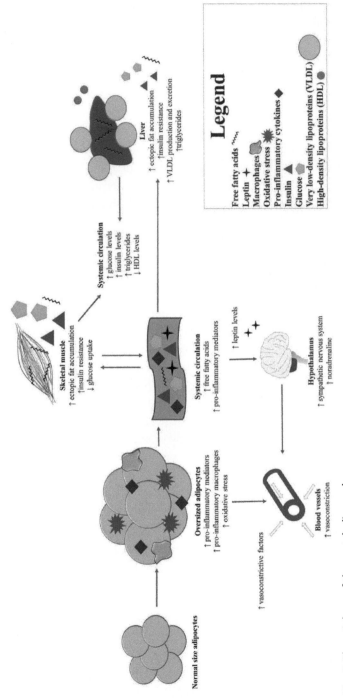

Fig. 2 The etiology of the metabolic syndrome.

(Vidal-Puig, 2014). In fact, the heritability of the MetS can be explained from 20% to 60% by genetic factors. To date, roughly 30 loci have been associated with the MetS according to recent GWAS (Fenwick et al., 2018). Some of the relevant genes associated with the MetS and theirs components are presented in Table 2.

4.1.1 Abdominal Obesity

Obesity is defined as an abnormal or excessive fat accumulation caused by chronic imbalance between energy intake and energy expenditure (World Health Organisation, 2015, 2018). Energy-dense diets and lack of physical activity bring a disequilibrium in energy expenditure and intake, which results in lipid deposit in the adipose tissue as energy storage (Chan & Woo, 2010). On the mid-long term, the adipose tissue expends until it reaches excessive proportions, obesogenic state. Fat accumulation around the abdomen, known as abdominal obesity, is considered the most hazardous site. Elevation of free fatty acids is observed in bloodstream when adipocytes get overloaded with fat. Increase in plasma free fatty acid levels enhance the accumulation of fat in other organs, such as the liver or the skeletal muscle, known as ectopic fat (Vidal-Puig, 2014). Overall, the expansion of the adipose tissue represents a major factor underlying the evolution of the MetS given that the translocation of fat in other organs is associated with many metabolic alterations.

Genetic background is known to greatly influence the prevalence of obesity in many populations. For some individuals, the expansion of the adipose tissue is not necessarily the origin of metabolic alterations since one might have normal bodyweight and still develop the MetS (Vidal-Puig, 2014). Further, approximately 30% of obese individuals remain metabolically healthy (Wildman, Muntner, Reynolds, et al., 2008). Overall, the heritability of obesity and related parameters is estimated at 50–80% (Fenwick et al., 2018). To date, GWAS identified up to 190 different gene loci associated with obesity and fat distribution, some of which have been widely studied for their potential interactions with diet (Hammad & Jones, 2017). Some of the genes associated with obesity and their specific role are presented in Table 2.

4.1.2 Low-Grade Chronic Inflammation

Inflammation is defined as a response to an injury or pathogens, which translates in increased blood flow, capillary dilatation, infiltration of leucocytes and locally produced inflammatory mediators in order to repair damaged

Table 2 Gene Variants Associated to the Metabolic Syndrome and Components

Gene	OMIM	Role[a]	Risk Factor	References
Lipoprotein lipase	*LPL*	Hydrolyse of triglycerides in lipoproteins (VLDL) and chylomicrons. Cellular uptake of free fatty acids, cholesterol-rich lipoproteins and chylomicron remnants	Incidence of the Mets	Brown and Walker (2016)
			Dyslipidemia	Knoblauch et al. (2004)
Apolipoprotein A1	*APOA1*	Major protein of HDL, key role in the reverse cholesterol transport process, co-factor of LCAT	Incidence of the Mets	Brown and Walker (2016)
			Dyslipidemia	Ziki and Mani (2016)
Apolipoprotein A4	*APO44*	Dietary fat absorption, chylomicron synthesis, activation of LCAT	Incidence of the Mets	Brown and Walker (2016)
			Obesity	Guclu-Geyik et al. (2012)
			Dyslipidemia	Ziki and Mani (2016)
Apolipoprotein A5	*APOA5*	Components of VLDL, HDL and chylomicrons. Important role in lipoprotein metabolism by interaction with LDL receptor gene	Incidence of the Mets	Brown and Walker (2016)
			Dyslipidemia	Aouizerat et al. (2003)
Apolipoprotein B	*APOB*	Major protein of VLDL, IDL, LDL and chylomicrons. Ligand of LDL receptors	Incidence of the Mets	Phillips et al. (2011)
			Dyslipidemia	Knoblauch et al. (2004)
Apolipoprotein C1	*APOC1*	Involved in changes of esterified cholesterol between lipoproteins. Role in the removal of cholesterol in tissues. Inhibition of CETP	Incidence of the Mets	Brown and Walker (2016)
			Obesity	Ziki and Mani (2016)
			Dyslipidemia	

		Function	Trait	References
Apolipoprotein E	APOE	Components of chylomicrons and IDLs. Role in the catabolism of triglyceride-rich lipoproteins and cholesterol metabolism	Obesity / Dyslipidemia	Ziki and Mani (2016) / Fallaize et al. (2017) / Knoblauch et al. (2004)
Cholesterol ester transfers protein	CETP	Transfer of esterified cholesterol from HDL to other lipoproteins	Incidence of the Mets / Dyslipidemia	Brown and Walker (2016) / Knoblauch et al. (2004)
Adrenergic receptor beta-3	ADRB3	Involved in lipolysis and thermogenesis in the adipose tissue	Obesity	Kurokawa, Nakai, Kameo, Liu, and Satoh (2012)
Cytochrome P450 family 27, subfamily A	CYP27A	Monooxygenases involved in the catalyzation of many drug metabolism and synthesis of cholesterol, steroids and other lipids	Obesity / Blood pressure	Ziki and Mani (2016)
LDL receptor	LDLR	Cell surface proteins, endocytosis of cholesterol form LDL. Receptor mediated by specific ligands	Dyslipidemia	Knoblauch et al. (2004)
Adiponectin	ADIPOQ	Express in the adipose tissue sclely. Adiponectin is involved in hormonal and metallic processes	Obesity / Inflammation / Insulin resistance	Skrypnik et al. (2017) / Perez, Moore-Carrasco, Gonzalez, Fuentes, and Palomo (2012)

Continued

Table 2 Gene Variants Associated to the Metabolic Syndrome and Components—cont'd

Gene	OMIM	Role	Risk Factor	References
Fat mass and obesity-related gene	*FTO*	Involved in the reversal of alkylated DNA and RNA damages by oxidative demethylation	Incidence of the Mets	Brown and Walker (2016)
			Obesity	Scuteri et al. (2007)
			Insulin resistance	Ziki and Mani (2016)
			Hyperinsulinemia	Merino, Udler, Leong, and Meigs (2017)
			Type 2 diabetes	
Peroxisome proliferator-activated receptor alpha	*PPARA*	Involved in the utilization and catabolism of fatty acids. Activated through ligand binding	Incidence of the MetS	Dong et al. (2015)
			Type 2 diabetes	
Peroxisome proliferator-activated receptor gamma	*PPARG*	Regulator of adipocyte differentiation	Obesity	Razquin, Martinez, Martinez-Gonzalez, Bes-Rastrollo, et al. (2009)
			Inflammation	Ziki and Mani (2016)
			Insulin resistance	Perez et al. (2012)
			Type 2 diabetes	Brown and Walker (2016)
				Merino et al. (2017)

			Condition	Reference
Insulin receptor substrate I	*IRS1*	Phosphorylated by insulin receptor tyrosine kinase. Involved in the signalization from insulin and insulin-like growth factor-1 receptors to intracellular pathways	Obesity	Ziki and Mani (2016)
			Insulin resistance	Brown and Walker (2016)
			Hyperinsulinemia	Merino et al. (2017)
			Type 2 diabetes	
Transcription factor 7 like 2	*TCF7L2*	Involved in the Wnt signaling pathway. Implicated in blood glucose homeostasis	Obesity	Ziki and Mani (2016)
			Insulin resistance	Brown and Walker (2016)
			Hyperglycemia	Merino et al. (2017)
			Type 2 diabetes	
Glucokinase regulator	*GCKR*	Regulatory protein that inhibits glucokinase in the liver and pancreatic islet cells	Incidence of the Mets	Brown and Walker (2016)
			Insulin resistance	Merino et al. (2017)
			Hyperglycemia	
			Type 2 diabetes	
Fatty acid binding protein 1	*FABP1*	Involved in fatty acid uptake, transport and metabolism in the liver	Dyslipidemia	Larifla et al. (2016)
				Turkovic, Pizent, Dodig, Pavlovic, and Pasalic (2012)
Melanocortin 4 receptor	*MC4R*	Interacts with adrenocorticotropic and MSH hormones. Mediated by G proteins	Obesity	Wang et al. (2012)
			Insulin resistance	Ziki and Mani (2016)
			Type 2 diabetes	Brown and Walker (2016)
				Merino et al. (2017)

Continued

Table 2 Gene Variants Associated to the Metabolic Syndrome and Components—cont'd

Gene	OMIM	Role	Risk Factor	References
Neuropeptide Y	*NPY*	Influence cortical excitability, stress response, food intake, circadian rhythms and cardiovascular function	Incidence of the Mets	de Luis, Izaola, de la Fuente, Primo, and Aller (2016)
			Obesity	O'Leary et al. (2016)
			Insulin resistance	
			Hyperglycemia	
			Dyslipidemia	
Gastric inhibitory polypeptide receptor	*GIPR*	Inhibits gastric acid secretion and gastrin release, stimulate insulin release in the presence of elevated glucose	Type 2 diabetes	O'Leary et al. (2016)
Leptin receptor	*LEPR*	Regulation of fat metabolism and body weight	Obesity	O'Leary et al. (2016)
Zinc transporter 8	*ZNT8*	Encodes for a zinc efflux transporter. Expressed in the islets of Langerhans in the pancreas. Involved in the secretory pathway of the insulin-secreting cells	Type 2 diabetes	O'Leary et al. (2016)
Free fatty acid receptor 4	*FFAR4*	Receptor for free fatty acids, including omega-3. Involved in suppressing anti-inflammatory responses and insulin sensitizing	Obesity	Waguri, Goda, Kasezawa, and Yamakawa-Kobayashi (2013)

Fatty acid desaturase 1 and 2	FADS1, FADS2	Desaturation of omega-3 and omega-6 PUFA at the delta-5 position (FADS1) and delta-6 position (FADS2)	Type 2 diabetes	Merino et al. (2017)
			Dyslipidemia	Fenwick et al. (2018)
				Ziki and Mani (2016)
Melatonin receptor 1B	MTNR1B	Receptor for melatonin. Involved in the neurobiological effects of melatonin and light-dependent functions in the retina	Hyperglycemia	Merino et al. (2017)
			Insulin resistance	Fenwick et al. (2018)
			Type 2 diabetes	Ziki and Mani (2016)
Lipase C, hepatic type	LIPC	Triglyceride hydrolase and ligand/bridging factor for the receptor-mediated lipoprotein uptake	Dyslipidemia	Fenwick et al. (2018)
ATP binding cassette A1	ABCA1	Cholesterol efflux pump in the cellular lipid removal pathway	Obesity	Ziki and Mani (2016)
			Dyslipidemia	Knoblauch et al. (2004)
Tumor necrosis factor-alpha	TNFA	Pro-inflammatory cytokine. Regulation of a multiple biological processes, such as cell proliferation, differentiation, apoptosis, lipid metabolism and coagulation	Incidence of the MetS	Joffe, Collins, and Goedecke (2013)
			Obesity	O'Leary et al. (2016)
			Insulin resistance	
			Dyslipidemia	
Angiotensin-converting enzyme	ACE	Vasopressor and aldosterone stimulating peptide, controls blood pressure and fluid-electrolyte balance. Key role in the renin-angiotensin system	Blood pressure homeostasis	Arnett and Claas (2018)

Continued

Table 2 Gene Variants Associated to the Metabolic Syndrome and Components—cont'd

Gene	OMIM	Role	Risk Factor	References
b-2 adrenergic receptor	ADRB2	Interacts with epinephrine. Lipolytic receptor in the adipose tissue	Obesity	Arnett and Claas (2018)
			Type 2 diabetes	Abete, Navas-Carretero, Marti, and Martinez (2012)
			Blood pressure	
			homeostasis	
Angiotensinogen	AGT	Involved in maintain blood pressure	Blood pressure homeostasis	Arnett and Claas (2018)
Nitric oxide synthase	NOS3	Biologic mediator in several processes such as neurotransmission, antimicrobial and antitumoral activities	Blood pressure	Ziki and Mani (2016)
Interleukin 6	IL-6	Cytokine involved in inflammatory processes and maturation of B cells	Obesity	Joffe et al. (2013)
			Inflammation	O'Leary et al. (2016)
			Dyslipidemia	Fernández-Real (2006)
			Blood pressure	

[a]From O'Leary et al. (2016).

HDL-C: high-density lipoprotein; IDL: intermediate-density lipoprotein; LCAT: lecithin-cholesterol acyltransferase, LDL-C: low-density lipoprotein; MetS: metabolic syndrome; MSH: melanocyte-stimulating hormone; PUFA: poly-unsaturated fatty acids; VLDL: very low-density lipoprotein.

tissues and eliminate pathogens (Minihane et al., 2015; Rocha, Bressan, & Hermsdorff, 2017). This definition mostly describes acute inflammation; yet, inflammation can also occur chronically, such as low-grade chronic inflammation. This low-grade chronic inflammation is usually asymptotic and results from metabolic perturbations such as abdominal obesity (Rocha et al., 2017). Low-grade chronic inflammation is a major player in the MetS etiology and evolution. The expansion of abdominal adipose tissue activates some transcription factors, such as the nuclear factor kappa-B (NF-κB), which regulates the expression of numerous pro-inflammatory genes (Lumeng & Saltiel, 2011). Specifically, the activation of pro-inflammatory cascades increases levels of pro-inflammatory cytokines and mediators, such as interleukin (IL)-1β, IL-6, tumor necrosis factor-alpha (TNF-α) or C-reactive protein (CRP) (Sanchez-Infantes, 2014). Further, the adipose tissue also produces its specific kind of endocrine and inflammatory mediators known as adipokines. For example, leptin is a pro-inflammatory adipokine which plasma levels correlate with the size of the adipocytes. Elevation of plasma leptin triggers the production of pro-inflammatory mediators (IL-1β, IL-6 or TNF-α) (Rocha et al., 2017). Leptin also have specific receptors in the hypothalamus which activates sympathetic nervous system and enhances thermogenesis (Sanchez-Infantes, 2014). Inversely with leptin, adiponectin is an anti-inflammatory adipokine linked with the resolution of inflammation. In low-grade chronic inflammation, plasma adiponectin levels are usually reduced (Rocha et al., 2017). In parallel, pro-inflammatory macrophages are attracted to the inflamed adipose tissue, which amplify the pro-inflammatory environment by producing a considerable amount of cytokines (Galic, Oakhill, & Steinberg, 2010). Further, excessive fat accumulation in adipocytes enhances ectopic fat deposition of some lipids, particularly diacylglycerols, ceramides and other phospholipids, in peripheral tissues and organs. The accumulation of these lipids activates oxidative stress mechanisms, known as lipotoxicity. Oxidative stress is induced by the presence of reactive oxygen species (ROS) which can cause damages to DNA and can impair various metabolic functions (Vidal-Puig, 2014). Some gene variants could explain the effect of chronic inflammation on other components of the MetS. Gene polymorphisms of the *tumor necrosis factor-alpha* (*TNFA*) gene or the *interleukin-6* (*IL-6*) gene have been associated with variable pro-inflammatory responses which subsequently affect glucose metabolism, lipid profile and blood pressure (Fernández-Real, 2006). Further, multiple gene variants of the *adiponectin* (*ADIPOQ*) gene have been well-documented in inflammation-related obesity and metabolic disorders. Obese subjects with metabolic alterations seem to have a lower expression of the *ADIPOQ* gene in comparison with metabolically healthy obese subjects. Along with *ADIPOQ* gene, polymorphisms in the *peroxisome proliferator-activated receptor gamma* (*PPARG*) gene are associated with variations in inflammatory response to obesity (Perez et al., 2012).

4.1.3 Insulin Resistance and Altered Glucose Metabolism

Glucose intolerance and insulin resistance are critical contributors of metabolic disturbances in the MetS etiology. Insulin resistance is defined as a decreased capacity of insulin to accomplish its normal physiological functions on insulin-sensitive tissues, causing hyperinsulinemia to maintain glucose levels in normal ranges (Perona, 2017). Insulin resistance usually precedes hyperglycemia, which is defined as elevated plasma glucose levels due to impaired uptake of glucose from the insulin-sensitive tissues. Elevation of blood glucose levels is the main parameter used in the definition of the MetS (Alberti et al., 2009). Hyperglycemia is responsible for damages in many tissues and increases the risk of many chronic diseases such as cardiovascular diseases, atherosclerosis, hypertension and type 2 diabetes (de la Iglesia et al., 2016).

Many hypotheses have been proposed to elucidate the key role of insulin resistance in the pathogenicity of the MetS. First, decreased membrane fluidity of the cells is observed in insulin resistance. This greater rigidity of membranes is associated with lower number of insulin receptors and reduced affinity to insulin, along with reduced translocation and insertion of glucose receptors in cell membranes (Perona, 2017). Second, decreased levels of adiponectin and increased pro-inflammatory cytokines have deleterious effects on glucose metabolism (Sanchez-Infantes, 2014). Pro-inflammatory cytokines also inhibit the gene expression of *glucose transporter-4* (*GLUT-4*), the major glucose transporter in the adipose tissue, and *insulin receptor substrate-1* (*IRS-1*), a protein involved in insulin signalization, resulting in decreased glucose uptake (Sanchez-Infantes, 2014). In addition, the elevation of free fatty acids in bloodstream affects the insulin signalization and decreases glucose uptake in the skeletal muscle (Perona, 2017). Since skeletal muscle usually utilizes around 70–80% of ingested glucose for energy and glycogen formation, the reduction of glucose uptake from the muscles has a key role in the development of hyperglycemia (Zorzano, Sebastian, & Monserrat, 2014). Furthermore, the hepatic production of glucose is not inhibited by insulin in the MetS and insulin resistance, resulting in elevation of plasma glucose levels and increased lipid production and accumulation in the liver. Hepatic accumulation of lipids creates a lipotoxic environment contributing to further impairment of lipid and glucose metabolisms (D'Amore, Palasciano, & Moschetta, 2014).

Genes have a major influence on the development of alterations to glucose metabolism. To date, near 140 loci have been associated with glucose homeostasis, mainly identified through GWAS (Merino et al., 2017; Mohlke & Boehnke, 2015). Some SNPs associated with altered glucose metabolism and their roles are presented in Table 2. Given the huge number of genes identified with an association to glucose metabolism, some GRSs have been elaborated to predict type 2 diabetes incidence, although the clinical utility or these GRSs still needs to be confirmed (Merino et al., 2017; Vassy et al., 2014).

4.1.4 Dyslipidemia
The MetS is characterized by major alterations of lipid metabolism. Dyslipidemia is defined as abnormalities in the composition and relative proportions of plasma lipoproteins. The role of lipoproteins is to ensure the transportation of fatty acids and cholesterol through the body (Bosomworth, 2013). In the MetS, the production of lipoproteins by the liver is impaired (D'Amore et al., 2014). Due to ectopic fat accumulation, the liver increases its production and secretion of large very low-density lipoproteins (VLDL) rich in triglycerides in bloodstream (Borén & Adiels, 2014). The augmentation of VLDL in circulation is tightly associated with increased apolipoprotein C-III (ApoCIII) levels, which are located on VLDL and HDL. ApoCIII inhibits the activity of the lipoprotein lipase reducing the clearance of triglyceride-rich lipoproteins (D'Amore et al., 2014). The augmentation of VLDL secretion alters HDL metabolism by increasing the catabolism of HDL particles. In parallel, the alterations of lipid profile in the MetS are often accompanied with increased small LDL levels and increased total-cholesterol (Total-C)/HDL ratio, which are markers associated with higher risk of atherosclerosis and cardiovascular diseases (Borén & Adiels, 2014). Changes in cells' membrane fluidity also alter exchanges between lipoproteins and cells, contributing to further impairment of lipid profile (Perona, 2017).

Dyslipidemia has a strong genetic component. Notably, triglycerides and HDL levels have an estimated 60–80% heritability. GWAS suggest that around 95 loci could explain 25–35% of plasma lipid levels (Fenwick et al., 2018). Some genes associated with dyslipidemia and their roles are presented in Table 2. Some important genes are the *apolipoprotein A (APOA)* genes, namely *APOA2*, *APOA4* and *APOA5* genes, which are mainly associated with triglyceride-rich lipoproteins. Likewise, *apolipoprotein E* (*APOE*) gene is associated with multiple lipoproteins such as VLDL, LDL and HDL, giving this gene a central role in the regulation of lipid metabolism (Peña-Romero, Navas-Carrillo, Marín, & Orenes-Piñero, 2017). The *APOE* gene has three distinct alleles known as ε2, ε3 and ε4, the ε3 being the most common (Fenwick et al., 2018). Individuals with the ε4 variant are associated with higher LDL levels, higher total-C levels and triglycerides, whereas ε3 have intermediate levels and ε2 have lower levels (Fenwick et al., 2018; Peña-Romero et al., 2017). Many other SNPs are associated with lipid profile, such as the *hepatic lipase* (*LIPC*) gene, the *cholesterylester transfer protein* (*CETP*) gene or the *fatty acid desaturase 2* (*FADS2*) gene to name a few (Fenwick et al., 2018).

4.1.5 Hypertension

High blood pressure is a major component of MetS and represents an important risk factor for cardiovascular diseases (de la Iglesia et al., 2016). Adipocytes possess the entire machinery for renin–angiotensin system although adipose tissue is not the main regulator of blood pressure (Mendizábal, Llorens, & Nava, 2013; Yanai et al., 2008). With excessive abdominal fat, the adipose tissue secretes angiotensin II and angiotensinogen, two hormones with vasoconstrictive effects on blood vessels (Yanai et al., 2008). In presence of low-grade inflammation, increased TNF-α levels stimulate the production of angiotensinogen and endothelin-1, another potent vasoconstrictive factor. Increased levels of leptin activate the sympathetic nervous system, stimulating noradrenaline secretion and contributing to the rise of blood pressure. The decline of adiponectin, which normally has a vasodilator effect, amplifies the vasoconstrictive profile. Further, insulin resistance impairs nitric oxide (NO) production and enhances the production of endothelin-I, promoting further vasoconstriction. Finally, glucose has a vasoconstrictive effect on the endothelium; thus, hyperglycemia contributes to worsen the situation (Mendizábal et al., 2013).

The heritability of blood pressure is estimated at around 30–60%. GWAS have identified around 90 loci associated with high blood pressure, some of which are presented in Table 2 (Fenwick et al., 2018). Yet, the variance of blood pressure can only be explained at 2% with the actual gene variants identified, suggesting a key role of environmental factors. In parallel, transcriptomic analyses warranted the discovery of different gene expression in subjects with or without high blood pressure. The expression of genes involved in insulin signaling and fatty acid metabolism from different tissues were increased with high blood pressure, with different expression signatures between individuals (Arnett & Claas, 2018).

In sum, the MetS is a complex cluster of risk factors which are linked together by multiples mechanisms not necessarily well understood. The genetic dimension of the MetS has an important role to play in the etiology of the disease and in the response to environmental factors. Since the expansion of abdominal adipose tissue is recognized as one of the main contributors of the Mets, most of the solutions proposed are often oriented on reduction of body weight, especially abdominal fat (Grundy et al., 2005). Indeed, weight loss could significantly reduce inflammation, plasma inflammatory markers, insulin resistance and insulin secretion and could improve lipid profile and blood pressure (de la Iglesia et al., 2016). Nonetheless, the identification of gene variants might be an important milestone in the personalization of therapeutic strategies in prevention or management of the MetS.

5. DIETARY FATTY ACIDS AND THEIR ROLE IN HEALTH

Dietary fatty acids have many roles in metabolic health. Lipids have functional and structural roles by composing the cell's membranes and are major determinants of membrane fluidity, ion permeability, receptor translocation and affinity with hormones and mediators (Heer & Egert, 2015). Dietary non-essential fatty acids (SFA and MUFA) have a limited effect on membrane fluidity since most of these lipids can be synthetized endogenously. On the counterpart, dietary omega-3 PUFA, omega-6 PUFA and TFA can have a significant effect on membrane fluidity (Perona, 2017).

Lipids also have many roles in cells activities, such as gene expression, regulatory effects on enzyme activities and inflammatory processes (Heer & Egert, 2015; Risérus, Willett, & Hu, 2009; Wang & Hu, 2017). Specifically, fatty acids can influence metabolic activities by binding to lipid sensitive transcription factors. From these transcription factors, the peroxisome proliferator-activated receptors (PPARs) have been widely studied due to their critical roles in energy metabolism. PPAR-α and PPAR-β are known for their role in lipid metabolism and inflammation, while PPAR-γ regulates adipocyte differentiation and is involved in lipid and glucose metabolism (Ordovas, Kaput, & Corella, 2007). The activation of PPARs is associated with upregulation of hundreds of genes involved in lipoprotein metabolism or lipid oxidation and downregulation of genes associated with pro-inflammatory mediators (Salter & Tarling, 2007).

Hence, the activation of PPARs is usually linked to resolution of inflammation, decreased lipotoxicity, increased insulin sensitivity and improved lipid profile (Georgiadi & Kersten, 2012; Salter & Tarling, 2007). On the counterpart, dysregulation of PPARs activity is tightly linked to the MetS, type 2 diabetes, inflammation and cardiovascular diseases (Azhar, 2010; Perez et al., 2012). Unsaturated fatty acids are natural ligands of PPARs, especially PUFA (Grygiel-Górniak, 2014). Omega-3 and omega-6 PUFA and derived metabolites can bind to all subtypes of PPARs, especially PPAR-α followed by PPAR-γ. Sterol regulatory binding proteins (SREBPs) are other transcription factors that can be modulated by fatty acids. SREBPs are composed of various proteins (SREBP1a, SREBP1c and SREBP2) involved in the expression of genes involved in lipid and lipoprotein metabolisms. Activation of SREBPs is associated with increased cholesterol synthesis, accumulation of cholesterol and triglycerides in the liver, but also reduction of hepatic insulin sensitivity. Unsaturated fatty acids, especially omega-6 and omega-3 PUFA, can efficiently modulate the activity of SREBPs. The specific actions of fatty acids binding on PPARs and SREBPs will be covered later on. Overall, changing the composition of fatty acids in diet may affect gene expression through the activation/repression of PPARs and SREBPs, but also other transcription factors or receptors (Salter & Tarling, 2007).

In sum, dietary fatty acids are involved in metabolic health through complex metabolic pathways which influence the fluidity of cell membranes, regulation of transcription factors, expression of multiple genes and inflammatory processes. Further, genetic background is known to greatly affect the response to dietary fatty acids, suggesting a more individualized approach could reduce the actual discrepancies surrounding dietary responses. The next section covers the roles of total and specific types of dietary fatty acids on the different components of the MetS. Recent knowledges about genomics studies surrounding dietary intake of fatty acids are also described.

5.1 Total Fatty Acids

Dietary fatty acids are commonly present in diets in conjunction with carbohydrates and proteins. Different balances between these three main macronutrients create different dietary patterns. Through the evolution of nutrition science, many dietary patterns have been identified to prevent or provoke metabolic alterations in the human body. Further, advances in nutrigenetics suggest a potential role of genetic background in the response to different macronutrient compositions of diets. The following paragraphs cover the metabolic effects of low-fat and high-fat diets on the components of the MetS in conjunction with gene expression and relevant genetic variants.

5.1.1 Obesity

Traditionally, low–fat diets have been a central dietary intervention in weight loss programs. The premise behind low-fat diets is to reduce energy intake by replacing fatty acids mainly with carbohydrates, which are less caloric (Perona, 2017). In an isocaloric context, a meta-analysis of random-ized controlled trials observed a slight reduction of body weight, body mass index and waist circumference by decreasing dietary fatty acids intake in an adult population in comparison with moderate fatty acid intake (Hooper et al., 2015). However, hypocaloric high fat, low–carbohydrate diets are becoming more and more popular due to their efficiency for weight loss. Lately, low-fat diets have been increasingly criticized through meta–analyses of randomized controlled trials which failed to see any advantage on weight loss in comparison with high–fat, low-carbohydrate diets (Johnston et al., 2014; Schwingshackl & Hoffmann, 2014a; Tobias et al., 2015). According to gene transcripts, differences in gene expression following a high–fat or a low–fat diet are minor; energy restriction seems to be the major regulator of gene expression, independently of fat composition in diet (Rudkowska & Pérusse, 2012).

Discrepancies surrounding the identification of the optimal diet to lose weight might be explained by genetic factors. Many gene-diet interactions have been identified in relation with dietary fat and obesity risk factors, such as *APOA2, fat mass and obesity-related gene (FTO), PPARG, IL6, TNFA, APOA5* and many others (Hammad & Jones, 2017). Some of these gene variants are presented in Table 3. For example, the Preventing Overweight Using Novel Dietary Strategies (POUNDS LOST) trial has largely contrib-uted to the identification of gene–diet interactions surrounding diets in a weight-loss context. Briefly, the POUNDS LOST study is a randomized controlled trial involving 811 overweight or obese adults assigned to one of four hypocaloric diets with different fat composition for a 2-year inter-vention. One of the genes identified in the POUNDS LOST study is the *Neuropeptide Y (NPY)* gene, which encoded protein has a key role in food intake, metabolic and cardiovascular functions (Table 1). The "C" carriers of the rs16147 on the *NPY* gene had a lower weight loss after 6 months in comparison with "T" carriers. However, the "T" carriers had smaller weight

Table 3 Dietary Fatty Acids and Diets, Gene Variants and the Components of the Metabolic Syndrome

Gene	OMIM	Polymorphisms	Metabolic Syndrome Component	Effects of Total Fat Intake	References
peroxisome proliferator-activated receptor gamma 2	PPARG2	Pro12Ala	Obesity	Ala carriers: Greater ↓ total weight loss with low-fat diet vs Pro12Pro and high-fat diet (hypocaloric diets). Higher waist circumference with low-fat diet after 2 years vs Pro12 carriers	Garaulet, Smith, Hernández-González, Lee, and Ordovás (2011)
				Ala12Ala genotype: Less important bodyweight reduction with high-fat diet vs CC genotype (hypocaloric diet)	Razquin, Martinez, Martinez-Gonzalez, Corella, et al. (2009)
Transcription factor 7 like 2	TCF7L2	Rs12255372	Obesity	TT: Greater ↓ BMI, total fat mass and trunk fat mass after 6 months with low-fat diet (hypocaloric diets)	Bodhini et al. (2017)
Protein phosphatase Mg2+/Mn2+ dependent 1K	PPM1K	Rs144081	Obesity	C carriers: Greater ↓ body weight with low-fat diet vs high-fat diet (hypocaloric diets)	Goni et al. (2017)
Fibroblast growth factor 21	FGF21	Rs838147	Obesity	C carriers: Greater ↓ of waist circumference, total body fat mass and trunk mass with low-fat diet vs high-fat diet (hypocaloric diets)	Heianza et al. (2016)

Gene	Symbol	SNP	Disease	Findings	Reference
Fat mass and obesity associated gene	*FTO*	Rs9939609	Obesity	A carriers: When >30% of total energy from fat, ↑ % body fat. Greater ↑ of body fat per risk allele (hypocaloric diets)	Labayen et al. (2016)
ATP binding cassette A1	*ABCA1*	Rs230C	Obesity	T carriers: ↑ visceral fat/subcutaneous fat ratio with low-fat diet, ↓ abdominal fat/subcutaneous fat ratio with high-fat diet	Jacobo-Albavera et al. (2015)
Neuropeptide Y	*NPY*	Rs16147	Obesity	T carriers: Smaller ↓ of body weight with high-fat diet vs CC genotype (hypocaloric diets)	Lin et al. (2015)
Gastric inhibitory polypeptide	*GIPR*	Rs2287019	Obesity	T carriers: Greater ↓ of bodyweight with low-fat diet vs high-fat, especially in white subjects (hypocaloric diets)	Qi, Bray, Hu, Sacks, and Qi (2012)
Apolipoprotein A5	*APOA5*	Rs3135506	Obesity	GG genotype: ↑ fat intake associated with ↑ risk of obesity	Domínguez-Reyes et al. (2015)
Apolipoprotein B	*APOB*	Rs1469513	Obesity	G carriers: High-fat intake is associated with ↑ risk of obesity vs A carriers	Doo, Won, and Kim (2015)
Apolipoprotein B	*APOB*	Rs512535	Obesity	GG genotype: High-fat intake ↑ BMI vs A carriers and low-fat diet	Phillips et al. (2011)
Apolipoprotein A1	*APOA1*	Rs670	Obesity	GG genotype: High-fat intake ↑ BMI vs A carriers and low-fat diet	Phillips et al. (2011)

Continued

Table 3 Dietary Fatty Acids and Diets, Gene Variants and the Components of the Metabolic Syndrome—cont'd

Gene	OMIM	Polymorphisms	Metabolic Syndrome Component	Effects of Total Fat Intake	References
Leptin receptor	LEPR	Rs1137101	Obesity	G carriers: ↑ fat intake is associated with ↑ risk of obesity	Domínguez-Reyes et al. (2015)
Interleukin-6	IL-6	IVS3 +281G>T	Obesity	G carriers: High dietary fat intake is associated with ↑ adiposity	Joffe et al. (2014)
Interleukin-6	IL-6	IVS4 +869A>G	Obesity	A carriers: High dietary fat intake is associated with ↑ adiposity	Joffe et al. (2014)
Hepatocyte nuclear factor 1α	HNF1A	Rs7957197	Obesity	T carriers: High-fat diet is associated with ↓ body weight, ↓ waist circumference vs non-carriers (hypocaloric diets)	Huang et al. (2018)
GRS with 16 SNPs associated with obesity			Obesity	High GRS associated with ↑ % body fat mass with ↑ fat intake	Goni, Cuervo, Milagro, and Martínez (2015)
GRS with 93 SNPs associated with increased BMI		—	Obesity	High GRS associated with ↑ BMI with high-fat diet vs low GRS. Smaller ↑ of BMI with low-fat diet	Celis-Morales et al. (2017)
Adiponectin gene	ADIPOQ	rs1501299 (+276G/T)	Inflammation	GG genotype: ↑ adiponectin with low-fat diet vs T carriers (hypocaloric diets)	de Luis, Izaola, Primo, and Aller (2017)
ATP binding cassette A1	ABCA1	Rs230C	Inflammation	T carriers: ↑ adiponectin levels with low-fat diet, ↓ adiponectin levels with high-fat diet in premenopausal women	Jacobo-Albavera et al. (2015)

GRS with five SNPs associated with adiponectin	—	—	Inflammation	High GRS: ↓ adiponectin with high-fat diet Low GRS: ↑ adiponectin with low-fat diet	Ma et al. (2017)
Apolipoprotein A1	APOA1	Rs670	Inflammation	GG genotype: High-fat intake ↑ CRP vs A carriers	Phillips et al. (2011)
Angiotensin-converting enzyme	ACE	Rs4343	Glucose metabolism	GG genotype: ↑ fasting insulin, ↑ HOMA-IR, ↑ AUC 2h post-ivGTT for insulin and glucose with high-fat diet vs A carriers. >37% of total energy from fat associated with 2.7 × ↑ risk of type 2 diabetes vs A carriers	Schüler, Osterhoff, Frahnow, Möhlig, et al. (2017)
Protein phosphatase Mg2+/Mn2+ dependent 1K	PPM1K	Rs144081	Glucose metabolism	T carriers: ↓ insulin levels and HOMA-B with high-fat diet vs low-fat diet (hypocaloric diets)	Goni et al. (2017)
				C carriers: Greater ↓ insulin levels and HOMA-IR with low-fat diet (hypocaloric diets)	Xu et al. (2013)
Proprotein convertase subtilisin/kexin type 7	PCSK7	Rs236908	Glucose metabolism	G carriers: ↓ fasting insulin and ↓ HOMA-IR on low-fat/high carbohydrate diet (hypocaloric diets)	Huang et al. (2015)
Adiponectin gene	ADIPOQ	rs1501299 (+276G/T)	Glucose metabolism	GG genotype: ↓ fasting insulin and ↓ HOMA-IR with low-fat diet vs T carriers (hypocaloric diets)	de Luis et al. (2017)

Continued

Table 3 Dietary Fatty Acids and Diets, Gene Variants and the Components of the Metabolic Syndrome—cont'd

Gene	OMIM	Polymorphisms	Metabolic Syndrome Component	Effects of Total Fat Intake	References
ATP binding cassette A1	ABCA1	Rs230C	Glucose metabolism	Risk allele carriers: ↓ HOMA–IR with high–fat diet	Jacobo–Albavera et al. (2015)
Insulin receptor substrate 1	IRS1	Rs2943641	Glucose metabolism	CC genotype: ↓ insulin levels and ↓ HOMA–IR after low–fat, high carbohydrate diet (hypocaloric diets)	Qi et al. (2011)
Gastric inhibitory polypeptide	GIPR	Rs2287019	Glucose metabolism	T carriers: Greater ↓ of HOMA–IR with low–fat diet vs high–fat diet, especially in white subjects (hypocaloric diets)	Qi et al. (2012)
Hepatocyte nuclear factor 1α	HNF1A	Rs7957197	Glucose metabolism	T carriers: High–fat diet is associated with ↓ fasting insulin and ↓ insulin resistance vs non–carriers (hypocaloric diets)	Huang et al. (2018)
Apolipoprotein A1	APOA1	Rs670	Glucose metabolism	GG genotype: High–fat intake ↑ fasting insulin and ↑ insulin resistance vs A carriers and low–fat diet	Phillips et al. (2011)
Apolipoprotein B	APOB	Rs512535	Glucose metabolism	GG genotype: High–fat intake ↑ fasting insulin and ↑ insulin resistance vs A carriers and low–fat diet	Phillips et al. (2011)

GRS of 14 SNPs associated with fasting glucose			Glucose metabolism	High GRS: ↑ fasting glucose with high-fat diet	Wang et al. (2016)
Transcription factor 7 like 2	TCF7L2	Rs12255372	Dyslipidemia	T carriers: ↑ HDL with lowest tertile of fat intake; ↓ HDL with highest tertile of fat intake	Mattei, Qi, Hu, Sacks, and Qi (2012)
Lipoprotein lipase	LPL	Rs1121923	Dyslipidemia	T carriers: ↑ HDL and ↓ risk of HDL with low-fat diet vs CC genotype	Ayyappa et al. (2017)
Melatonin receptor 1B	MTNR1B	Rs10830963	Dyslipidemia	G carriers: Greater reduction of total-C and LDL with a low-fat diet vs high-fat diet (hypocaloric diets)	Goni et al. (2018)
Hepatic lipase	LIPC	Rs1800588	Dyslipidemia	C carriers: ↑ HDL levels with high-fat diet vs low-fat and TT genotype	Smith et al. (2017)
Hepatic lipase	LIPC	Rs2070895	Dyslipidemia	A carriers: ↓ total-C and LDL and ↑ HDL with low-fat diet vs G carriers, ↑ total-cholesterol and LDL with high-fat diet (hypocaloric diets)	Xu et al. (2015)
Apolipoprotein A5	APOA5	Rs964184	Dyslipidemia	G carriers: Greater ↓ of total-C, ↓ LDL vs non-carriers with low-fat diet, greater ↑ of HDL with a high-fat diet vs non-carriers (hypocaloric diets)	Zhang, Qi, Bray, et al. (2012)

Continued

Table 3 Dietary Fatty Acids and Diets, Gene Variants and the Components of the Metabolic Syndrome—cont'd

Gene	OMIM	Polymorphisms	Metabolic Syndrome Component	Effects of Total Fat Intake	References
Apolipoprotein A1	APOA1	Rs670	Dyslipidemia	GG genotype: High-fat intake ↑ ApoCIII levels and ↑ LDL/HDL ratio vs "A" carriers	Phillips et al. (2011)
Cholesterol ester transfer protein	CETP	Rs3764261	Dyslipidemia	CC genotype: Greater ↑ HDL and ↓ triglycerides with high-fat diet vs low-fat diet (hypocaloric diets)	Qi et al. (2015)
ATP binding assette A1	ABCA1	rs9282541	Dyslipidemia	"T" carriers: ↓ HDL with low-fat diet, ↑ HDL with high-fat diet	Jacobo-Albavera et al. (2015)
Leptin receptor	LEPR	Rs1137101	Dyslipidemia	"G" carriers: ↑ fat intake is associated with ↑ risk of hypercholesterolemia	Domínguez-Reyes et al. (2015)
Interleukin-6	IL-6	IVS4 +869A > G	Dyslipidemia	"G" carriers: High dietary fat intake is associated with ↑ triglycerides and ↑ total-C/HDL ratio in black women	Joffe et al. (2014)
Neuropeptide Y	NPY	Rs16147	Blood pressure	C carriers: Greater ↓ of blood pressure with low-fat diet vs high-fat diet. Greater changes in hypertensive subjects (hypocaloric diets)	Zhang, Qi, Liang, et al. (2012)

Apolipoprotein A1	APOA1	Rs670	Metabolic syndrome	GG genotype: High-fat intake ↑ risk of the MetS, while low-fat diet did not increase the risk of MetS, especially in males	Phillips et al. (2011)
Apolipoprotein B	APOB	Rs512535	Metabolic syndrome	GG genotype: High-fat intake ↑ risk of the MetS, while low-fat diet did not increase the risk of MetS, especially in males	Phillips et al. (2011)
Insulin receptor substrate 1	IRS1	Rs1522813	Metabolic syndrome	A carriers: reversion of the Mets with high-fat diet vs low-fat diet (hypocaloric diets)	Qi et al. (2013)

ApoCIII: apolipoprotein C-III; AUC: area under the curve; BMI: body mass index; CRP: C-reactive protein; GRS: genetic risk score; HDL: high-density lipoproteins; HOMA-β: homeostatic model assessment of β-cells; HOMA-IR: homeostatic model assessment of insulin resistance; ivGTT: intravenous glucose tolerance test; LDL: low-density lipoproteins.

loss on a high-fat diet in comparison with "C" homozygous carriers (Lin et al., 2015). The *peroxisome proliferator-activated receptor gamma 2 (PPARG2)* Pro12Ala also presents a gene-diet interaction with fat intake. The Ala12 carriers observed a greater reduction of bodyweight on a low-fat diet compared to the Pro12Pro genotype or AlA12 carriers on a high-fat diet in the POUNDS LOST trial (Garaulet et al., 2011). In contrast, Ala12 carriers presented a higher waist circumference on a low-fat diet without energy restriction in comparison with Pro12 carriers in another study (Razquin, Martinez, Martinez-Gonzalez, Corella, et al., 2009). Celis-Morales et al. (2017) proposed a GRS with 93 SNPs associated with body mass index (BMI); in subjects with high GRS, a high-fat diet was associated with increased BMI in comparison with individuals with low GRS (Celis-Morales et al., 2017).

5.1.2 Chronic Inflammation

Low-fat diets have been associated with decreased inflammatory state (Delgado-Lista et al., 2016; Jonasson, Guldbrand, Lundberg, & Nystrom, 2014). In contrast, hypercaloric high-fat diets are associated with pro-inflammatory profile and increased free fatty acids flux to insulin-sensitive tissues, which enhance the accumulation of diglycerides and ceramides and the production of ROS (Perona, 2017; Rocha et al., 2017). However, high-fat, low-carbohydrate diets in a hypocaloric context were shown to increase adiponectin levels, suggesting improvement in the inflammatory profile (Heer & Egert, 2015; Ruiz-Núñez et al., 2016). Some gene variants are involved in the interaction between inflammatory markers with total fat intake (Table 3). For example, the carriers of T allele of the rs230C on the *ATP binding cassette A1 (ABCA1)* gene had higher adiponectin levels when following a low-fat diet; yet, the carriers of the T allele had decreased adiponectin levels with a high-fat diet (Jacobo-Albavera et al., 2015). Likewise, the carriers of the GG genotype of the rs1501299 on the *ADIPOQ* gene had increased levels of adiponectin with a low-fat diet in comparison with T carriers in a hypocaloric context (de Luis et al., 2018). Within the POUNDS LOST study, a GRS composed of five SNPs associated with adiponectin levels was designed; subjects with high GRS had decreased levels of adiponectin with a high-fat diet, whereas increased levels of adiponectin were observed in subjects with low GRS following a low-fat diet (Ma et al., 2017). Overall, numerous gene variants showed interactions with fat composition of diets in inflammatory responses, suggesting low-fat diets might greater improve adiponectin levels in individuals with specific polymorphisms in comparison with others.

5.1.3 Glucose Metabolism

Low-fat, high carbohydrate diets have been shown efficient to improve glucose metabolism; however, some individuals seem to be more susceptible to hyperglycemia and hyperinsulinemia with a diet reduced in fatty acids (Gow, Garnett, Baur, & Lister, 2016). High-fat, low-carbohydrate diets for weight loss are also associated with improvement in glucose metabolism (Gow et al., 2016; Heer & Egert, 2015; Ruiz-Núñez et al., 2016). The differences in individual responses on glucose metabolism and fat intake have been assessed within nutrigenetic studies and multiple genetic variants have been identified, some of which are presented in Table 3. The Nutrient Gene Interactions in Human Obesity: implications for dietary guidelines (NUGENOB) study contributed to the identification of gene variants like the POUNDS LOST study. The NUGENOB is a randomized clinical trial in which 757 subjects from seven different European countries were randomized to either a low-fat or high-fat hypocaloric diet for a 10-week dietary intervention. *Protein phosphate Mg2+/Mn2+ Dependent 1K (PPM1K)* gene, which encoded protein is involved in cell survival and mitochondria permeability, was shown to interact with dietary fat composition on glucose metabolism. The T carriers of the rs1440581 in the *PPM1K* gene had limited improvement of glucose levels after the intervention in comparison with C allele carriers. The T carriers had lower fasting insulin levels and decreased β-cell functionality following a high-fat diet in comparison with low-fat diet and other genotypes (Goni et al., 2017). Complementary, the carriers of the C allele had greater reduction in insulin resistance and insulin levels on a low-fat diet after 1 year in the POUNDS LOST trial, with milder improvement in the high-fat diet (Xu et al., 2013). Using the data from POUNDS LOST trial, a GRS based on 14 SNPs involved in glucose metabolism was associated with dietary fat intake, especially high-fat diets, and changes in parameters associated with glucose metabolism in obese individuals (Wang et al., 2016). Overall, the great differences in glycemic and insulinemic responses to high-fat or low-fat diets between individuals might be due to some gene variants.

5.1.4 Dyslipidemia

Both high-fat and low-fat diets have unique effects on dyslipidemia when combined with energy restriction. Low-fat diets are associated with reduction on total-C and LDL levels; however, low-fat diets were also associated with decreased HDL levels and increased triglycerides levels in some individuals (Ordovas et al., 2007). In contrast, high-fat, low-carbohydrate diets with energy restriction are associated with reduced triglyceride levels

(Heer & Egert, 2015; Ruiz-Núñez et al., 2016). Genetic background might have a key role in the magnitude of response to one diet or another. Some gene-diet interactions between lipid profile and fat intake are presented in Table 3. In the POUNDS LOST trial, the A carriers of the rs2070895 on the *LIPC* gene had lower total-C and LDL levels and a small increase in HDL levels on the low-fat diet, while the A carriers on the high-fat diet saw their total-C and LDL levels increased (Xu et al., 2015). In contrast, the CC genotype of the rs3764261 on the *CETP* gene had greater improvement in the HDL levels and decreased triglycerides following a high-fat diet compared to a low-fat diet in the POUNDS LOST trial (Qi et al., 2015). Many other SNPs from the *APOA5* gene, the *IL-6* gene, the *leptin receptor* (*LEPR*) gene or the *ABCA1* gene also present some gene-diet interactions with fatty acid composition of diet on lipid profile (Table 3).

5.1.5 Blood Pressure

Hypocaloric high-fat, low-carbohydrate diets are linked to decreased diastolic blood pressure (Heer & Egert, 2015; Ruiz-Núñez et al., 2016). Low-fat diets could have potential effects on blood pressure in combination with weight loss; however, reducing fat content did not sustainably reduce blood pressure and the risk of hypertension in 48,835 post-menopausal women after an 8.3-year follow-up (Allison et al., 2016). Gene-diet interactions regarding fat composition of diet and blood pressure are presented in Table 3. Within the POUNDS LOST trial, the C carriers of the rs16147 on the *NPY* gene had greater reduction of systolic and diastolic blood pressure following a low-fat diet in comparison with a high-fat diet, especially in hypertensive subjects (Zhang, Qi, Liang, et al., 2012).

Finally, some gene variants have been shown to have gene-diet interactions with fat intake in diet and the incidence of the MetS, as presented in Table 3. For example, carriers of GG genotype of both rs670 of *APOA1* and rs521535 of *apolipoprotein B* (*APOB*) had increased risk of developing the MetS in comparison with low-fat diets especially in males (Phillips et al., 2011). On the contrary, A carriers of the rs1522813 on the *insulin receptor substrate 1* (*IRS-1*) gene had a reversion of the Mets on a hypocaloric high-fat diet in comparison with a low-fat diet (Qi et al., 2013). These results illustrate the contrasting variability in response to fat content between individuals. In conclusion, the optimal fat composition of diet in the prevention of the MetS is far from being a "*one size fits all.*" Genetic profiling in personalized nutrition might be a useful tool in the identification of the optimal macronutrient composition for individuals.

6. TYPES OF FATTY ACIDS

Although some changes in micronutrient composition of diets show interesting results on metabolic health, studying total fatty acid intake gives only a limited glimpse on the impact of different types of fatty acids. The different types of fatty acids have different saturation degrees, lengths and concentrations in foods, which all potentially affect metabolic health in a different way (Perona, 2017). The following sections examine the role of different types of fatty acids, namely SFA, MUFA, omega-6 and omega-3 PUFA and TFA on the various components of the MetS. Nutrigenetics and nutrigenomics evidences on each type of fatty acids are also covered.

6.1 Saturated Fatty Acids

SFA are important sources of energy and are involved in many metabolic pathways, from gene transcription to structural functions of the cells (German & Dillard, 2004). The liver and the adipose tissue can both synthesize and store SFA mostly in the form of palmitic acid (16:0). In the human body, roughly 30–40% of total fatty acids are SFA (Ruiz-Núñez et al., 2016). In the past decades, dietary SFA have been associated with a negative impact on health, built from evidences suggesting that diets rich in SFA increase the risk of cardiometabolic diseases, MetS and its components (Melanson, Astrup, & Donahoo, 2009; Wang & Hu, 2017). However, metabolic differences have been observed according to the specific length of SFA types (Poudyal & Brown, 2015).

6.1.1 Obesity

Prospective studies suggest SFA intake is positively associated with body weight (Field, Willett, Lissner, & Colditz, 2012). The obesogenic effect of SFA-rich diets might be linked to upregulation of lipogenic genes and increased hepatic lipogenesis. In contrast, MCFA have been associated with increased fat oxidation, thermogenesis and satiety together with reduced accumulation of fat in adipocytes (Hammad & Jones, 2017). The consumption of MCFA-rich milk was associated with decreased total fat percentage, total body weight and visceral fat percentage compared to a low-MCFA milk (Matualatupauw, Bohl, Gregersen, Hermansen, & Afman, 2017). The difference between long-chain fatty acids and MCFA might be due to the rapid absorption and metabolization of MCFA in comparison with long-chain SFA (Hammad & Jones, 2017). Further, MCFA could potentially upregulate gene expression related to energy metabolism in the adipose tissue (Matualatupauw et al., 2017).

Despite the detrimental effects attributed to SFA, great variations between ethnic groups and individuals are observed (Raatz, Conrad, Johnson, Picklo, & Jahns, 2017). Numerous SNPs have been identified for having gene-diet interactions with SFA intake, some of which are presented in Table 4. For example, the CC genotype of the rs5082 on *APOA2* gene and GG genotype of the rs3135506 of *APOA5* gene have been consistently associated with increased risk of obesity (Corella et al., 2009; Domínguez-Reyes et al., 2015). Some SNPs from the *PPARG* gene were associated with different outcomes on obesity parameters, where the carriers of the minor allele of the rs2938395 of *PPARG* had increased body weight, in comparison with the minor allele carriers of rs2972162 and rs709157 who had decreased bodyweight compared to homozygous wild type (Larsen et al., 2016). Obesity-related GRS showed positive associations between BMI and waist circumference with high SFA intake, confirming actual dietary requirements for SFA are effective for populations with high obesity-related GRS (Casas-Agustench et al., 2014; Celis-Morales et al., 2017).

6.1.2 Inflammation

SFA could have a pro-inflammatory effect according to in vitro and cross-sectional studies (Minihane et al., 2015). The pro-inflammatory effects of SFA are attributed to the activation of NF-κB and subsequent upregulation of pro-inflammatory genes (Georgiadi & Kersten, 2012). Transcriptomics analyses showed that high intake of SFA from animal sources was associated with increased *TNF-α* and *IL-6* gene expression in healthy subjects (Jiménez-Gómez et al., 2008). Similarly, SFA consumption upregulated post-prandial pro-inflammatory genes and downregulated anti-inflammatory gene expressions in comparison with high-MUFA diet in individuals with the MetS (Cruz-Teno et al., 2012; Perez-Martinez et al., 2010; Rangel-Zúñiga et al., 2015). In addition, SFA are not efficient ligands of PPARs; therefore, high SFA intake could decrease PPARs activity and contribute to a pro-inflammatory environment (Afman & Müller, 2012; Perez et al., 2012). The length of SFA can also have different effects on inflammatory responses. High levels of palmitic acid (16:0) was associated with increased TNF-α and IL-1β plasma concentrations and decreased *adiponectin* gene expression (Da Silva & Rudkowska, 2015; Kien et al., 2015; Rocha et al., 2017). Myristic acid

Table 4 Saturated Fatty Acids, Gene Variants and the Components of the Metabolic Syndrome

Gene	OMIM	Polymorphisms	Metabolic Syndrome Component	Effects of Saturated Fatty Acids	Reference
Phosphoenolpyruvate carboxykinase 2	*PCK2*	Rs9783666	Obesity	Minor allele carriers: High SFA intake is associated with ↓ body weight vs homozygous wild type	Larsen et al. (2016)
Apolipoprotein A2	*APOA2*	Rs5082	Obesity	CC genotype: High SFA intake is associated with ↑ obesity prevalence	Corella et al. (2009)
Apolipoprotein A5	*APOA5*	Rs3135506	Obesity	GG genotype: High SFA intake is associated with ↑ risk of obesity	Domínguez-Reyes et al. (2015)
Leptin receptor	*LEPR*	Rs1137101	Obesity	G carriers: High SFA intake is associated with ↑ risk of obesity	Domínguez-Reyes et al. (2015)
CCAAT/enhancer binding protein B	*CEBPB*	Rs4253449	Obesity	A carriers: High SFA intake is associated with higher risk of ↑ body weight vs TT wild type (not replicated)	Larsen et al. (2016)
Peroxisome proliferator-activated receptor gamma	*PPARG*	Rs2938395	Obesity	Minor allele carriers: High SFA intake is associated with ↑ body weight vs homozygous wild type (not replicated)	Larsen et al. (2016)
Peroxisome proliferator-activated receptor gamma	*PPARG*	Rs2972162 and rs709157	Obesity	Minor allele carriers: High SFA intake is associated with ↓ body weight vs homozygous wild type (not replicated)	Larsen et al. (2016)

Continued

Table 4 Saturated Fatty Acids, Gene Variants and the Components of the Metabolic Syndrome—cont'd

Gene	OMIM	Polymorphisms	Metabolic Syndrome Component	Effects of Saturated Fatty Acids	Reference
GRS with 93 SNPs associated with higher BMI			Obesity	High GRS with high SFA intake is associated with ↑ BMI vs low GRS	Celis-Morales et al. (2017)
GRS with 63 SNPs associated with obesity			Obesity	High GRS with high SFA intake is associated with ↑ BMI and waist circumference vs low SFA intake	Casas-Agustench et al. (2014)
Apolipoprotein A2	APOA2	Rs5082	Inflammation	T carriers: High SFA intake is associated with ↑ CRP levels	Keramat et al. (2017)
S100 calcium-binding protein A9	S100A9	Rs2014866	Glucose metabolism	CC genotype: High SFA/carbohydrate ratio is associated with ↑ HOMA-IR. No effect with low SFA/carbohydrate ratio	Blanco-Rojo et al. (2016)
Scavenger receptor class B type 1 protein	SCARB1	Exon 1 variant (G > A)	Glucose metabolism	A carriers: Greater ↓ in LDL by replacing SFA by CHO–rich diet vs GG genotype	Pérez-Martínez et al. (2003)
Retinoid X receptor Alpha	RXRA	rs11185660	Dyslipidemia	CC genotype: High SFA intake ↑ triglycerides levels after fish oil supplementation. Low SFA intake ↓ triglycerides levels after fish oil supplementation	Bouchard-Mercier, Rudkowska, Lemieux, Couture, and Vohl (2014)

Retinoid X receptor Alpha	RXRA	Rs12339187	Dyslipidemia	G carriers: Low SFA intake ↓ triglycerides after fish oil supplementation vs high SFA and A-A genotype	Bouchard-Mercier, Rudkowska, Lemieux, Couture, and Vohl (2014)
Tumor necrosis factor-alpha	TNFA	−308G/A	Dyslipidemia	GG genotype: High SFA intake is associated with ↓ total-C; A carriers: High SFA intake is associated with ↑ total-C	Joffe et al. (2011)
Angiotensin-converting enzyme	ACE	Rs4343	Blood pressure	GG genotype: High SFA intake associated with ↑ concentrations of ACE and ↑ systolic blood pressure vs AA/GG	Schüler, Osterhoff, Frahnow, Seltmann, et al. (2017)
Melanocortin 4 receptor	MC4R	Rs12970134	Metabolic syndrome	A carriers: ↑ risk of the MetS in the highest quartile of SFA consumption vs lowest quartile	Koochakpoor et al. (2016)

ACE: angiotensin-converting enzyme; BMI: body mass index; GRS: genetic risk score; HOMA-IR: homeostatic model assessment of insulin resistance, LDL; low-density lipoproteins; MetS: metabolic syndrome; SFA: saturated fatty acids; Total-C: total-cholesterol.

(14:0) was associated with an increased expression of pro-inflammatory genes in adipocytes, such as *IL-6* (Dordevic, Konstantopoulos, & Cameron-Smith, 2014; Yeop et al., 2010).

Conversely, the intake of MCFA was associated with downregulation of *NF-κB* and the upregulation *of PPAR-α*, promoting anti-inflammatory profile (Matualatupauw et al., 2017). In contrast with long-chain SFA, lauric acid (12:0) does not increase gene expression of *IL-6* or TNF-α in adipocytes (Da Silva & Rudkowska, 2015).

To study effects of dietary fatty acids on the MetS, a large clinical trial called LIPGENE was performed, including 417 subjects with the MetS which were randomized to four distinct diets for 12 weeks: a high-fat diet rich in SFA (38% of total energy); high-fat diet rich in MUFA; a low-fat diet rich in complex carbohydrates and a low-fat, high carbohydrate diet enriched with omega-3 PUFA (Caswell, 2009). Reducing SFA intake did not affect the levels of CRP, IL-6, TNF-α, adiponectin or leptin, suggesting no benefit of reducing SFA on chronic inflammation in the MetS (Tierney et al., 2010). Although evidences supporting the role of genes in the stimulation of inflammation by SFA remain limited, gene-diet interactions could explain the heterogeneity in the response to SFA (Table 4). As an example, the rs5082 of the *APOA2* gene has been associated with inflammation and SFA, where the T carriers had higher CRP levels with high SFA intake (Keramat et al., 2017).

6.1.3 Glucose Metabolism

SFA-rich diets have been associated with worsen glucose and insulin homeostasis when compared with MUFA-rich diets (Vessby et al., 2001; Wang & Hu, 2017). Some mechanisms propose that SFA-rich diets may repress PPARs activation, which could contribute to insulin resistance. In parallel, insulin resistance can be triggered by increased pro-inflammatory cytokines in response to high SFA diets. Regarding specific SFA, palmitic acid (16:0) has been associated with decreased insulin signalization of IRS-1 (Bjørnshave & Hermansen, 2014). In contrast, lauric acid (12:0) intake was associated with improved insulin sensitivity, increased insulin secretion and decreased glucose levels (Poudyal & Brown, 2015). However, in the LIPGENE trial, reducing SFA intake did not improve insulin sensitivity, fasting insulin or glucose levels (Tierney et al., 2010). Genetic background could also impact the response to SFA on glucose metabolism although the number of identified gene variants remains scarce (Table 4). For example, the *S100 calcium-binding protein A9* (*S100A9*) gene, which is involved in the recruitment of lymphocytes in inflammatory sites, had a gene-diet interaction with SFA intake on glucose metabolism (O'Leary et al., 2016). Specifically, carriers of the CC genotype of the rs2014866 on the *S100A9* gene are more likely to improve insulin resistance by adopting a diet low in SFA in replacement of carbohydrates (Blanco-Rojo et al., 2016).

6.1.4 Dyslipidemia

Actual evidence suggests replacing dietary SFA by PUFA or MUFA is associated with decreased total-C, LDL, HDL levels and total-C/HDL ratio. Further, when SFA are substituted for carbohydrates, total-C, LDL and HDL levels are decreased; in addition, no difference is observed in total-C/HDL ratio. Mechanistically, SFA have a unique effect of suppressing *LDL receptor* gene expression, promoting elevation of LDL levels in plasma. SFA-rich diets are also associated with increased expression of *SRBEP-1c* resulting in increased cholesterol synthesis and accumulation of triglycerides and cholesterol in the liver (Salter & Tarling, 2007). Further, high SFA diets are associated with increased liberation of free fatty acid levels in bloodstream (Camargo et al., 2014). Regarding specific SFA, lauric acid (12:0), myristic acid (14:0) and palmitic acid (16:0) are associated with increased LDL, elevated HDL, increased total-C levels together with lowered triglyceride levels (Mensink, Zock, Kester, & Katan, 2003). In counterpart, stearic acid (18:0) intake was associated with decreased LDL levels, lower total-C/HDL ratio when compared with other types of SFA and TFA. However, HDL levels were lower and total-C/HDL ratio was increased with stearic acid (18:0) intake in comparison with unsaturated fatty acid intake (Poudyal & Brown, 2015). These data suggest stearic acid (18:0) might have neutral or beneficial effects on lipid profile when replacing other types of long-chain SFA.

Although altered lipid profile is linked to detrimental effects on cardiovascular health, recent meta-analyses of cohort studies found no association between SFA intake and increased cardiovascular risk (Chowdhury et al., 2014; de Souza et al., 2015; Siri-Tarino, Sun, Hu, & Krauss, 2010). Despite the elevation of LDL levels, dietary SFA seem to mostly increase large LDL particles, while small and dense LDL are the one known to have greater detrimental effects on cardiovascular health (Ruiz-Núñez et al., 2016). Hence, the reduction of SFA solely does not seem efficient to reduce cardiovascular risk, which questions the relevance of actual recommendations about SFA on cardiometabolic health (Wang & Hu, 2017). However, interpretation of studies can be strongly biased by absence of consideration for the quality of nutrients replacing SFA (Sacks et al., 2017). For example, replacing SFA with low-glycemic index carbohydrates or processed complex carbohydrates seems to have a neutral effect of cardiometabolic health (Briggs, Petersen, & Kris-Etherton, 2017; Sacks et al., 2017). However, high-glycemic index carbohydrates increase the risk of cardiovascular events in comparison with SFA, while the replacement of SFA with non-processed fiber-rich carbohydrates reduces the risk of cardiovascular diseases, decreased LDL levels and increased HDL levels in comparison with refined, high-glycemic index carbohydrates (Briggs et al., 2017).

Despite epidemiological evidences, great disparities exist in individual responses to SFA intake. Gene variants of the *APOA* gene family (*APOA1*, *APOA2*, *APOA5*), the *APOE* gene and many others have shown gene-diet interactions with SFA intake and lipid profile and have been extensively described previously (Rudkowska & Vohl, 2010; Sotos-Prieto & Penalvo, 2013). Some other gene variants presenting gene-diet interactions with SFA intake and dyslipidemia are presented in Table 4. For example, the −308G/A of the *TNFA* gene was associated with changes in lipid profile with SFA consumption. The GG genotype carriers had lower total-C levels with a high SFA diet, while the A allele carriers had higher total-C levels with high SFA diet (Joffe et al., 2011).

6.1.5 Blood Pressure

SFA consumption does not seem to affect blood pressure according to the LIPGENE intervention trial and a systematic review of randomized controlled trials (Hooper et al., 2011; Tierney et al., 2010). Yet, gene-diet interactions between SFA intake and blood pressure were observed with the *angiotensin-converting enzyme gene* (*ACE*) (Table 4). Specifically, high SFA intake was associated with increased concentrations of angiotensin-converting enzyme in bloodstream and elevation of systolic blood pressure in carriers of the GG genotype in comparison with A carriers (Schüler, Osterhoff, Frahnow, Seltmann, et al., 2017).

Regarding the incidence of the MetS, the *melanocortin 4 receptor gene* (*MC4R*) rs1290134 was shown to have a gene-diet interaction with SFA intake (Table 4). Notably, the A allele carriers of the rs12970134 with the highest quartile of SFA intake had an increased risk of the MetS in comparison with A carriers in the lowest quartile of SFA intake (Koochakpoor et al., 2016).

Overall, SFA intake is likely to have detrimental effects of the MetS components in comparison with MUFA, PUFA or low-glycemic index carbohydrates. Although many SNPs have been identified in relation with SFA consumption and the different components of the MetS, the combination of relevant gene variants will better identify those who would really benefit from SFA reduction.

6.2 Mono-Unsaturated Fatty Acids

MUFA have been widely studied for their potential beneficial effects on metabolic health. The most studied MUFA-rich food source is surely olive oil, which contains a huge amount of MUFA and bioactive elements such as phenols, tocopherols, carotenoids or chlorophyll (Rocha et al., 2017). MUFA intake and olive oil are also often studied as part of the MedDiet.

6.2.1 Obesity

Unsaturated fatty acids such as MUFA have been previously linked with downregulation of genes related to lipogenesis and increased expression of genes associated with fatty acid oxidation, which might limit fat deposition in the adipose tissue (Hammad & Jones, 2017). In some clinical trials, a MUFA-rich diet compared to SFA-rich or carbohydrate diet decreased abdominal adiposity, body mass and fat mass in subjects with normal or excess body weight (Gillingham, Harris-Janz, & Jones, 2011). Regarding gene-diet interactions, some genes variants of *PPARG* were associated with MUFA intake and obesity outcomes. Specifically, the Ala12 carriers of the Pro12Ala (rs1801282) of the *PPARG2* gene and the minor allele carriers of the rs10865710 of *PPARG* had decreased obesity with high-MUFA intake in comparison with homozygous wild types (Garaulet et al., 2011; Larsen et al., 2016). Many gene-diet interactions involving MUFA intake, obesity, *APOA* genes family and many other SNP have been reviewed previously (Hammad & Jones, 2017). Some other relevant genes variants are presented in Table 5. These gene-diet interactions suggest the expression of genes in response to MUFA intake on bodyweight might be regulated by some genes variations.

6.2.2 Inflammation

Dietary MUFA could potentially affect inflammatory response. In the LIP-GENE trial, MUFA-rich diet was associated reduction of post-prandial NF-κB activity and TNF-α levels compared to a SFA-rich diet and diets rich in carbohydrates with or without n-3 PUFA in individuals with the MetS

Table 5 Mono-Unsaturated Fatty Acids, Gene Variants and the Components of the Metabolic Syndrome

Gene	OMIM	Polymorphisms	Metabolic Syndrome Component	Effects of Mono-Unsaturated Fatty Acids	References
Apolipoprotein A1	APOA1	Rs670	Metabolic syndrome	GG genotype: High-MUFA intake associated with ↑ risk of MetS especially in males	Phillips et al. (2011)
Apolipoprotein B	APOB	Rs512535	Metabolic syndrome	GG genotype: High-MUFA intake associated with ↑ risk of MetS especially in males	Phillips et al. (2011)
CCAAT/ enhancer binding protein B	CEBPB	Rs4253449	Obesity	A carriers: High-MUFA intake is associated with ↓ body weight with each additional A allele (not replicated)	Larsen et al. (2016)
Peroxisome proliferator-activated receptor gamma	PPARG	Rs10865710	Obesity	Minor allele carriers: High-MUFA intake is associated with ↓ body weight vs homozygous wild type	Larsen et al. (2016)
Peroxisome proliferator-activated receptor gamma 2	PPARG2	Pro12Alal (rs1801282)	Obesity	Ala12 carriers (G): Decreased obesity with high-MUFA intake vs Pro12Pro homozygous. No difference between genotype with low-MUFA intake	Garaulet et al. (2011)
Apolipoprotein A2	APOA2	Rs5082	Inflammation	CC genotype: High-MUFA intake is associated with ↓ Il-8 and CRP levels	Keramat et al. (2017)
Telomerase RNA component	TERC	Rs1269630	Inflammation	CC genotype: High-MUFA intake is associated with ↓ CRP levels	Gomez-Delgado et al. (2018)
Fatty acid desaturase 2	FASD2	Ala54Thr	Glucose metabolism	Thr54 carriers: Replacing olive oil by sunflower oil is associated with ↑ insulin resistance	Morcillo et al. (2007)

Gene	OMIM	Polymorphisms	Metabolic Syndrome Component	Effects of the Mediterranean Diet	References
Peroxisome proliferator-activated receptor gamma 2	PPARG2	Pro12Ala1 (rs1801282)	Glucose metabolism	Ala12 carriers (G): Low-MUFA intake is associated with ↑ HOMA-IR in obese subjects	Soriguer et al. (2006)
Scavenger receptor class B type 1 protein	SCARB1	Exon 1 variant (G>A)	Glucose metabolism	GG genotype: High-MUFA diet was associated with ↑ steady-state plasma glucose and ↑ free fatty acids vs GA genotype A carriers: High-MUFA diet was associated with ↑ insulin sensitivity	Pérez-Martínez et al. (2005)
Apolipoprotein E	APOE	Rs1064725	Dyslipidemia	TT genotype: ↓ total-C with high-MUFA diet vs high SFA or high PUFA diet	Shatwan, Weech, Jackson, Lovegrove, and Vimaleswaran (2017)
Adiponectin gene	ADIPOQ	−11377C/G	Dyslipidemia	G carriers: MUFA/SFA ratio <1 is associated with ↑ total-C and ↑ LDL	Park et al. (2014)
Peroxisome proliferator-activated receptor gamma 2	PPARG2	Pro12Ala1 (rs1801282)	Obesity	Ala12 carriers (G): Decreased waist circumference with the MedDiet in subjects with type 2 diabetes	Razquin, Martinez, Martinez-Gonzalez, Corella, et al. (2009)
Interleukin 6	IL-6	−174G>C	Obesity	CC genotype: Greater ↓ bodyweight with a MedDiet enriched in virgin olive oil	Razquin, Martinez, Martinez-Gonzalez, Bes-Rastrollo, et al. (2010)

Continued

Table 5 Mono-Unsaturated Fatty Acids, Gene Variants and the Components of the Metabolic Syndrome—cont'd

Gene	OMIM	Polymorphisms	Metabolic Syndrome Component	Effects of the Mediterranean Diet	References
Fat mass and obesity-related gene	FTO	Rs9939609	Obesity	AA genotype: Less weight gain with the MedDiet vs low-fat diet	Razquin, Martinez, Martinez-Gonzalez, Bes-Rastrollo, et al. (2009)
Telomerase RNA component	TERC	Rs1269630	Inflammation	CC genotype: Greater ↓ CRP levels with the MedDiet vs G allele carriers	Gomez-Delgado et al. (2018)
Melanocortin 4 receptor and fat mass and obesity-related gene	MC4R + FTO	Rs17782313 Rs9939609	Glucose metabolism	Risk alleles carriers: Low adherence to the MedDiet is associated with ↑ T2D vs wild alleles. With a good adherence to the MedDiet, no association with T2D is detected	Ortega-Azorín et al. (2012)
Transcription factor 7 like 2	TCF7L2	Rs7903146	Glucose metabolism	TT genotype: Low adherence to the MedDiet is associated with ↑ fasting glucose levels vs high adherence	Corella et al. (2013)
Adiponectin	ADIPOQ	276G/T (rs1501299)	Glucose metabolism	G carriers: ↓ glucose levels,↓ insulin levels and ↓ HOMA-IR with a MedDiet vs T carriers	de Luis et al. (2017)
NOD-like receptor pyrin domain containing-3″ inflammasome	NLRP3	Rs10733113	Glucose metabolism	A carriers: ↑ insulin sensibility with the MedDiet in non-diabetic subjects	Roncero-Ramos et al. (2018)

Adiponectin	*ADIPOQ*	276G/T (rs1501299)	Dyslipidemia	G carriers: ↓ total-Cl levels and ↓ LDL levels with a MedDiet vs T carriers	de Luis et al. (2017)
Cholesterol ester transfer protein	*CETP*	Rs3764261	Dyslipidemia	T carriers: ↑ HDL and ↓ triglycerides with a MedDiet vs GG genotype	Garcia-Rios et al. (2018)
Transcription factor 7 like 2	*TCF7L2*	Rs7903146	Dyslipidemia	TT genotype: Low adherence to the MedDiet is associated with ↑ total-C, ↑ LDL and ↑ triglyceride levels vs high adherence. With a good adherence, ↓ risk of stroke vs low-fat diet	Corella et al. (2013)
Serpin peptidase inhibitor clade E, member 1 (plasminogen activator inhibitor 1)	*SERPINE1 (PAI-I)*	Rs6950982	Dyslipidemia	GG genotype: Good adherence to the MedDiet is associated with ↓ triglyceride levels	Sotos-Prieto et al. (2013)

CRP: C-reactive protein; HDL: high-density lipoprotein; HOMA-IR: homeostatic model assessment of insulin resistance, IL-8: interleukin-8; LDL: low-density lipo-proteins; MedDiet: Mediterranean diet; MetS: metabolic syndrome; MUFA: mono-unsaturated fatty acids; PUFA: poly-unsaturated fatty acids; SFA: saturated fatty acids, T2D: type 2 diabetes; Total-C: total-cholesterol.

(Cruz-Teno et al., 2012; Perez-Martinez et al., 2010). Through trans-criptomics analysis, increased dietary palmitic acid/oleic acid ratio (SFA/MUFA) lowered the expression of pro-inflammatory genes in peripheral blood mononuclear cells (PBMC) and decreased pro-inflammatory cyto-kines secretion from leukocytes (Kien et al., 2015). Oleic acid intake was associated with increased *adiponectin* gene expression and decreased excre-tion of *TNF-α* and *IL-6* in adipocytes (Da Silva & Rudkowska, 2015). High-MUFA diet in comparison with high SFA diet was associated with downregulation in the expression of genes involved in oxidative phosphor-ylation pathway, thereby reducing metabolic stress (van Dijk et al., 2012). However, expression of anti-inflammatory genes remained unchanged when phenols were removed olive oil, supporting the hypothesis that phe-nols modify pro-inflammatory gene expression to a greater extent than MUFA in olive oil (Camargo et al., 2010; Smith, 2012). Some gene-diet interactions have been identified with variants of the *APOA2* gene and the *telomerase RNA component* (*TERC*) genes (Table 5). The CC genotype of the rs5082 of *APOA2* gene and the CC genotype of the rs1269630 of the *TERC* gene were associated with decreased levels of CRP with MUFA-rich diet (Gomez-Delgado et al., 2018; Keramat et al., 2017).

6.2.3 Glucose Metabolism

Increased MUFA consumption has been associated with improved β-cells function and increased insulin sensitivity in replacement of SFA (Heer & Egert, 2015; Perona, 2017). A meta-analysis of long-term clinical trials observed an association between high-MUFA diets and decreased glycated hemoglobin, a biochemical marker often used in the assessment of type 2 diabetes evolution, in comparison with low-MUFA and low-fat diets. However, no association was observed for fasting glucose, fasting insulin and insulin resistance (Schwingshackl, Strasser, & Hoffmann, 2011). In con-trast, in the LIPGENE trial, the consumption of high-MUFA diet decreased insulin sensitivity in females. These contrasting results could be explained by methodological and population differences between studies, but also genetic background (Perona, 2017; Tierney et al., 2010). Despite the potential role of genes in the heterogeneity of responses to MUFA, the identification of gene variants associated with glucose metabolism and MUFA intake remain limited (Table 5). As an example, the carriers of the Thr54 allele of the Ala54Thr on the *fatty acid desaturase 2* (*FADS2*) had increased insulin resis-tance when replacing olive oil by sunflower oil, the latter being composed with less MUFA than olive oil (Morcillo et al., 2007). Some gene-diet interaction was observed between the Pro12Ala located on the *PPARG* gene, peripheral insulin sensitivity, type 2 diabetes and high oleic acid intake in observational studies (Ordovas et al., 2007). Notably, the Ala12 obese carriers had increased insulin resistance with a low–MUFA diet (Soriguer et al., 2006).

6.2.4 Dyslipidemia

Replacing SFA with MUFA is associated with decreased total-C, LDL levels and total-C/HDL ratio together with mild increase of HDL levels, suggesting an overall improvement of lipid profile. When MUFA replace carbohydrates, triglyceride levels are reduced in healthy subjects, but also in individuals suffering from the MetS and type 2 diabetes. Substitution of 1% of total energy from carbohydrates by 1% of total energy from MUFA could reduce LDL levels and total-C/HDL levels ratio (Mensink et al., 2003). Transcriptomics studies observed that intake of unsaturated fatty acids such as MUFA was associated with downregulation genes involved in lipogenesis (Hammad & Jones, 2017). Despite these positive effects of MUFA, triglycerides and LDL-lowering effects of MUFA-rich diets seem to be less effective or comparable to PUFA-rich diets (Gillingham et al., 2011). Some gene variants interacting with MUFA intake on lipid profile outcomes are enumerated in Table 5 and many others have been detailed elsewhere (Rudkowska & Vohl, 2010; Sotos-Prieto & Penalvo, 2013). An important gene variant is the rs1064725 of the *APOE* gene, for which the TT genotype was associated with decreased total-C levels following a high–MUFA diet in comparison with high SFA or PUFA-rich diet (Shatwan et al., 2017). Similarly, a low–MUFA/SFA ratio is associated with increased total-C and LDL levels in G carriers of the −11377 C/G of the *ADIPOQ* gene (Park et al., 2014).

6.2.5 Blood Pressure

MUFA intake, especially oleic acid, was associated with reduction of blood pressure in comparison with SFA-rich and carbohydrate-rich diets (Delgado-Lista et al., 2016; Gillingham et al., 2011; Turner & Spatz, 2016). Some mechanisms propose that oleic acid could inhibit angiotensin receptors and enhance NO upregulation (Turner & Spatz, 2016). Olive oil is also associated with decreased systolic blood pressure (Hohmann et al., 2015). Mechanisms proposed are that olive oil can change the cells' membrane phospholipid compositions in subjects with hypertension. Similar with inflammation, the effects of olive oil on blood pressure seem to be mostly attributable of minor components of olive oil such as phenolic compounds (Perona, 2017). Regarding nutrigenetics, gene variants associated with blood pressure and MUFA intake are scarce.

Overall, actual evidences propose that replacing SFA by MUFA could be either neutral or protective for cardiometabolic health (Bowen, Sullivan, Kris-Etherton, & Petersen, 2018; Briggs et al., 2017; Sacks et al., 2017). Some beneficial effects of MUFA on the components of the MetS have been observed; however, studying the individual effect of MUFA on health can be very challenging since MUFA dietary sources vary greatly between populations. Many cofounding nutrients that can have beneficial or detrimental effects on health are present in MUFA-rich foods, such as SFA and TFA in meat or phenolic compounds in olive oil for example (Bowen et al., 2018; Briggs et al., 2017; Imamura et al., 2016). Further, some nutrigenetic studies demonstrate that not all subjects respond positively to a MUFA-rich diet. For example, the male carriers of the GG genotype of the rs670 on *APOA1* gene and the rs512535 on *APOB* gene had increased risk of the MetS with high-MUFA intake (Phillips et al., 2011). In conclusion, genetic background plays a key role in the response to MUFA intake. Despite the potential aforementioned gene variants, GRS might be the optimal solution to clearly identify the specific responders to high-MUFA intake in prevention of the MetS.

6.3 Mediterranean Diet

The MedDiet is characterized as a primary plant-based diet which includes high proportions of fat content (between 30% and 40% of total energy intake), with high-MUFA/SFA ratio. The MedDiet is mainly composed with vegetables, especially green and leafy vegetables, fruits, nuts and seeds, legumes, accompanied with moderate to high intake of fish and sea foods, low to moderate intake of dairy products and moderate intake of red wine during meals. Red meat should be consumed occasionally in small portions. A unique characteristic of the MedDiet is the presence of extra virgin olive oil as the main source of dietary fat (Bloomfield et al., 2015). The overall diet is composed with anti-inflammatory foods rich in anti-oxidants, such as phenol-rich olive oil and omega-3 sources, all have interesting contributions to the overall beneficial effect of the diet (Delgado-Lista et al., 2016).

A good adherence to the MedDiet is associated with positive effects on the MetS and its components in comparison with low-fat diet (Esposito, Maiorino, Bellastella, Panagiotakos, & Giugliano, 2017). A large multicenter randomized controlled trial on the MedDiet named PREDIMED was undertaken from 2003 to 2010. This clinical trial randomized 7447 individuals at risk of cardiovascular diseases to various versions of MedDiet for 4.8 years (Estruch et al., 2013). One of the MedDiet was supplemented in virgin olive oil (1 L per week), the other variant of the MedDiet was supplemented with nuts (30 g/day) and the control diet was a traditional low-fat diet. Subjects who had a MetS diagnosis before the trial and who followed the MedDiet enriched with virgin olive oil had reversal of the MetS after 4.8 years (Babio et al., 2015; Babio, Bulló, & Salas-Salvadó, 2009). In addition, a meta-analysis including 50 observational and clinical studies observed a reduction of the risk of MetS up to 50% with a good adherence to the MedDiet (Kastorini et al., 2011).

6.3.1 Obesity
The high content of fat of the MedDiet has been a topic of concern for weight management. However, some meta-analyses of clinical trials reported no increased weight gain with the MedDiet, but observed greater weight loss in comparison with controlled diet (Ajala, English, & Pinkney, 2013; Esposito, Kastorini, Panagiotakos, & Giugliano, 2010). Some gene-diet interactions including the MedDiet and obesity parameters are presented in Table 5. For example, the carriers of the AA genotype of the rs9939609 on the *FTO* gene were less likely to gain weight with the MedDiet in comparison with a low-fat diet (Razquin, Martinez, Martinez-Gonzalez, Bes-Rastrollo, et al., 2009). The CC genotype of the $-174G > C$ SNP on the *IL-6* gene had greater reduction of bodyweight by following a MedDiet enriched in virgin oil (Razquin et al., 2010).

6.3.2 Inflammation
Then MedDiet is associated with anti–inflammatory effects through the reduction of pro-inflammatory cytokines and increased adiponectin levels (Schwingshackl & Hoffmann, 2014b). Many pro-inflammatory genes are downregulated with the MedDiet, such as the *TNF-α* gene or the *interleukin-1β* gene (Corella, Coltell, Macian, & Ordovás, 2018). The PREDIMED trial observed a reduction of IL-6, CRP and pro-inflammatory chemokines plasma levels with a long-term adherence to the MedDiet, which on the contrary were increased with the low-fat diet (Estruch, 2010; Fitó et al., 2016). The MedDiet is also associated with reduced oxidative stress through reduction of gene expression associated with oxidative stress and enhanced antioxidant actions (Delgado-Lista et al., 2016; Fitó et al., 2016). Phenols

from olive oil, red wine and whole plant-based foods seem to be the major players in the antioxidant power of the MedDiet (Delgado-Lista et al., 2016). Regarding potential gene-diet interactions between the MedDiet and inflammation, the carriers of the CC genotype of the rs1269630 on *TERC* gene had greater decreased in CRP levels with the MedDiet in comparison with G carriers (Table 5) (Gomez-Delgado et al., 2018).

6.3.3 Glucose Metabolism

Glucose metabolism integrity could highly benefit from the anti-inflammatory effects of the MedDiet. In the PREDIMED trial, participants in the MedDiet supplemented with olive oil showed a 40% reduction of type 2 diabetes risk in comparison with the low-fat diet (Esposito et al., 2009). In another long-term trial, subjects with type 2 diabetes de novo consuming a MedDiet for 8.1 years observed improved inflammatory markers, which were associated with reduction of insulin resistance and improved glucose metabolism in comparison with a low-fat diet (Maiorino et al., 2016a, 2016b). Results from the PREDIMED trial suggest the MedDiet could change phospholipid composition and cholesterol content in cell membranes and could consequently improve membrane receptors functionality (Barceló et al., 2009). The glycemic response to the MedDiet is also modulated by gene variants, some of which are presented in Table 5. Notably, the carriers of the TT genotype of the rs7903146 of the *transcription factor 7 like 2 gene* (*TCF7L2*) gene could avoid alterations in glucose levels with a good adherence to the MedDiet in contrast with a low adherence (Corella et al., 2013). Similarly, the carriers of the G allele of the 276G/T (rs1501299) of the *ADIPOQ* gene had lower glucose levels, lower insulin levels and improved insulin sensitivity with a MedDiet in comparison with T carriers (de Luis et al., 2017). The combination of the rs17782313 of the *MC4R* gene and the rs9939609 of the *FTO* gene also presented a diet-gene interaction with the adherence to the MedDiet. When the adherence to the MedDiet was low, the risk alleles of these SNPs were associated with increased risk of type 2 diabetes; however, when the adherence to the diet is high, the risk of type 2 diabetes disappeared (Fitó et al., 2016; Ortega-Azorín et al., 2012).

6.3.4 Dyslipidemia

The MedDiet has interesting effects on lipid profile. The atherogenic total-C/HDL ratio and LDL levels were improved with the MedDiet in comparison with high SFA diet (Delgado-Lista et al., 2016). In the

PREDIMED trial, circulating triglycerides in VLDL were decreased after 3 months of consuming the MedDiet rich in olive oil (Perona et al., 2010). Further, a meta-analysis of 50 observational and randomized controlled trials in which the MedDiet was associated with increased levels of HDL (1.17 mg/dL) and decreased triglycerides (−6.14 mg/dL) (Kastorini et al., 2011). Some gene-diet interactions have been identified between the MedDiet and lipid profile, some of which are presented in Table 5. For example, carriers of the TT genotype of the rs7903146 on the *TCF7L2* gene observed greater alterations of lipid profile when adherence to the MedDiet was low, in comparison with a good adherence. In contrast, a good adherence to the MedDiet reduced the risk of stroke in TT genotype compared to a low-fat diet (Corella et al., 2013).

6.3.5 Blood Pressure

A systematic review of observational and clinical studies suggested an improvement of both systolic and diastolic blood pressure (−2.35 and −1.58 mmHg, respectively) with the MedDiet (Kastorini et al., 2011). Transcriptomics analysis identified many pathways modulated by the MedDiet enriched with olive oil, such as renin-angiotensin pathway, while NO synthase signaling pathways were modified by the two variants of the MedDiet of the PREDIMED trial (Castaner et al., 2013). These results suggest virgin olive oil may have an important role in the hypotensive effects on the MedDiet. Further studies are required to better understand the role of genetic variations in the response to the MedDiet on blood pressure.

In conclusion, the MedDiet seems to have greater beneficial attributes than other dietary patterns on the MetS and its components. Although some individuals might not respond to the same level to the diet, the adoption of the MedDiet is publicly encouraged for the prevention of cardiometabolic diseases (U.S. Department of Health and Human Services and U.S. Department of Agriculture, 2015). Nutrigenetic studies suggest that some carriers of gene variants might greater benefit from the diet. Taken together, these results suggest that the MedDiet is one of the best dietary patterns to prevent the MetS and its components.

6.4 Poly-Unsaturated Fatty Acids

Both omega-6 and omega-3 PUFA showed interesting effects on the components of the MetS. Despites often studied together, omega-6 and omega-3 PUFA have distinct actions on metabolic health.

6.4.1 Obesity

The consumption of PUFA is associated with downregulation of many genes associated with lipogenesis by the repression of SREBPs activities. Yet, omega-6 PUFA intake is linked to promote fat deposition, while omega-3 PUFA seems to have opposite effects. Therefore, omega-3 PUFA might counteracts the lipogenic actions of omega-6 PUFA. These data suggest that an increased omega-6/omega-3 ratio, similarly to what is observed in Western Diets, is more likely to increase fat deposition (Hammad & Jones, 2017). Regarding omega-3 specifically, fish oil and fish consumption have been associated with mild reduction of bodyweight, BMI and waist circumference in meta-analysis of randomized controlled trials (Bender et al., 2014).

Many gene variants are associated with variations in response of bodyweight to PUFA intake, some of which are presented in Table 6. For example, carriers of the A allele of the rs4253449 on the *CCAAT/enhancer binding protein* (*CEBPB*) gene and the carriers of the minor allele of rs709157 of *PPARG* gene, which are two genes involved in adipocyte differentiation, had increased bodyweight with high PUFA intake compared to homozygous wild type. These data suggest activation or repression of transcription factors by PUFA is modulated by genetic variants, which can change the affinity to dietary lipids as ligands (Larsen et al., 2016). Regarding omega-6 linoleic acid, the T carriers of the rs174547 of the *fatty acid desaturase 1* (*FADS1*) gene consuming low intake of LA had lower BMI and lower waist circumference compared to high LA consumers (Dumont et al., 2017).

6.4.2 Inflammation

Omega-6 and omega-3 PUFA are essential for the production of eicosanoids, which influence inflammatory and immune responses through the body. Dietary LA is transformed in ARA to be utilized for the production of some eicosanoids mostly associated with pro-inflammatory effects. In contrast, omega-3 PUFA are involved in the production of anti-inflammatory eicosanoids. Eicosanoids from omega-3 PUFA also enhance the production of mediators involved in resolution of inflammation compared to ARA (Calder, 2015; Serhan, Chiang, & Van Dyke, 2008). Eicosanoids from omega-6 PUFA are usually more powerful than those derived from omega-3 PUFA; therefore, the ratio omega-6/omega-3 is often used to understand the inflammatory response following the consumption of dietary PUFA (Rustan & Drevon, 2005). Further, PUFA can efficiently activate PPARs. When activated, PPAR-α and PPAR-γ increase the expression of *adiponectin* gene, *adiponectin receptor* gene and many other genes, which induce anti-inflammatory processes

Table 6 Poly-Unsaturated Fatty Acids, Gene Variants and the Components of the Metabolic Syndrome

Gene	OMIM	Polymorphisms	Metabolic Syndrome Component	Effects of Poly-Unsaturated Fatty Acids	References
Lipoprotein lipase	LPL	Rs320 (HindIII)	Obesity	TT genotype: High PUFA intake associated with ↓ BMI and waist circumference vs G carriers	Ma et al. (2014)
Fat mass and obesity associated gene	FTO	Rs9939609	Obesity	A carriers: High PUFA intake is associated with ↓ BMI, ↓ bodyweight, ↓ fat mass vs TT genotype (hypocaloric diets)	De Luis, Aller, Izaola, and Pacheco (2015)
CCAAT/enhancer binding protein B	CEBPB	Rs4253449	Obesity	Minor allele carriers: High PUFA intake is associated with increased bodyweight vs homozygous wild type	Larsen et al. (2016)
Peroxisome proliferator-activated receptor gamma	PPARG	Rs709157	Obesity	Minor allele carriers: High PUFA intake is associated with ↑ body weight vs homozygous wild type (not replicated)	Larsen et al. (2016)
Fatty acid desaturase 1	FADS1	Rs174547	Obesity	T carriers: Low consumption of LA is associated with ↓ BMI, ↓ waist circumference for each additional minor allele vs high LA consumers	Dumont et al. (2017)

Continued

Table 6 Poly-Unsaturated Fatty Acids, Gene Variants and the Components of the Metabolic Syndrome—cont'd

Gene	OMIM	Polymorphisms	Metabolic Syndrome Component	Effects of Poly-Unsaturated Fatty Acids	References
Zinc transporter 8	*ZNT8*	Rs1326634	Obesity	CC genotype: High fish intake is associated with ↓ risk of abdominal obesity vs T carriers	Hosseini-Esfahani et al. (2017)
Interleukin-6	*IL-6*	−174G>C	Obesity	C carriers: High omega-3 intake or high omega-3/omega-6 ratio is associated with ↓ BMI in white women	Joffe et al. (2014)
Interleukin-6	*IL-6*	281G>T	Obesity	C carriers: High omega-3 intake or high omega-3/omega-6 ratio is associated with ↓ BMI in white women	Joffe et al. (2014)
Interleukin-6	*IL-6*	869A>G	Obesity	AG genotype: High omega-3 intake or high omega-3/omega-6 ratio is associated with ↓ BMI in white women	Joffe et al. (2014)
Zinc transporter 8	*ZNT8*	Rs1326634	Metabolic syndrome	CC genotype: High omega-3 intake is associated with ↓ risk of Mets vs T carriers	Hosseini-Esfahani et al. (2017).

Fat mass and obesity associated gene	*FTO*	Rs9939609	Inflammation	A carriers: High PUFA intake is associated with ↓ leptin vs TT genotype (hypocaloric diets)	De Luis et al. (2015)
Apolipoprotein A2	*APOA2*	Rs5082	Inflammation	CC genotype: High omega-3 intake is associated with ↓ IL-1β and ↓ CRP levels	Keramat et al. (2017)
Sterol regulatory element binding protein 1c	*SREBF1*	Rs2297508	Glucose metabolism	CC genotype: High PUFA intake is associated with ↑ insulin sensitivity	Bouchard-Mercier, Rudkowska, Lemieux, Couture, Pérusse, et al. (2014)
Fat mass and obesity associated gene	*FTO*	Rs9939609	Glucose metabolism	A carriers: High PUFA intake is associated with ↓ fasting insulin and ↓ HOMA-IR vs TT genotype (hypocaloric diets)	De Luis et al. (2015)
Zing transporter 8	*ZNT8*	Rs13266634	Glucose metabolism	CC genotype: High PUFA intake is associated with ↓ risk of high fasting glucose levels vs CT/TT	Hosseini-Esfahani et al. (2017)
Free fatty acid receptor or G protein-coupled receptor 120	*FFAR4 or GPR120*	Rs17108973 Rs11187537 Rs7081686 Rs717484310	Glucose metabolism	Minor allele carriers: ↑ HOMA-IR and ↑ fasting insulin levels after omega-3 supplementation vs homozygous wild type had ↓ fasting insulin avec omega-3 supplementation	Vallée Marcotte et al. (2017)

Continued

Table 6 Poly-Unsaturated Fatty Acids, Gene Variants and the Components of the Metabolic Syndrome—cont'd

Gene	OMIM	Polymorphisms	Metabolic Syndrome Component	Effects of Poly-Unsaturated Fatty Acids	References
Phosphatidylinositol-4,5-Bisphosphate 3-kinase Catalytic Subunit Alpha—Potassium Calcium-Activated Channel Subfamily M Regulatory Beta Subunit 3 region	*P1K3CA-KCNMB3*	Rs7645550	Glucose metabolism	TT genotype: Low omega-3/omega-6 ratio is associated with ↓ HOMA-IR and ↓ insulin levels *vs* C carriers, but not with high omega-3/omega-6 ratio	Zheng et al. (2013)
Phosphatidylinositol-4,5-Bisphosphate 3-kinase Catalytic Subunit Alpha—Potassium Calcium-Activated Channel Subfamily M Regulatory Beta Subunit 3 region	*P1K3CA-KCNMB3*	Rs1183319	Glucose metabolism	G carriers: High omega-3/omega-6 ratio intake is associated with ↑ HbA1c *vs* AA genotype, but not when the omega-3/omega-6 ratio was low. Low omega-3/omega-6 ratio was associated with ↑ HOMA-IR, by not when the ratio was high	Zheng et al. (2013)
Peroxisome proliferator-activated receptor alpha	*PPARA*	Leu162Val	Dyslipidemia	Val162 carriers: When PUFA intake is low, ↑ triglycerides and ↑ ApoCIII levels. When PUFA intake is high, ↓ ApoCIII	Tai et al. (2005)

Transcription factor 7 like 2	TCF7L2	Rs12255372	Dyslipidemia	T carriers: Lowest tertile of PUFA intake is associated with ↑ HDL, highest tertile of PUFA intake associated with ↓ HDL vs GG genotype	Bodhini et al. (2017)
Apolipoprotein A1	APOA1	−75G/A	Dyslipidemia	A carriers: High PUFA intake is associated with ↑ HDL levels G carriers: high PUFA intake is associated with ↓ HDL levels	Ordovas et al. (2002)
Tumor necrosis factor-alpha	TNFA	238G/A	Dyslipidemia	GG genotype: High PUFA intake is associated with ↓ HDL, greater interaction when combined with −308G/A SNP A carriers: High PUFA intake is associated with ↑ HDL levels	Fontaine–Bisson et al. (2007)
Tumor necrosis factor-alpha	TNFA	−308G/A	Dyslipidemia	A carriers: High PUFA intake is associated with ↓ HDL levels, greater interaction when combined with 238G/A SNP	Fontaine–Bisson et al. (2007)
Fatty acid desaturase 1	FADS1	Rs174547	Dyslipidemia	T carriers: High linoleic acid intake is associated with ↓ HDL	Dumont et al. (2017)

Continued

Table 6 Poly-Unsaturated Fatty Acids, Gene Variants and the Components of the Metabolic Syndrome—cont'd

Gene	OMIM	Polymorphisms	Metabolic Syndrome Component	Effects of Poly-Unsaturated Fatty Acids	References
Apolipoprotein A5	APOA5	Rs662799	Dyslipidemia	C carriers: High n-6 PUFA is associated with ↑ triglycerides, ↑ VLDL size and ↓ LDL size	Lai et al. (2006)
Zinc transporter 8	ZNT8	Rs1326634	Dyslipidemia	CC genotype: High n-3 intake is associated with ↓ risk of dyslipidemia vs T carriers	Hosseini-Esfahani et al. (2017)
Fatty acid desaturase 1	FADS1	Rs174546	Dyslipidemia	T carriers: Greater ↓ of triglycerides levels after EPA and DHA supplementation vs CC genotype	AlSaleh et al. (2014)
Interleukin-6	IL-6	−174G>C	Dyslipidemia	C carriers: High omega-3 intake or high omega-3/omega-6 ratio is associated with ↓ triglycerides and ↓ total-C/HDL ratio	Joffe et al. (2014)
Cholesterol ester transfer protein	CETP	Rs5882	Dyslipidemia	T carriers: Increased omega-3 levels in red blood cells membrane is associated with ↓ total-cholesterol, ↓ LDL and ↑ HDL	Rudkowska et al. (2013)

Cholesterol ester transfer protein	CETP	Rs183130	Dyslipidemia	T carriers: Increased omega-3 levels in red blood cells membrane is associated with ↓ total–cholesterol, ↓ LDL and ↑ HDL	Rudkowska et al. (2013)
Cholesterol ester transfer protein	CETP	Rs4783961	Dyslipidemia	AG genotype: Increased omega-3 levels in red blood cells membrane is associated with ↓ total–cholesterol, ↓ LDL and ↑ HDL	Rudkowska et al. (2013)
Angiotensinogen	AGT	Rs699	Dyslipidemia	TT genotype: Increased omega-3 levels in red blood cells membrane is associated with ↓ total–cholesterol, ↓ LDL and ↑ HDL	Rudkowska et al. (2013)
Peroxisome proliferator-activated receptor gamma 2	PPARG2	Pro12Ala (rs1801282)	Dyslipidemia	Pro12 carriers: With omega-3 supplementation, especially fish oil, greater ↑ LDL vs non-carriers not receiving omega-3	Zheng et al. (2018)
CD36 molecule	CD36	Rs1527483	Dyslipidemia	G carriers: With omega-3 fish oil supplementation, greater ↓ triglycerides vs non-carriers	Zheng et al. (2018)

Continued

Table 6 Poly-Unsaturated Fatty Acids, Gene Variants and the Components of the Metabolic Syndrome—cont'd

Gene	OMIM	Polymorphisms	Metabolic Syndrome Component	Effects of Poly-Unsaturated Fatty Acids	References
Nitrite oxide synthase	*NOS3*	Rs1799983	Dyslipidemia	A carriers: Low omega-3 is associated with ↑ triglycerides, high omega-3 intake is associated with ↓ triglyceride levels A carriers without omega-3 supplementation had ↑ triglycerides vs CC genotype, while A carriers with omega-3 supplementation had no change in triglyceride levels	Ferguson et al. (2010) Zheng et al. (2018)
GRS of 3 SNPs associated with omega-3 and lipid profile	*CD36 NOS3 PPARG2*	rs1527483-G allele, rs1799983-A allele, rs1801282-G allele	Dyslipidemia	High GRS subjects who received omega-3 supplementation, especially fish oil had ↓ triglycerides levels vs high GRS subjects without omega-3	Zheng et al. (2018)
Cytochrome P450 family 4 subfamily F member 2	*CYP4F2*	V433M	Blood pressure	V433V genotype: High omega-3 intake is associated with greater ↓ in blood pressure in time	Tagetti et al. (2015)

ApoCIII: apolipoprotein C-III; BMI: body mass index; DHA: docohexaenoic acid; EPA: eicosapentaenoic acid; GRS: genetic risk score; HbA1c: glycated hemoglobin A; HDL: high-density lipoprotein; HOMA-IR: homeostatic model assessment insulin resistance; IL: Interleukin; LDL: low-density lipoprotein; MetS: metabolic syndrome; PUFA: poly-unsaturated fatty acids; VLDL: very low-density lipoprotein.

(Perez et al., 2012). Omega-6 and omega-3 PUFA are both potent ligands; yet; their individual effects on inflammation differ partially due to their unique derivative bioactive compounds from the eicosanoid pathway. Further, PUFA intake is associated with the repression of NF-κB and pro-inflammatory gene expression (Georgiadi & Kersten, 2012).

Despite the aforementioned hypotheses, human clinical trials failed to see any increase of inflammatory cytokines with LA intake (Harris et al., 2009). The binding of omega-6 PUFA on PPAR-α may counteracts with pro-inflammatory eicosanoids. Taken together, omega-6 PUFA consumption does not seem to lead to significant increased inflammation (Innes & Calder, 2018). In parallel, omega-3 PUFA are more potent ligands to PPAR-α than LA; therefore, omega-3 PUFA have other anti-inflammatory properties than eicosanoid production (Salter & Tarling, 2007). DHA and EPA consumption downregulates pro-inflammatory cytokine production via reduced expression of *NF-κB*, *IL-1β* and *IL-6* genes (Minihane et al., 2015; Shah et al., 2017). Interestingly, the consumption of EPA and DHA can effectively reduce gene expression of pro-inflammatory cytokines on a long-time period (Afman & Müller, 2012). Further, modifications in the proportions of EPA and DHA result in variations in gene expression, suggesting EPA and DHA have specific anti-inflammatory actions (Rudkowska & Pérusse, 2012). Despite these beneficial effects, high inter-individual variability is observed in the inflammatory response to PUFA intake. Some gene variants have been shown to have gene-diet interactions with PUFA intake on the response of inflammation, some of which are presented in Table 6. For example, the CC genotype of the rs5082 of the *APOA2* gene was associated with reduction of IL-1β and CRP levels after omega-3 intake (Keramat et al., 2017). Gene variants of the *IL-6* gene were associated with variations in the response to omega-3 PUFA on obesity risk factors. The C allele carriers of the −174G > C and the 281G > A on the *IL-6* gene with high omega-3 PUFA intake or high omega-3/omega-6 ratio had decreased BMI, especially in white women (Joffe et al., 2014). These results suggest omega-3 PUFA could potentially modulate *IL-6* gene expression differently according to the presence/absence of risk alleles and improve phenotypes of other components of the MetS related to inflammation, such as obesity.

6.4.3 Glucose Metabolism
PUFA intake could also be beneficial for glucose metabolism. First, PUFA intake could modulate PPAR-γ expression resulting in increased transcription

of *GLUT-4*, improved insulin resistance and decreased pro-inflammatory state (Perona, 2017). However, high concentrations of PUFA and eicosanoids are necessary to fully activate PPARs (Grygiel-Górniak, 2014). PUFA intake is also associated with decreased expression of SREBPs, which could potentially promote insulin sensitivity (D'Amore et al., 2014; Salter & Tarling, 2007). In parallel, dietary PUFA could increase membrane fluidity, improve receptor translocation and consequently improve insulin sensitivity (Imamura et al., 2016). Regarding omega-6 PUFA, increased omega-6/omega-3 ratio could lead to reduced insulin sensitivity in skeletal muscle and adipocytes. The detrimental effects of omega-6 PUFA are linked to ARA, which increases insulin resistance (Simopoulos, 2016). On the contrary, a diet rich in omega-3 PUFA increases EPA and DHA content in cells membranes, which enhances membrane fluidity and facilitates cells membrane functionality (Rustan & Drevon, 2005). In contrast with mechanistic hypotheses, results from randomized controlled trials failed to confirm a consistent effect of omega-3 PUFA on insulin resistance in subjects with the MetS (Hartweg et al., 2008; Lopez-Huertas, 2012; Wu et al., 2012). Conversely, some studies observed detrimental effects of marine omega-3 on blood glucose (Heer & Egert, 2015). These contrasting results are partially explained by the great heterogeneity in responses to PUFA intake regarding glucose metabolism. Some gene variants associated with glucose metabolism and PUFA intake are presented in Table 6. For example, the carriers of the CC genotype of the rs2297508 on the *sterol regulatory element binding transcription factor 1 gene (SREBF1)* had increased insulin sensitivity with high PUFA intake. Since the *SREBF1* gene encodes for SREBP1 transcription factor, these results suggest PUFA intake might repress SBREPs more efficiently in some individuals according to their genetic profile. Regarding omega-3 PUFA, the discrepancies in epidemiological studies might be partially due to gene variants, some of which are presented in Table 6. For example, minor allele carriers of some SNPs on the *free fatty acid receptor gene (FFAR4)* (also known also the *G protein-coupled receptor 120 gene (GPR120)*) had deteriorated insulin metabolism after the supplementation of omega-3 PUFA, while homozygous wild type had decreased insulin levels after omega-3 PUFA supplementation (Vallée Marcotte et al., 2017). *FFAR* gene encodes for the receptors of free fatty acids, including omega-3 PUFA (O'Leary et al., 2016). These results highlight the great inter-variability between subjects in the response to omega-3 PUFA and confirm the relevance of future personalized dietary advices.

6.4.4 Dyslipidemia

Replacing SFA with PUFA is associated with decreased total-C, LDL, total-C/HDL ratio and mild increase of HDL levels (Mensink et al., 2003; Sacks et al., 2017). The beneficial effects of PUFA intake have been observed with both omega-6 PUFA and omega-3 PFUA (Imamura et al., 2016). Evidence suggests that replacing SFA or carbohydrates with LA is associated with decreased LDL and total-C/HDL ratio (Mensink et al., 2003). This beneficial effect of LA on lipid profile could be due to the activation of PPAR-α, which has cholesterol-lowering actions (Pandey, Renwick, Misquith, Sokoll, & Sparks, 2008). In addition to PPARs activation, PUFA intake was shown to decrease SREBPs signaling, resulting in modulation of lipogenesis (Rudkowska & Pérusse, 2012). Regarding omega-3 PUFA, DHA is known as a strong repressor of *SREBPs* gene expression in the liver, which inhibits lipogenesis (Afman & Müller, 2012). The activation of PPAR-α in the liver by omega-3 PUFA is another potential mechanism, which may lead to increased hepatic fatty acid oxidation and decreasing endogenous lipogenesis in the liver (de la Iglesia et al., 2016). Likewise, the LIPGENE trial observed a reduction of plasma triglyceride levels, ApoCIII levels and free fatty acids levels with a low-fat diet enriched in omega-3 PUFA (Tierney et al., 2010).

Despite these promising results of PUFA intake on lipid profile, metabolic responses differ greatly between individuals. The heterogeneity of responses to PUFA intake has been extensively studied with gene variants on *APOA1*, *APOA4*, *APOA5* or *CETP* genes and has been detailed previously (Rudkowska & Vohl, 2010; Sotos-Prieto & Penalvo, 2013). Some other gene variants are presented in Table 6. For example, the Val162 carriers of the Leu162Val of the *PPARA* gene had increased triglycerides and ApoCIII levels with low PUFA intake; yet, ApoCIII levels decreased with high PUFA intake (Tai et al., 2005). These results support that gene variations in PPARs might modulate the affinity of fatty acids as ligands for the activation of these transcription factors. Regarding omega-6 PUFA, T carriers and C carriers of the *FADS1* (rs174547) and *APOA5* (rs662799), respectively, promoted deterioration of lipid profile with high omega-6 PUFA intake or LA intake, while non-carriers were not affected (Dumont et al., 2017; Lai et al., 2006). Individual responses to omega-3 PUFA on lipid profile are also explained by multiple gene variants, some of which are presented in Table 6. For example, some gene variants of the *CETP* gene (rs5882, rs183130 and rs4783961) and the rs699 of the *AGT* gene had greater improvement of lipid profile when omega-3 PUFA levels in red blood cells were increased (Rudkowska et al., 2013). A high GRS combining SNPs of the *cluster of differentiation 36 gene* (*CD36*), *nitric oxide synthase gene* (*NOS3*) and *PPARG* was associated with decreased triglycerides levels with supplementation of fish oil compared to subjects with a low GRS (Zheng et al., 2018).

6.4.5 Blood Pressure

Omega-3 PUFA intake was associated with improved blood pressure risk factors (Delgado-Lista et al., 2016; Heer & Egert, 2015). A meta-analysis of randomized controlled trials assessing the supplementation of EPA and DHA noted a reduction of both systolic and diastolic blood pressure, with stronger effect among untreated hypertensive subjects (Miller, Van Elswyk, & Alexander, 2014). Possible mechanisms include repression of vasoconstrictive eicosanoids, increased production and liberation of NO, reduction of noradrenaline in plasma, suppression of angiotensin-converting enzyme activity and inhibited angiotensin II formation (Perona, 2017). The beneficial effects of omega-3 on blood pressure are more important with EPA and DHA, but ALA also have an influence to a lesser extend (Khalesi, Irwin, & Schubert, 2015). Gene-diet interactions involving PUFA intake and blood pressure are limited (Table 6). Regarding omega-3 PUFA, the V433V genotype of the V433M polymorphism of the *Cytochrome P450 Family 4 Subfamily F Member 2 gene* (*CYP4F2*) was associated with greater improvement in blood pressure in time with high omega-3 PUFA (Tagetti et al., 2015).

Overall, both omega-6 and omega-3 PUFA seem to have beneficial effects in the prevention of the MetS; yet, the great variance in response to PUFA intake suggests a personalized approach in dietary advices might be more adapted in order to prevent or manage the different components of the Mets.

6.5 Trans Fatty Acids

TFA have well-known detrimental effects on cardiovascular health and metabolic diseases. However, actual evidence usually focuses on total TFA intake despite the fact that iTFA and rTFA have respective unique actions on metabolic health. This section covers the various effects of total TFA and iTFA intake on the components of the MetS, followed by a subsection focusing exclusively on rTFA. Regarding genetic background related to dietary TFA, data of one trial studying gene variants and the components of the MetS was identified; additional studies are thus required to identify further gene-diet interactions with TFA.

6.5.1 Obesity

The association between long-term TFA consumption and bodyweight has not been studied in clinical trials; yet, some observational evidences suggest total TFA intake is associated with weight gain, especially abdominal fat (Field et al., 2012; Micha & Mozaffarian, 2009; Mozaffarian, Aro, & Willett, 2009). High consumption of total TFA in diet has been linked to increased lipogenesis in the liver, elevating the risk of excessive fat deposition (Hammad & Jones, 2017). Total TFA could increase fat deposit in abdominal adipose tissue and increase body weight in comparison with isocaloric diets without TFA (Dhaka et al., 2011; Micha & Mozaffarian, 2009). Nonetheless, further studies are required to better understand the potential role of iTFA on weight parameters (Micha & Mozaffarian, 2009).

6.5.2 Inflammation

iTFA have been associated with accentuated systemic inflammation. Evidences from various randomized controlled trials noted increased IL-6 levels, TNF-α and CRP activities with a consumption of 6.7–8% total TFA in comparison with a minimal intake. Total TFA intake also increases the production of pro-inflammatory cytokines in macrophages and monocytes, perpetuating the pro-inflammatory environment. The pro-inflammatory effects of total TFA consumption could be related to the development of further complications linked to the MetS (Micha & Mozaffarian, 2009).

6.5.3 Glucose Metabolism

Evidences from observational studies suggest increased risk of insulin resistance and type 2 diabetes with iTFA consumption (Dhaka et al., 2011). Reducing total TFA intake from 3% to 2% of total energy intake and replacing by the original non-hydrogenated oil could reduce type 2 diabetes incidence by 40% (Salmerón et al., 2001). Mechanistically, iTFA have been shown to increase free fatty acid liberation from adipocytes by reducing triglyceride uptake, contributing to insulin resistance. Further, iTFA incorporation into cells' membranes affects the functionality of membrane receptors of glucose and insulin (Micha & Mozaffarian, 2009). Further, iTFA might bind to transcription factors such as PPARs or SREBPs and alter the expression of genes involved in glucose metabolism (Mozaffarian, Katan, Ascherio, Stampfer, & Willett, 2006). However, detrimental effects of total TFA and iTFA intake remain controversial in clinical trials (Micha & Mozaffarian, 2009; Mozaffarian et al., 2009; Salmerón et al., 2001). A meta-analysis of clinical trials failed to observe any change in glucose or insulin levels with various TFA doses (Aronis, Khan, & Mantzoros, 2012).

The effects of total TFA and iTFA are probably mediated by baseline insulin resistance or susceptibility of subjects to insulin resistance (Micha & Mozaffarian, 2009).

6.5.4 Dyslipidemia

Since the integration of iTFA in processed foods, evidence leads toward detrimental effects on lipid profile, with increased total-C, LDL levels and total-C/HDL ratio and decreased HDL levels (Micha & Mozaffarian, 2009; Wang & Hu, 2017). iTFA consumption is also associated with increased triglyceride levels and reduced size of LDL (Dhaka et al., 2011; Mensink et al., 2003; Micha & Mozaffarian, 2009; Mozaffarian & Clarke, 2009). Increased total TFA and iTFA consumption has been consistently and strongly associated with increased risk of cardiovascular diseases, even when compared with SFA (de Souza et al., 2015; Sacks et al., 2017; Wang & Hu, 2017). However, mechanisms underlying total TFA and iTFA actions on lipid profile remain poorly understood (Dhaka et al., 2011). Some hypotheses suggest that iTFA could alter cell membrane functionality by integrating phospholipid membranes (Micha & Mozaffarian, 2009). Total TFA intake has been associated with alterations in the expression of the *lipoprotein lipase* gene (Mozaffarian et al., 2006). Total TFA and iTFA intake could also alter hepatic lipoprotein secretion and thus, increases accumulation of lipids in hepatocytes. Most of the evidences on iTFA are based on elaïdic acid, yet the role of other iTFA in the deterioration of metabolic health is less understood (Micha & Mozaffarian, 2009). A potential gene-diet interaction has been identified with the *fatty acid binding protein-2 gene* (*FABP2*), which is involved in fatty acid uptake and metabolism. In a post-prandial study, the carriers of the Thr54 of the Ala54Thr on *FABP2* had increased fractional triacylglycerol synthetic rates with a high total TFA diet in comparison with high oleic acid diet of non-carriers (Lefevre et al., 2005).

6.5.5 Blood Pressure

Regarding blood pressure, at first glance total TFA and iTFA consumption does not seem to affect blood pressure in comparison with SFA, MUFA and PUFA. Human intervention studies do not give further evidence of a potential link between iTFA consumption and blood pressure (Mozaffarian et al., 2009).

Overall, iTFA have detrimental effects on the different components of the MetS which seem to be worse than SFA intake. The well-known adverse effects of iTFA encouraged many health institutions to support the elimination of iTFA in processed foods. Despite these convincing results, data on gene-diet interactions between iTFA intake and the MetS are scarce. However, given the actual knowledge on iTFA, choosing to avoid iTFA in processed foods might be the best choice for the population.

6.6 Ruminant *trans* Fatty Acids

rTFA are natural TFA present in ruminant food sources. rTFA such as TPA (*trans* 16:1, n-7) and TVA (*trans* 18:1, n-7) do not seem to have the detrimental effects observed with elaïdic acid (*trans* 18:1, n-9) on metabolic health (Mozaffarian et al., 2009). Notably, rTFA from dairy products could have beneficial effects on health. CLA raised interest for their potential beneficial roles in obesity, type 2 diabetes and hypertension; however, actual evidence showed controversial results which are still not clarified (Kim et al., 2016; Viladomiu, Hontecillas, & Bassaganya-Riera, 2016). Regarding gene-diet interactions, one study examining gene variants associated with rTFA consumption and the various components of the MetS was identified. Additional studies are required to better understand the interindividual variability in the response to rTFA.

6.6.1 Obesity

Data surrounding the role of TPA and TVA on obesity risk factors are limited. A prospective analysis observed lower adiposity in subjects with high TPA levels in plasma phospholipids. However, it worth mentioning that TPA intake was highly correlated to high whole-fat dairy products; hence, the effect might be modulated by other nutrients in the dairy matrix (Mozaffarian et al., 2010). CLA supplementation is associated with a reduction of body fat in humans. Further studies support that CLA supplementation, especially *trans*-10, *cis*-12 isomer could help reduce BMI, waist circumference and abdominal fat mass, which is promising in prevention of the MetS (Dilzer & Park, 2012). *Trans-10, cis-12* could reduce lipid storage and adipogenesis in the adipose tissue, but also increase fatty acid oxidation in skeletal muscles. However, additional studies are required to confirm the anti-obesogenic effects of CLA supplementation (Kim et al., 2016).

6.6.2 Inflammation

Available reports suggest TPA and TVA have no effect or modest effect on inflammatory cytokines (Tremblay & Rudkowska, 2017). Recent in vitro study suggests both TPA and TVA can reduce the gene expression of

TNF-α in human endothelial cells and lower *TNF-α* and *IL-8* gene expression in hepatocytes induced with TNF-α. Interestingly, the effect of rTFA was independent of PPAR-γ activation, suggesting rTFA might modulate cells' membranes functionality (Da Silva, Julien, Bilodeau, Barbier, & Rudkowska, 2017). Furthermore, high rTFA levels in plasma phospholipids were associated with higher adiponectin levels in a cross-sectional study with Canadian subjects (Da Silva, Julien, Pérusse, Vohl, & Rudkowska, 2015). Despite these promising results, clinical trials in humans are scarce. Regarding CLA supplementation, the isomers 9*cis*-11*trans*, 10*trans*-12*cis* or their combination could have specific effects on inflammation. The isomer 9*cis*-11*trans* has been associated with beneficial effects on inflammation in macrophages and in the adipose tissue through the activation of PPAR-γ, downregulation of NF-κB and decreased *TNF-α* gene expression (Moloney et al., 2007; Tremblay & Rudkowska, 2017). In contrast, in adipocytes, the isomer 10*trans*-12*cis* CLA was associated with increased expression of pro-inflammatory genes, increased cytokine liberation and activation of NF-κB (Chung, Brown, Provo, Hopkins, & McIntosh, 2005). 10*trans*-12*cis* CLA was also associated with decreased adiponectin and leptin levels (Tremblay & Rudkowska, 2017).

6.6.3 Glucose Metabolism

Despite the negative effects of total TFA and iTFA on glucose metabolism, rTFA seem to have neutral effect on type 2 diabetes. In observational evidence, high TPA levels in plasma phospholipids were associated with lower insulin resistance and lower risk of developing type 2 diabetes (Mozaffarian et al., 2010). Mechanisms proposed that TPA could stimulate lipid oxidation and inhibit lipogenesis in the liver, thereby improving muscle insulin sensitivity and reducing hepatic fat content. TPA could also have similar actions as *cis*-palmitoleic acid, which is a regulator of insulin sensitivity and glucose tolerance in animal model (Tremblay & Rudkowska, 2017). Yet, isolating the effect of TPA can be quite challenging due to other bioactive components in food matrix (Wang & Hu, 2017). Other mechanisms suggest TVA could promote β-cell replication, insulin biosynthesis and secretion through increased expression of *PPAR-γ* and *GLUT-2*, the main glucose receptor on the surface of β-cells. Regarding CLA, the isomer 9*cis*-11*trans* was associated with increased expression of *GLUT-4* and *PPAR-γ* genes, suggesting enhanced glucose uptake and insulin sensitivity (Tremblay & Rudkowska, 2017). In contrast, the consumption of 1% of total energy from the isomer 10*trans*-12*cis* showed alterations of insulin sensitivity in obese men with abdominal obesity and suffering from the MetS, suggesting a potential diabetogenic effect (Dhaka et al., 2011; Heer & Egert, 2015; Risérus et al., 2009). Gene variants might explain the heterogeneity between individual responses to CLA supplementation. A gene-diet interaction was identified between CLA supplementation and the *PPARG* Pro12Ala polymorphism. The Pro12Pro carriers had decreased expression of *PPARG* gene after CLA supplementation in comparison with Ala12Ala genotype.

6.6.4 Dyslipidemia

A prospective study observed that high TPA levels in plasma phospholipids were associated with lower triglyceride levels, lower total-C/HDL ratio and increased HDL levels (Mozaffarian et al., 2013). TPA and TVA consumption has been associated with reduced hepatic fat accumulation in the liver and reduced secretion of triglycerides in animal models (Tremblay & Rudkowska, 2017). Yet, according to a meta-analysis of randomized controlled trials, rTFA consumption up to 4.19% of total energy intake does not have any effect of lipid profile (Gayet-Boyer et al., 2014). Regarding the role of CLA on lipid profile, the evidence is controversial. Some studies observed beneficial effects on lipid profile, suggesting that 10*trans*-12*cis* could reduce triglycerides and cholesterol content in the liver by reducing SREBPs activity (Kim et al., 2016). Conversely, CLA supplementation was also associated with reduced levels of HDL levels and increased LDL/HDL ratio in other clinical studies, supporting a potential deleterious effect on lipid profile (Dilzer & Park, 2012). These discrepancies might have a genetic dimension which has not been fully explored.

6.6.5 Blood Pressure

High rTFA incorporated in plasma phospholipids is associated with lower systolic and diastolic blood pressure in a cross-sectional study with Canadian subjects (Da Silva et al., 2015). Similarly, CLA supplementation was associated with improvement of blood pressure in hypertensive subjects (Koba & Yanagita, 2014). Further studies are needed to confirm the effectiveness of rTFA on blood pressure.

In conclusion, rTFA seem to have either beneficial or neutral effects on the various components of the MetS; yet, additional studies are required to better understand the unique role of rTFA in metabolic health. Regarding CLA supplementation, both beneficial and detrimental effects have been identified on the different components of the MetS. However, most studies using CLA supplementation administered higher doses than what is found naturally in natural food sources. In addition, the metabolic differences observed between iTFA and rTFA could be due to the very small quantities of rTFA consumed in average diet (<0.5% of total energy) (Micha et al., 2014). These small amounts consumed could thus explain the absence of association with cardiovascular diseases (Dhaka et al., 2011).

7. GAPS AND CHALLENGES IN THE IMPLEMENTATION OF PERSONALIZED NUTRITION

Nutrigenomics studies have opened new paths toward a better understanding of optimal health with nutrition in relation with individual's specific needs. Despite promising results in the identification of responders and non-responders to dietary treatments, many gaps and challenges remain before the implementation of personalized nutrition in public health care services.

Regarding nutrigenomics research, most studies available to date focus on a small handful of SNPs, usually one or two gene variants at the time. However, since chronic diseases like the MetS are complex multifactorial and multigenetic conditions, the elaboration of additional GRS are essential to better evaluate gene–diet interactions. Furthermore, identified SNPs associated with food must be replicated in different populations, which is not the case for most aforementioned gene–diet interactions (Merched & Chan, 2013). In addition, other OMICs technologies such as epigenomics, metabolomics, proteomics and metagenomics are growing in popularity; hence, their integration in nutrition studies will be more and more important in other to eventually elaborate personalized dietary advices (Fig. 1). More globally, personalized nutrition research often focuses primarily on genetic profiling; however, further importance must be given to the global environment including physical activity, socio-economic status, sleep, stress, ethnicity and many other factors for future personalized dietary advices (Comerford & Pasin, 2017).

The integration of personalized nutrition in public health priorities would be a great step toward a better health. Personalized nutrition would have many benefits, such as early prevention of chronic diseases and reduction of utilization of resources. Genetic profiling in personalized nutrition could also increase motivation to make behavioral changes and adopting healthier lifestyle. However, early identification of genetic risk would put more responsibility on individuals for their health and could have different psychological effects depending on the person (de Roos, 2013). For example, genotype information might have negative effects on some people and promote the adoption of adverse behaviors (de Roos, 2013; Heianza & Qi, 2017). Therefore, more studies should be performed to better understand social implications of genetic testing in dietary advices. Further, the success of genetic testing depends greatly on a profound and complete understanding of implications related to genetic information (de Roos, 2013). Education of the public, but also of medical health providers is essential to limit unexpected adverse consequences on genotype testing, especially regarding informed consent. In order to implement personalized nutrition in public health care services, consensus of experts on a well-defined code of practice must be elaborated, but also respected and trusted. Ethical considerations must be considered such as discrimination based on genotype, especially in relation with insurance and employment, but also the freedom of choice in the uptake of genotype testing or not (Gibney & Walsh, 2013). Further, there is a need for legislative implementation regarding the accessibility and protection of genetic information (de Roos, 2013).

In summary, actual dietary requirements are integrating some levels of personalization by classifying individuals in different groups according to age, sex, health status (e.g., pregnancy) or medical conditions. The utilization of genetic profiling in the future will ensure a higher level of precision in the classification of individuals according to their optimal needs and thus, optimization of nutritional interventions. Despite the great advances in nutrigenomics, many milestones remain on the path to personalized nutrition using genetic profiling (de Toro-Martín et al., 2017).

8. CONCLUSION

In conclusion, dietary fatty acids have many purposes in the human body and their relative intake can be either beneficial or detrimental on metabolic health. Actual evidences on the effects of different dietary fatty acids on the components of the MetS are summarized in Table 7. In sum, the main issues on dietary fatty acids are not necessarily about total intake per se, but rather on the nature of the different fatty acids in relation to the others. SFA and iTFA intake are associated with detrimental effects on the components of the MetS, while MUFA and PUFA intake are associated with improvement of metabolic health when replacing SFA. In contrast with iTFA, rTFA seem to have neutral or beneficial effects on the MetS. Given these evidences, the MedDiet, which has optimal proportions of the

Table 7 Evidences and Conclusions of the Different Fatty Acids on the Components of the Metabolic Syndrome

Dietary Fatty Acids	Dietary Guidelines	Evidences	Conclusions
Total fatty acids	Between 20% and 35% of total energy intake	Obesity: Both low-fat and high-fat diets can be efficient for the management of obesity in a hypocaloric context Inflammation: Hypercaloric high-fat diets are associated with pro-inflammatory profile, while low-fat diets are associated with decreased inflammation. In a hypocaloric context, both low-fat and high-fat diets are efficient to decrease inflammation Glucose metabolism: Improvement of glucose metabolism with both high-fat and low-fat diets in weight loss context. Great difference in individual responses Dyslipidemia: Low-fat diets are associated with ↓ total-C and LDL, but ↓ HDL and ↑ triglycerides in some individuals. High-fat diet in weight loss context is associated with ↓ triglycerides Blood pressure: No sustainable changes in blood pressure with low-fat diet. High-fat diet is associated with ↓ diastolic blood pressure	Hypercaloric high-fat diets should be avoided Both high-fat and low-fat diets can be appropriated in the prevention of management of the MetS, especially in a hypocaloric context Great disparities between individual responses to dietary fat can be explained by gene variants

| Saturated fatty acids | Replacement of SFA by unsaturated fatty acids. Maximum 7–10% | Obesity: SFA intake associated with ↑ body weight, while MCFA intake is associated with ↓ bodyweight and ↓ visceral fat
Inflammation: SFA is associated with pro-inflammatory profile. MCFA is associated with downregulation or no effects on pro-inflammatory genes
Glucose metabolism: SFA is associated with worsen glucose homeostasis and insulin resistance
Dyslipidemia: SFA intake is associated with ↑ total-C, HDL, LDL and total-C/HDL ratio when replacing MUFA or PUFA, but no increased risk of cardiovascular diseases. When SFA replace high-glycemic index carbohydrates, ↓ risk of cardiovascular diseases. When SFA replace low-glycemic, non-processed high-fiber carbohydrates, the ↑ risk of cardiovascular diseases
Blood pressure: SFA intake is not associated with blood pressure | High SFA intake is likely to contribute to the evolution of the various components of the MetS excepted blood pressure
SFA should be replaced by MUFA, PUFA or high-fiber, non-processed and low-glycemic carbohydrates
MCFA could prevent and manage obesity and related pro-inflammatory profile |

Continued

Table 7 Evidences and Conclusions of the Different Fatty Acids on the Components of the Metabolic Syndrome—cont'd

Dietary Fatty Acids	Dietary Guidelines	Evidences	Conclusions
Mono-unsaturated fatty acids	Replacement of SFA by unsaturated fatty acids	Obesity: MUFA intake is associated with ↓ body mass, fat mass and abdominal adiposity when replacing SFA Inflammation: MUFA-rich diet is associate with ↓ inflammation and ↓ oxidative stress vs SFA-rich diet Glucose metabolism: MUFA-rich diet is associated with ↓ total-C, LDL and total-C/HDL ratio and ↑ HbA1c vs low-fat diets and low-MUFA diets, while fasting glucose, insulin and insulin resistance were not affected by MUFA. ↓ insulin sensitivity in females with MUFA-rich diet Dyslipidemia: MUFA intake is associated with ↓ total-C, LDL and total-C/HDL ratio and ↑ HDL when replacing SFA. MUFA intake is associated with ↓ triglycerides, ↓ LDL and ↓ total-C/HDL when replacing carbohydrates Blood pressure: MUFA intake (oleic acid specifically) is associated with ↓ blood pressure vs SFA and carbohydrates. Olive oil is associated with ↓ blood pressure, probably due to phenolic compounds	High-MUFA intake is likely to prevent the evolution of the various components of the MetS when replacing SFA The effects of MUFA can be influenced by other elements in the food matrix The replacement of SFA or TFA by olive oil is likely to have beneficial effects on the MetS

Mediterranean Diet	Adopt a Mediterranean-style diet	Obesity: The MedDiet is not associated with weight gain despite high-fat content Inflammation: The MedDiet is associated with reduction of pro-inflammatory profile and oxidative stress. The antioxidant and phenolic compounds seem to be main contributor Glucose metabolism: the MedDiet is associated with improved insulin sensitivity. The MedDiet rich in olive oil is associated with reduction of type 2 diabetes vs low-fat diet Dyslipidemia: The MedDiet is associated with ↓ LDL, total-C/HDL ratio, ↓ HDL and ↓ triglycerides Blood pressure: The MedDiet is associated with ↓ systolic and diastolic blood pressure	Adopting a MedDiet is likely to prevent and reverse the MetS Good adherence of the MedDiet is associated with ↓ risk of type 2 diabetes and ↓ cardiovascular diseases
Omega-6 poly-unsaturated fatty acids	5–10% of total energy intake Replacement of SFA by unsaturated fatty acids	Obesity: Omega-6 PUFA is associated with ↑ fat deposition. ↑ omega-6/omega-3 PUFA ratio is associated with ↑ obesity Inflammation: Omega-6 PUFA intake do not seem to ↑ inflammation Glucose metabolism: ↑ Omega-6/omega-3 ratio is associated with ↓ insulin sensitivity Dyslipidemia: LA intake is associated with ↓ LDL and ↓ total-C/HDL ratio when replacing SFA or carbohydrates. Blood pressure: No evidence suggesting omega-6 PUFA can affect blood pressure	Westernized dietary patterns usually have higher omega-6/omega-3 ratio, which is associated with ↑ obesity and worsen insulin resistance LA intake is associated with improved lipid profile

Continued

Table 7 Evidences and Conclusions of the Different Fatty Acids on the Components of the Metabolic Syndrome—cont'd

Dietary Fatty Acids	Dietary Guidelines	Evidences	Conclusions
Omega-3 poly-unsaturated fatty acids	0.6–1.2% of total energy intake 2 servings of fish/week	Obesity: Omega-3 PUFA is associated with ↓ fat deposition. Decreased omega-6/omega-3 PUFA ratio is associated with ↓ obesity Inflammation: Omega-3 PUFA are associated with ↓ pro-inflammatory profile. EPA and DHA have the strongest effects vs ALA. DHA and EPA have distinct effects on inflammation Glucose metabolism: Omega-3 PUFA effects on glucose metabolism remains contradictory. Great heterogeneity in the individual response to omega-3 PUFA Dyslipidemia: Omega-3 PUFA intake is associated with ↓ triglycerides, ↓ ApoCIII and ↓ free fatty acids in plasma vs low-fat diets. Great variability in the response to omega-3 Blood pressure: Omega-3 PUFA, especially EPA and DHA, are associated with ↓ systolic and diastolic blood pressure	Decreasing omega-6/omega-3 by increasing omega-3 PUFA in diet is likely to better prevent of manage the MetS EPA and DHA seem to be the main contributors in the improvement of the components of the MetS Eating fish and sea foods is encouraged to consume the right amount of EPA and DHA Great variability in response to omega-3 PUFA on the components of the MetS can be explained by gene variants
Trans fatty acids	<1% of total energy intake	Obesity: TFA intake is associated with ↑ abdominal fat. High levels of TPA in phospholipids are associated with ↓ adiposity. CLA supplementation could ↓ BMI, ↓ waist circumference and ↓ abdominal fat mass Inflammation: iTFA intake is associated with ↑ pro-inflammatory profile. rTFA seems to have neutral of beneficial	iTFA intake is likely to accelerate the evolution of the MetS. Consumption of iTFA should be avoided rTFA seems to have either neutral or beneficial effects on the various components of the MetS. Additional studies are required before any further recommendations CLA supplementation could have beneficial effects on some of the components of the

effects of inflammation. CLA supplementation might have either beneficial or detrimental effects on inflammation according to specific isomers

Glucose metabolism: iTFA intake is associated with ↑ insulin resistance and ↑ risk of type 2 diabetes.

rTFA seem to have either neutral or beneficial effects on insulin resistance and type 2 diabetes. CLA supplementation might have either beneficial or detrimental effects on glucose metabolism according to specific isomers

Dyslipidemia: TFA intake is associated with ↑ total-C, ↑ LDL, ↑ total-C/HDL and ↓ HDL, ↑ triglycerides and ↓ size of LDL. TFA intake is associated with ↑ cardiovascular risk in comparison with SFA.

High levels of TPA in plasma were associated with ↓ triglyceride levels, ↓ total-C/HDL ratio and ↑ HDL levels. Data on CLA supplementation and lipid profile are controversial

Blood pressure: No evidence suggesting TFA can affect blood pressure.

High rTFA in plasma is associated with ↓ systolic and diastolic blood pressure

MetS; yet, results are highly controversial

Further studies examining gene–diet interactions are needed regarding TFA intake and the MetS

ALA: alpha-linolenic acid; ApoCIII: apolipoprotein C–III; ARA: arachidonic acid; CLA: conjugated linoleic acid; DHA: docohexaenoic acid; EPA: eicosapentaenoic acid; HbA1c: glycated hemoglobin; HDL: high-density lipoproteins; iTFA: industrial *trans* fatty acids; LA: linoleic acid; LDL: low-density lipoproteins; MCFA: medium–chain fatty acids; Med-Diet: Mediterranean diet; MetS: metabolic syndrome; MUFA: mono-unsaturated fatty acids; PUFA: poly-unsaturated fatty acids; rTFA: ruminant *trans* fatty acids; SFA: saturated fatty acids; TFA: *trans* fatty acids; total-C: total-cholesterol.

aforementioned dietary fatty acids, is recognized as one of the best dietary patterns in prevention of the MetS and related chronic diseases. In parallel, given the fact that great interindividual variability exists in the metabolic response to dietary fatty acids, many gene variants have been identified showing relevant gene-diet interactions with dietary fatty acids intake. Despite these promising discoveries, many gaps remain before the implementation of personalized nutrition in public health care systems. Additional studies combining different OMICs technologies are needed to achieve a comprehensive understanding behind the complexity of the MetS and eventually, to develop tailored dietary advices with a more global approach.

REFERENCES

Abete, I., Navas-Carretero, S., Marti, A., & Martinez, J. A. (2012). Nutrigenetics and nutrigenomics of caloric restriction. In C. Bouchard & J. M. Ordovas (Eds.), *Progress in molecular biology and translational science: Vol. 108* (pp. 323–346). Academic Press. https://doi.org/10.1016/B978-0-12-398397-8.00013-7.

Afman, L. A., & Müller, M. (2012). Human nutrigenomics of gene regulation by dietary fatty acids. *Progress in Lipid Research, 51*(1), 63–70. https://doi.org/10.1016/j.plipres.2011.11.005.

Ajala, O., English, P., & Pinkney, J. (2013). Systematic review and meta-analysis of different dietary approaches to the management of type 2 diabetes. *The American Journal of Clinical Nutrition, 97*(3), 505–516. https://doi.org/10.3945/ajcn.112.042457.

Alberti, K. G. M. M., Eckel, R. H., Grundy, S. M., Zimmet, P. Z., Cleeman, J. I., Donato, K. A., et al. (2009). Harmonizing the metabolic syndrome: A joint interim statement of the International Diabetes Federation Task Force on Epidemiology and Prevention; National Heart, Lung, and Blood Institute; American Heart Association; World Heart Federation; International Atherosclerosis Society; and International Association for the Study of Obesity. *Circulation, 120*(16), 1640–1645. https://doi.org/10.1161/CIRCULATIONAHA.109.192644.

Allison, M. A., Aragaki, A. K., Ray, R. M., Margolis, K. L., Beresford, S. A. A., Kuller, L., et al. (2016). A randomized trial of a low-fat diet intervention on blood pressure and hypertension: Tertiary analysis of the WHI dietary modification trial. *American Journal of Hypertension, 29*(8), 959–968. https://doi.org/10.1093/ajh/hpv196.

AlSaleh, A., Maniou, Z., Lewis, F. J., Hall, W. L., Sanders, T. A. B., & O'Dell, S. D. (2014). Genetic predisposition scores for dyslipidaemia influence plasma lipid concentrations at baseline, but not the changes after controlled intake of n-3 polyunsaturated fatty acids. *Genes & Nutrition, 9*(4), 412. https://doi.org/10.1007/s12263-014-0412-8.

Aouizerat, B. E., Kulkarni, M., Heilbron, D., Drown, D., Raskin, S., Pullinger, C. R., et al. (2003). Genetic analysis of a polymorphism in the human apoA-V gene: Effect on plasma lipids. *Journal of Lipid Research, 44*(6), 1167–1173.

Arnett, D. K., & Claas, S. A. (2018). Omics of blood pressure and hypertension. *Circulation Research, 122*(10), 1409. https://doi.org/10.1161/CIRCRESAHA.118.311342.

Aronis, K. N., Khan, S. M., & Mantzoros, C. S. (2012). Effects of trans fatty acids on glucose homeostasis: A meta-analysis of randomized, placebo-controlled clinical trials. *The American Journal of Clinical Nutrition, 96*(5), 1093–1099. https://doi.org/10.3945/ajcn.112.040576.

Australian Government. (2015). *Fat*. Retrieved May 17, 2018. Canberra, Australia: National Health and Medical Research Council. from https://www.eatforhealth.gov.au/food-essentials/fat-salt-sugars-and-alcohol/fat.

Ayyappa, K. A., Shatwan, I., Bodhini, D., Bramwell, L. R., Ramya, K., Sudha, V., et al. (2017). High fat diet modifies the association of lipoprotein lipase gene polymorphism with high density lipoprotein cholesterol in an Asian Indian population. *Nutrition & Metabolism*, *14*, 8. https://doi.org/10.1186/s12986-016-0155-1.

Azhar, S. (2010). Peroxisome proliferator-activated receptors, metabolic syndrome and cardiovascular disease. *Future Cardiology*, *6*(5), 657–691. https://doi.org/10.2217/fca.10.86.

Babio, N., Becerra-Tomás, N., Martínez-González, M. Á., Corella, D., Estruch, R., Ros, E., et al. (2015). Consumption of yogurt, low-fat milk, and other low-fat dairy products is associated with lower risk of metabolic syndrome incidence in an elderly Mediterranean population. *The Journal of Nutrition*, *145*(10), 2308–2316. https://doi.org/10.3945/jn.115.214593.

Babio, N., Bulló, M., & Salas-Salvadó, J. (2009). Mediterranean diet and metabolic syndrome: The evidence. *Public Health Nutrition*, *12*(9A), 1607–1617. https://doi.org/10.1017/S1368980009990449.

Barceló, F., Perona, J. S., Prades, J., Funari, S. S., Gomez-Gracia, E., Conde, M., et al. (2009). Mediterranean-style diet effect on the structural properties of the erythrocyte cell membrane of hypertensive patients. *Hypertension*, *54*(5), 1143. https://doi.org/10.1161/HYPERTENSIONAHA.109.137471.

Bender, N., Portmann, M., Heg, Z., Hofmann, K., Zwahlen, M., & Egger, M. (2014). Fish or n3-PUFA intake and body composition: A systematic review and meta-analysis. *Obesity Reviews*, *15*(8), 657–665. https://doi.org/10.1111/obr.12189.

Benkeblia, N. (2012). Nutrition science and "omics" technologies: Ethical aspects in global health. In *OMICs technologies: Tools for food science* (1st ed.). Boca Raton: CRC Press.

Bjørnshave, A., & Hermansen, K. (2014). Effects of dairy protein and fat on the metabolic syndrome and type 2 diabetes. *The Review of Diabetic Studies*, *11*(2), 153–166. https://doi.org/10.1900/RDS.2014.11.153.

Blanco-Rojo, R., Delgado-Lista, J., Lee, Y.-C., Lai, C.-Q., Perez-Martinez, P., Rangel-Zuñiga, O., et al. (2016). Interaction of an S100A9 gene variant with saturated fat and carbohydrates to modulate insulin resistance in 3 populations of different ancestries. *The American Journal of Clinical Nutrition*, *104*(2), 508–517. https://doi.org/10.3945/ajcn.116.130898.

Bloomfield, H. E., Kane, R., Koeller, E., Greer, N., MacDonald, R., & Wilt, T. (2015). *Benefits and harms of the Mediterranean diet compared to other diets*. Washington, DC: Department of Veterans Affairs (US). Retrieved from http://www.ncbi.nlm.nih.gov/books/NBK379574/.

Bodhini, D., Gaal, S., Shatwan, I., Ramya, K., Ellahi, B., Surendran, S., et al. (2017). Interaction between TCF7L2 polymorphism and dietary fat intake on high density lipoprotein cholesterol. *PLoS One*, *12*(11), e0188382. https://doi.org/10.1371/journal.pone.0188382.

Borén, J., & Adiels, M. (2014). Lipid metabolism in metabolic syndrome. In M. Orešič, A. Vidal-Puig, & J. M. Stephens (Eds.), *A systems biology approach to study metabolic syndrome* (pp. 157–170). Switzerland: Springer International Publishing [Chapter 8].

Bosomworth, N. J. (2013). Approach to identifying and managing atherogenic dyslipidemia. *Canadian Family Physician*, *59*(11), 1169.

Bouchard-Mercier, A., Rudkowska, I., Lemieux, S., Couture, P., Pérusse, L., & Vohl, M.-C. (2014). SREBF1 gene variations modulate insulin sensitivity in response to a fish oil supplementation. *Lipids in Health and Disease*, *13*, 152. https://doi.org/10.1186/1476-511X-13-152.

Bouchard-Mercier, A., Rudkowska, I., Lemieux, S., Couture, P., & Vohl, M.-C. (2014). Polymorphisms in genes involved in fatty acid β-oxidation interact with dietary fat intakes to modulate the plasma TG response to a fish oil supplementation. *Nutrients*, *6*(3), 1145–1163. https://doi.org/10.3390/nu6031145.

Bowen, K. J., Sullivan, V. K., Kris-Etherton, P. M., & Petersen, K. S. (2018). Nutrition and cardiovascular disease—An update. *Current Atherosclerosis Reports*, *20*(2), 8. https://doi.org/10.1007/s11883-018-0704-3.

Briggs, M. A., Petersen, K. S., & Kris-Etherton, P. M. (2017). Saturated fatty acids and cardiovascular disease: Replacements for saturated fat to reduce cardiovascular risk. *Healthcare (Basel, Switzerland)*, *5*(2), 29. https://doi.org/10.3390/healthcare5020029.

Brown, A. E., & Walker, M. (2016). Genetics of insulin resistance and the metabolic syndrome. *Current Cardiology Reports*, *18*, 75. https://doi.org/10.1007/s11886-016-0755-4.

Calder, P. C. (2015). Marine omega-3 fatty acids and inflammatory processes: Effects, mechanisms and clinical relevance, Oxygenated metabolism of PUFA: Analysis and biological relevance. *Biochimica et Biophysica Acta*, *1851*(4), 469–484. https://doi.org/10.1016/j.bbalip.2014.08.010.

Camargo, A., Meneses, M. E., Pérez-Martínez, P., Delgado-Lista, J., Rangel-Zúñiga, O. A., Marín, C., et al. (2014). Dietary fat modifies lipid metabolism in the adipose tissue of metabolic syndrome patients. *Genes & Nutrition*, *9*(4), 409. https://doi.org/10.1007/s12263-014-0409-3.

Camargo, A., Ruano, J., Fernandez, J. M., Parnell, L. D., Jimenez, A., Santos-Gonzalez, M., et al. (2010). Gene expression changes in mononuclear cells in patients with metabolic syndrome after acute intake of phenol-rich virgin olive oil. *BMC Genomics*, *11*, 253. https://doi.org/10.1186/1471-2164-11-253.

Casas-Agustench, P., Arnett, D. K., Smith, C. E., Lai, C.-Q., Parnell, L. D., Borecki, I. B., et al. (2014). Saturated fat intake modulates the association between a genetic risk score of obesity and BMI in two US populations. *Journal of the Academy of Nutrition and Dietetics*, *114*(12), 1954–1966. https://doi.org/10.1016/j.jand.2014.03.014.

Castaner, O., Corella, D., Covas, M.-I., Sorli, J. V., Subirana, I., Flores-Mateo, G., et al. (2013). In vivo transcriptomic profile after a Mediterranean diet in high-cardiovascular risk patients: A randomized controlled trial. *The American Journal of Clinical Nutrition*, *98*(3), 845–853. https://doi.org/10.3945/ajcn.113.060582.

Caswell, H. (2009). A summary of findings from the 5-year Lipgene project. *Nutrition Bulletin*, *34*(1), 92–96. https://doi.org/10.1111/j.1467-3010.2008.01728.x.

Celis-Morales, C. A., Lyall, D. M., Gray, S. R., Steell, L., Anderson, J., Iliodromiti, S., et al. (2017). Dietary fat and total energy intake modifies the association of genetic profile risk score on obesity: Evidence from 48170 UK Biobank participants. *International Journal of Obesity*, *41*, 1761.

Chan, R. S., & Woo, J. (2010). Prevention of overweight and obesity: How effective is the current public health approach. *International Journal of Environmental Research and Public Health*, *7*(3), 765–783. https://doi.org/10.3390/ijerph7030765.

Chowdhury, R., Warnakula, S., Kunutsor, S., Crowe, F., Ward, H. A., Johnson, L., et al. (2014). Association of dietary, circulating, and supplement fatty acids with coronary risk: A systematic review and meta-analysis. *Annals of Internal Medicine*, *160*(6), 398–406. https://doi.org/10.7326/M13-1788.

Chung, S., Brown, J. M., Provo, J. N., Hopkins, R., & McIntosh, M. K. (2005). Conjugated linoleic acid promotes human adipocyte insulin resistance through NFκB-dependent cytokine production. *The Journal of Biological Chemistry*, *280*(46), 38445–38456. https://doi.org/10.1074/jbc.M508159200.

Comerford, K. B., & Pasin, G. (2017). Gene–dairy food interactions and health outcomes: A review of nutrigenetic studies. *Nutrients*, *9*(7), 710. https://doi.org/10.3390/nu9070710.

Corella, D., Carrasco, P., Sorlí, J. V., Estruch, R., Rico-Sanz, J., Martínez-González, M. Á., et al. (2013). Mediterranean diet reduces the adverse effect of the TCF7L2-rs7903146 polymorphism on cardiovascular risk factors and stroke incidence: A randomized controlled trial in a high-cardiovascular-risk population. *Diabetes Care*, *36*(11), 3803–3811. https://doi.org/10.2337/dc13-0955.

Corella, D., Coltell, O., Macian, F., & Ordovás, J. M. (2018). Advances in understanding the molecular basis of the Mediterranean diet effect. *Annual Review of Food Science and Technology*, *9*(1), 227–249. https://doi.org/10.1146/annurev-food-032217-020802.

Corella, D., Coltell, O., Mattingley, G., Sorlí, J. V., & Ordovas, J. M. (2017). Utilizing nutritional genomics to tailor diets for the prevention of cardiovascular disease: A guide for upcoming studies and implementations. *Expert Review of Molecular Diagnostics*, *17*(5), 495–513. https://doi.org/10.1080/14737159.2017.1311208.

Corella, D., Peloso, G., Arnett, D. K., Demissie, S., Cupples, L. A., Tucker, K., et al. (2009). APOA2, dietary fat and body mass index: Replication of a gene-diet interaction in three independent populations. *Archives of Internal Medicine*, *169*(20), 1897–1906. https://doi.org/10.1001/archinternmed.2009.343.

Cruz-Teno, C., Pérez-Martínez, P., Delgado-Lista, J., Yubero-Serrano, E. M., García-Ríos, A., Marín, C., et al. (2012). Dietary fat modifies the postprandial inflammatory state in subjects with metabolic syndrome: The LIPGENE study. *Molecular Nutrition & Food Research*, *56*(6), 854–865. https://doi.org/10.1002/mnfr.201200096.

D'Amore, S., Palasciano, G., & Moschetta, A. (2014). Chapter 3: The liver in metabolic syndrome. In M. Orešič & A. Vidal-Puig (Eds.), *A systems biology approach to study metabolic syndrome* (pp. 27–61). Switzerland: Springer International Publishing [Chapter 3].

Da Silva, M. S., Julien, P., Bilodeau, J.-F., Barbier, O., & Rudkowska, I. (2017). Trans fatty acids suppress TNF-α-induced inflammatory gene expression in endothelial (HUVEC) and hepatocellular carcinoma (HepG2) cells. *Lipids*, *52*(4), 315–325. https://doi.org/10.1007/s11745-017-4243-4.

Da Silva, M. S., Julien, P., Pérusse, L., Vohl, M.-C., & Rudkowska, I. (2015). Natural rumen-derived trans fatty acids are associated with metabolic markers of cardiac health. *Lipids*, *50*(9), 873–882. https://doi.org/10.1007/s11745-015-4055-3.

Da Silva, M. S., & Rudkowska, I. (2015). Dairy nutrients and their effect on inflammatory profile in molecular studies. *Molecular Nutrition & Food Research*, *59*(7), 1249–1263. https://doi.org/10.1002/mnfr.201400569.

de la Iglesia, R., Loria-Kohen, V., Zulet, M. A., Martinez, J. A., Reglero, G., & Ramirez de Molina, A. (2016). Dietary strategies implicated in the prevention and treatment of metabolic syndrome. *International Journal of Molecular Sciences*, *17*(11), 1877–1898. https://doi.org/10.3390/ijms17111877.

De Luis, D. A., Aller, R., Izaola, O., & Pacheco, D. (2015). Role of rs9939609 FTO gene variant in weight loss, insulin resistance and metabolic parameters after a high monounsaturated vs a high polyunsaturated fat hypocaloric diets. *Nutrición Hospitalaria*, *32*(1), 175–181. https://doi.org/10.3305/nh.2015.32.1.9169.

de Luis, D. A., Izaola, O., de la Fuente, B., Primo, D., & Aller, R. (2016). Association of neuropeptide Y Gene rs16147 polymorphism with cardiovascular risk factors, adipokines, and metabolic syndrome in patients with obesity. *Journal of Nutrigenetics and Nutrigenomics*, *9*(5–6), 213–221.

de Luis, D. A., Izaola, O., Primo, D., & Aller, R. (2017). Role of rs1501299 variant in the adiponectin gene on total adiponectin levels, insulin resistance and weight loss after a Mediterranean hypocaloric diet. *Diabetes Research and Clinical Practice* (in press). https://doi.org/10.1016/j.diabres.2017.11.007.

de Luis, D. A., Izaola, O., Primo, D., Aller, R., Ortola, A., Gómez, E., et al. (2018). The association of SNP276G > T at adiponectin gene with insulin resistance and circulating adiponectin in response to two different hypocaloric diets. *Diabetes Research and Clinical Practice*, *137*, 93–99. https://doi.org/10.1016/j.diabres.2018.01.003.

de Roos, B. (2013). Personalised nutrition: Ready for practice? *Proceedings of the Nutrition Society*, *72*(1), 48–52. https://doi.org/10.1017/S0029665112002844.

de Souza, R. J., Mente, A., Maroleanu, A., Cozma, A. I., Ha, V., Kishibe, T., et al. (2015). Intake of saturated and trans unsaturated fatty acids and risk of all cause mortality, cardiovascular disease, and type 2 diabetes: Systematic review and meta-analysis of observational studies. *British Medical Journal*, *351*, h3978. https://doi.org/10.1136/bmj.h3978.

de Toro-Martín, J., Arsenault, B. J., Després, J.-P., & Vohl, M.-C. (2017). Precision nutrition: A review of personalized nutritional approaches for the prevention and management of metabolic syndrome. *Nutrients*, *9*(8), 913. https://doi.org/10.3390/nu9080913.

Delgado-Lista, J., Perez-Martinez, P., Garcia-Rios, A., Perez-Caballero, A. I., Perez-Jimenez, F., & Lopez-Miranda, J. (2016). Mediterranean diet and cardiovascular risk: Beyond traditional risk factors. *Critical Reviews in Food Science and Nutrition*, *56*(5), 788–801. https://doi.org/10.1080/10408398.2012.726660.

Dhaka, V., Gulia, N., Ahlawat, K. S., & Khatkar, B. S. (2011). Trans fats—Sources, health risks and alternative approach—A review. *Journal of Food Science and Technology*, *48*(5), 534–541. https://doi.org/10.1007/s13197-010-0225-8.

Dilzer, A., & Park, Y. (2012). Implication of conjugated linoleic acid (CLA) in human health. *Critical Reviews in Food Science and Nutrition*, *52*(6), 488–513. https://doi.org/10.1080/10408398.2010.501409.

Domínguez-Reyes, T., Astudillo-López, C. C., Salgado-Goytia, L., Muñoz-Valle, J. F., Salgado-Bernabé, A. B., Guzmán-Guzmán, I. P., et al. (2015). Interaction of dietary fat intake with APOA2, APOA5 and LEPR polymorphisms and its relationship with obesity and dyslipidemia in young subjects. *Lipids in Health and Disease*, *14*, 106. https://doi.org/10.1186/s12944-015-0112-4.

Dong, C., Zhou, H., Shen, C., Yu, L.-G., Ding, Y., Zhang, Y.-H., et al. (2015). Role of peroxisome proliferator-activated receptors gene polymorphisms in type 2 diabetes and metabolic syndrome. *World Journal of Diabetes*, *6*(4), 654–661. https://doi.org/10.4239/wjd.v6.i4.654.

Doo, M., Won, S., & Kim, Y. (2015). Association between the APOB rs1469513 polymorphism and obesity is modified by dietary fat intake in Koreans. *Nutrition*, *31*(5), 653–658. https://doi.org/10.1016/j.nut.2014.10.007.

Dordevic, A. L., Konstantopoulos, N., & Cameron-Smith, D. (2014). 3T3-L1 preadipocytes exhibit heightened monocyte-chemoattractant protein-1 response to acute fatty acid exposure. *PLoS One*, *9*(6), e99382. https://doi.org/10.1371/journal.pone.0099382.

Dumont, J., Goumidi, L., Grenier-Boley, B., Cottel, D., Maréaux, N., Montaye, M., et al. (2017). Dietary linoleic acid interacts with FADS1 genetic variability to modulate HDL-cholesterol and obesity-related traits. *Clinical Nutrition* (in press). https://doi.org/10.1016/j.clnu.2017.07.012.

Eckel, R. H., Jakicic, J. M., Ard, J. D., Hubbard, V. S., de Jesus, J. M., Lee, I.-M., et al. (2013). AHA/ACC guideline on lifestyle management to reduce cardiovascular risk. *Circulation*, *129*, S76–S99. https://doi.org/10.1161/01.cir.0000437740.48606.d1.

Esposito, K., Kastorini, C.-M., Panagiotakos, D. B., & Giugliano, D. (2010). Mediterranean diet and weight loss: Meta-analysis of randomized controlled trials. *Metabolic Syndrome and Related Disorders*, *9*(1), 1–12. https://doi.org/10.1089/met.2010.0031.

Esposito, K., Maiorino, M. I., Bellastella, G., Panagiotakos, D. B., & Giugliano, D. (2017). Mediterranean diet for type 2 diabetes: Cardiometabolic benefits. *Endocrine*, *56*(1), 27–32. https://doi.org/10.1007/s12020-016-1018-2.

Esposito, K., Maiorino, M. I., Ciotola, M., Di Palo, C., Scognamiglio, P., Gicchino, M., et al. (2009). Effects of a Mediterranean-style diet on the need for antihyperglycemic drug therapy in patients with newly diagnosed type 2 diabetes: A randomized trial. *Annals of Internal Medicine*, *151*(5), 306–314.

Estruch, R. (2010). Anti-inflammatory effects of the Mediterranean diet: The experience of the PREDIMED study. *Proceedings of the Nutrition Society, 69*(3), 333–340. https://doi.org/10.1017/S0029665110001539.

Estruch, R., Ros, E., Salas-Salvadó, J., Covas, M.-I., Corella, D., Arós, F., et al. (2013). Primary prevention of cardiovascular disease with a Mediterranean diet. *New England Journal of Medicine, 368*(14), 1279–1290. https://doi.org/10.1056/NEJMoa1200303.

Fallaize, R., Carvalho-Wells, A. L., Tierney, A. C., Marin, C., Kieć-Wilk, B., Dembińska-Kieć, A., et al. (2017). APOE genotype influences insulin resistance, apolipoprotein CII and CIII according to plasma fatty acid profile in the metabolic syndrome. *Scientific Reports, 7*(1), 6274. https://doi.org/10.1038/s41598-017-05802-2.

Fenwick, P. H., Jeejeebhoy, K., Dhaliwal, R., Royall, D., Brauer, P., Tremblay, A., et al. (2018). Lifestyle genomics and the metabolic syndrome: A review of genetic variants that influence response to diet and exercise interventions. *Critical Reviews in Food Science and Nutrition*, 1–12. https://doi.org/10.1080/10408398.2018.1437022.

Ferguson, J. F., Phillips, C. M., McMonagle, J., Pérez-Martínez, P., Shaw, D. I., Lovegrove, J. A., et al. (2010). NOS3 gene polymorphisms are associated with risk markers of cardiovascular disease and interact with omega-3 polyunsaturated fatty acids. *Atherosclerosis, 211*(2), 539–544. https://doi.org/10.1016/j.atherosclerosis.2010.03.027.

Fernández-Real, J.-M. (2006). Genetic predispositions to low-grade inflammation and type 2 diabetes. *Diabetes Technology & Therapeutics, 8*(1), 55–66. https://doi.org/10.1089/dia.2006.8.55.

Field, A. E., Willett, W. C., Lissner, L., & Colditz, G. A. (2012). Dietary fat and weight gain among women in the nurses' health study. *Obesity, 15*(4), 967–976. https://doi.org/10.1038/oby.2007.616.

Fitó, M., Melander, O., Martínez, J. A., Toledo, E., Carpéné, C., & Corella, D. (2016). Advances in integrating traditional and omic biomarkers when analyzing the effects of the Mediterranean diet intervention in cardiovascular prevention. *International Journal of Molecular Sciences, 17*(9), 1469. https://doi.org/10.3390/ijms17091469.

Fontaine-Bisson, B., Wolever, T. M., Chiasson, J.-L., Rabasa-Lhoret, R., Maheux, P., Josse, R. G., et al. (2007). Genetic polymorphisms of tumor necrosis factor-α modify the association between dietary polyunsaturated fatty acids and fasting HDL-cholesterol and apo A-I concentrations. *The American Journal of Clinical Nutrition, 86*(3), 768–774. https://doi.org/10.1093/ajcn/86.3.768.

Galic, S., Oakhill, J. S., & Steinberg, G. R. (2010). Adipose tissue as an endocrine organ, Endocrine aspects of obesity. *Molecular and Cellular Endocrinology, 316*(2), 129–139. https://doi.org/10.1016/j.mce.2009.08.018.

Garaulet, M., Smith, C. E., Hernández-González, T., Lee, Y.-C., & Ordovás, J. M. (2011). PPARγ Pro12Ala interacts with fat intake for obesity and weight loss in a behavioural treatment based on the Mediterranean diet. *Molecular Nutrition & Food Research, 55*(12), 1771–1779. https://doi.org/10.1002/mnfr.201100437.

Garcia-Rios, A., Alcala-Diaz, J. F., Gomez-Delgado, F., Delgado-Lista, J., Marin, C., Leon-Acuña, A., et al. (2018). Beneficial effect of CETP gene polymorphism in combination with a Mediterranean diet influencing lipid metabolism in metabolic syndrome patients: CORDIOPREV study. *Clinical Nutrition, 37*(1), 229–234. https://doi.org/10.1016/j.clnu.2016.12.011.

Gayet-Boyer, C., Tenenhaus-Aziza, F., Prunet, C., Marmonier, C., Malpuech-Brugère, C., Lamarche, B., et al. (2014). Is there a linear relationship between the dose of ruminant trans-fatty acids and cardiovascular risk markers in healthy subjects: Results from a systematic review and meta-regression of randomised clinical trials. *The British Journal of Nutrition, 112*(12), 1914–1922. https://doi.org/10.1017/S0007114514002578.

Georgiadi, A., & Kersten, S. (2012). Mechanisms of gene regulation by fatty acids. *Advances in Nutrition, 3*(2), 127–134. https://doi.org/10.3945/an.111.001602.

German, J. B., & Dillard, C. J. (2004). Saturated fats: What dietary intake? *The American Journal of Clinical Nutrition, 80*(3), 550–559. https://doi.org/10.1093/ajcn/80.3.550.

Gerster, H. (1998). Can adults adequately convert alpha-linolenic acid (18:3n-3) to eicosapentaenoic acid (20:5n-3) and docosahexaenoic acid (22:6n-3)? *International Journal for Vitamin and Nutrition Research. Internationale Zeitschrift für Vitamin- und Ernährungsforschung. Journal International de Vitaminologie et de Nutrition, 68*(3), 159–173.

Gibney, M. J., & Walsh, M. C. (2013). The future direction of personalised nutrition: My diet, my phenotype, my genes. *Proceedings of the Nutrition Society, 72*(2), 219–225. https://doi.org/10.1017/S0029665112003436.

Gillingham, L. G., Harris-Janz, S., & Jones, P. J. H. (2011). Dietary monounsaturated fatty acids are protective against metabolic syndrome and cardiovascular disease risk factors. *Lipids, 46*(3), 209–228. https://doi.org/10.1007/s11745-010-3524-y.

Gomez-Delgado, F., Delgado-Lista, J., Lopez-Moreno, J., Rangel-Zuñiga, O. A., Alcala-Diaz, J. F., Leon-Acuña, A., et al. (2018). Telomerase RNA component genetic variants interact with the Mediterranean diet modifying the inflammatory status and its relationship with aging: CORDIOPREV study. *The Journals of Gerontology. Series A, Biological Sciences and Medical Sciences, 73*(3), 327–332. https://doi.org/10.1093/gerona/glw194.

Goni, L., Cuervo, M., Milagro, F. I., & Martínez, J. A. (2015). A genetic risk tool for obesity predisposition assessment and personalized nutrition implementation based on macronutrient intake. *Genes & Nutrition, 10*(1), 445. https://doi.org/10.1007/s12263-014-0445-z.

Goni, L., Qi, L., Cuervo, M., Milagro, F. I., Saris, W. H., MacDonald, I. A., et al. (2017). Effect of the interaction between diet composition and the PPM1K genetic variant on insulin resistance and β cell function markers during weight loss: Results from the nutrient gene interactions in human obesity: Implications for dietary guidelines (NUGENOB) randomized trial. *The American Journal of Clinical Nutrition, 106*(3), 902–908. https://doi.org/10.3945/ajcn.117.156281.

Goni, L., Sun, D., Heianza, Y., Wang, T., Huang, T., Cuervo, M., et al. (2018). Macronutrient-specific effect of the MTNR1B genotype on lipid levels in response to 2 year weight-loss diets. *Journal of Lipid Research, 59*(1), 155–161. https://doi.org/10.1194/jlr.P078634.

Gow, M. L., Garnett, S. P., Baur, L. A., & Lister, N. B. (2016). The effectiveness of different diet strategies to reduce type 2 diabetes risk in youth. *Nutrients, 8*(8), 486. https://doi.org/10.3390/nu8080486.

Grundy, S. M., Cleeman, J. I., Daniels, S. R., Donato, K. A., Eckel, R. H., Franklin, B. A., et al. (2005). Diagnosis and management of the metabolic syndrome: An American Heart Association/National Heart, Lung, and Blood Institute scientific statement. *Circulation, 112*(17), 2735–2752. https://doi.org/10.1161/CIRCULATIONAHA.105.169404.

Grygiel-Górniak, B. (2014). Peroxisome proliferator-activated receptors and their ligands: Nutritional and clinical implications—A review. *Nutrition Journal, 13*, 17. –17. https://doi.org/10.1186/1475-2891-13-17.

Guclu-Geyik, F., Onat, A., Coban, N., Komurcu-Bayrak, E., Sansoy, V., Can, G., et al. (2012). Minor allele of the APOA4 gene T347S polymorphism predisposes to obesity in postmenopausal Turkish women. *Molecular Biology Reports, 39*(12), 10907–10914. https://doi.org/10.1007/s11033-012-1990-4.

Hammad, S. S., & Jones, P. J. (2017). Dietary fatty acid composition modulates obesity and interacts with obesity-related genes. *Lipids, 52*(10), 803–822. https://doi.org/10.1007/s11745-017-4291-9.

Harika, R. K., Eilander, A., Alssema, M., Osendarp, S. J. M., & Zock, P. L. (2013). Intake of fatty acids in general populations worldwide does not meet dietary recommendations to prevent coronary heart disease: A systematic review of data from 40 countries. *Annals of Nutrition and Metabolism, 63*(3), 229–238.

Harris, W. S., Mozaffarian, D., Rimm, E., Kris-Etherton, P., Rudel, L. L., Appel, L. J., et al. (2009). Omega-6 fatty acids and risk for cardiovascular disease. *Circulation, 119*(6), 902. https://doi.org/10.1161/CIRCULATIONAHA.108.191627.

Hartweg, J., Perera, R., Montori, V., Dinneen, S., Neil, H. A. W., & Farmer, A. (2008). Omega-3 polyunsaturated fatty acids (PUFA) for type 2 diabetes mellitus. *The Cochrane Database of Systematic Reviews, 1*, CD003205. https://doi.org/10.1002/14651858. CD003205.pub2.

Health Canada. (2007). *Eating well with Canada's food guide*. Ottawa, Ontario, Canada: HC Pub. 4651. Queen's Printer.

Heer, M., & Egert, S. (2015). Nutrients other than carbohydrates: Their effects on glucose homeostasis in humans. *Diabetes/Metabolism Research and Reviews, 31*(1), 14–35. https://doi.org/10.1002/dmrr.2533.

Heianza, Y., Ma, W., Huang, T., Wang, T., Zheng, Y., Smith, S. R., et al. (2016). Macronutrient intake-associated FGF21 genotype modifies effects of weight-loss diets on 2-year changes of central adiposity and body composition: The POUNDS Lost trial. *Diabetes Care, 39*(11), 1909–1914. https://doi.org/10.2337/dc16-1111.

Heianza, Y., & Qi, L. (2017). Gene–diet interaction and precision nutrition in obesity. *International Journal of Molecular Sciences, 18*(4), 787. https://doi.org/10.3390/ijms18040787.

Hohmann, C. D., Cramer, H., Michalsen, A., Kessler, C., Steckhan, N., Choi, K., et al. (2015). Effects of high phenolic olive oil on cardiovascular risk factors: A systematic review and meta-analysis. *Phytomedicine, 22*(6), 631–640. https://doi.org/10.1016/j.phymed.2015.03.019.

Hooper, L., Abdelhamid, A., Bunn, D., Brown, T., Summerbell, C. D., & Skeaff, C. M. (2015). Effects of total fat intake on body weight. *The Cochrane Database of Systematic Reviews, 8*, CD011834. https://doi.org/10.1002/14651858.CD011834.

Hooper, L., Summerbell, C. D., Thompson, R., Sills, D., Roberts, F. G., Moore, H., et al. (2011). Reduced or modified dietary fat for preventing cardiovascular disease. *The Cochrane Database of Systematic Reviews*, (5), CD002137. https://doi.org/10. 1002/14651858.CD002137.pub2.

Hosseini-Esfahani, F., Mirmiran, P., Koochakpoor, G., Daneshpour, M. S., Guity, K., & Azizi, F. (2017). Some dietary factors can modulate the effect of the zinc transporters 8 polymorphism on the risk of metabolic syndrome. *Scientific Reports, 7*, 1649. https://doi.org/10.1038/s41598-017-01762-9.

Huang, T., Huang, J., Qi, Q., Li, Y., Bray, G. A., Rood, J., et al. (2015). PCSK7 genotype modifies effect of a weight-loss diet on 2-year changes of insulin resistance: The POUNDS LOST trial. *Diabetes Care, 38*(3), 439–444. https://doi.org/10.2337/dc14-0473.

Huang, T., Wang, T., Heianza, Y., Sun, D., Ivey, K., Durst, R., et al. (2018). HNF1A variant, energy-reduced diets and insulin resistance improvement during weight loss: The POUNDS Lost trial and DIRECT. *Diabetes, Obesity and Metabolism, 20*(6), 1445–1452. https://doi.org/10.1111/dom.13250.

Imamura, F., Micha, R., Wu, J. H. Y., de Oliveira Otto, M. C., Otite, F. O., Abioye, A. I., et al. (2016). Effects of saturated fat, polyunsaturated fat, monounsaturated fat, and carbohydrate on glucose-insulin homeostasis: A systematic review and meta-analysis of randomised controlled feeding trials. *PLoS Medicine, 13*(7), e1002087. https://doi.org/10.1371/journal.pmed.1002087.

Innes, J. K., & Calder, P. C. (2018). Omega-6 fatty acids and inflammation. *Prostaglandins, Leukotrienes, and Essential Fatty Acids, 132*, 41–48. https://doi.org/10.1016/j.plefa.2018.03.004.

Institute of Medicine. (2006). *Dietary reference intakes: The essential guide to nutrient requirements*. Washington, DC: The National Academies Press.

International Diabetes Federation. (2006). *The IDF consensus worldwide definition of the metabolic syndrome*. Belgium: IDF Communications.

Jacobo-Albavera, L., Posadas-Romero, C., Vargas-Alarcón, G., Romero-Hidalgo, S., Posadas-Sánchez, R., González-Salazar, M. d. C., et al. (2015). Dietary fat and carbohydrate modulate the effect of the ATP-binding cassette A1 (ABCA1) R230C variant on metabolic risk parameters in premenopausal women from the genetics of atherosclerotic disease (GEA) study. *Nutrition & Metabolism*, *12*, 45. https://doi.org/10.1186/s12986-015-0040-3.

Jacobson, T. A., Maki, K. C., Orringer, C. E., Jones, P. H., Kris-Etherton, P., Sikand, G., et al. (2015). National lipid association recommendations for patient-centered management of dyslipidemia: Part 2. *Journal of Clinical Lipidology*, *9*(6) S1–S122.e1. https://doi.org/10.1016/j.jacl.2015.09.002.

Jiménez-Gómez, Y., López-Miranda, J., Blanco-Colio, L. M., Marín, C., Pérez-Martínez, P., Ruano, J., et al. (2008). Olive oil and walnut breakfasts reduce the postprandial inflammatory response in mononuclear cells compared with a butter breakfast in healthy men. *Atherosclerosis*, *204*(2), e70–e76. https://doi.org/10.1016/j.atherosclerosis.2008.09.011.

Joffe, Y. T., Collins, M., & Goedecke, J. H. (2013). The relationship between dietary fatty acids and inflammatory genes on the obese phenotype and serum lipids. *Nutrients*, *5*(5), 1672–1705. https://doi.org/10.3390/nu5051672.

Joffe, Y. T., van der Merwe, L., Collins, M., Carstens, M., Evans, J., Lambert, E. V., et al. (2011). The -308 G/A polymorphism of the tumour necrosis factor-α gene modifies the association between saturated fat intake and serum total cholesterol levels in white South African women. *Genes & Nutrition*, *6*(4), 353–359. https://doi.org/10.1007/s12263-011-0213-2.

Joffe, Y. T., van der Merwe, L., Evans, J., Collins, M., Lambert, E. V., September, A., et al. (2014). Interleukin-6 gene polymorphisms, dietary fat intake, obesity and serum lipid concentrations in black and white south African women. *Nutrients*, *6*(6), 2436–2465. https://doi.org/10.3390/nu6062436.

Johnston, B. C., Kanters, S., Bandayrel, K., Wu, P., Naji, F., Siemieniuk, R. A., et al. (2014). Comparison of weight loss among named diet programs in overweight and obese adults: A meta-analysis. *JAMA*, *312*(9), 923–933. https://doi.org/10.1001/jama.2014.10397.

Jonasson, L., Guldbrand, H., Lundberg, A. K., & Nystrom, F. H. (2014). Advice to follow a low-carbohydrate diet has a favourable impact on low-grade inflammation in type 2 diabetes compared with advice to follow a low-fat diet. *Annals of Medicine*, *46*(3), 182–187. https://doi.org/10.3109/07853890.2014.894286.

Kastorini, C.-M., Milionis, H. J., Esposito, K., Giugliano, D., Goudevenos, J. A., & Panagiotakos, D. B. (2011). The effect of Mediterranean diet on metabolic syndrome and its components: A meta-analysis of 50 studies and 534,906 individuals. *Journal of the American College of Cardiology*, *57*(11), 1299–1313. https://doi.org/10.1016/j.jacc.2010.09.073.

Keramat, L., Sadrzadeh-Yeganeh, H., Sotoudeh, G., Zamani, E., Eshraghian, M., Mansoori, A., et al. (2017). Apolipoprotein A2 −265 T > C polymorphism interacts with dietary fatty acids intake to modulate inflammation in type 2 diabetes mellitus patients. *Nutrition*, *37*, 86–91. https://doi.org/10.1016/j.nut.2016.12.012.

Khalesi, S., Irwin, C., & Schubert, M. (2015). Flaxseed consumption may reduce blood pressure: A systematic review and meta-analysis of controlled trials. *The Journal of Nutrition*, *145*(4), 758–765. https://doi.org/10.3945/jn.114.205302.

Kien, C. L., Bunn, J. Y., Fukagawa, N. K., Anathy, V., Matthews, D. E., Crain, K. I., et al. (2015). Lipidomic evidence that lowering the typical dietary palmitate to oleate ratio in humans decreases the leukocyte production of proinflammatory cytokines and muscle expression of redox-sensitive genes. *The Journal of Nutritional Biochemistry*, *26*(12), 1599–1606. https://doi.org/10.1016/j.jnutbio.2015.07.014.

Kim, J. H., Kim, Y., Kim, Y. J., & Park, Y. (2016). Conjugated linoleic acid: Potential health benefits as a functional food ingredient. *Annual Review of Food Science and Technology*, 7(1), 221–244. https://doi.org/10.1146/annurev-food-041715-033028.

Knoblauch, H., Bauerfeind, A., Toliat, M. R., Becker, C., Luganskaja, T., Günther, U. P., et al. (2004). Haplotypes and SNPs in 13 lipid-relevant genes explain most of the genetic variance in high-density lipoprotein and low-density lipoprotein cholesterol. *Human Molecular Genetics*, 13(10), 993–1004. https://doi.org/10.1093/hmg/ddh119.

Koba, K., & Yanagita, T. (2014). Health benefits of conjugated linoleic acid (CLA). *Obesity Research & Clinical Practice*, 8(6), e525–e532. https://doi.org/10.1016/j.orcp.2013. 10.001.

Koochakpoor, G., Daneshpour, M. S., Mirmiran, P., Hosseini, S. A., Hosseini-Esfahani, F., Sedaghatikhayat, B., et al. (2016). The effect of interaction between melanocortin-4 receptor polymorphism and dietary factors on the risk of metabolic syndrome. *Nutrition & Metabolism*, 13, 35. https://doi.org/10.1186/s12986-016-0092-z.

Kurokawa, N., Nakai, K., Kameo, S., Liu, Z.-M., & Satoh, H. (2012). Association of BMI with the β3-adrenergic receptor gene polymorphism in Japanese: Meta-analysis. *Obesity Research*, 9(12), 741–745. https://doi.org/10.1038/oby.2001.102.

Labayen, I., Ruiz, J. R., Huybrechts, I., Ortega, F. B., Arenaza, L., González-Gross, M., et al. (2016). Dietary fat intake modifies the influence of the FTO rs9939609 polymorphism on adiposity in adolescents: The HELENA cross-sectional study. *Nutrition, Metabolism, and Cardiovascular Diseases*, 26(10), 937–943. https://doi.org/10.1016/j.numecd.2016. 07.010.

Lai, C.-Q., Corella, D., Demissie, S., Cupples, L. A., Adiconis, X., Zhu, Y., et al. (2006). Dietary intake of n-6 fatty acids modulates effect of apolipoprotein A5 gene on plasma fasting triglycerides, remnant lipoprotein concentrations, and lipoprotein particle size. *Circulation*, 113(17), 2062. https://doi.org/10.1161/CIRCULATIONAHA.105. 577296.

Larifla, L., Rambhojan, C., Joannes, M.-O., Maimaitiming-Madani, S., Donnet, J.-P., Marianne-Pépin, T., et al. (2016). Gene polymorphisms of FABP2, ADIPOQ and ANP and risk of hypertriglyceridemia and metabolic syndrome in Afro-Caribbeans. *PLoS One*, 11(9), e0163421. https://doi.org/10.1371/journal.pone.0163421.

Larsen, S. C., Ängquist, L., Østergaard, J. N., Ahluwalia, T. S., Vimaleswaran, K. S., Roswall, N., et al. (2016). Intake of total and subgroups of fat minimally affect the associations between selected single nucleotide polymorphisms in the PPARγ pathway and changes in anthropometry among European adults from cohorts of the DiOGenes study. *The Journal of Nutrition*, 146(3), 603–611. https://doi.org/10.3945/ jn.115.219675.

Lefevre, M., Lovejoy, J. C., Smith, S. R., DeLany, J. P., Champagne, C., Most, M. M., et al. (2005). Comparison of the acute response to meals enriched with cis- or trans-fatty acids on glucose and lipids in overweight individuals with differing FABP2 genotypes. *Metabolism—Clinical and Experimental*, 54(12), 1652–1658. https://doi.org/10.1016/ j.metabol.2005.06.015.

Lin, X., Qi, Q., Zheng, Y., Huang, T., Lathrop, M., Zelenika, D., et al. (2015). Neuropeptide Y genotype, central obesity, and abdominal fat distribution: The POUNDS LOST trial. *The American Journal of Clinical Nutrition*, 102(2), 514–519. https://doi.org/10.3945/ajcn.115.107276.

Lopez-Huertas, E. (2012). The effect of EPA and DHA on metabolic syndrome patients: A systematic review of randomised controlled trials. *British Journal of Nutrition*, 107(S2), S185–S194. https://doi.org/10.1017/S0007114512001572.

Lumeng, C. N., & Saltiel, A. R. (2011). Inflammatory links between obesity and metabolic disease. *The Journal of Clinical Investigation*, 121(6), 2111–2117. https://doi.org/10.1172/ JCI57132.

Ma, W., Huang, T., Heianza, Y., Wang, T., Sun, D., Tong, J., et al. (2017). Genetic variations of circulating adiponectin levels modulate changes in appetite in response to weight-loss diets. *The Journal of Clinical Endocrinology and Metabolism, 102*(1), 316–325. https://doi.org/10.1210/jc.2016-2909.

Ma, Y., Tucker, K., Smith, C., Lee, Y., Huang, T., Richardson, K., et al. (2014). Lipoprotein lipase variants interact with polyunsaturated fatty acids for obesity traits in women: Replication in two populations. *Nutrition, Metabolism, and Cardiovascular Diseases, 24*(12), 1323–1329. https://doi.org/10.1016/j.numecd.2014.07.003.

Maiorino, M. I., Bellastella, G., Petrizzo, M., Scappaticcio, L., Giugliano, D., & Esposito, K. (2016a). Anti-inflammatory effect of Mediterranean diet in type 2 diabetes is durable: 8-year follow-up of a controlled trial. *Diabetes Care, 39*(3), e44–e45. https://doi.org/10.2337/dc15-2356.

Maiorino, M. I., Bellastella, G., Petrizzo, M., Scappaticcio, L., Giugliano, D., & Esposito, K. (2016b). Mediterranean diet cools down the inflammatory milieu in type 2 diabetes: The MÉDITA randomized controlled trial. *Endocrine, 54*(3), 634–641. https://doi.org/10.1007/s12020-016-0881-1.

Marten, B., Pfeuffer, M., & Schrezenmeir, J. (2006). Medium-chain triglycerides, Technological and health aspects of bioactive components of milk. *International Dairy Journal, 16*(11), 1374–1382. https://doi.org/10.1016/j.idairyj.2006.06.015.

Mattei, J., Qi, Q., Hu, F. B., Sacks, F. M., & Qi, L. (2012). TCF7L2 genetic variants modulate the effect of dietary fat intake on changes in body composition during a weight-loss intervention. *The American Journal of Clinical Nutrition, 96*(5), 1129–1136. https://doi.org/10.3945/ajcn.112.038125.

Matualatupauw, J. C., Bohl, M., Gregersen, S., Hermansen, K., & Afman, L. A. (2017). Dietary medium-chain saturated fatty acids induce gene expression of energy metabolism-related pathways in adipose tissue of abdominally obese subjects. *International Journal of Obesity, 41*, 1348.

Melanson, E. L., Astrup, A., & Donahoo, W. T. (2009). The relationship between dietary fat and fatty acid intake and body weight, diabetes, and the metabolic syndrome. *Annals of Nutrition & Metabolism, 55*(1–3), 229–243. https://doi.org/10.1159/000229004.

Mendizábal, Y., Llorens, S., & Nava, E. (2013). Hypertension in metabolic syndrome: Vascular pathophysiology. *International Journal of Hypertension, 2013*, 230868. https://doi.org/10.1155/2013/230868.

Mensink, R. P., Zock, P. L., Kester, A. D., & Katan, M. B. (2003). Effects of dietary fatty acids and carbohydrates on the ratio of serum total to HDL cholesterol and on serum lipids and apolipoproteins: A meta-analysis of 60 controlled trials. *The American Journal of Clinical Nutrition, 77*(5), 1146–1155. https://doi.org/10.1093/ajcn/77.5.1146.

Merched, A. J., & Chan, L. (2013). Nutrigenetics and nutrigenomics of atherosclerosis. *Current Atherosclerosis Reports, 15*(6), 328. https://doi.org/10.1007/s11883-013-0328-6.

Merino, J., Udler, M. S., Leong, A., & Meigs, J. B. (2017). A decade of genetic and metabolomic contributions to type 2 diabetes risk prediction. *Current Diabetes Reports, 17*(12), 135. –135. https://doi.org/10.1007/s11892-017-0958-0.

Micha, R., Khatibzadeh, S., Shi, P., Fahimi, S., Lim, S., Andrews, K. G., et al. (2014). Global, regional, and national consumption levels of dietary fats and oils in 1990 and 2010: A systematic analysis including 266 country-specific nutrition surveys. *British Medical Journal, 348*, g2272. https://doi.org/10.1136/bmj.g2272.

Micha, R., & Mozaffarian, D. (2009). Trans fatty acids: Effects on metabolic syndrome, heart disease and diabetes. *Nature Reviews Endocrinology, 5*(6), 335–344.

Miller, P. E., Van Elswyk, M., & Alexander, D. D. (2014). Long-chain Omega-3 fatty acids eicosapentaenoic acid and docosahexaenoic acid and blood pressure: A meta-analysis of randomized controlled trials. *American Journal of Hypertension, 27*(7), 885–896. https://doi.org/10.1093/ajh/hpu024.

Mine, Y., Miyashita, K., & Shahidi, F. (2009). Nutrigenomics and proteomics in health and disease: An overview. In Y. Mine, K. Miyashita, & F. Shahidi (Eds.), *Nutrigenomics and proteomics in health and disease food factors and gene interactions* (pp. 3–10). Iowa: Wiley-Blackwell [Chapter 1].

Minihane, A. M., Vinoy, S., Russell, W. R., Baka, A., Roche, H. M., Tuohy, K. M., et al. (2015). Low-grade inflammation, diet composition and health: Current research evidence and its translation. *The British Journal of Nutrition, 114*(7), 999–1012. https://doi.org/10.1017/S0007114515002093.

Mohlke, K. L., & Boehnke, M. (2015). Recent advances in understanding the genetic architecture of type 2 diabetes. *Human Molecular Genetics, 24*(R1), R85–R92. https://doi.org/10.1093/hmg/ddv264.

Moloney, F., Toomey, S., Noone, E., Nugent, A., Allan, B., Loscher, C. E., et al. (2007). Antidiabetic effects of cis-9, trans-11-conjugated linoleic acid may be mediated via anti-inflammatory effects in white adipose tissue. *Diabetes, 56*(3), 574. https://doi.org/10.2337/db06-0384.

Monteiro, J., Leslie, M., Moghadasian, M. H., Arendt, B. M., Allard, J. P., & Ma, D. W. L. (2014). The role of n-6 and n-3 polyunsaturated fatty acids in the manifestation of the metabolic syndrome in cardiovascular disease and non-alcoholic fatty liver disease. *Food & Function, 5*(3), 426–435. https://doi.org/10.1039/C3FO60551E.

Morcillo, S., Rojo-Martínez, G., Cardona, F., de la Cruz Almaraz, M., de la Soledad Ruiz de Adana, M., Esteva, I., et al. (2007). Effect of the interaction between the fatty acid-binding protein 2 gene Ala54Thr polymorphism and dietary fatty acids on peripheral insulin sensitivity: A cross-sectional study. *The American Journal of Clinical Nutrition, 86*(4), 1232–1237. https://doi.org/10.1093/ajcn/86.4.1232.

Mozaffarian, D., Aro, A., & Willett, W. C. (2009). Health effects of trans-fatty acids: Experimental and observational evidence. *European Journal of Clinical Nutrition, 63*(Suppl. 2), S5–21. https://doi.org/10.1038/sj.ejcn.1602973.

Mozaffarian, D., Cao, H., King, I. B., Lemaitre, R. N., Song, X., Siscovick, D. S., et al. (2010). Trans-palmitoleic acid, metabolic risk factors, and new-onset diabetes in US adults. *Annals of Internal Medicine, 153*(12), 790–799. https://doi.org/10.1059/0003-4819-153-12-201012210-00005.

Mozaffarian, D., & Clarke, R. (2009). Quantitative effects on cardiovascular risk factors and coronary heart disease risk of replacing partially hydrogenated vegetable oils with other fats and oils. *European Journal of Clinical Nutrition, 63*, S22.

Mozaffarian, D., de Oliveira Otto, M. C., Lemaitre, R. N., Fretts, A. M., Hotamisligil, G., Tsai, M. Y., et al. (2013). trans-Palmitoleic acid, other dairy fat biomarkers, and incident diabetes: The multi-ethnic study of atherosclerosis (MESA). *The American Journal of Clinical Nutrition, 97*(4), 854–861. https://doi.org/10.3945/ajcn.112.045468.

Mozaffarian, D., Katan, M. B., Ascherio, A., Stampfer, M. J., & Willett, W. C. (2006). Trans fatty acids and cardiovascular disease. *New England Journal of Medicine, 354*(15), 1601–1613. https://doi.org/10.1056/NEJMra054035.

National Human Genome Research Institute. (2016). *What are genetics and genomics?* Retrieved May 30, 2018. Rockville Pike, Bethesda, MD: National Human Genome Research Institute. from https://www.genome.gov/19016904/.

Ning, B., & Kaput, J. (2009). Toward personalized nutrition and medicine: Promises and challenges. In Y. Mine, K. Miyashita, & F. Shahidi (Eds.), *Nutrigenomics and proteomics in health and disease: Food factors and gene interactions* (pp. 33–46). Iowa: Wiley-Blackwell [Chapter 3].

Nishida, C., & Uauy, R. (2009). WHO scientific update on health consequences of trans fatty acids: Introduction. *European Journal of Clinical Nutrition, 63*, S1.

O'Leary, N. A., Wright, M. W., Brister, J. R., Ciufo, S., Haddad, D., McVeigh, R., et al. (2016). Reference sequence (RefSeq) database at NCBI: Current status, taxonomic expansion, and functional annotation. *Nucleic Acids Research, 44*(Database issue), D733–D745. https://doi.org/10.1093/nar/gkv1189.

O'Neill, S., & O'Driscoll, L. (2015). Metabolic syndrome: A closer look at the growing epidemic and its associated pathologies. *Obesity Reviews: An Official Journal of the International Association for the Study of Obesity, 16*(1), 1–12. https://doi.org/10.1111/obr.12229.

Ordovas, J. M., Corella, D., Cupples, L. A., Demissie, S., Kelleher, A., Coltell, O., et al. (2002). Polyunsaturated fatty acids modulate the effects of the APOA1 G-A polymorphism on HDL-cholesterol concentrations in a sex-specific manner: The Framingham study. *The American Journal of Clinical Nutrition, 75*(1), 38–46. https://doi.org/10.1093/ajcn/75.1.38.

Ordovas, J. M., Kaput, J., & Corella, D. (2007). Nutrition in the genomics era: Cardiovascular disease risk and the Mediterranean diet. *Molecular Nutrition & Food Research, 51*(10), 1293–1299. https://doi.org/10.1002/mnfr.200700041.

Ortega-Azorín, C., Sorlí, J. V., Asensio, E. M., Coltell, O., Martínez-González, M. Á., Salas-Salvadó, J., et al. (2012). Associations of the FTO rs9939609 and the MC4R rs17782313 polymorphisms with type 2 diabetes are modulated by diet, being higher when adherence to the Mediterranean diet pattern is low. *Cardiovascular Diabetology, 11*, 137. https://doi.org/10.1186/1475-2840-11-137.

Pandey, N. R., Renwick, J., Misquith, A., Sokoll, K., & Sparks, D. L. (2008). Linoleic acid-enriched phospholipids act through peroxisome proliferator-activated receptors α to stimulate hepatic apolipoprotein A-I secretion. *Biochemistry, 47*(6), 1579–1587. https://doi.org/10.1021/bi702148f.

Park, J. Y., Lee, H.-J., Jang, H. B., Hwang, J.-Y., Kang, J. H., Han, B.-G., et al. (2014). Interactions between ADIPOQ gene variants and dietary monounsaturated: Saturated fatty acid ratio on serum lipid levels in Korean children. *Nutrition, Metabolism, and Cardiovascular Diseases, 24*(1), 83–90. https://doi.org/10.1016/j.numecd.2013.04.007.

Peña-Romero, A. C., Navas-Carrillo, D., Marín, F., & Orenes-Piñero, E. (2017). The future of nutrition: Nutrigenomics and nutrigenetics in obesity and cardiovascular diseases. *Critical Reviews in Food Science and Nutrition*, 1–12. https://doi.org/10.1080/10408398.2017.1349731.

Perez, P. M., Moore-Carrasco, R., Gonzalez, D. R., Fuentes, E. Q., & Palomo, I. G. (2012). Gene expression of adipose tissue, endothelial cells and platelets in subjects with metabolic syndrome (review). *Molecular Medicine Reports, 5*(5), 1135–1140. https://doi.org/10.3892/mmr.2012.785.

Perez-Martinez, P., Garcia-Quintana, J. M., Yubero-Serrano, E. M., Tasset-Cuevas, I., Tunez, I., Garcia-Rios, A., et al. (2010). Postprandial oxidative stress is modified by dietary fat: Evidence from a human intervention study. *Clinical Science, 119*(6), 251. https://doi.org/10.1042/CS20100015.

Pérez-Martínez, P., Ordovás, J. M., López-Miranda, J., Gómez, P., Marín, C., Moreno, J., et al. (2003). Polymorphism exon 1 variant at the locus of the scavenger receptor class B type I gene: Influence on plasma LDL cholesterol in healthy subjects during the consumption of diets with different fat contents. *The American Journal of Clinical Nutrition, 77*(4), 809–813. https://doi.org/10.1093/ajcn/77.4.809.

Pérez-Martínez, P., Pérez-Jiménez, F., Bellido, C., Ordovás, J. M., Moreno, J. A., Marín, C., et al. (2005). A polymorphism exon 1 variant at the locus of the scavenger receptor class B type I (SCARB1) gene is associated with differences in insulin sensitivity in healthy people during the consumption of an olive oil-rich diet. *The Journal of Clinical Endocrinology & Metabolism, 90*(4), 2297–2300. https://doi.org/10.1210/jc.2004-1489.

Perona, J. S. (2017). Membrane lipid alterations in the metabolic syndrome and the role of dietary oils. *Biochimica et Biophysica Acta, 1859*(9 Pt. B), 1690–1703. https://doi.org/10.1016/j.bbamem.2017.04.015.

Perona, J. S., Covas, M.-I., Fitó, M., Cabello-Moruno, R., Aros, F., Corella, D., et al. (2010). Reduction in systemic and VLDL triacylglycerol concentration after a 3-month Mediterranean-style diet in high-cardiovascular-risk subjects. *The Journal of Nutritional Biochemistry*, *21*(9), 892–898. https://doi.org/10.1016/j.jnutbio.2009.07.005.

Phillips, C. M., Goumidi, L., Bertrais, S., Field, M. R., McManus, R., Hercberg, S., et al. (2011). Gene–nutrient interactions and gender may modulate the association between ApoA1 and ApoB gene polymorphisms and metabolic syndrome risk. *Atherosclerosis*, *214*(2), 408–414. https://doi.org/10.1016/j.atherosclerosis.2010.10.029.

Poudyal, H., & Brown, L. (2015). Should the pharmacological actions of dietary fatty acids in cardiometabolic disorders be classified based on biological or chemical function? *Progress in Lipid Research*, *59*, 172–200. https://doi.org/10.1016/j.plipres.2015.07.002.

Qi, Q., Bray, G. A., Hu, F. B., Sacks, F. M., & Qi, L. (2012). Weight-loss diets modify glucose-dependent insulinotropic polypeptide receptor rs2287019 genotype effects on changes in body weight, fasting glucose, and insulin resistance: The preventing over-weight using novel dietary strategies trial. *The American Journal of Clinical Nutrition*, *95*(2), 506–513. https://doi.org/10.3945/ajcn.111.025270.

Qi, Q., Bray, G. A., Smith, S. R., Hu, F. B., Sacks, F. M., & Qi, L. (2011). Insulin receptor substrate 1 (IRS1) gene variation modifies insulin resistance response to weight-loss diets in a two-year randomized trial. *Circulation*, *124*(5), 563–571. https://doi.org/10.1161/CIRCULATIONAHA.111.025767.

Qi, Q., Durst, R., Schwarzfuchs, D., Leitersdorf, E., Shpitzen, S., Li, Y., et al. (2015). CETP genotype and changes in lipid levels in response to weight-loss diet intervention in the POUNDS LOST and DIRECT randomized trials. *Journal of Lipid Research*, *56*(3), 713–721. https://doi.org/10.1194/jlr.P055715.

Qi, Q., Xu, M., Wu, H., Liang, L., Champagne, C. M., Bray, G. A., et al. (2013). IRS1 genotype modulates metabolic syndrome reversion in response to 2-year weight-loss diet intervention: The POUNDS LOST trial. *Diabetes Care*, *36*(11), 3442–3447. https://doi.org/10.2337/dc13-0018.

Raatz, S. K., Conrad, Z., Johnson, L. K., Picklo, M. J., & Jahns, L. (2017). Relationship of the reported intakes of fat and fatty acids to body weight in US adults. *Nutrients*, *9*(5), 438. https://doi.org/10.3390/nu9050438.

Rangel-Zúñiga, O. A., Camargo, A., Marin, C., Peña-Orihuela, P., Pérez-Martínez, P., Delgado-Lista, J., et al. (2015). Proteome from patients with metabolic syndrome is regulated by quantity and quality of dietary lipids. *BMC Genomics*, *16*, 509. https://doi.org/10.1186/s12864-015-1725-8.

Razquin, C., Martinez, J. A., Martinez-Gonzalez, M. A., Bes-Rastrollo, M., Fernández-Crehuet, J., & Marti, A. (2009). A 3-year intervention with a Mediterranean diet modified the association between the rs9939609 gene variant in FTO and body weight changes. *International Journal of Obesity*, *34*, 266.

Razquin, C., Martinez, J. A., Martinez-Gonzalez, M. A., Corella, D., Santos, J. M., & Marti, A. (2009). The Mediterranean diet protects against waist circumference enlargement in 12Ala carriers for the PPARγ gene: 2 years' follow-up of 774 subjects at high cardiovascular risk. *British Journal of Nutrition*, *102*(5), 672–679. https://doi.org/10.1017/S0007114509289008.

Razquin, C., Martinez, J. A., Martinez-Gonzalez, M. A., Fernández-Crehuet, J., Santos, J. M., & Marti, A. (2010). A Mediterranean diet rich in virgin olive oil may reverse the effects of the -174G/C IL6 gene variant on 3-year body weight change. *Molecular Nutrition & Food Research*, *54*(S1), S75–S82. https://doi.org/10.1002/mnfr.200900257.

Risérus, U., Willett, W. C., & Hu, F. B. (2009). Dietary fats and prevention of type 2 diabetes. *Progress in Lipid Research*, *48*(1), 44–51. https://doi.org/10.1016/j.plipres.2008.10.002.

Rocha, D. M., Bressan, J., & Hermsdorff, H. H. (2017). The role of dietary fatty acid intake in inflammatory gene expression: A critical review. *São Paulo Medical Journal = Revista Paulista de Medicina*, *135*(2), 157–168. https://doi.org/10.1590/1516-3180. 2016.008607072016.

Roncero-Ramos, I., Rangel-Zuñiga, O. A., Lopez-Moreno, J., Alcala-Diaz, J. F., Perez-Martinez, P., Jimenez-Lucena, R., et al. (2018). Mediterranean diet, glucose homeostasis, and Inflammasome genetic variants: The CORDIOPREV study. *Molecular Nutrition & Food Research*, *62*(9), 1700960. https://doi.org/10.1002/mnfr.201700960.

Ros, E. (2003). Dietary cis-monounsaturated fatty acids and metabolic control in type 2 diabetes. *The American Journal of Clinical Nutrition*, *78*(3 Suppl), 617S–625S.

Rudkowska, I., Ouellette, C., Dewailly, E., Hegele, R. A., Boiteau, V., Dubé-Linteau, A., et al. (2013). Omega-3 fatty acids, polymorphisms and lipid related cardiovascular disease risk factors in the Inuit population. *Nutrition & Metabolism*, *10*, 26. https://doi.org/10. 1186/1743-7075-10-26.

Rudkowska, I., & Pérusse, L. (2012). Individualized weight management: What can be learned from nutrigenomics and nutrigenetics? In C. Bouchard & J. M. Ordovas (Eds.), *Progress in molecular biology and translational science: Vol. 108* (pp. 347–382). Academic Press. https://doi.org/10.1016/B978-0-12-398397-8.00014-9.

Rudkowska, I., & Vohl, M.-C. (2010). Interaction between diets, polymorphisms and plasma lipid levels. *Clinical Lipidology*, *5*(3), 421–438. https://doi.org/10.2217/clp.10.26.

Ruiz-Núñez, B., Dijck-Brouwer, D. A. J., & Muskiet, F. A. J. (2016). The relation of saturated fatty acids with low-grade inflammation and cardiovascular disease. *The Journal of Nutritional Biochemistry*, *36*, 1–20. https://doi.org/10.1016/j.jnutbio.2015.12.007.

Rustan, A. C., & Drevon, C. A. (2005). *Fatty acids: Structures and properties. In eLS*. American Cancer Society *https://doi.org/10.1038/npg.els.0003894*.

Sacks, F. M., Lichtenstein, A. H., Wu, J. H. Y., Appel, L. J., Creager, M. A., Kris-Etherton, P. M., et al. (2017). Dietary fats and cardiovascular disease: A presidential advisory from the American Heart Association. *Circulation*, *136*(3), e1–e23. https://doi. org/10.1161/CIR.0000000000000510.

Salmerón, J., Hu, F. B., Manson, J. E., Stampfer, M. J., Colditz, G. A., Rimm, E. B., et al. (2001). Dietary fat intake and risk of type 2 diabetes in women. *The American Journal of Clinical Nutrition*, *73*(6), 1019–1026. https://doi.org/10.1093/ajcn/73.6.1019.

Salter, A. M., & Tarling, E. J. (2007). Regulation of gene transcription by fatty acids. *Animal*, *1*(9), 1314–1320. https://doi.org/10.1017/S1751731107000675.

Sanchez-Infantes, D. (2014). Role of adipose tissue in the pathogenesis and treatment of metabolic syndrome. In M. Orešič & A. Vidal-Puig (Eds.), *A systems biology approach to study metabolic syndrome* (pp. 63–83). Switzerland: Springer International Publishing [Chapter 4].

Schaefer, E. J., Lamon-Fava, S., Ausman, L. M., Ordovas, J. M., Clevidence, B. A., Judd, J. T., et al. (1997). Individual variability in lipoprotein cholesterol response to National Cholesterol Education Program Step 2 diets. *The American Journal of Clinical Nutrition*, *65*(3), 823–830. https://doi.org/10.1093/ajcn/65.3.823.

Schüler, R., Osterhoff, M. A., Frahnow, T., Möhlig, M., Spranger, J., Stefanovski, D., et al. (2017). Dietary fat intake modulates effects of a frequent ACE gene variant on glucose tolerance with association to type 2 diabetes. *Scientific Reports*, *7*, 9234. https://doi.org/ 10.1038/s41598-017-08300-7.

Schüler, R., Osterhoff, M. A., Frahnow, T., Seltmann, A., Busjahn, A., Kabisch, S., et al. (2017). High-saturated-fat diet increases circulating angiotensin-converting enzyme, which is enhanced by the rs4343 polymorphism defining persons at risk of nutrient-dependent increases of blood pressure. *Journal of the American Heart Association: Cardiovascular and Cerebrovascular Disease*, *6*(1), e004465. https://doi.org/10.1161/ JAHA.116.004465.

Schwingshackl, L., & Hoffmann, G. (2014a). Comparison of the long-term effects of high-fat v. low-fat diet consumption on cardiometabolic risk factors in subjects with abnormal glucose metabolism: A systematic review and meta-analysis. *British Journal of Nutrition, 111*(12), 2047–2058. https://doi.org/10.1017/S0007114514000464.

Schwingshackl, L., & Hoffmann, G. (2014b). Mediterranean dietary pattern, inflammation and endothelial function: A systematic review and meta-analysis of intervention trials. *Nutrition, Metabolism, and Cardiovascular Diseases, 24*(9), 929–939. https://doi.org/10.1016/j.numecd.2014.03.003.

Schwingshackl, L., Strasser, B., & Hoffmann, G. (2011). Effects of monounsaturated fatty acids on glycaemic control in patients with abnormal glucose metabolism: A systematic review and meta-analysis. *Annals of Nutrition and Metabolism, 58*(4), 290–296.

Scuteri, A., Sanna, S., Chen, W.-M., Uda, M., Albai, G., Strait, J., et al. (2007). Genome-wide association scan shows genetic variants in the FTO gene are associated with obesity-related traits. *PLoS Genetics, 3*(7), e115. https://doi.org/10.1371/journal.pgen.0030115.

Serhan, C. N., Chiang, N., & Van Dyke, T. E. (2008). Resolving inflammation: Dual anti-inflammatory and pro-resolution lipid mediators. *Nature Reviews. Immunology, 8*(5), 349–361. https://doi.org/10.1038/nri2294.

Shah, R. D., Xue, C., Zhang, H., Tuteja, S., Li, M., Reilly, M. P., et al. (2017). Expression of calgranulin genes S100A8, S100A9 and S100A12 is modulated by n-3 PUFA during inflammation in adipose tissue and mononuclear cells. *PLoS One, 12*(1), e0169614. https://doi.org/10.1371/journal.pone.0169614.

Shatwan, I. M., Weech, M., Jackson, K. G., Lovegrove, J. A., & Vimaleswaran, K. S. (2017). Apolipoprotein E gene polymorphism modifies fasting total cholesterol concentrations in response to replacement of dietary saturated with monounsaturated fatty acids in adults at moderate cardiovascular disease risk. *Lipids in Health and Disease, 16*, 222. https://doi.org/10.1186/s12944-017-0606-3.

Simopoulos, A. P. (2016). An increase in the omega-6/omega-3 fatty acid ratio increases the risk for obesity. *Nutrients, 8*(3), 128. https://doi.org/10.3390/nu8030128.

Siri-Tarino, P. W., Sun, Q., Hu, F. B., & Krauss, R. M. (2010). Meta-analysis of prospective cohort studies evaluating the association of saturated fat with cardiovascular disease. *The American Journal of Clinical Nutrition, 91*(3), 535–546. https://doi.org/10.3945/ajcn.2009.27725.

Skrypnik, K., Suliburska, J., Skrypnik, D., Pilarski, L., Regula, J., & Bogdanski, P. (2017). The genetic basis of obesity complications. *Acta Scientiarum Polonorum. Technologia Alimentaria, 16*(1), 83–91.

Smith, C. E. (2012). Plant oils and cardiometabolic risk factors: The role of genetics. *Current Nutrition Reports, 1*(3), 161–168. https://doi.org/10.1007/s13668-012-0018-y.

Smith, C. E., Van Rompay, M. I., Mattei, J., Garcia, J. F., Garcia-Bailo, B., Lichtenstein, A. H., et al. (2017). Dietary fat modulation of hepatic lipase variant −514 C/T for lipids: A crossover randomized dietary intervention trial in Caribbean hispanics. *Physiological Genomics, 49*(10), 592–600. https://doi.org/10.1152/physiolgenomics.00036.2017.

Soriguer, F., Morcillo, S., Cardona, F., Rojo-Martínez, G., de la Cruz Almaráz, M., de la Soledad Ruiz de Adana, M., et al. (2006). Pro12Ala polymorphism of the PPARG2 gene is associated with type 2 diabetes mellitus and peripheral insulin sensitivity in a population with a high intake of oleic acid. *The Journal of Nutrition, 136*(9), 2325–2330. https://doi.org/10.1093/jn/136.9.2325.

Sotos-Prieto, M., Guillén, M., Portolés, O., Sorlí, J. V., González, J. I., Asensio, E. M., et al. (2013). Association between the rs6950982 polymorphism near the SERPINE1 gene and blood pressure and lipid parameters in a high-cardiovascular-risk population: Interaction with Mediterranean diet. *Genes & Nutrition, 8*(4), 401–409. https://doi.org/10.1007/s12263-012-0327-1.

Sotos-Prieto, M., & Penalvo, J. L. (2013). Genetic variation of apolipoproteins, diet and other environmental interactions; an updated review. *Nutrición Hospitalaria, 28*(4), 999–1009. https://doi.org/10.3305/nh.2013.28.4.6475.

Tagetti, A., Ericson, U., Montagnana, M., Danese, E., Almgren, P., Nilsson, P., et al. (2015). Intakes of omega-3 polyunsaturated fatty acids and blood pressure change over time: Possible interaction with genes involved in 20-HETE and EETs metabolism. *Prostaglandins & Other Lipid Mediators, 120*, 126–133. https://doi.org/10.1016/j.prostaglandins. 2015.05.003.

Tai, E. S., Corella, D., Demissie, S., Cupples, L. A., Coltell, O., Schaefer, E. J., et al. (2005). Polyunsaturated fatty acids interact with the PPARA-L162V polymorphism to affect plasma triglyceride and apolipoprotein C-III concentrations in the Framingham heart study. *The Journal of Nutrition, 135*(3), 397–403. https://doi.org/10.1093/jn/135.3.397.

Tardy, A.-L., Morio, B., Chardigny, J.-M., & Malpuech-Brugère, C. (2011). Ruminant and industrial sources of trans-fat and cardiovascular and diabetic diseases. *Nutrition Research Reviews, 24*(1), 111–117. https://doi.org/10.1017/S0954422411000011.

Tierney, A. C., McMonagle, J., Shaw, D. I., Gulseth, H. L., Helal, O., Saris, W. H. M., et al. (2010). Effects of dietary fat modification on insulin sensitivity and on other risk factors of the metabolic syndrome—LIPGENE: A European randomized dietary intervention study. *International Journal of Obesity, 35*, 800.

Tobias, D. K., Chen, M., Manson, J. E., Ludwig, D. S., Willett, W., & Hu, F. B. (2015). Effect of low-fat vs. other diet interventions on long-term weight change in adults: A systematic review and meta-analysis. *The Lancet. Diabetes & Endocrinology, 3*(12), 968–979. https://doi.org/10.1016/S2213-8587(15)00367-8.

Tremblay, B. L., & Rudkowska, I. (2017). Nutrigenomic point of view on effects and mechanisms of action of ruminant trans fatty acids on insulin resistance and type 2 diabetes. *Nutrition Reviews, 75*(3), 214–223. https://doi.org/10.1093/nutrit/nuw066.

Turkovic, L. F., Pizent, A., Dodig, S., Pavlovic, M., & Pasalic, D. (2012). FABP 2 gene polymorphism and metabolic syndrome in elderly people of Croatian descent. *Biochemia Medica, 22*(2), 217–224.

Turner, J. M., & Spatz, E. S. (2016). Nutritional supplements for the treatment of hypertension: A practical guide for clinicians. *Current Cardiology Reports, 18*(12), 126. https://doi.org/10.1007/s11886-016-0806-x.

U.S. Department of Health and Human Services and U.S. Department of Agriculture. (2015). *Dietary guidelines for Americans 2015–2010* (8th ed.). Retrieved from http://health.gov/dietaryguidelines/2015/guidelines/.

Vallée Marcotte, B., Cormier, H., Rudkowska, I., Lemieux, S., Couture, P., & Vohl, M.-C. (2017). Polymorphisms in FFAR4 (GPR120) gene modulate insulin levels and sensitivity after fish oil supplementation. *Journal of Personalized Medicine, 7*(4), 15. https://doi.org/ 10.3390/jpm7040015.

van Dijk, S. J., Feskens, E. J. M., Bos, M. B., de Groot, L. C. P. G. M., de Vries, J. H. M., Müller, M., et al. (2012). Consumption of a high monounsaturated fat diet reduces oxidative phosphorylation gene expression in peripheral blood mononuclear cells of abdominally overweight men and women. *The Journal of Nutrition, 142*(7), 1219–1225. https://doi.org/10.3945/jn.111.155283.

Vannice, G., & Rasmussen, H. (2014). Position of the academy of nutrition and dietetics: Dietary fatty acids for healthy adults. *Journal of the Academy of Nutrition and Dietetics, 114*(1), 136–153. https://doi.org/10.1016/j.jand.2013.11.001.

Vassy, J. L., Hivert, M.-F., Porneala, B., Dauriz, M., Florez, J. C., Dupuis, J., et al. (2014). Polygenic type 2 diabetes prediction at the limit of common variant detection. *Diabetes, 63*(6), 2172–2182. https://doi.org/10.2337/db13-1663.

Vessby, B., Uusitupa, M., Hermansen, K., Riccardi, G., Rivellese, A. A., Tapsell, L. C., et al. (2001). Substituting dietary saturated for monounsaturated fat impairs insulin sensitivity in healthy men and women: The KANWU study. *Diabetologia, 44*(3), 312–319.

Vidal-Puig, A. (2014). The metabolic syndrome and its complex pathophysiology. In M. Orešič & A. Vidal-Puig (Eds.), *A systems biology approach to study metabolic syndrome* (pp. 3–16). Switzerland: Springer International Publishing [Chapter 1].

Viladomiu, M., Hontecillas, R., & Bassaganya-Riera, J. (2016). Modulation of inflammation and immunity by dietary conjugated linoleic acid, Immunopharmacology of fatty acids. *European Journal of Pharmacology, 785,* 87–95. https://doi.org/10.1016/j.ejphar. 2015.03.095.

Waguri, T., Goda, T., Kasezawa, N., & Yamakawa-Kobayashi, K. (2013). The combined effects of genetic variations in the GPR120 gene and dietary fat intake on obesity risk. *Biomedical Research, 34*(2), 69–74. https://doi.org/10.2220/biomedres.34.69.

Wang, D. D., & Hu, F. B. (2017). Dietary fat and risk of cardiovascular disease: Recent controversies and advances. *Annual Review of Nutrition, 37*(1), 423–446. https://doi. org/10.1146/annurev-nutr-071816-064614.

Wang, T., Huang, T., Zheng, Y., Rood, J., Bray, G. A., Sacks, F. M., et al. (2016). Genetic variation of fasting glucose and changes in glycemia in response to 2-year weight-loss diet intervention: The POUNDS LOST trial. *International Journal of Obesity (2005), 40*(7), 1164–1169. https://doi.org/10.1038/ijo.2016.41.

Wang, D., Ma, J., Zhang, S., Hinney, A., Hebebrand, J., Wang, Y., et al. (2012). Association of the MC4R V103I polymorphism with obesity: A Chinese case–control study and meta-analysis in 55,195 individuals. *Obesity, 18*(3), 573–579. https://doi.org/10.1038/ oby.2009.268.

Wildman, R. P., Muntner, P., Reynolds, K., et al. (2008). The obese without cardiometabolic risk factor clustering and the normal weight with cardiometabolic risk factor clustering: Prevalence and correlates of 2 phenotypes among the US population (nhanes 1999–2004). *Archives of Internal Medicine, 168*(15), 1617–1624. https://doi. org/10.1001/archinte.168.15.1617.

World Health Organisation. (2015). *Healthy diet.* Retrieved April 8, 2018, from http://www. who.int/mediacentre/factsheets/fs394/en/.

World Health Organisation. (2018). *Obesity and overweight.* Retrieved March 27, 2018, from, http://www.who.int/mediacentre/factsheets/fs311/en/.

Wu, J. H., Micha, R., Imamura, F., Pan, A., Biggs, M. L., Ajaz, O., et al. (2012). Omega-3 fatty acids and incident type 2 diabetes: A systematic review and meta-analysis. *The British Journal of Nutrition, 107*(2), S214–S227. https://doi.org/10.1017/S0007114512001602.

Xu, M., Ng, S. S., Bray, G. A., Ryan, D. H., Sacks, F. M., Ning, G., et al. (2015). Dietary fat intake modifies the effect of a common variant in the LIPC gene on changes in serum lipid concentrations during a long-term weight-loss intervention trial. *The Journal of Nutrition, 145*(6), 1289–1294. https://doi.org/10.3945/jn.115.212514.

Xu, M., Qi, Q., Liang, J., Bray, G. A., Hu, F. B., Sacks, F. M., et al. (2013). Genetic determinant for amino acid metabolites and changes in body weight and insulin resistance in response to weight-loss diets: The POUNDS LOST trial. *Circulation, 127*(12), 1283–1289. https://doi.org/10.1161/CIRCULATIONAHA.112.000586.

Yanai, H., Tomono, Y., Ito, K., Furutani, N., Yoshida, H., & Tada, N. (2008). The underlying mechanisms for development of hypertension in the metabolic syndrome. *Nutrition Journal, 7,* 10. https://doi.org/10.1186/1475-2891-7-10.

Yeop, H. C., Kargi, A. Y., Omer, M., Chan, C. K., Wabitsch, M., O'Brien, K. D., et al. (2010). Differential effect of saturated and unsaturated free fatty acids on the generation of monocyte adhesion and chemotactic factors by adipocytes: Dissociation of adipocyte hypertrophy from inflammation. *Diabetes, 59*(2), 386–396. https://doi.org/10.2337/ db09-0925.

Zhang, X., Qi, Q., Bray, G. A., Hu, F. B., Sacks, F. M., & Qi, L. (2012). APOA5 genotype modulates 2-y changes in lipid profile in response to weight-loss diet intervention: The Pounds Lost trial. *The American Journal of Clinical Nutrition, 96*(4), 917–922. https://doi. org/10.3945/ajcn.112.040907.

Zhang, X., Qi, Q., Liang, J., Hu, F. B., Sacks, F. M., & Qi, L. (2012). Neuropeptide Y promoter polymorphism modifies effects of a weight-loss diet on 2-year changes of blood pressure: The Pounds Lost trial. *Hypertension, 60*(5), 1169–1175. https://doi.org/10.1161/hypertensionaha.112.197855.

Zheng, J.-S., Arnett, D. K., Parnell, L. D., Lee, Y.-C., Ma, Y., Smith, C. E., et al. (2013). Polyunsaturated fatty acids modulate the association between PIK3CA-KCNMB3 genetic variants and insulin resistance. *PLoS One, 8*(6), e67394. https://doi.org/10.1371/journal.pone.0067394.

Zheng, J.-S., Chen, J., Wang, L., Yang, H., Fang, L., Yu, Y., et al. (2018). Replication of a gene-diet interaction at CD36, NOS3 and PPARG in response to omega-3 fatty acid supplements on blood lipids: A double-blind randomized controlled trial. *eBioMedicine, 31*, 150–156. https://doi.org/10.1016/j.ebiom.2018.04.012.

Ziki, M. D. A., & Mani, A. (2016). Metabolic syndrome: Genetic insights into disease pathogenesis. *Current Opinion in Lipidology, 27*(2), 162–171. https://doi.org/10.1097/MOL.0000000000000276.

Zorzano, A., Sebastian, D., & Monserrat, R. (2014). The skeletal muscle in metabolic syndrome. In M. Orešič & A. Vidal-Puig (Eds.), *A systems biology approach to study metabolic syndrome* (pp. 137–156). Switzerland: Springer International Publishing [Chapter 6].

CHAPTER THREE

Microbial Ecology of Fermented Vegetables and Non-Alcoholic Drinks and Current Knowledge on Their Impact on Human Health

Laura Lavefve*,†, Daya Marasini*, Franck Carbonero*,1

*Department of Food Science and Center for Human Nutrition, University of Arkansas, Fayetteville, AR, United States
†Direction des Etudes Et Prestations (DEEP), Institut Polytechnique UniLaSalle, Beauvais, France
[1]Corresponding author: e-mail address: fgcarbon@uark.edu

Contents

Abstract

Fermented foods are currently experiencing a re-discovery, largely driven by numerous health benefits claims. While fermented dairy, beer, and wine (and other alcoholic

Advances in Food and Nutrition Research, Volume 87
ISSN 1043-4526
https://doi.org/10.1016/bs.afnr.2018.09.001

147

fermented beverages) have been the subject of intensive research, other plant-based fermented foods that are in some case widely consumed (kimchi/sauerkraut, pickles, kombucha) have received less scientific attention. In this chapter, the current knowledge on the microbiology and potential health benefits of such plant-based fermented foods are presented. Kimchi is the most studied, characterized by primarily acidic fermentation by lactic acid bacteria. Anti-obesity and anti-hypertension properties have been reported for kimchi and other pickled vegetables. Kombucha is the most popular non-alcoholic fermented drink. Kombucha's microbiology is remarkable as it involves all fermenters described in known fermented foods: lactic acid bacteria, acetic acid bacteria, fungi, and yeasts. While kombucha is often hyped as a "super-food," only antioxidant and antimicrobial properties toward foodborne pathogens are well established; and it is unknown if these properties incur beneficial impact, even *in vitro* or in animal models. The mode of action that has been studied and demonstrated the most is the probiotic one. However, it can be expected that fermentation metabolites may be prebiotic, or influence host health directly. To conclude, plant-based fermented foods and drinks are usually safe products; few negative reports can be found, but more research, especially human dietary intervention studies, are warranted to substantiate any health claim.

1. INTRODUCTION

The consumption of fermented food has long been associated with health benefits. Many of these claims derived from anecdotal evidence, but more recently, food and nutrition scientists have begun to provide more robust evidence for the potential positive impact of the consumption of fermented products on several chronic diseases (Gille, Schmid, Walther, & Vergeres, 2018). These benefits have been attributed to the microorganisms themselves and the metabolites they produce in the food. Moreover, light has been shed on the importance of the human microbiota on the host's health and the way they interact with each other, mostly *via* the gut microbes (Marco et al., 2017).

However, it should be noted that very few controlled human dietary intervention studies have been performed so far to demonstrate these potential health benefits (Marco et al., 2017), and there is a notable difference in scientific knowledge between the more popular fermented products and other plant–based fermented foods. Specifically, fermented dairy products have been studied extensively (Aryana & Olson, 2017; Fernandez, Panahi, Daniel, Tremblay, & Marette, 2017; Gille et al., 2018) and are also commonly used as probiotic/prebiotic vehicles. Wine, bread, and to some extent beer have also received considerable attention (De Vuyst, Vrancken, Ravyts, Rimaux, & Weckx, 2009; Fragopoulou, Choleva, Antonopoulou, &

Demopoulos, 2018; Pastor et al., 2017; Rodhouse & Carbonero, 2017). This chapter will summarize the current knowledge on less-studied plant-based foods, thereby excluding wine and beer. We will describe fermented vegetables first with focus on the most studied sauerkraut and kimchi, and then fermented plant-based beverages with focus on kombucha.

2. FOOD FERMENTATION

2.1 Historical Perspective and Current Economic Importance

Fermentation, together with drying, is the oldest known food preservation method. Wine-like beverages have been produced since the Neolithic period (between 8500 and 4000 BC), and there are evidences that milk fermentation were common processes in 6000 BC in India and Egypt (McGovern et al., 2004; Sicard & Legras, 2011). It is commonly believed that food fermentation represented an accidental discovery, where humans realized that food could be preserved longer under specific conditions, without realizing the processes allowing preservation. All along history, new fermented products were developed and perfected: soy sauce was developed in Asia about 2500 years ago, trade of cheese was common during Antiquity in Greece and the Rome, fermented sausages have been reported to be consumed by Roman soldiers (Prajapati & Nair, 2003); and yogurt become part of regular diet in Western countries because of its purported health benefits at the beginning of the 20th century (Fisberg & Machado, 2015).

It is only fairly recently (end of the 18th century) that scientists were able to decipher the microbial processes responsible for food fermentation. Lavoisier and Gay-Lussac were the first to describe the chemistry of the alcoholic fermentation, without identification of the role played by microorganisms. Louis Pasteur eventually demonstrated that fermentation would not be possible without microorganisms and hence was "a living process." Following this discovery, scientific knowledge of food fermentation evolved quite rapidly; with identification and isolation of microbial species and strains involved in different fermentation and optimization of the fermentation parameters (Prajapati & Nair, 2003).

While originally the main purpose of fermentation was shelf life extension, scientific investigation revealed different processes involved in food safety but also food quality and organoleptic properties. Microorganisms in fermented foods help prevent the growth of spoilers or pathogens either by direct competitive exclusion (Arroyo-Lopez, Perez-Traves, Querol, &

Barrio, 2011; Zhu, Xiao, Shen, & Hao, 2010) or by production of antimicrobial compounds, with the two main antimicrobial compounds being lactic (and to some extent other organic acids) acid produced by lactic acid bacteria (LAB) (Ozogul & Hamed, 2018) and ethanol produced by yeasts (Hatoum, Labrie, & Fliss, 2012). More recently, specific LAB strains have been shown to produce a range of other antimicrobial compounds (Bali, Panesar, Bera, & Kennedy, 2016; Cleveland, Montville, Nes, & Chikindas, 2001; Corr et al., 2007; Mokoena, Mutanda, & Olaniran, 2016). Fermentation can also help to remove or convert toxic components from food like phytic acid in plants (Garcia-Mantrana, Monedero, & Haros, 2015; Garcia-Mantrana, Yebra, Haros, & Monedero, 2016), and improve food digestibility with for example cheeses and yogurts that can be typically consumed by lactose-intolerant individuals (Marco et al., 2017) or kimchi and sauerkraut which will be described in more detail (Kim, Noh, & Song, 2018; Raak, Ostermann, Boehm, & Molsberger, 2014). Microorganisms also release hundreds of metabolites that incur desirable organoleptic properties to the final fermented food (Akissoe et al., 2015; Altay, Karbancioglu-Guler, Daskaya-Dikmen, & Heperkan, 2013; Blanco, Andres-Iglesias, & Montero, 2016).

During the 20th century, plant-based fermented foods (excluding bread, wine, and beer) popularity has somewhat declined with the industrialization of food production, particularly in Western countries. However, in the last 10–15 years, plant-based fermented foods have become part of the new food trends. For example, the market of kefir was estimated around USD 130 million in 2014, while it was negligible in the 1990s. This renewed interest is explained by consumers' demand for healthy, locally-sourced food with original and creative organoleptic properties. Fermentation allows labeling the food as entirely natural, allowing the development of novel foods inspired from traditional foods. For example, non-alcoholic fermented beverages, such as kombucha (Greenwalt, Steinkraus, & Ledford, 2000), have evident advantages for consumers looking for similar taste without alcohol side effects.

2.2 Microorganisms Involved in Food Fermentation and Their Metabolic Pathways

Food fermentation can be defined as "a controlled microbial growth and enzymatic conversions of major and minor food components" (Marco et al., 2017). The diversity of microorganisms that can ferment food products is very important and usually the fermentation of one specific product results from the presence of several types of microbes. Still food fermentation can be generally classified by acidic fermentation on one hand and alcoholic fermentation on the other hand. These two types of fermentation have been reviewed extensively (Hill et al., 2017; Shiby & Mishra, 2013; Song, In, Lim, & Rahim, 2017; Tamang, Watanabe, & Holzapfel, 2016), and a very brief overview is provided in this chapter.

The main phylum of bacteria involved in acidic fermentation is the firmicutes, which contains the LAB; with *Lactobacillus*, *Leuconostoc*, *Lactococcus*, and *Streptococcus* being the most common starter cultures. Fermentation of dairy, cereal, vegetables, and meat involve LAB, with or without the intervention of other bacteria or yeasts. LAB is classified as homo or hetero-fermentative, depending on their metabolic abilities. Homo-fermentative LAB, such as *Lactococcus*, produce exclusively lactic acid; while hetero-fermentative LAB like *Leuconostoc* produce a variety of other metabolites leading to more complex organoleptic properties (Coda, Cagno, Gobbetti, & Rizzello, 2014; Fiorda et al., 2017; Randazzo, Caggia, & Neviani, 2009; Schroeter & Klaenhammer, 2009). Another phylum of bacteria playing a role in acidic fermentation is Proteobacteria with *Acetobacter*, *Gluconobacter*, and *Gluconacetobacter* genera. They mainly produce acetic acid through the oxidation of ethanol, but can also use sugars and other compounds (De Roos & De Vuyst, 2018). These genera are mainly known to ferment wine into vinegar (Mas, Torija, Garcia-Parrilla Mdel, & Troncoso, 2014), but have more recently been identified as playing important roles in tea, cocoa, and coffee fermentation (Coton et al., 2017; De Bruyn et al., 2016; Illeghems, Pelicaen, De Vuyst, & Weckx, 2016). Finally, bacteria from the Actinobacteria phylum can also play a role in food fermentation. *Bifidobacterium*, which also produce lactic acid, are not usually involved in the fermentation of food itself but are often, added to dairy products for their purported probiotics properties (Eales et al., 2017; Shiby & Mishra, 2013). *Corynebacterium glutamicum* and *Propionibacterium* are two other Actinobacteria members used in specific food fermentation for their release of glutamate and propionate, respectively (Denis & Irlinger, 2008; Meile, Le Blay, & Thierry, 2008; Moslemi, Mazaheri Nezhad Fard, Hosseini, Homayouni-Rad, & Mortazavian, 2016).

Alcoholic food fermentation is only performed by yeast, leading to the release of ethanol and carbon dioxide. *Saccharomyces* are by far the most commonly used yeast in food fermentation, and a large number of species and strains have been domesticated and selected for bread, beer, wine, and other products fermentation (Chen et al., 2016; Sicard & Legras, 2011).

Saccharomyces convert simple sugars and some polysaccharides to ethanol and carbon dioxide, in widely different proportion depending on the strains selected for specific purposes (Canonico, Comitini, & Ciani, 2014; Gonzalez-Perez & Alcalde, 2014; Marongiu et al., 2015). Recently, other yeasts that used to be considered spoilage or wild fermenters such as *Brettanomyces/Dekkera*, *Toluraspora*, and *Pichia* have begun to be used purposely in food fermentation (Tamang et al., 2016).

2.3 Safety Concerns

While food fermentation is generally expected to improve food safety, it is not an entirely risk-free process. The risks of contamination of fermented food increase when spontaneous fermentation are used instead of well-defined starter cultures (Hondrodimou, Kourkoutas, & Panagou, 2011; Panagou, Tassou, Vamvakoula, Saravanos, & Nychas, 2008), if the quality of the raw material was questionable and if deviation from the optimal fermentation conditions occur. The two main safety issues with fermented foods are potential pathogens development and production of toxic compounds.

There is ample evidence that standard concentrations of lactic acid and/or ethanol provide strong antimicrobial properties toward all known foodborne pathogens. However, in a study comparing low and normal alcohol beers, *Escherichia coli*, *Salmonella Typhimurium*, and *Listeria monocytogenes* were all found to grow into beer with ethanol levels below 4% (Menz, Vriesekoop, Zarei, Zhu, & Aldred, 2010). Similarly, deviations in salt concentration and temperature were found to enhance the survival of *Escherichia coli* and *Listeria monocytogenes* in comparison with traditionally produced sauerkraut, though they were not found in unsafe amounts in the final product (Niksic et al., 2005). Further, an experimentation made on fermented cauliflower reported the survival of *Listeria monocytogenes* and *Salmonella* Typhimurium after the fermentation and despite being in a normal pH and total titrable acidity levels (Paramithiotis, Doulgeraki, Tsilikidis, Nychas, & Drosinos, 2012).

Biogenic amines represent the main problematic metabolites in fermented foods. Members of the *Enterobacteriaceae* or lactic acid bacteria produce those compounds from free amino acids through the activity of decarboxylase enzymes to increase their resistance to acidic environment; and the main biogenic amines found in food are histamine, tyramine, putrescin, and cadaverine derived from the amino acids histidine, tyrosine, ornithine, and lysine respectively (Linares et al., 2016; Pessione & Cirrincione, 2016). While biogenic amines are metabolized by members of the gut microbiota, high intakes can lead to several detrimental health impact (Pugin et al., 2017). To limit the final content of biogenic amines in food, treatments such as heat or the use of specific starters checked for lack of decarboxylase activity can be used (Kung et al., 2017; Ozogul & Hamed, 2018). Initiating the fermentation of sauerkraut with *Lactobacillus plantarum*, *L. casei*, or *L. curvatus* reduced the total content of biogenic amines by seven folds after 45 days of storage compared to a spontaneous fermentation of sauerkraut (Rabie, Siliha, el-Saidy, el-Badawy, & Malcata, 2011).

3. FERMENTED VEGETABLES

3.1 Introduction

Fermented plant foods include fruits and vegetables, but fruits are often too high in sugars or acidic to be fermented other than by alcoholic fermentation; therefore, fruits are out of the scope of this chapter. Notable exception is olives and cocoa beans which require acidic fermentation for human consumption (Abriouel, Benomar, Lucas, & Galvez, 2011; D'Antuono et al., 2018; Illeghems et al., 2016; Montoro et al., 2016), but they will be referred as vegetables for the purpose of this chapter. The first cucumbers were fermented around 2000 BC in the Middle East, and the first reported cabbage fermentation was kimchi in Korea (Jung, Lee, & Jeon, 2014). The technology of fermenting cabbage was brought from China to Europe during the 13th century, where it was called the sauerkraut (Pfohl, 1983). Nowadays, the most popular fermented vegetables in Western countries are cucumber pickles, with a market of USD 2 billion in the United States, olives and sauerkraut, even if the consumption of the last one tends to decrease in the United States. However, sauerkraut and kimchi are the fermented vegetables for which the most microbiology and human health research has been performed (Kim, Yang, Kim, Lee, & Lee, 2018; Park, Jeong, Lee, & Daily, 2014; Patra, Das, Paramithiotis, & Shin, 2016; Raak et al., 2014), so they will be used as the main examples in this section.

3.2 Fermentation of Cabbage: Sauerkraut and Kimchi

Two species of cabbage are fermented worldwide. The most popular is *Brassica rapa* which is used to produce kimchi in Asia (Patra et al., 2016). In Europe and in the United States, sauerkraut results from the fermentation of *Brassica oleracea* (Peñas, Martinez-Villaluenga, & Frias, 2017). Before fermentation, the exterior leaves are removed to limit the contamination by undesirable microbiota, as well as the core that contains too much sucrose. Leaves are then shredded and placed into water with salt that prevents growth of undesirable microorganisms (Peñas et al., 2017). Fermentation is conducted at room temperature until the desirable level of titratable acidity is reached (at least 1% of lactic acid formed), usually after 2–3 weeks (Jung et al., 2014; Peñas et al., 2017). The sauerkraut can then be stored at refrigeration temperature before packaging and distribution for extended times, and sauerkraut/kimchi shelf life is typically between 6 and 12 months, and can reach up to 2 years if canned (Plengvidhya, Breidt, Lu, & Fleming, 2007).

Cabbage fermentation is most commonly conducted without starter cultures by taking advantage of the naturally occurring lactic acid bacteria (LAB) present on the cabbage (Jung et al., 2014; Peñas et al., 2017) which initiate a spontaneous fermentation. On fresh cabbage, the microbial population is greater on the exterior leaves and decreases as it gets closest to the core. Shredded cabbage contains about 5×10^6 cfu/g of total aerobes and less than 10^2 cfu/g of fungi (Plengvidhya, Breidt, & Fleming, 2004). This spontaneous fermentation most commonly consists of the succession of two periods. The first step (3–7 days) is the hetero-fermentative stage: the LAB ferment glucose into several products such as acids, alcohol, and carbon dioxide. Then, in the homo-fermentative step (10–14 days) other LAB ferment glucose into lactic acid only (Jeong et al., 2013; Jeong, Jung, Lee, Jin, & Jeon, 2013; Jung et al., 2012; Lee, Jung, & Jeon, 2015). The bacterial composition varies greatly from one step to the other (Lu, Breidt, Plengvidhya, & Fleming, 2003).

Sauerkraut/kimchi microbial ecology and succession were initially characterized through culture-based methods. These techniques were successful in identifying the primary LAB species responsible for the sauerkraut fermentation: *Leuconostoc mesenteroides* and *Lactobacillus brevis* for the heterofermentative step, and *Pediococcus cerevisiae* and *Lactobacillus plantarum* for the homo-fermentative step (Stamer, Stoyla, & Dunckel, 1971; Xiong, Guan, Song, Hao, & Xie, 2012). However, new culture-independent methods have allowed gaining deeper knowledge of the microbial diversity and successions during cabbage fermentation. A study with a wide array of kimchi samples demonstrated that salt concentration, fermentation duration, and manufacturing processes (home–made vs industrial) all contribute to microbiota profiles and successions (Lee, Song, Jung, Lee, & Chang,

2017). In particular, this and other studies revealed Weissella as a common bacterial genus involved in cabbage fermentation (Hong, Choi, Lee, Yang, & Lee, 2016; Kwak, Cho, Noh, & Om, 2014; Plengvidhya et al., 2007). Examples of microorganisms reported to be present in Kimchi, Sauerkraut, and other fermented vegetables are shown in Table 1.

3.3 Other Fermented Vegetables

When harvested for human consumption as table olives, these fruits need to be treated to reduce bitterness, a result that is almost exclusively achieved by fermentation (Coton, Coton, Levert, Casaregola, & Sohier, 2006; Garrido-Fernandez & Vaughn, 1978). In this case, the main objective of the fermentation is to remove the bitter phenolic oleuropein (Garrido-Fernandez & Vaughn, 1978). Different fermentation processes are carried out depending on local traditions and ripeness of the olives, but all processes include the use of salted brine where spontaneous lactic acid fermentation occurs (Argyri et al., 2013; Montoro et al., 2016). Not surprisingly, the main taxa of LAB described for sauerkraut are the dominant fermenters (Argyri et al., 2013; Ciafardini, Marsilio, Lanza, & Pozzi, 1994; Montoro et al., 2016; Zaragoza et al., 2017). However, some recent culture-independent studies have revealed the additional presence of halophilic and alkaliphilic bacteria (De Angelis et al., 2015; Lucena-Padros & Ruiz-Barba, 2016; Randazzo et al., 2017) and a diversity of yeasts (Leventdurur et al., 2016) in fermenting olives. In addition, there is a recent increased interest in identifying and developing LAB strains with improved oleuropein and phenolic degradation abilities (Bevilacqua et al., 2015; Johnson, Melliou, Zweigenbaum, & Mitchell, 2018; Kaltsa, Papaliaga, Papaioannou, & Kotzekidou, 2015; Pistarino et al., 2013; Tataridou & Kotzekidou, 2015; Zago et al., 2013).

Pickling is a vernacular term that encompasses fermentation or immersion in vinegar of a variety of foods (vegetables, but also fruits, eggs, and meat) to extend their shelf life. Again, microbiologically driven pickling is very similar to sauerkraut fermentation with use of salted brine and lactic acid bacteria performing the acidic fermentation (Perez-Diaz et al., 2017; Perez-Diaz & McFeeters, 2011). Small cucumbers are the most popular pickled vegetables worldwide (and are hence commonly referred as "pickles"). Cucumber fermentation is conducted by, in order of prevalence, *Lactobacillus pentosus*, *Lb. plantarum*, *Lb. brevis*, *Weissella* spp., *Pediococcus ethanolidurans*, *Leuconostoc* spp., and *Lactococcus* spp. (Perez-Diaz et al., 2017). Many other pickled vegetables have been studied for their bacterial

Table 1 Microorganisms Present in Kimchi, Sauerkraut, and Other Fermented Vegetables

	Microorganisms	References
Kimchi Sauerkraut	Lactic acid bacteria (LAB)	Xiong et al. (2012) and Stamer et al. (1971)
	Leuconostoc mesenteroides	
	Lactobacillus brevis	
	Pediococcus cerevisiae	
	Lactobacillus plantarum	
	Weissella spp.	Kwak et al. (2014), Hong et al. (2016), and Plengvidhya et al. (2007)
Olives	*Halophilic* and *Alkaliphilic* bacteria	De Angelis et al. (2015), Lucena-Padros and Ruiz-Barba (2016), and Randazzo et al. (2017)
	Yeasts	Leventdurur et al. (2016)
Cucumber	*Lactobacillus pentosus*	Perez-Diaz et al. (2017)
	Lb. plantarum	
	Lb. brevis	
	Weissella spp.	
	Pediococcus ethanolidurans	
	Leuconostoc spp.	
	Lactococcus spp.	
Fermented peppers	*Lactobacillus* spp.	Gonzalez-Quijano et al. (2014)
	Leuconostoc citreum	
	Weissella cibaria	
	Lactobacillus plantarum	
	Lactobacillus paraplantarum	
	Hanseniaspora pseudoguilliermondii	
	Kodamaea ohmeri	
Soy sauce	*Bacillus, Aspergillus, Klebsiella, Cladosporium,* and *Shimwellia*	

Table 1 Microorganisms Present in Kimchi, Sauerkraut, and Other Fermented Vegetables—cont'd

	Microorganisms	References
Tofu and Tempeh	*Bacillus* or LAB species	Yang et al. (2017)
	Lactic Acid Bacteria	Lee et al. (2017) and Nam, Lee, and Lim (2012)
Cocoa	Firmicutes, acetic acid bacteria, and yeast	Illeghems et al. (2016), Camu et al. (2007), Visintin et al. (2017), Ouattara et al. (2017), Mahazar, Zakuan, Norhayati, MeorHussin, and Rukayadi (2017), Samagaci, Ouattara, Niamke, and Lemaire (2016), and Ho, Fleet, and Zhao (2018)

composition (Bao et al., 2016; Liu, Kuda, Takahashi, & Kimura, 2018; Liu et al., 2011; Ono et al., 2014; Suzuki, Honda, Suganuma, Saito, & Yajima, 2014; Tang, Hu, Wang, Wang, & Wang, 2016; Yu et al., 2012), with similar LAB profiles but also detection of yeasts. Fermented peppers (*Capsicum annuum*) require significant levels of acetic acid, which potentially explains differences in microbial profiles (Gonzalez-Quijano et al., 2014). While *Lactobacillus* species were predominant during the fermentation, *Leuconostoc citreum* and *Weissella cibaria* were present at the beginning of the fermentation process and later in the fermentation, *Lactobacillus plantarum* and *Lactobacillus paraplantarum* became predominant. In addition, several yeasts were also identified such as *Hanseniaspora pseudoguilliermondii* and *Kodamaea ohmeri* (Gonzalez-Quijano et al., 2014). Soybeans are traditionally fermented to soy sauce, tofu, and tempeh. It has been shown that long-term fermentation of soy sauce involved uncommon microorganisms such as *Bacillus*, *Aspergillus*, *Klebsiella*, *Cladosporium*, and *Shimwellia* (Yang et al., 2017). Tofu and tempeh fermentation may involve primarily *Bacillus* or LAB species (Lee, Li, et al., 2017; Nam et al., 2012).

Cocoa refers to the dried and fermented seed of *Theobroma cacao* that are used to prepare different chocolate presentations (Ho et al., 2018). Cocoa fermentation is almost exclusively spontaneous (Camu et al., 2007), which is suspected to lead to large variation in quality and organoleptic properties based on local climate and environmental factors (De Vuyst & Weckx, 2016; Visintin, Alessandria, Valente, Dolci, & Cocolin, 2016). Cocoa fermentation involves a wide variety of LAB and other Firmicutes, acetic acid bacteria, and yeast (Camu et al., 2007; Ho et al., 2018; Illeghems et al., 2016; Mahazar et al., 2017; Ouattara, Reverchon, Niamke, & Nasser, 2017; Samagaci et al., 2016; Visintin et al., 2016) and these fermenting microbiota appear to be associated to geography (Bortolini, Patrone, Puglisi, & Morelli, 2016). There have been studies to test potential starter cultures to obtain more homogeneous results (De Vuyst & Weckx, 2016; Mahazar et al., 2017; Papalexandratou & Nielsen, 2016; Visintin et al., 2017).

3.4 The Consumption of Fermented Vegetables and Effects on Health

Fermented vegetables are believed to have a positive impact on health. Kimchi and sauerkraut are probably the most popular fermented vegetable when these come to purported health beneficial properties (Raak et al., 2014). Studies have reported that regular consumption of kimchi is associated with lower obesity incidence (Kim, Lee, & Jung, 2012), and potential direct anti-obesogenic effects have been described using rodents and *in vitro* models (Lee et al., 2015; Park et al., 2012) and obese human subjects (Kim et al., 2011). Potential antidiabetic properties were also reported with improvements of certain parameters in a rat (Islam & Choi, 2009) and a human study (An et al., 2013). Epidemiological studies in Asian countries' human cohorts have reported kimchi consumption as the most significant parameter negatively correlated with cancer risk (Patra et al., 2016; Wie et al., 2017). Further, kimchi was shown to prevent *Helicobacter pylori* induced cancer symptoms and inflammatory response in mice (Jeong et al., 2015); and kimchi starter cultures, especially *Lactobacillus plantarum* and *Weissella cibaria* strains, have been suggested to incur cancer preventive properties (Kwak et al., 2014; Lee, Kim, Lee, & Park, 2016). Kimchi has also been reported to incur potential beneficial effects toward atherosclerosis and atopic dermatitis, but only based on *in vitro* or epidemiology studies (Kim, Noh, & Song, 2018; Park & Bae, 2016). Sauerkraut and kimchi have also been reported to have remarkable antioxidant, anti-inflammatory, and immunomodulatory properties (Choi et al., 2017; Jeong et al., 2015; Kusznierewicz, Smiechowska, Bartoszek, & Namiesnik, 2008; Lee, Kim, Lee, Jang, & Choue, 2014; Lee, Song, Jang, & Han, 2014; Park, Joe, Rhee, Jeong, & Jeong, 2017; van Dijk et al., 2000).

Only a few studies on health effects have been conducted on other fermented vegetables. In a rat study, fermented soybeans consumption lead to increased insulinotropic response compared to unfermented soybeans

(Kwon et al., 2007). Similarly, fermented buckwheat sprouts were found to incur significantly increased blood pressure level reduction in hypertensive rats models (Nakamura, Naramoto, & Koyama, 2013), a response that was explained by the presence of specific antihypertensive peptides (Koyama, Hattori, Amano, Watanabe, & Nakamura, 2014; Koyama et al., 2013). Nukazuke, newly developed Japanese pickles, was found to incur similar response in hypertensive rats (Oda, Imanishi, Yamane, Ueno, & Mori, 2014; Oda, Nagai, Ueno, & Mori, 2015).

4. PLANT-BASED FERMENTED DRINKS
4.1 Introduction

While beer and wine are among the oldest known fermented foods (Rodhouse & Carbonero, 2017), fermented tea and coffee quickly joined their ranks. Fermented tea (kombucha) consumption was first reported in 220 BC from northeast China and then expanded to Korea and Japan (Greenwalt et al., 2000). With the expansion of commercial trades between countries, kombucha became popular in Russia, and later in Germany. After World War II, its consumption spread to the rest of Western Europe (Rasu, Malbaša, Lončar, Vitas, & Muthuswamy, 2014). Today, kombucha is particularly popular in the United States and other western countries partly because of its unverified reputation to strengthen the immune system (Greenwalt et al., 2000; Holbourn & Hurdman, 2017). The development of coffee as a popular drink worldwide required the development of appropriate processes, including the crucial initial step that was later identified as a fermentation step (Silva, Schwan, Sousa Dias, & Wheals, 2000). In this chapter, those two popular drinks are given as the main examples.

4.2 Kombucha

Kombucha differs strikingly with the other fermented foods discussed in this chapter in that it is often home-brewed, and industrial fermentation is not as common. Kombucha is essentially a fermented sweet tea, with very little amounts of alcohol and presence of carbon dioxide, but fermentation processes are very variable (Rasu et al., 2014; Villarreal-Soto, Beaufort, Bouajila, Souchard, & Taillandier, 2018). The common denominator of all kombucha production is the use of a SCOBY (Symbiotic Colony of Bacteria and Yeast) that ferments the tea (Rasu et al., 2014), which is recycled and re-used several times. The fermentation is conducted during 6–14 days under aerobic conditions and at room temperature. In modern iterations, a secondary fermentation with fruit juice or other sugary liquid is often conducted in the final container to obtain stronger taste and increased effervescence.

Kombucha is the result of a semi-spontaneous fermentation of the tea by yeasts and bacteria, and the SCOBY quality is crucial in obtaining a safe drink with desired quality. Unlike wine or beer, pure starter cultures which could improve consistency in fermentation process have not been identified. Therefore, Kombucha microbiota composition and successions are influenced by multiple factors such as environment, vessel used, fermentation parameters; but the main driver remains the SCOBY composition (Coton et al., 2017). When a secondary composition is conducted, only microorganisms present in the fermented tea are present, but the acidification and carbon dioxide production prove that both acetic bacteria and yeasts are involved in the secondary fermentation.

Interestingly, the microbial quality of kombucha had been very sparsely studied by culture-dependent methods, where mainly acetobacteraceae members and yeast were detected (Chen & Liu, 2000; Greenwalt et al., 2000; Teoh, Heard, & Cox, 2004). The first culture-independent study on kombucha revealed a higher diversity compared to other fermented foods with five abundant bacterial phyla: *Actinobacteria*, *Bacteroidetes*, *Deinococcus-Thermus*, *Firmicutes*, and *Proteobacteria*, with variations according to the geographic origin of the samples (Marsh, O'Sullivan, Hill, Ross, & Cotter, 2014). Subsequent studies reported either dominance of LAB or acetic acid bacteria among the bacteria (Chakravorty et al., 2016; Coton et al., 2017; De Filippis, Troise, Vitaglione, & Ercolini, 2018; Podolich et al., 2017). Yeast and fungal profiles appear to be extremely variable, with dominance of *Zygosaccharomyces* in one case (Marsh et al., 2014), other *Saccharomycetaceae* (Chakravorty et al., 2016), and *Dekkera* and *Hanseniaspora* (Coton et al., 2017). Microorganisms present in Kombucha are shown in Table 2.

4.3 Other Plant-Based Fermented Drinks

Coffee is a widely consumed drink made from beans of several species in the *Coffea* genus. While fermentation of the beans is required to remove the mucilage layer, the next heat-intensive steps (roasting and brewing) result

Table 2 Microorganisms Present in Kombucha

Microorganisms	Mostly Involved	References
Bacteria	*Actinobacteria*	Marsh et al. (2014)
	Bacteroidetes	
	Deinococcus-Thermus	
	Firmicutes	
	Proteobacteria	
	Lactic acid bacteria (LAB)	Chakravorty et al. (2016) and Coton et al. (2017)
	Acetic acid bacteria	De Filippis et al. (2018) and Podolich et al. (2017)
Fungi and yeasts	*Zygosaccharomyces*	Coton et al. (2017)
	Saccharomycetaceae	
	Dekkera	
	Hanseniaspora	

in a drink that is virtually sterile (Waters, Arendt, & Moroni, 2017). Since fermentation is conducted for the specific purpose of mucilage removal, spontaneous fermentation is almost universally used (Waters et al., 2017). While it is suspected that fermentation affects the final quality products, there has been relatively little research in determining the microbial composition and successions (De Bruyn et al., 2016; Ludlow et al., 2016). It has been reported that wet fermentation is conducted by LAB primarily but also Enterobacteriaceae, and yeast while dry fermentation is driven by acetic acid bacteria and *Pichia* yeasts (De Bruyn et al., 2016). Since the mid-1900s, numerous species of microorganisms have been isolated from the fermentation phase of wet processing (Agate & Bhat, 1966; Avallone, Guyot, Brillouet, Olguin, & Guiraud, 2001; de Melo Pereira et al., 2014; Masoud, Cesar, Jespersen, & Jakobsen, 2004; Masoud & Kaltoft, 2006); these microorganisms are reported in Table 3.

Many traditional drinks are obtained by plant fermentation worldwide. Chicha is made from maize and other cereals, nuts, and fruits in South and Central America (Freire, Zapata, Mosquera, Mejia, & Trueba, 2016). Chicha's production often includes human mastication and maintenance of worst outdoor in uncleaned containers. As a result, culture-independent

Table 3 Microorganisms in Coffee

Type	Mostly Involved	References
Wet fermentation	LAB, Enterobacteriaceae, and Yeast	De Bruyn et al. (2016)
	Klebsiella ozaenae, K. oxytoca, Erwinia herbicola, E. dissolvens Hafnia spp., *Enterobacter aerogenes, Leuconostoc mesenteroides, Lactobacillus brevis Kloeckera apis apiculata, Candida guilliermondii, C. tropicalis, C. parapsilosis, Cryptococcus albidus, C. laurentii, Pichia kluyveri, P. anomala, Hanseniaspora uvarum, Saccharomyces cerevisiae, Debaryomyces hansenii, Torulaspora delbrueckii,* and *Rhodotorula mucilaginosa*	Agate and Bhat (1966), Avallone et al. (2001), Avallone, Brillouet, Guyot, Olguin, and Guiraud (2002), de Melo Pereira et al. (2014), and Vilela, Pereira, Silva, Batista, and Schwan (2010)
	Aspergillus, Penicillium, and *Fusarium* species	Silva et al. (2000)
Dry fermentation	acetic acid bacteria and *Pichia* yeasts	De Bruyn et al. (2016)

analyses have demonstrated that oral bacteria (Elizaquivel et al., 2015; Freire et al., 2016; Puerari, Magalhães-Guedes, & Schwan, 2015; Resende et al., 2018) and environmental yeasts (Mendoza, Neef, Vignolo, & Belloch, 2017; Rodriguez et al., 2017) are involved in chicha's fermentation. Similarly, cassava–based fermented drinks from South America and Africa were shown to primarily harbor LAB and yeasts (Almeida, Rachid, & Schwan, 2007; Colehour et al., 2014; Crispim et al., 2013; Ramos et al., 2010, 2015; Wilfrid Padonou et al., 2009).

4.4 Health Properties of Plant-Based Fermented Drinks

Kombucha is known for its numerous, sometimes outrageous, health claims including anything from its detoxifying power, reducing blood pressure, increasing the resistance to cancer and even reducing the appearance of gray hair. Of course, the majority of these claims are completely unsubstantiated, and it is therefore important to distinguish kombucha's demonstrated potential health-promoting properties (Holbourn & Hurdman, 2017; Rasu et al., 2014; Vina, Semjonovs, Linde, & Denina, 2014).

There are substantial evidences that fermentation increases kombucha antioxidant potential (Gamboa-Gomez et al., 2016, 2017; Gharib, 2009; Sun, Li, & Chen, 2015; Velicanski, Cvetkovic, Markov, Saponjac, & Vulic, 2014), which is hypothesized to be the result of microbial metabolism of tea polyphenols (Chakravorty et al., 2016). This antioxidant potential has been suggested to incur health benefits, but only in *in vitro* studies (Bhattacharya, Manna, Gachhui, & Sil, 2011; Gamboa-Gomez et al., 2016, 2017; Vazquez-Cabral et al., 2017) or animal models (Aloulou et al., 2012; Banerjee et al., 2010; Bellassoued et al., 2015; Bhattacharya, Gachhui, & Sil, 2013; Gamboa-Gomez et al., 2017; Gharib, 2009; Hartmann, Burleson, Holmes, & Geist, 2000; Pakravan et al., 2017; Salafzoon, Mahmoodzadeh Hosseini, & Halabian, 2017). Antimicrobial properties of kombucha are also well documented toward several known human bacterial and fungal pathogens (Bhattacharya et al., 2016, 2018; Mahmoudi et al., 2016; Sknepnek et al., 2018; Velicanski et al., 2014), and intriguingly toward some viruses (Fu, Wu, Lv, He, & Jiang, 2015). However, it is not known to what extent antimicrobial properties translate *in vivo* in the human gastro-intestinal tract. Interestingly, isolated cases of adverse reaction due to excessive consumption, or consumption of compromised kombucha, have also been reported (Gedela, Potu, Gali, Alyamany, & Jha, 2016; SungHee Kole, Jones, Christensen, & Gladstein, 2009). Studies on chicha were motivated by purported health benefits (Freire et al., 2016), however no report on such properties is found in the literature.

While the impact of coffee on human health has been recently reevaluated, and now considered as neutral or potentially beneficial (Berretta et al., 2018; Carlstrom & Larsson, 2018; Xie et al., 2018), any potential benefits have been associated with its bioactive compounds (polyphenols and melanoidins) content (Godos et al., 2014; Lopez-Barrera, Vazquez-Sanchez, Loarca-Pina, & Campos-Vega, 2016; Moreira, Nunes, Domingues, & Coimbra, 2012) as reported in Table 4. Whether fermentation could be used as a mean to optimize coffee's bioactives profiles is unclear (De Bruyn et al., 2016). There could be a risk of mycotoxins contamination from fungi contamination during fermentation, but a study found very low levels in processed coffee beans (Jeszka-Skowron, Zgola-Grzeskowiak, Waskiewicz, Stepien, & Stanisz, 2017).

Table 4 Reported Potential Health Benefits From Coffee Consumption
References

Antioxidant activity	Martini et al. (2016)
Antimutagenic effects	Nehlig and Debry (1994)
Type 2 diabetes	Akash, Rehman, and Chen (2014)
Cardiovascular diseases	Godos et al. (2014)
Parkinson's disease	Scheperjans, Pekkonen, Kaakkola, and Auvinen (2015)
Bladder cancer	Demirel, Cakan, Yalcinkaya, Topcuoglu, and Altug (2008) and Villanueva et al. (2009)
Breast cancer	Ganmaa et al. (2008) and Tang, Zhou, Wang, and Yu (2009)
Gastro-intestinal cancers	Larsson, Giovannucci, and Wolk (2006), Larsson, Bergkvist, Giovannucci, and Wolk (2006), and Botelho, Lunet, and Barros (2006)
Other cancers	Zvrko, Gledovic, and Ljaljevic (2008), Tang, Wu, Ma, Wang, and Yu (2010), Wigle, Turner, Gomes, and Parent (2008), and Larsson, Giovannucci, and Wolk (2017)

5. POTENTIAL MECHANISMS INVOLVED IN HEALTH BENEFITS

5.1 Introduction

In 2015, Chilton et al. suggested that fermented foods should be part of national nutritional guides. However, there are few examples of national nutritional guidelines explicitly citing fermented foods as a category. One explanation could be the recent industrialization of fermented foods, which used to be home-made and therefore potentially unsafe (Chilton et al., 2015). The consumption of fermented foods has been claimed to exert many positive effects on health, but these claims often stem from *in vitro* or animal studies, thereby reducing the relevance to human health. These properties are considered to vary according to the food type and the type of microorganisms responsible for the fermentation. A general overview of the potential health benefits associated with fermented foods consumption, which have been reviewed extensively is presented (Marco et al., 2017).

Since many fermented foods are known to harbor an abundant and viable microbiota (Lang, Eisen, & Zivkovic, 2014; Marco et al., 2017), they are often branded as probiotics, even if data supporting such claims is often conflicting or incomplete. It is commonly accepted that a significant number of viable bacteria from fermented foods can reach the lower intestinal tract, but the extent to which those potential probiotics are able to colonize and outcompete the resident microbiota is still unclear (Derrien & van Hylckama Vlieg, 2015; Plé, Breton, Daniel, & Foligné, 2015). Fermentative LAB and propionic acid producers, in particular strains of *Lactobacillus*, have long been selected for their potential health benefits based on *in vitro* or animal studies (Choi et al., 2017; Corr et al., 2007; Ple et al., 2016; Vasiee, Alizadeh Behbahani, Tabatabaei Yazdi, Mortazavi, & Noorbakhsh, 2017), but there is still limited evidence to support translation of such observations to consistent human health modulation (Kolmeder et al., 2016; Zhang et al., 2016).

The other main mechanism by which fermented foods are considered to be potentially beneficial to health is by the altered nutritive and bioactive profiles through enzymatic and microbial activities. Suppression of biogenic amines and phytic acid are beneficial properties as described earlier. Moreover, sourdough fermentation can reduce the amount of fermentable oligosaccharides, disaccharides, monosaccharides, and polyols (FODMAPS), which may in turn incur improved digestive health (Laatikainen et al., 2016). Yeast-fermented foods are known to be enriched in beneficial macro- and micronutrients derived from dead cells (Rodhouse & Carbonero, 2017). Native plant bioactive phytochemicals are also known to be converted to a variety of potentially more beneficial metabolites, which will be described in more details later in this chapter.

Epidemiological studies have reported positive association between consumption of fermented dairy foods and type 2 diabetes risk (Brouwer-Brolsma, Sluik, Singh-Povel, & Feskens, 2018; Gijsbers et al., 2016; Salas-Salvado, Guasch-Ferre, Diaz-Lopez, & Babio, 2017). Other health conditions for which a beneficial impact of fermented foods have been reported include hypertension, osteoporosis, and irritable bowel syndrome (Fekete, Givens, & Lovegrove, 2015; Laatikainen et al., 2016; Lim et al., 2015; Tu et al., 2015). Fermented dairy products have also been suggested to have potential anti-cancer properties, but only based on *in vitro* studies (Ghoneum & Gimzewski, 2014; Khoury et al., 2014).

5.2 Impact on the Gut Microbiota

The gut microbiota is a term used to refer as the collection of microorganisms residing in the human gastro-intestinal tract, particularly in the colon, and has been the subject of numerous studies in the last 2 decades (Candela et al., 2014; Carbonero, Benefiel, Alizadeh-Ghamsari, & Gaskins, 2012; Carbonero, Benefiel, & Gaskins, 2012; Everard & Cani, 2013; Faith et al., 2013; Flint, 2011; Sheflin, Melby, Carbonero, & Weir, 2017; Turnbaugh et al., 2007). The adult human gut microbiota has been shown to be rather stable (Faith et al., 2013; Martinez, Muller, & Walter, 2013; Turroni et al., 2017) and only drastic dietary changes (O'Keefe et al., 2015; Ou et al., 2013; Scott, Gratz, Sheridan, Flint, & Duncan, 2013; Sheflin et al., 2017), significant intrinsic challenges (Gibson, Crofts, & Dantas, 2015; Maurice, Haiser, & Turnbaugh, 2013), or health conditions (Chow & Mazmanian, 2010; Davis-Richardson & Triplett, 2015; Kinross, Darzi, & Nicholson, 2011; Liou et al., 2013; Turnbaugh et al., 2009) are known to induce significant modulation. Pre- and probiotics are popular food additives that targets the gut microbiota (Davis, Martinez, Walter, Goin, & Hutkins, 2011; Martinez, Kim, Duffy, Schlegel, & Walter, 2010; Martinez et al., 2013), and the open question is whether fermented foods do exhibit pre- and/or probiotic properties (Marco et al., 2017).

Fermented dairy products have been the preferred model to test the potential probiotic properties of fermented foods, with conflicting findings in the ability of foodborne probiotics to establish long-term stable populations in the colon (Derrien & van Hylckama Vlieg, 2015; Hill et al., 2017; Shiby & Mishra, 2013). In general, plant-based fermented foods are branded as probiotic on the basis that microorganisms supposedly probiotic have been isolated or detected in the fermentation process or the final product (Argyri et al., 2013; Bansal, Mangal, Sharma, & Gupta, 2016; Duangjitcharoen, Kantachote, Ongsakul, Poosaran, & Chaiyasut, 2008; Greppi et al., 2017; Hatoum et al., 2012; Yu et al., 2013). Only in some cases have these potential probiotic properties been partially confirmed through *in vitro* (Arasu & Al-Dhabi, 2017; Argyri et al., 2013; Yu et al., 2013) or animal studies (Jo et al., 2016; Li et al., 2016; Yu et al., 2013); but no human study has been performed.

In theory, microbial fermentation of plant phytochemicals should result in the production of a diversity of metabolites, thereby increasing the chances for prebiotic compounds to be present. Therefore, a new criterion for starter cultures strains could be their ability to release significant amounts of prebiotic. For example, specific strains of Leuconostoc sugars were successfully included in kimchi fermentation to slightly increase the concentration of bifidogenic isomaltooligosaccharides (Cho et al., 2014; Eom, Seo, & Han, 2007). LAB-derived exopolysaccharides (Caggianiello, Kleerebezem, & Spano, 2016; Korcz, Kerenyi, & Varga, 2018; Lynch, Zannini, Coffey, & Arendt, 2018; Salazar, Gueimonde, de Los Reyes-Gavilan, & Ruas-Madiedo, 2016; Suzuki et al., 2014; Zeidan et al., 2017) or yeast cell walls (More & Vandenplas, 2018) are other classes of compounds with potential prebiotic properties that could be preferentially produced in fermented plant products.

5.3 Other Potential Mechanisms

Metabolites present in fermented foods, including dairy (Zheng et al., 2015; Zhu, Wang, Hollis, & Jacques, 2015), sourdough bread (Bondia-Pons et al., 2011), wine (Regueiro, Vallverdu-Queralt, Simal-Gandara, Estruch, & Lamuela-Raventos, 2013), and beer (Gurdeniz et al., 2016; Sandoval-Ramirez et al., 2017) have been shown to be directly transferred to the human metabolome. Remarkably, while metabolomics of the plant-based fermented food covered in this chapter have been studied to some extent, and shown to be uniquely distinct from these more commonly fermented foods (De Angelis et al., 2015; Kim, Jung, Bong, Lee, & Hwang, 2012; Randazzo et al., 2017; Roda et al., 2017; Tomita, Saito, Nakamura, Sekiyama, & Kikuchi, 2017), there has been no documented attempt at determining the fate of these metabolites neither in animal nor in human studies.

Fermented dairy is often associated with somewhat misleading claims of improved immune system modulation. While there has been extensive research on the direct impact of probiotic strains from fermented dairy on the immune system (Bumgardner et al., 2018; Khan et al., 2012; Walter, Britton, & Roos, 2011), it is difficult to conclude if the effects are really beneficial. One study reported potential anti-inflammatory action of a kimchi-derived Lactobacillus, but under extremely specific conditions that are unlikely to be translatable to human health (Choi et al., 2017). Further, a randomized controlled dietary intervention study in health young adults showed no significant impact on immune system biomarkers from short-term kimchi consumption (Lee, Kim, et al., 2014).

6. CONCLUSION

Plant-based fermented foods outside of bread, beer, and wine represent a significant and increasing portion of human dietary intakes; and harbor higher numbers of viable microorganisms. Fermentation is primarily conducted by acetic and lactic acid bacteria and yeasts, but other taxa can be found and microbial composition and dynamics vary considerably depending on the type of food and fermentation processes.

The common belief that plant-based fermented foods are healthy still requires extensive research to be scientifically substantiated. Still, limited evidences do exist in favor of beneficial effects of fermented foods consumption, and their general microbiological safety is also an argument for recommendation in dietary guidelines. The extent to which microorganisms involved in fermentation can establish themselves in the human colon is open for debate, thus probiotics claims are still impossible to substantiate. However, the majority of the few studies on the impact of plant-based fermented foods point to neutral or potentially positive impact, and these foods are typically safe, thus considering them as part of a balanced diet appears reasonable.

REFERENCES

Abriouel, H., Benomar, N., Lucas, R., & Galvez, A. (2011). Culture-independent study of the diversity of microbial populations in brines during fermentation of naturally-fermented Alorena green table olives. *International Journal of Food Microbiology*, *144*, 487–496.

Agate, A. D., & Bhat, J. V. (1966). Role of pectinolytic yeasts in the degradation of mucilage layer of Coffea robusta cherries. *Applied Microbiology*, *14*, 256–260.

Akash, M. S., Rehman, K., & Chen, S. (2014). Effects of coffee on type 2 diabetes mellitus. *Nutrition*, *30*, 755–763.

Akissoe, N. H., Sacca, C., Declemy, A. L., Bechoff, A., Anihouvi, V. B., Dalode, G., et al. (2015). Cross-cultural acceptance of a traditional yoghurt-like product made from fermented cereal. *Journal of the Science of Food and Agriculture*, *95*, 1876–1884.

Almeida, E. G., Rachid, C. C. T. C., & Schwan, R. F. (2007). Microbial population present in fermented beverage 'cauim' produced by Brazilian Amerindians. *International Journal of Food Microbiology*, *120*, 146–151.

Aloulou, A., Hamden, K., Elloumi, D., Ali, M. B., Hargafi, K., Jaouadi, B., et al. (2012). Hypoglycemic and antilipidemic properties of kombucha tea in alloxan-induced diabetic rats. *BMC Complementary and Alternative Medicine*, *12*, 63. https://doi.org/10.1186/1472-6882-12-63.

Altay, F., Karbancioglu-Guler, F., Daskaya-Dikmen, C., & Heperkan, D. (2013). A review on traditional Turkish fermented non-alcoholic beverages: Microbiota, fermentation process and quality characteristics. *International Journal of Food Microbiology*, *167*, 44–56.

An, S. Y., Lee, M. S., Jeon, J. Y., Ha, E. S., Kim, T. H., Yoon, J. Y., et al. (2013). Beneficial effects of fresh and fermented kimchi in prediabetic individuals. *Annals of Nutrition & Metabolism*, *63*, 111–119.

Arasu, M. V., & Al-Dhabi, N. A. (2017). In vitro antifungal, probiotic, and antioxidant functional of a novel *Lactobacillus paraplantarum* isolated from fermented dates in Saudi Arabia. *Journal of the Science of Food and Agriculture*, *97*, 5287–5295.

Argyri, A. A., Zoumpopoulou, G., Karatzas, K. A., Tsakalidou, E., Nychas, G. J., Panagou, E. Z., et al. (2013). Selection of potential probiotic lactic acid bacteria from fermented olives by in vitro tests. *Food Microbiology*, *33*, 282–291.

Arroyo-Lopez, F. N., Perez-Traves, L., Querol, A., & Barrio, E. (2011). Exclusion of *Saccharomyces kudriavzevii* from a wine model system mediated by *Saccharomyces cerevisiae*. *Yeast*, *28*, 423–435.

Aryana, K. J., & Olson, D. W. (2017). A 100-year review: Yogurt and other cultured dairy products. *Journal of Dairy Science*, *100*, 9987–10013.

Avallone, S., Brillouet, J. M., Guyot, B., Olguin, E., & Guiraud, J. P. (2002). Involvement of pectolytic micro-organisms in coffee fermentation. *International Journal of Food Science and Technology*, *37*, 191–198.

Avallone, S., Guyot, B., Brillouet, J. M., Olguin, E., & Guiraud, J. P. (2001). Microbiological and biochemical study of coffee fermentation. *Current Microbiology*, *42*, 252–256.

Bali, V., Panesar, P. S., Bera, M. B., & Kennedy, J. F. (2016). Bacteriocins: Recent trends and potential applications. *Critical Reviews in Food Science and Nutrition*, *56*, 817–834.

Banerjee, D., Hassarajani, S. A., Maity, B., Narayan, G., Bandyopadhyay, S. K., & Chattopadhyay, S. (2010). Comparative healing property of kombucha tea and black tea against indomethacin-induced gastric ulceration in mice: Possible mechanism of action. *Food & Function*, *1*, 284–293.

Bansal, S., Mangal, M., Sharma, S. K., & Gupta, R. K. (2016). Non-dairy based probiotics: A healthy treat for intestine. *Critical Reviews in Food Science and Nutrition*, *56*, 1856–1867.

Bao, Q., Song, Y., Xu, H., Yu, J., Zhang, W., Menghe, B., et al. (2016). Multilocus sequence typing of *Lactobacillus casei* isolates from naturally fermented foods in China and Mongolia. *Journal of Dairy Science*, *99*, 5202–5213.

Bellassoued, K., Ghrab, F., Makni-Ayadi, F., Van Pelt, J., Elfeki, A., & Ammar, E. (2015). Protective effect of kombucha on rats fed a hypercholesterolemic diet is mediated by its antioxidant activity. *Pharmaceutical Biology*, *53*, 1699–1709.

Berretta, M., Micek, A., Lafranconi, A., Rossetti, S., Di Francia, R., De Paoli, P., et al. (2018). Coffee consumption is not associated with ovarian cancer risk: A dose-response meta-analysis of prospective cohort studies. *Oncotarget*, *9*, 20807–20815.

Bevilacqua, A., de Stefano, F., Augello, S., Pignatiello, S., Sinigaglia, M., & Corbo, M. R. (2015). Biotechnological innovations for table olives. *International Journal of Food Sciences and Nutrition*, *66*, 127–131.

Bhattacharya, D., Bhattacharya, S., Patra, M. M., Chakravorty, S., Sarkar, S., Chakraborty, W., et al. (2016). Antibacterial activity of polyphenolic fraction of kombucha against enteric bacterial pathogens. *Current Microbiology*, *73*, 885–896.

Bhattacharya, S., Gachhui, R., & Sil, P. C. (2013). Effect of Kombucha, a fermented black tea in attenuating oxidative stress mediated tissue damage in alloxan induced diabetic rats. *Food and Chemical Toxicology*, *60*, 328–340.

Bhattacharya, D., Ghosh, D., Bhattacharya, S., Sarkar, S., Karmakar, P., Koley, H., et al. (2018). Antibacterial activity of polyphenolic fraction of Kombucha against *Vibrio cholerae*: Targeting cell membrane. *Letters in Applied Microbiology*, *66*, 145–152.

Bhattacharya, S., Manna, P., Gachhui, R., & Sil, P. C. (2011). Protective effect of kombucha tea against tertiary butyl hydroperoxide induced cytotoxicity and cell death in murine hepatocytes. *Indian Journal of Experimental Biology*, *49*, 511–524.

Blanco, C. A., Andres-Iglesias, C., & Montero, O. (2016). Low-alcohol beers: Flavor compounds, defects, and improvement strategies. *Critical Reviews in Food Science and Nutrition*, *56*, 1379–1388.

Bondia-Pons, I., Nordlund, E., Mattila, I., Katina, K., Aura, A. M., Kolehmainen, M., et al. (2011). Postprandial differences in the plasma metabolome of healthy Finnish subjects after intake of a sourdough fermented endosperm rye bread versus white wheat bread. *Nutrition Journal*, *10*, 116. https://doi.org/10.1186/1475-2891-10-116.

Bortolini, C., Patrone, V., Puglisi, E., & Morelli, L. (2016). Detailed analyses of the bacterial populations in processed cocoa beans of different geographic origin, subject to varied fermentation conditions. *International Journal of Food Microbiology*, *236*, 98–106.

Botelho, F., Lunet, N., & Barros, H. (2006). Coffee and gastric cancer: Systematic review and meta-analysis. *Cadernos de Saúde Pública*, *22*, 889–900.

Brouwer-Brolsma, E. M., Sluik, D., Singh-Povel, C. M., & Feskens, E. J. M. (2018). Dairy product consumption is associated with pre-diabetes and newly diagnosed type 2 diabetes in the Lifelines Cohort Study. *The British Journal of Nutrition*, *119*, 442–455.

Bumgardner, S. A., Zhang, L., LaVoy, A. S., Andre, B., Frank, C. B., Kajikawa, A., et al. (2018). Nod2 is required for antigen-specific humoral responses against antigens orally delivered using a recombinant Lactobacillus vaccine platform. *PLoS One*, *13*, e0196950.

Caggianiello, G., Kleerebezem, M., & Spano, G. (2016). Exopolysaccharides produced by lactic acid bacteria: From health-promoting benefits to stress tolerance mechanisms. *Applied Microbiology and Biotechnology*, *100*, 3877–3886.

Camu, N., Winter, T. D., Verbrugghe, K., Cleenwerck, I., Vandamme, P., Takrama, J. S., et al. (2007). Dynamics and biodiversity of populations of lactic acid bacteria and acetic acid bacteria involved in spontaneous heap fermentation of cocoa beans in Ghana. *Applied and Environmental Microbiology*, *73*, 1809–1824.

Candela, M., Turroni, S., Biagi, E., Carbonero, F., Rampelli, S., Fiorentini, C., et al. (2014). Inflammation and colorectal cancer, when microbiota-host mutualism breaks. *World Journal of Gastroenterology*, *20*, 908–922.

Canonico, L., Comitini, F., & Ciani, M. (2014). Dominance and influence of selected *Saccharomyces cerevisiae* strains on the analytical profile of craft beer refermentation. *Journal of the Institute of Brewing*, *120*, 262–267.

Carbonero, F., Benefiel, A. C., Alizadeh-Ghamsari, A. H., & Gaskins, H. R. (2012). Microbial pathways in colonic sulfur metabolism and links with health and disease. *Frontiers in Physiology*, *3*, 448.

Carbonero, F., Benefiel, A. C., & Gaskins, H. R. (2012). Contributions of the microbial hydrogen economy to colonic homeostasis. *Nature Reviews. Gastroenterology & Hepatology*, *9*, 504–518.

Carlstrom, M., & Larsson, S. C. (2018). Coffee consumption and reduced risk of developing type 2 diabetes: A systematic review with meta-analysis. *Nutrition Reviews*, *76*, 395–417.

Chakravorty, S., Bhattacharya, S., Chatzinotas, A., Chakraborty, W., Bhattacharya, D., & Gachhui, R. (2016). Kombucha tea fermentation: Microbial and biochemical dynamics. *International Journal of Food Microbiology*, *220*, 63–72.

Chen, P., Dong, J., Yin, H., Bao, X., Chen, L., He, Y., et al. (2016). Genome comparison and evolutionary analysis of different industrial lager yeasts (*Saccharomyces pastorianus*). *Journal of the Institute of Brewing*, *122*, 42–47.

Chen, C., & Liu, B. Y. (2000). Changes in major components of tea fungus metabolites during prolonged fermentation. *Journal of Applied Microbiology*, *89*, 834–839.

Cho, S. K., Eom, H. J., Moon, J. S., Lim, S. B., Kim, Y. K., Lee, K. W., et al. (2014). An improved process of isomaltooligosaccharide production in kimchi involving the addition of a Leuconostoc starter and sugars. *International Journal of Food Microbiology*, *170*, 61–64.

Choi, C. Y., Kim, Y. H., Oh, S., Lee, H. J., Kim, J. H., Park, S. H., et al. (2017). Anti-inflammatory potential of a heat-killed Lactobacillus strain isolated from Kimchi on house dust mite-induced atopic dermatitis in NC/Nga mice. *Journal of Applied Microbiology, 123*, 535–543.

Chow, J., & Mazmanian, S. K. (2010). A pathobiont of the microbiota balances host colonization and intestinal inflammation. *Cell Host & Microbe, 7*, 265–276.

Ciafardini, G., Marsilio, V., Lanza, B., & Pozzi, N. (1994). Hydrolysis of oleuropein by *Lactobacillus plantarum* strains associated with olive fermentation. *Applied and Environmental Microbiology, 60*, 4142–4147.

Cleveland, J., Montville, T. J., Nes, I. F., & Chikindas, M. L. (2001). Bacteriocins: Safe, natural antimicrobials for food preservation. *International Journal of Food Microbiology, 71*, 1–20.

Coda, R., Cagno, R. D., Gobbetti, M., & Rizzello, C. G. (2014). Sourdough lactic acid bacteria: Exploration of non-wheat cereal-based fermentation. *Food Microbiology, 37*, 51–58.

Colehour, A. M., Meadow, J. F., Liebert, M. A., Cepon-Robins, T. J., Gildner, T. E., Urlacher, S. S., et al. (2014). Local domestication of lactic acid bacteria via cassava beer fermentation. *PeerJ, 2*, e479.

Corr, S. C., Li, Y., Riedel, C. U., O'Toole, P. W., Hill, C., & Gahan, C. G. (2007). Bacteriocin production as a mechanism for the antiinfective activity of *Lactobacillus salivarius* UCC118. *Proceedings of the National Academy of Sciences of the United States of America, 104*, 7617–7621.

Coton, M., Coton, E., Levert, D., Casaregola, S., & Sohier, D. (2006). Yeast ecology in French cider and black olive natural fermentations. *International Journal of Food Microbiology, 108*, 130–135.

Coton, M., Pawtowski, A., Taminiau, B., Burgaud, G., Deniel, F., Coulloumme-Labarthe, L., et al. (2017). Unraveling microbial ecology of industrial-scale Kombucha fermentations by metabarcoding and culture-based methods. *FEMS Microbiology Ecology, 93*. https://doi.org/10.1093/femsec/fix048.

Crispim, S. M., Nascimento, A. M., Costa, P. S., Moreira, J. L., Nunes, A. C., Nicoli, J. R., et al. (2013). Molecular identification of *Lactobacillus* spp. associated with puba, a Brazilian fermented cassava food. *Brazilian Journal of Microbiology, 44*, 15–21.

D'Antuono, I., Bruno, A., Linsalata, V., Minervini, F., Garbetta, A., Tufariello, M., et al. (2018). Fermented Apulian table olives: Effect of selected microbial starters on polyphenols composition, antioxidant activities and bioaccessibility. *Food Chemistry, 248*, 137–145.

Davis, L. M., Martinez, I., Walter, J., Goin, C., & Hutkins, R. W. (2011). Barcoded pyrosequencing reveals that consumption of galactooligosaccharides results in a highly specific bifidogenic response in humans. *PLoS One, 6*, e25200.

Davis-Richardson, A. G., & Triplett, E. W. (2015). A model for the role of gut bacteria in the development of autoimmunity for type 1 diabetes. *Diabetologia, 58*, 1386–1393.

De Angelis, M., Campanella, D., Cosmai, L., Summo, C., Rizzello, C. G., & Caponio, F. (2015). Microbiota and metabolome of un-started and started Greek-type fermentation of Bella di Cerignola table olives. *Food Microbiology, 52*, 18–30.

De Bruyn, F., Zhang, S. J., Pothakos, V., Torres, J., Lambot, C., Moroni, A. V., et al. (2016). Exploring the impacts of postharvest processing on the microbiota and metabolite profiles during green coffee bean production. *Applied and Environmental Microbiology, 83*. https://doi.org/10.1128/AEM.02398-16. Print 2017 Jan 1.

De Filippis, F., Troise, A. D., Vitaglione, P., & Ercolini, D. (2018). Different temperatures select distinctive acetic acid bacteria species and promotes organic acids production during Kombucha tea fermentation. *Food Microbiology, 73*, 11–16.

de Melo Pereira, G. V., Soccol, V. T., Pandey, A., Medeiros, A. B., Andrade Lara, J. M., Gollo, A. L., et al. (2014). Isolation, selection and evaluation of yeasts for use in fermentation of coffee beans by the wet process. *International Journal of Food Microbiology, 188,* 60–66.

De Roos, J., & De Vuyst, L. (2018). Acetic acid bacteria in fermented foods and beverages. *Current Opinion in Biotechnology, 49,* 115–119.

De Vuyst, L., Vrancken, G., Ravyts, F., Rimaux, T., & Weckx, S. (2009). Biodiversity, ecological determinants, and metabolic exploitation of sourdough microbiota. *Food Microbiology, 26,* 666–675.

De Vuyst, L., & Weckx, S. (2016). The cocoa bean fermentation process: From ecosystem analysis to starter culture development. *Journal of Applied Microbiology, 121,* 5–17.

Demirel, F., Cakan, M., Yalcinkaya, F., Topcuoglu, M., & Altug, U. (2008). The association between personal habits and bladder cancer in Turkey. *International Urology and Nephrology, 40,* 643–647.

Denis, C., & Irlinger, F. (2008). Safety assessment of dairy microorganisms: Aerobic coryneform bacteria isolated from the surface of smear-ripened cheeses. *International Journal of Food Microbiology, 126,* 311–315.

Derrien, M., & van Hylckama Vlieg, J. E. (2015). Fate, activity, and impact of ingested bacteria within the human gut microbiota. *Trends in Microbiology, 23,* 354–366.

Duangjitcharoen, Y., Kantachote, D., Ongsakul, M., Poosaran, N., & Chaiyasut, C. (2008). Selection of probiotic lactic acid bacteria isolated from fermented plant beverages. *Pakistan Journal of Biological Sciences, 11,* 652–655.

Eales, J., Gibson, P., Whorwell, P., Kellow, J., Yellowlees, A., Perry, R. H., et al. (2017). Systematic review and meta-analysis: The effects of fermented milk with *Bifidobacterium lactis* CNCM I-2494 and lactic acid bacteria on gastrointestinal discomfort in the general adult population. *Therapeutic Advances in Gastroenterology, 10,* 74–88.

Elizaquivel, P., Perez-Cataluna, A., Yepez, A., Aristimuno, C., Jimenez, E., Cocconcelli, P. S., et al. (2015). Pyrosequencing vs. culture-dependent approaches to analyze lactic acid bacteria associated to chicha, a traditional maize-based fermented beverage from Northwestern Argentina. *International Journal of Food Microbiology, 198,* 9–18.

Eom, H. J., Seo, D. M., & Han, N. S. (2007). Selection of psychrotrophic *Leuconostoc* spp. producing highly active dextransucrase from lactate fermented vegetables. *International Journal of Food Microbiology, 117,* 61–67.

Everard, A., & Cani, P. D. (2013). Diabetes, obesity and gut microbiota. *Best Practice & Research. Clinical Gastroenterology, 27,* 73–83.

Faith, J. J., Guruge, J. L., Charbonneau, M., Subramanian, S., Seedorf, H., Goodman, A. L., et al. (2013). The long-term stability of the human gut microbiota. *Science, 341,* 1237439.

Fekete, A. A., Givens, D. I., & Lovegrove, J. A. (2015). Casein-derived lactotripeptides reduce systolic and diastolic blood pressure in a meta-analysis of randomised clinical trials. *Nutrients, 7,* 659–681.

Fernandez, M. A., Panahi, S., Daniel, N., Tremblay, A., & Marette, A. (2017). Yogurt and cardiometabolic diseases: A critical review of potential mechanisms. *Advances in Nutrition, 8,* 812–829.

Fiorda, F. A., de Melo Pereira, G. V., Thomaz-Soccol, V., Rakshit, S. K., Pagnoncelli, M. G. B., Vandenberghe, L. P. S., et al. (2017). Microbiological, biochemical, and functional aspects of sugary kefir fermentation—A review. *Food Microbiology, 66,* 86–95.

Fisberg, M., & Machado, R. (2015). History of yogurt and current patterns of consumption. *Nutrition Reviews, 73*(Suppl. 1), 4–7.

Flint, H. J. (2011). Obesity and the gut microbiota. *Journal of Clinical Gastroenterology, 45 Suppl,* S128–S132.

Fragopoulou, E., Choleva, M., Antonopoulou, S., & Demopoulos, C. A. (2018). Wine and its metabolic effects. A comprehensive review of clinical trials. *Metabolism, 83*, 102–119.

Freire, A. L., Zapata, S., Mosquera, J., Mejia, M. L., & Trueba, G. (2016). Bacteria associated with human saliva are major microbial components of Ecuadorian indigenous beers (chicha). *PeerJ, 4*, e1962.

Fu, N., Wu, J., Lv, L., He, J., & Jiang, S. (2015). Anti-foot-and-mouth disease virus effects of Chinese herbal kombucha in vivo. *Brazilian Journal of Microbiology, 46*, 1245–1255.

Gamboa-Gomez, C. I., Gonzalez-Laredo, R. F., Gallegos-Infante, J. A., Perez, M. D., Moreno-Jimenez, M. R., Flores-Rueda, A. G., et al. (2016). Antioxidant and angiotensin-converting enzyme inhibitory activity of *Eucalyptus camaldulensis* and *Litsea glaucescens* infusions fermented with Kombucha consortium. *Food Technology and Biotechnology, 54*, 367–374.

Gamboa-Gomez, C. I., Simental-Mendia, L. E., Gonzalez-Laredo, R. F., Alcantar-Orozco,-E. J., Monserrat-Juarez, V. H., Ramirez-Espana, J. C., et al. (2017). In vitro and in vivo assessment of anti-hyperglycemic and antioxidant effects of Oak leaves (*Quercus convallata* and *Quercus arizonica*) infusions and fermented beverages. *Food Research International, 102*, 690–699.

Ganmaa, D., Willett, W. C., Li, T. Y., Feskanich, D., van Dam, R. M., Lopez-Garcia, E., et al. (2008). Coffee, tea, caffeine and risk of breast cancer: A 22-year follow-up. *International Journal of Cancer, 122*, 2071–2076.

Garcia-Mantrana, I., Monedero, V., & Haros, M. (2015). Reduction of phytate in soy drink by fermentation with *Lactobacillus casei* expressing phytases from bifidobacteria. *Plant Foods for Human Nutrition, 70*, 269–274.

Garcia-Mantrana, I., Yebra, M. J., Haros, M., & Monedero, V. (2016). Expression of bifidobacterial phytases in *Lactobacillus casei* and their application in a food model of whole-grain sourdough bread. *International Journal of Food Microbiology, 216*, 18–24.

Garrido-Fernandez, A., & Vaughn, R. H. (1978). Utilization of oleuropein by microorganisms associated with olive fermentations. *Canadian Journal of Microbiology, 24*, 680–684.

Gedela, M., Potu, K. C., Gali, V. L., Alyamany, K., & Jha, L. K. (2016). A case of hepatotoxicity related to Kombucha tea consumption. *South Dakota Medicine, 69*, 26–28.

Gharib, O. A. (2009). Effects of Kombucha on oxidative stress induced nephrotoxicity in rats. *Chinese Medicine, 4*, 23. https://doi.org/10.1186/1749-8546-4-23.

Ghoneum, M., & Gimzewski, J. (2014). Apoptotic effect of a novel kefir product, PFT, on multidrug-resistant myeloid leukemia cells via a hole-piercing mechanism. *International Journal of Oncology, 44*, 830–837.

Gibson, M. K., Crofts, T. S., & Dantas, G. (2015). Antibiotics and the developing infant gut microbiota and resistome. *Current Opinion in Microbiology, 27*, 51–56.

Gijsbers, L., Ding, E. L., Malik, V. S., de Goede, J., Geleijnse, J. M., & Soedamah-Muthu, S. S. (2016). Consumption of dairy foods and diabetes incidence: A dose-response meta-analysis of observational studies. *The American Journal of Clinical Nutrition, 103*, 1111–1124.

Gille, D., Schmid, A., Walther, B., & Vergeres, G. (2018). Fermented food and non-communicable chronic diseases: A review. *Nutrients, 10*(4), 448. https://doi.org/10.3390/nu10040448.

Godos, J., Pluchinotta, F. R., Marventano, S., Buscemi, S., Li Volti, G., Galvano, F., et al. (2014). Coffee components and cardiovascular risk: Beneficial and detrimental effects. *International Journal of Food Sciences and Nutrition, 65*, 925–936.

Gonzalez-Perez, D., & Alcalde, M. (2014). Assembly of evolved ligninolytic genes in *Saccharomyces cerevisiae*. *Bioengineered, 5*, 254–263.

Gonzalez-Quijano, G. K., Dorantes-Alvarez, L., Hernandez-Sanchez, H., Jaramillo-Flores,-M. E., de, J., Perea-Flores, M., et al. (2014). Halotolerance and survival kinetics of lactic acid bacteria isolated from jalapeno pepper (*Capsicum annuum* L.) fermentation. *Journal of Food Science, 79*, M1545–M1553.

Greenwalt, C. J., Steinkraus, K. H., & Ledford, R. A. (2000). Kombucha, the fermented tea: Microbiology, composition, and claimed health effects. *Journal of Food Protection, 63*, 976–981.

Greppi, A., Saubade, F., Botta, C., Humblot, C., Guyot, J. P., & Cocolin, L. (2017). Potential probiotic *Pichia kudriavzevii* strains and their ability to enhance folate content of traditional cereal-based African fermented food. *Food Microbiology, 62*, 169–177.

Gurdeniz, G., Jensen, M. G., Meier, S., Bech, L., Lund, E., & Dragsted, L. O. (2016). Detecting beer intake by unique metabolite patterns. *Journal of Proteome Research, 15*, 4544–4556.

Hartmann, A. M., Burleson, L. E., Holmes, A. K., & Geist, C. R. (2000). Effects of chronic kombucha ingestion on open-field behaviors, longevity, appetitive behaviors, and organs in c57-bl/6 mice: A pilot study. *Nutrition, 16*, 755–761.

Hatoum, R., Labrie, S., & Fliss, I. (2012). Antimicrobial and probiotic properties of yeasts: From fundamental to novel applications. *Frontiers in Microbiology, 3*, 421.

Hill, D., Sugrue, I., Arendt, E., Hill, C., Stanton, C., & Ross, R. P. (2017). Recent advances in microbial fermentation for dairy and health. *F1000Research, 6*, 751.

Ho, V. T. T., Fleet, G. H., & Zhao, J. (2018). Unravelling the contribution of lactic acid bacteria and acetic acid bacteria to cocoa fermentation using inoculated organisms. *International Journal of Food Microbiology, 279*, 43–56.

Holbourn, A., & Hurdman, J. (2017). Kombucha: Is a cup of tea good for you? *BML Case Reports, 2017*. https://doi.org/10.1136/bcr-2017-221702.

Hondrodimou, O., Kourkoutas, Y., & Panagou, E. Z. (2011). Efficacy of natamycin to control fungal growth in natural black olive fermentation. *Food Microbiology, 28*, 621–627.

Hong, S. W., Choi, Y. J., Lee, H. W., Yang, J. H., & Lee, M. A. (2016). Microbial community structure of Korean cabbage kimchi and ingredients with denaturing gradient gel electrophoresis. *Journal of Microbiology and Biotechnology, 26*, 1057–1062.

Illeghems, K., Pelicaen, R., De Vuyst, L., & Weckx, S. (2016). Assessment of the contribution of cocoa-derived strains of *Acetobacter ghanensis* and *Acetobacter senegalensis* to the cocoa bean fermentation process through a genomic approach. *Food Microbiology, 58*, 68–78.

Islam, M. S., & Choi, H. (2009). Antidiabetic effect of Korean traditional Baechu (Chinese cabbage) kimchi in a type 2 diabetes model of rats. *Journal of Medicinal Food, 12*, 292–297.

Jeong, J. W., Choi, I. W., Jo, G. H., Kim, G. Y., Kim, J., Suh, H., et al. (2015). Anti-inflammatory effects of 3-(4'-hydroxyl-3',5'-dimethoxyphenyl)propionic acid, an active component of Korean cabbage Kimchi, in lipopolysaccharide-stimulated BV2 microglia. *Journal of Medicinal Food, 18*, 677–684.

Jeong, S. H., Jung, J. Y., Lee, S. H., Jin, H. M., & Jeon, C. O. (2013). Microbial succession and metabolite changes during fermentation of dongchimi, traditional Korean watery kimchi. *International Journal of Food Microbiology, 164*, 46–53.

Jeong, S. H., Lee, H. J., Jung, J. Y., Lee, S. H., Seo, H. Y., Park, W. S., et al. (2013). Effects of red pepper powder on microbial communities and metabolites during kimchi fermentation. *International Journal of Food Microbiology, 160*, 252–259.

Jeong, M., Park, J. M., Han, Y. M., Park, K. Y., Lee, D. H., Yoo, J. H., et al. (2015). Dietary prevention of Helicobacter pylori-associated gastric cancer with kimchi. *Oncotarget, 6*, 29513–29526.

Jeszka-Skowron, M., Zgola-Grzeskowiak, A., Waskiewicz, A., Stepien, L., & Stanisz, E. (2017). Positive and negative aspects of green coffee consumption—antioxidant activity versus mycotoxins. *Journal of the Science of Food and Agriculture, 97*, 4022–4028.

Jo, S. G., Noh, E. J., Lee, J. Y., Kim, G., Choi, J. H., Lee, M. E., et al. (2016). *Lactobacillus curvatus* WiKim38 isolated from kimchi induces IL-10 production in dendritic cells and alleviates DSS-induced colitis in mice. *Journal of Microbiology, 54,* 503–509.

Johnson, R., Melliou, E., Zweigenbaum, J., & Mitchell, A. E. (2018). Quantitation of oleuropein and related phenolics in cured Spanish-style green, California-style black Ripe, and Greek-style natural fermentation olives. *Journal of Agricultural and Food Chemistry, 66,* 2121–2128.

Jung, J. Y., Lee, S. H., & Jeon, C. O. (2014). Kimchi microflora: History, current status, and perspectives for industrial kimchi production. *Applied Microbiology and Biotechnology, 98,* 2385–2393.

Jung, J. Y., Lee, S. H., Lee, H. J., Seo, H. Y., Park, W. S., & Jeon, C. O. (2012). Effects of *Leuconostoc mesenteroides* starter cultures on microbial communities and metabolites during kimchi fermentation. *International Journal of Food Microbiology, 153,* 378–387.

Kaltsa, A., Papaliaga, D., Papaioannou, E., & Kotzekidou, P. (2015). Characteristics of oleuropeinolytic strains of *Lactobacillus plantarum* group and influence on phenolic compounds in table olives elaborated under reduced salt conditions. *Food Microbiology, 48,* 58–62.

Khan, M. W., Zadeh, M., Bere, P., Gounaris, E., Owen, J., Klaenhammer, T., et al. (2012). Modulating intestinal immune responses by lipoteichoic acid-deficient *Lactobacillus acidophilus*. *Immunotherapy, 4,* 151–161.

Khoury, N., El-Hayek, S., Tarras, O., El-Sabban, M., El-Sibai, M., & Rizk, S. (2014). Kefir exhibits antiproliferative and proapoptotic effects on colon adenocarcinoma cells with no significant effects on cell migration and invasion. *International Journal of Oncology, 45,* 2117–2127.

Kim, E. K., An, S. Y., Lee, M. S., Kim, T. H., Lee, H. K., Hwang, W. S., et al. (2011). Fermented kimchi reduces body weight and improves metabolic parameters in overweight and obese patients. *Nutrition Research, 31,* 436–443.

Kim, J., Jung, Y., Bong, Y. S., Lee, K. S., & Hwang, G. S. (2012). Determination of the geographical origin of kimchi by (1)H NMR-based metabolite profiling. *Bioscience, Biotechnology, and Biochemistry, 76,* 1752–1757.

Kim, J. H., Lee, J. E., & Jung, I. K. (2012). Dietary pattern classifications and the association with general obesity and abdominal obesity in Korean women. *Journal of the Academy of Nutrition and Dietetics, 112,* 1550–1559.

Kim, H. J., Noh, J. S., & Song, Y. O. (2018). Beneficial effects of Kimchi, a Korean fermented vegetable food, on pathophysiological factors related to atherosclerosis. *Journal of Medicinal Food, 21,* 127–135.

Kim, M. S., Yang, H. J., Kim, S. H., Lee, H. W., & Lee, M. S. (2018). Effects of Kimchi on human health: A protocol of systematic review of controlled clinical trials. *Medicine (Baltimore), 97,* e0163.

Kinross, J. M., Darzi, A. W., & Nicholson, J. K. (2011). Gut microbiome-host interactions in health and disease. *Genome Medicine, 3,* 14.

Kolmeder, C. A., Salojarvi, J., Ritari, J., de Been, M., Raes, J., Falony, G., et al. (2016). Faecal metaproteomic analysis reveals a personalized and stable functional microbiome and limited effects of a probiotic intervention in adults. *PLoS One, 11,* e0153294.

Korcz, E., Kerenyi, Z., & Varga, L. (2018). Dietary fibers, prebiotics, and exopolysaccharides produced by lactic acid bacteria: Potential health benefits with special regard to cholesterol-lowering effects. *Food & Function, 9,* 3057–3068.

Koyama, M., Hattori, S., Amano, Y., Watanabe, M., & Nakamura, K. (2014). Blood pressure-lowering peptides from neo-fermented buckwheat sprouts: A new approach to estimating ACE-inhibitory activity. *PLoS One, 9,* e105802.

Koyama, M., Naramoto, K., Nakajima, T., Aoyama, T., Watanabe, M., & Nakamura, K. (2013). Purification and identification of antihypertensive peptides from fermented buckwheat sprouts. *Journal of Agricultural and Food Chemistry, 61,* 3013–3021.

Kung, H. F., Lee, Y. C., Huang, Y. L., Huang, Y. R., Su, Y. C., & Tsai, Y. H. (2017). Degradation of histamine by *Lactobacillus plantarum* isolated from miso products. *Journal of Food Protection, 80,* 1682–1688.

Kusznierewicz, B., Smiechowska, A., Bartoszek, A., & Namiesnik, J. (2008). The effect of heating and fermenting on antioxidant properties of white cabbage. *Food Chemistry, 108,* 853–861.

Kwak, S. H., Cho, Y. M., Noh, G. M., & Om, A. S. (2014). Cancer preventive potential of kimchi lactic acid bacteria (*Weissella cibaria, Lactobacillus plantarum*). *Journal of Cancer Prevention, 19,* 253–258.

Kwon, D. Y., Jang, J. S., Hong, S. M., Lee, J. E., Sung, S. R., Park, H. R., et al. (2007). Long-term consumption of fermented soybean-derived Chungkookjang enhances insulinotropic action unlike soybeans in 90% pancreatectomized diabetic rats. *European Journal of Nutrition, 46,* 44–52.

Laatikainen, R., Koskenpato, J., Hongisto, S. M., Loponen, J., Poussa, T., Hillila, M., et al. (2016). Randomised clinical trial: Low-FODMAP rye bread vs. regular rye bread to relieve the symptoms of irritable bowel syndrome. *Alimentary Pharmacology & Therapeutics, 44,* 460–470.

Lang, J. M., Eisen, J. A., & Zivkovic, A. M. (2014). The microbes we eat: Abundance and taxonomy of microbes consumed in a day's worth of meals for three diet types. *PeerJ, 2,* e659.

Larsson, S. C., Bergkvist, L., Giovannucci, E., & Wolk, A. (2006). Coffee consumption and incidence of colorectal cancer in two prospective cohort studies of Swedish women and men. *American Journal of Epidemiology, 163,* 638–644.

Larsson, S. C., Giovannucci, E., & Wolk, A. (2006). Coffee consumption and stomach cancer risk in a cohort of Swedish women. *International Journal of Cancer, 119,* 2186–2189.

Larsson, S. C., Giovannucci, E. L., & Wolk, A. (2017). Coffee consumption and risk of gallbladder cancer in a prospective study. *Journal of the National Cancer Institute, 109,* 1–3.

Lee, S. H., Jung, J. Y., & Jeon, C. O. (2015). Source tracking and succession of kimchi lactic acid bacteria during fermentation. *Journal of Food Science, 80,* M1871–M1877.

Lee, H., Kim, D. Y., Lee, M. A., Jang, J. Y., & Choue, R. (2014). Immunomodulatory effects of *kimchi* in chinese healthy college students: A randomized controlled trial. *Clinical Nutrition Research, 3,* 98–105.

Lee, H. A., Kim, H., Lee, K. W., & Park, K. Y. (2016). Dietary nanosized *Lactobacillus plantarum* enhances the anticancer effect of kimchi on azoxymethane and dextran sulfate sodium-induced colon cancer in C57BL/6J mice. *Journal of Environmental Pathology, Toxicology and Oncology, 35,* 147–159.

Lee, M. H., Li, F. Z., Lee, J., Kang, J., Lim, S. I., & Nam, Y. D. (2017). Next-generation sequencing analyses of bacterial community structures in soybean pastes produced in Northeast China. *Journal of Food Science, 82,* 960–968.

Lee, H. A., Song, Y. O., Jang, M. S., & Han, J. S. (2014). Effect of baechu kimchi added ecklonia cava extracts on high glucose-induced oxidative stress in human umbilical vein endothelial cells. *Preventive Nutrition and Food Science, 19,* 170–177.

Lee, M., Song, J. H., Jung, M. Y., Lee, S. H., & Chang, J. Y. (2017). Large-scale targeted metagenomics analysis of bacterial ecological changes in 88 kimchi samples during fermentation. *Food Microbiology, 66,* 173–183.

Lee, K. H., Song, J. L., Park, E. S., Ju, J., Kim, H. Y., & Park, K. Y. (2015). Anti-obesity effects of starter fermented kimchi on 3T3-L1 adipocytes. *Preventive Nutrition and Food Science, 20,* 298–302.

Leventdurur, S., Sert-Aydin, S., Boyaci-Gunduz, C. P., Agirman, B., Ben Ghorbal, A., Francesca, N., et al. (2016). Yeast biota of naturally fermented black olives in different brines made from cv. Gemlik grown in various districts of the Cukurova region of Turkey. *Yeast, 33,* 289–301.

Li, J., Sung, C. Y., Lee, N., Ni, Y., Pihlajamaki, J., Panagiotou, G., et al. (2016). Probiotics modulated gut microbiota suppresses hepatocellular carcinoma growth in mice. *Proceedings of the National Academy of Sciences of the United States of America, 113,* E1306–E1315.

Lim, J. H., Jung, E. S., Choi, E. K., Jeong, D. Y., Jo, S. W., Jin, J. H., et al. (2015). Supplementation with *Aspergillus oryzae*-fermented kochujang lowers serum cholesterol in subjects with hyperlipidemia. *Clinical Nutrition, 34,* 383–387.

Linares, D. M., del Rio, B., Redruello, B., Ladero, V., Martin, M. C., Fernandez, M., et al. (2016). Comparative analysis of the in vitro cytotoxicity of the dietary biogenic amines tyramine and histamine. *Food Chemistry, 197,* 658–663.

Liou, A. P., Paziuk, M., Luevano, J. M., Jr., Machineni, S., Turnbaugh, P. J., & Kaplan, L. M. (2013). Conserved shifts in the gut microbiota due to gastric bypass reduce host weight and adiposity. *Science Translational Medicine, 5,* 178ra41.

Liu, X., Kuda, T., Takahashi, H., & Kimura, B. (2018). Bacterial and fungal microbiota of spontaneously fermented Chinese products, Rubing milk cake and Yan-cai vegetable pickles. *Food Microbiology, 72,* 106–111.

Liu, P., Shen, S. R., Ruan, H., Zhou, Q., Ma, L. L., & He, G. Q. (2011). Production of conjugated linoleic acids by Lactobacillus plantarum strains isolated from naturally fermented Chinese pickles. *Journal of Zhejiang University. Science. B, 12,* 923–930.

Lopez-Barrera, D. M., Vazquez-Sanchez, K., Loarca-Pina, M. G., & Campos-Vega, R. (2016). Spent coffee grounds, an innovative source of colonic fermentable compounds, inhibit inflammatory mediators in vitro. *Food Chemistry, 212,* 282–290.

Lu, Z., Breidt, F., Plengvidhya, V., & Fleming, H. P. (2003). Bacteriophage ecology in commercial sauerkraut fermentations. *Applied and Environmental Microbiology, 69,* 3192–3202.

Lucena-Padros, H., & Ruiz-Barba, J. L. (2016). Diversity and enumeration of halophilic and alkaliphilic bacteria in Spanish-style green table-olive fermentations. *Food Microbiology, 53,* 53–62.

Ludlow, C. L., Cromie, G. A., Garmendia-Torres, C., Sirr, A., Hays, M., Field, C., et al. (2016). Independent origins of yeast associated with coffee and cacao fermentation. *Current Biology, 26,* 965–971.

Lynch, K. M., Zannini, E., Coffey, A., & Arendt, E. K. (2018). Lactic acid bacteria exopolysaccharides in foods and beverages: Isolation, properties, characterization, and health benefits. *Annual Review of Food Science and Technology, 9,* 155–176.

Mahazar, N. H., Zakuan, Z., Norhayati, H., MeorHussin, A. S., & Rukayadi, Y. (2017). Optimization of culture medium for the growth of Candida sp. and Blastobotrys sp. as starter culture in fermentation of cocoa beans (Theobroma cacao) using response surface methodology (RSM). *Pakistan Journal of Biological Sciences, 20,* 154–159.

Mahmoudi, E., Saeidi, M., Marashi, M. A., Moafi, A., Mahmoodi, V., & Zeinolabedini Zamani, M. (2016). In vitro activity of kombucha tea ethyl acetate fraction against Malassezia species isolated from seborrhoeic dermatitis. *Current Medical Mycology, 2,* 30–36.

Marco, M. L., Heeney, D., Binda, S., Cifelli, C. J., Cotter, P. D., Foligne, B., et al. (2017). Health benefits of fermented foods: Microbiota and beyond. *Current Opinion in Biotechnology, 44,* 94–102.

Marongiu, A., Zara, G., Legras, J., Del Caro, A., Mascia, I., Fadda, C., et al. (2015). Novel starters for old processes: Use of *Saccharomyces cerevisiae* strains isolated from artisanal sourdough for craft beer production at a brewery scale. *Journal of Industrial Microbiology & Biotechnology, 42,* 85–92.

Marsh, A. J., O'Sullivan, O., Hill, C., Ross, R. P., & Cotter, P. D. (2014). Sequence-based analysis of the bacterial and fungal compositions of multiple kombucha (tea fungus) samples. *Food Microbiology*, *38*, 171–178.

Martinez, I., Kim, J., Duffy, P. R., Schlegel, V. L., & Walter, J. (2010). Resistant starches types 2 and 4 have differential effects on the composition of the fecal microbiota in human subjects. *PLoS One*, *5*, e15046.

Martinez, I., Lattimer, J. M., Hubach, K. L., Case, J. A., Yang, J., Weber, C. G., et al. (2013). Gut microbiome composition is linked to whole grain-induced immunological improvements. *The ISME Journal*, *7*, 269–280.

Martinez, I., Muller, C. E., & Walter, J. (2013). Long-term temporal analysis of the human fecal microbiota revealed a stable core of dominant bacterial species. *PLoS One*, *8*, e69621.

Martini, D., Del Bo', C., Tassotti, M., Riso, P., Del Rio, D., Brighenti, F., et al. (2016). Coffee consumption and oxidative stress: A review of human intervention studies. *Molecules*, *21*(8), 979. https://doi.org/10.3390/molecules21080979.

Mas, A., Torija, M. J., Garcia-Parrilla Mdel, C., & Troncoso, A. M. (2014). Acetic acid bacteria and the production and quality of wine vinegar. *Scientific World Journal*, *2014*, 394671.

Masoud, W., Cesar, L. B., Jespersen, L., & Jakobsen, M. (2004). Yeast involved in fermentation of *Coffea arabica* in East Africa determined by genotyping and by direct denaturing gradient gel electrophoresis. *Yeast*, *21*, 549–556. https://doi.org/10.1002/yea.1124.

Masoud, W., & Kaltoft, C. H. (2006). The effects of yeasts involved in the fermentation of Coffea arabica in East Africa on growth and ochratoxin A (OTA) production by Aspergillus ochraceus. *International Journal of Food Microbiology*, *106*(2), 229–234.

Maurice, C. F., Haiser, H. J., & Turnbaugh, P. J. (2013). Xenobiotics shape the physiology and gene expression of the active human gut microbiome. *Cell*, *152*, 39–50.

McGovern, P. E., Zhang, J., Tang, J., Zhang, Z., Hall, G. R., Moreau, R. A., et al. (2004). Fermented beverages of pre- and proto-historic China. *Proceedings of the National Academy of Sciences of the United States of America*, *101*, 17593–17598.

Meile, L., Le Blay, G., & Thierry, A. (2008). Safety assessment of dairy microorganisms: Propionibacterium and Bifidobacterium. *International Journal of Food Microbiology*, *126*, 316–320.

Mendoza, L. M., Neef, A., Vignolo, G., & Belloch, C. (2017). Yeast diversity during the fermentation of Andean chicha: A comparison of high-throughput sequencing and culture-dependent approaches. *Food Microbiology*, *67*, 1–10.

Menz, G., Vriesekoop, F., Zarei, M., Zhu, B., & Aldred, P. (2010). The growth and survival of food-borne pathogens in sweet and fermenting brewers' wort. *International Journal of Food Microbiology*, *140*, 19–25.

Mokoena, M. P., Mutanda, T., & Olaniran, A. O. (2016). Perspectives on the probiotic potential of lactic acid bacteria from African traditional fermented foods and beverages. *Food & Nutrition Research*, *60*, 29630.

Montoro, B. P., Benomar, N., Lavilla Lerma, L., Castillo Gutierrez, S., Galvez, A., & Abriouel, H. (2016). Fermented alorena table olives as a source of potential probiotic *Lactobacillus pentosus* strains. *Frontiers in Microbiology*, *7*, 1583.

More, M. I., & Vandenplas, Y. (2018). *Saccharomyces boulardii* CNCM I-745 improves intestinal enzyme function: A trophic effects review. *Clinical Medicine Insights. Gastroenterology*, *11*, 1179552217752679.

Moreira, A. S., Nunes, F. M., Domingues, M. R., & Coimbra, M. A. (2012). Coffee melanoidins: Structures, mechanisms of formation and potential health impacts. *Food & Function*, *3*, 903–915.

Moslemi, M., Mazaheri Nezhad Fard, R., Hosseini, S. M., Homayouni-Rad, A., & Mortazavian, A. M. (2016). Incorporation of propionibacteria in fermented milks as a probiotic. *Critical Reviews in Food Science and Nutrition, 56*, 1290–1312.

Nakamura, K., Naramoto, K., & Koyama, M. (2013). Blood-pressure-lowering effect of fermented buckwheat sprouts in spontaneously hypertensive rats. *Journal of Functional Foods, 5*, 406–415. https://doi.org/10.1016/j.jff.2012.11.013.

Nam, Y. D., Lee, S. Y., & Lim, S. I. (2012). Microbial community analysis of Korean soybean pastes by next-generation sequencing. *International Journal of Food Microbiology, 155*, 36–42.

Nehlig, A., & Debry, G. (1994). Potential genotoxic, mutagenic and antimutagenic effects of coffee: A review. *Mutation Research, 317*, 145–162.

Niksic, M., Niebuhr, S. E., Dickson, J. S., Mendonca, A. F., Koziczkowski, J. J., & Ellingson, J. L. (2005). Survival of *Listeria monocytogenes* and *Escherichia coli* O157:H7 during sauerkraut fermentation. *Journal of Food Protection, 68*, 1367–1374.

Oda, K., Imanishi, T., Yamane, Y., Ueno, Y., & Mori, Y. (2014). Bio-functional pickles that reduce blood pressure of rats. *Bioscience, Biotechnology, and Biochemistry, 78*, 882–890.

Oda, K., Nagai, T., Ueno, Y., & Mori, Y. (2015). Further evidence that a new type of Japanese pickles reduce the blood pressure of spontaneously hypertensive rats. *Bioscience, Biotechnology, and Biochemistry, 79*, 318–324.

O'Keefe, S. J., Li, J. V., Lahti, L., Ou, J., Carbonero, F., Mohammed, K., et al. (2015). Fat, fibre and cancer risk in African Americans and rural Africans. *Nature Communications, 6*, 6342.

Ono, H., Nishio, S., Tsurii, J., Kawamoto, T., Sonomoto, K., & Nakayama, J. (2014). Monitoring of the microbiota profile in nukadoko, a naturally fermented rice bran bed for pickling vegetables. *Journal of Bioscience and Bioengineering, 118*, 520–525.

Ou, J., Carbonero, F., Zoetendal, E. G., DeLany, J. P., Wang, M., Newton, K., et al. (2013). Diet, microbiota, and microbial metabolites in colon cancer risk in rural Africans and African Americans. *The American Journal of Clinical Nutrition, 98*, 111–120.

Ouattara, H. D., Ouattara, H. G., Droux, M., Reverchon, S., Nasser, W., & Niamke, S. L. (2017). Lactic acid bacteria involved in cocoa beans fermentation from Ivory Coast: Species diversity and citrate lyase production. *International Journal of Food Microbiology, 256*, 11–19.

Ouattara, H. G., Reverchon, S., Niamke, S. L., & Nasser, W. (2017). Regulation of the synthesis of pulp degrading enzymes in Bacillus isolated from cocoa fermentation. *Food Microbiology, 63*, 255–262.

Ozogul, F., & Hamed, I. (2018). The importance of lactic acid bacteria for the prevention of bacterial growth and their biogenic amines formation: A review. *Critical Reviews in Food Science and Nutrition, 58*, 1660–1670.

Pakravan, N., Mahmoudi, E., Hashemi, S. A., Kamali, J., Hajiaghayi, R., Rahimzadeh, M., et al. (2017). Cosmeceutical effect of ethyl acetate fraction of Kombucha tea by intradermal administration in the skin of aged mice. *Journal of Cosmetic Dermatology, 00*, 1–9.

Panagou, E. Z., Tassou, C. C., Vamvakoula, P., Saravanos, E. K., & Nychas, G. J. (2008). Survival of *Bacillus cereus* vegetative cells during Spanish-style fermentation of conservolea green olives. *Journal of Food Protection, 71*, 1393–1400.

Papalexandratou, Z., & Nielsen, D. S. (2016). It's gettin' hot in here: Breeding robust yeast starter cultures for cocoa fermentation. *Trends in Microbiology, 24*, 168–170.

Paramithiotis, S., Doulgeraki, A. I., Tsilikidis, I., Nychas, G. E., & Drosinos, E. H. (2012). Fate of *Listeria monocytogenes* and *Salmonella typhimurium* during spontaneous cauliflower fermentation. *Food Control, 27*, 178–183. https://doi.org/10.1016/j.foodcont.2012.03.022.

Park, S., & Bae, J. H. (2016). Fermented food intake is associated with a reduced likelihood of atopic dermatitis in an adult population (Korean National Health and Nutrition Examination Survey 2012-2013). *Nutrition Research, 36,* 125–133.

Park, K. Y., Jeong, J. K., Lee, Y. E., & Daily, J. W., 3rd. (2014). Health benefits of kimchi (Korean fermented vegetables) as a probiotic food. *Journal of Medicinal Food, 17,* 6–20.

Park, J. S., Joe, I., Rhee, P. D., Jeong, C. S., & Jeong, G. (2017). A lactic acid bacterium isolated from kimchi ameliorates intestinal inflammation in DSS-induced colitis. *Journal of Microbiology, 55,* 304–310.

Park, J. A., Tirupathi Pichiah, P. B., Yu, J. J., Oh, S. H., Daily, J. W., 3rd, & Cha, Y. S. (2012). Anti-obesity effect of kimchi fermented with Weissella koreensis OK1-6 as starter in high-fat diet-induced obese C57BL/6J mice. *Journal of Applied Microbiology, 113,* 1507–1516.

Pastor, R. F., Restani, P., Di Lorenzo, C., Orgiu, F., Teissedre, P. L., Stockley, C., et al. (2017). Resveratrol, human health and winemaking perspectives. *Critical Reviews in Food Science and Nutrition, 5,* 1–19.

Patra, J. K., Das, G., Paramithiotis, S., & Shin, H. S. (2016). Kimchi and other widely consumed traditional fermented foods of Korea: A review. *Frontiers in Microbiology, 7,* 1493.

Peñas, E., Martinez-Villaluenga, C., & Frias, J. (2017). Sauerkraut: Production, composition, and health benefits. In *Fermented foods in health and disease prevention* (pp. 557–576). Academic Press. chapter 24.

Perez-Diaz, I. M., Hayes, J., Medina, E., Anekella, K., Daughtry, K., Dieck, S., et al. (2017). Reassessment of the succession of lactic acid bacteria in commercial cucumber fermentations and physiological and genomic features associated with their dominance. *Food Microbiology, 63,* 217–227.

Perez-Diaz, I. M., & McFeeters, R. F. (2011). Preparation of a *Lactobacillus plantarum* starter culture for cucumber fermentations that can meet kosher guidelines. *Journal of Food Science, 76,* M120–M123.

Pessione, E., & Cirrincione, S. (2016). Bioactive molecules released in food by lactic acid bacteria: Encrypted peptides and biogenic amines. *Frontiers in Microbiology, 7,* 876.

Pfohl, G. (1983). Sauerkraut or liberty cabbage? Native country and reality. *Die Medizinische Welt, 34,* 536–541.

Pistarino, E., Aliakbarian, B., Casazza, A. A., Paini, M., Cosulich, M. E., & Perego, P. (2013). Combined effect of starter culture and temperature on phenolic compounds during fermentation of Taggiasca black olives. *Food Chemistry, 138,* 2043–2049.

Plé, C., Breton, J., Daniel, C., & Foligné, B. (2015). Maintaining gut ecosystems for health: Are transitory food bugs stowaways or part of the crew? *International Journal of Food Microbiology, 213,* 139–143. https://doi.org/10.1016/j.ijfoodmicro.2015.03.015.

Ple, C., Breton, J., Richoux, R., Nurdin, M., Deutsch, S. M., Falentin, H., et al. (2016). Combining selected immunomodulatory *Propionibacterium freudenreichii* and *Lactobacillus delbrueckii* strains: Reverse engineering development of an anti-inflammatory cheese. *Molecular Nutrition & Food Research, 60,* 935–948.

Plengvidhya, V., Breidt, F., J., & Fleming, H. P. (2004). Use of RAPD-PCR as a method to follow the progress of starter cultures in sauerkraut fermentation. *International Journal of Food Microbiology, 93,* 287–296.

Plengvidhya, V., Breidt, F., Jr., Lu, Z., & Fleming, H. P. (2007). DNA fingerprinting of lactic acid bacteria in sauerkraut fermentations. *Applied and Environmental Microbiology, 73,* 7697–7702.

Podolich, O., Zaets, I., Kukharenko, O., Orlovska, I., Reva, O., Khirunenko, L., et al. (2017). Kombucha multimicrobial community under simulated spaceflight and martian conditions. *Astrobiology, 17,* 459–469.

Prajapati, J., & Nair, B. (2003). The history of fermented foods. In *Handbook of fermented functional foods.* CRC Press.

Puerari, C., Magalhães-Guedes, K. T., & Schwan, R. F. (2015). Physicochemical and micro-biological characterization of chicha, a rice-based fermented beverage produced by Umutina Brazilian Amerindians. *Food Microbiology, 46*, 210–217.

Pugin, B., Barcik, W., Westermann, P., Heider, A., Wawrzyniak, M., Hellings, P., et al. (2017). A wide diversity of bacteria from the human gut produces and degrades biogenic amines. *Microbial Ecology in Health and Disease, 28*, 1353881.

Raak, C., Ostermann, T., Boehm, K., & Molsberger, F. (2014). Regular consumption of sauerkraut and its effect on human health: A bibliometric analysis. *Global Advances in Health and Medicine, 3*, 12–18.

Rabie, M. A., Siliha, H., el-Saidy, S., el-Badawy, A. A., & Malcata, F. X. (2011). Reduced biogenic amine contents in sauerkraut via addition of selected lactic acid bacteria. *Food Chemistry, 129*, 1778–1782. https://doi.org/10.1016/j.foodchem.2011.05.106.

Ramos, C. L., de Almeida, E. G., Pereira, G. V., Cardoso, P. G., Dias, E. S., & Schwan, R. F. (2010). Determination of dynamic characteristics of microbiota in a fermented beverage produced by Brazilian Amerindians using culture-dependent and culture-independent methods. *International Journal of Food Microbiology, 140*, 225–231.

Ramos, C. L., de Sousa, E. S., Ribeiro, J., Almeida, T. M., Santos, C. C., Abegg, M. A., et al. (2015). Microbiological and chemical characteristics of taruba, an indigenous beverage produced from solid cassava fermentation. *Food Microbiology, 49*, 182–188.

Randazzo, C. L., Caggia, C., & Neviani, E. (2009). Application of molecular approaches to study lactic acid bacteria in artisanal cheeses. *Journal of Microbiological Methods, 78*, 1–9.

Randazzo, C. L., Todaro, A., Pino, A., Pitino, I., Corona, O., & Caggia, C. (2017). Micro-biota and metabolome during controlled and spontaneous fermentation of Nocellara Etnea table olives. *Food Microbiology, 65*, 136–148.

Rasu, J., Malbaša, R. V., Lončar, E. S., Vitas, J. S., & Muthuswamy, S. (2014). A review on kombucha tea—Microbiology, composition, fermentation, beneficial effects, toxicity, and tea fungus. *Comprehensive Reviews in Food Science and Food Safety, 13*, 538–550.

Regueiro, J., Vallverdu-Queralt, A., Simal-Gandara, J., Estruch, R., & Lamuela-Raventos,-R. (2013). Development of a LC-ESI-MS/MS approach for the rapid quantification of main wine organic acids in human urine. *Journal of Agricultural and Food Chemistry, 61*, 6763–6768.

Resende, L. V., Pinheiro, L. K., Miguel, M. G. D. C. P., Ramos, C. L., Vilela, D. M., & Schwan, R. F. (2018). Microbial community and physicochemical dynamics during the production of 'Chicha', a traditional beverage of Indigenous people of Brazil. *World Journal of Microbiology and Biotechnology, 34*, 46.

Roda, A., Lucini, L., Torchio, F., Dordoni, R., De Faveri, D. M., & Lambri, M. (2017). Metabolite profiling and volatiles of pineapple wine and vinegar obtained from pineapple waste. *Food Chemistry, 229*, 734–742.

Rodhouse, L., & Carbonero, F. (2017). Overview of craft brewing specificities and poten-tially associated microbiota. *Critical Reviews in Food Science and Nutrition*, 1–12.

Rodriguez, M. E., Perez-Traves, L., Sangorrin, M. P., Barrio, E., Querol, A., & Lopes, C. A. (2017). Saccharomyces uvarum is responsible for the traditional fermentation of apple chicha in Patagonia. *FEMS Yeast Research, 17*(1). https://doi.org/10.1093/femsyr/fow109.

Salafzoon, S., Mahmoodzadeh Hosseini, H., & Halabian, R. (2017). Evaluation of the anti-oxidant impact of ginger-based kombucha on the murine breast cancer model. *Journal of Complementary & Integrative Medicine, 15*(1). https://doi.org/10.1515/jcim-2017-0071.

Salas-Salvado, J., Guasch-Ferre, M., Diaz-Lopez, A., & Babio, N. (2017). Yogurt and dia-betes: Overview of recent observational studies. *The Journal of Nutrition, 147*, 1452S–1461S.

Salazar, N., Gueimonde, M., de Los Reyes-Gavilan, C. G., & Ruas-Madiedo, P. (2016). Exopolysaccharides produced by lactic acid bacteria and bifidobacteria as fermentable substrates by the intestinal microbiota. *Critical Reviews in Food Science and Nutrition, 56*, 1440–1453.

Samagaci, L., Ouattara, H., Niamke, S., & Lemaire, M. (2016). Pichia kudrazevii and Candida nitrativorans are the most well-adapted and relevant yeast species fermenting cocoa in Agneby-Tiassa, a local Ivorian cocoa producing region. *Food Research International, 89*, 773–780.

Sandoval-Ramirez, B. A., M Lamuela-Raventos, R., Estruch, R., Sasot, G., Domenech, M., & Tresserra-Rimbau, A. (2017). Beer polyphenols and menopause: Effects and mechanisms—A review of current knowledge. *Oxidative Medicine and Cellular Longevity, 2017*, 4749131.

Scheperjans, F., Pekkonen, E., Kaakkola, S., & Auvinen, P. (2015). Linking smoking, coffee, urate, and Parkinson's disease—A role for gut microbiota? *Journal of Parkinsons Disease, 5*, 255–262.

Schroeter, J., & Klaenhammer, T. (2009). Genomics of lactic acid bacteria. *FEMS Microbiology Letters, 292*, 1–6.

Scott, K. P., Gratz, S. W., Sheridan, P. O., Flint, H. J., & Duncan, S. H. (2013). The influence of diet on the gut microbiota. *Pharmacological Research, 69*, 52–60.

Sheflin, A. M., Melby, C. L., Carbonero, F., & Weir, T. L. (2017). Linking dietary patterns with gut microbial composition and function. *Gut Microbes, 8*, 113–129.

Shiby, V. K., & Mishra, H. N. (2013). Fermented milks and milk products as functional foods—A review. *Critical Reviews in Food Science and Nutrition, 53*, 482–496.

Sicard, D., & Legras, J. (2011). Bread, beer and wine: Yeast domestication in the Saccharomyces sensu stricto complex. *Comptes Rendus Biologies, 334*, 229–236.

Silva, C. F., Schwan, R. F., Sousa Dias, Ë., & Wheals, A. E. (2000). Microbial diversity during maturation and natural processing of coffee cherries of *Coffea arabica* in Brazil. *International Journal of Food Microbiology, 60*, 251–260. https://doi.org/10.1016/S0168-1605(00)00315-9.

Sknepnek, A., Pantic, M., Matijasevic, D., Miletic, D., Levic, S., Nedovic, V., et al. (2018). Novel Kombucha beverage from Lingzhi or Reishi medicinal mushroom, Ganoderma lucidum, with antibacterial and antioxidant effects. *International Journal of Medicinal Mushrooms, 20*, 243–258.

Song, A. A., In, L. L. A., Lim, S. H. E., & Rahim, R. A. (2017). A review on Lactococcus lactis: From food to factory. *Microbial Cell Factories, 16*, 55.

Stamer, J. R., Stoyla, B. O., & Dunckel, B. A. (1971). Growth rates and fermentation patterns of lactic acid bacteria associated with the sauerkraut fermentation. *Journal of Milk and Food Technology, 34*, 521–525.

Sun, T. Y., Li, J. S., & Chen, C. (2015). Effects of blending wheatgrass juice on enhancing phenolic compounds and antioxidant activities of traditional kombucha beverage. *Journal of Food and Drug Analysis, 23*, 709–718.

SungHee Kole, A., Jones, H. D., Christensen, R., & Gladstein, J. (2009). A case of Kombucha tea toxicity. *Journal of Intensive Care Medicine, 24*, 205–207.

Suzuki, S., Honda, H., Suganuma, H., Saito, T., & Yajima, N. (2014). Growth and bile tolerance of *Lactobacillus brevis* strains isolated from Japanese pickles in artificial digestive juices and contribution of cell-bound exopolysaccharide to cell aggregation. *Canadian Journal of Microbiology, 60*, 139–145.

Suzuki, S., Kimoto-Nira, H., Suganuma, H., Suzuki, C., Saito, T., & Yajima, N. (2014). Cellular fatty acid composition and exopolysaccharide contribute to bile tolerance in *Lactobacillus brevis* strains isolated from fermented Japanese pickles. *Canadian Journal of Microbiology, 60*, 183–191.

Tamang, J. P., Watanabe, K., & Holzapfel, W. H. (2016). Review: Diversity of microorganisms in global fermented foods and beverages. *Frontiers in Microbiology*, 7, 377.

Tang, W., Hu, W., Wang, J., Wang, J., & Wang, Y. (2016). Identification of a new probiotic *Lactobacillus alimentarius* W369 from Chinese traditional pickles. *Wei Sheng Wu Xue Bao*, 56, 932–942.

Tang, N., Wu, Y., Ma, J., Wang, B., & Yu, R. (2010). Coffee consumption and risk of lung cancer: A meta-analysis. *Lung Cancer*, 67, 17–22.

Tang, N., Zhou, B., Wang, B., & Yu, R. (2009). Coffee consumption and risk of breast cancer: A metaanalysis. *American Journal of Obstetrics and Gynecology*, 200, 290. e1–290.e9.

Tataridou, M., & Kotzekidou, P. (2015). Fermentation of table olives by oleuropeinolytic starter culture in reduced salt brines and inactivation of *Escherichia coli* O157:H7 and Listeria monocytogenes. *International Journal of Food Microbiology*, 208, 122–130.

Teoh, A. L., Heard, G., & Cox, J. (2004). Yeast ecology of Kombucha fermentation. *International Journal of Food Microbiology*, 95, 119–126.

Tomita, S., Saito, K., Nakamura, T., Sekiyama, Y., & Kikuchi, J. (2017). Rapid discrimination of strain-dependent fermentation characteristics among Lactobacillus strains by NMR-based metabolomics of fermented vegetable juice. *PLoS One*, 12, e0182229.

Tu, M. Y., Chen, H. L., Tung, Y. T., Kao, C. C., Hu, F. C., & Chen, C. M. (2015). Short-term effects of kefir-fermented milk consumption on bone mineral density and bone metabolism in a randomized clinical trial of osteoporotic patients. *PLoS One*, 10, e0144231.

Turnbaugh, P. J., Hamady, M., Yatsunenko, T., Cantarel, B. L., Duncan, A., Ley, R. E., et al. (2009). A core gut microbiome in obese and lean twins. *Nature*, 457, 480–484.

Turnbaugh, P. J., Ley, R. E., Hamady, M., Fraser-Liggett, C. M., Knight, R., & Gordon, J. I. (2007). The human microbiome project. *Nature*, 449, 804–810.

Turroni, S., Rampelli, S., Biagi, E., Consolandi, C., Severgnini, M., Peano, C., et al. (2017). Temporal dynamics of the gut microbiota in people sharing a confined environment, a 520-day ground-based space simulation, MARS500. *Microbiome*, 5, 39. https://doi.org/10.1186/s40168-017-0256-8.

van Dijk, C., Ebbenhorst-Selles, T., Ruisch, H., Stolle-Smits, T., Schijvens, E., van Deelen, W., et al. (2000). Product and redox potential analysis of sauerkraut fermentation. *Journal of Agricultural and Food Chemistry*, 48, 132–139.

Vasiee, A., Alizadeh Behbahani, B., Tabatabaei Yazdi, F., Mortazavi, S. A., & Noorbakhsh, H. (2017). Diversity and probiotic potential of lactic acid bacteria isolated from horreh, a traditional Iranian fermented food. *Probiotics and Antimicrobial Proteins*, 10(2), 258–268.

Vazquez-Cabral, B. D., Larrosa-Perez, M., Gallegos-Infante, J. A., Moreno-Jimenez, M. R., Gonzalez-Laredo, R. F., Rutiaga-Quinones, J. G., et al. (2017). Oak kombucha protects against oxidative stress and inflammatory processes. *Chemico-Biological Interactions*, 272, 1–9.

Velicanski, A. S., Cvetkovic, D. D., Markov, S. L., Saponjac, V. T., & Vulic, J. J. (2014). Antioxidant and antibacterial activity of the beverage obtained by fermentation of sweetened lemon balm(*Melissa officinalis* L.) tea with symbiotic consortium of bacteria and yeasts. *Food Technology and Biotechnology*, 52, 420–429.

Vilela, D. M., Pereira, G. V., Silva, C. F., Batista, L. R., & Schwan, R. F. (2010). Molecular ecology and polyphasic characterization of the microbiota associated with semi-dry processed coffee (*Coffea arabica* L.). *Food Microbiology*, 27, 1128–1135.

Villanueva, C. M., Silverman, D. T., Murta-Nascimento, C., Malats, N., Garcia-Closas, M., Castro, F., et al. (2009). Coffee consumption, genetic susceptibility and bladder cancer risk. *Cancer Causes & Control*, 20, 121–127.

Villarreal-Soto, S. A., Beaufort, S., Bouajila, J., Souchard, J. P., & Taillandier, P. (2018). Understanding kombucha tea fermentation: A review. *Journal of Food Science, 83,* 580–588.

Vina, I., Semjonovs, P., Linde, R., & Denina, I. (2014). Current evidence on physiological activity and expected health effects of kombucha fermented beverage. *Journal of Medicinal Food, 17,* 179–188.

Visintin, S., Alessandria, V., Valente, A., Dolci, P., & Cocolin, L. (2016). Molecular identification and physiological characterization of yeasts, lactic acid bacteria and acetic acid bacteria isolated from heap and box cocoa bean fermentations in West Africa. *International Journal of Food Microbiology, 216,* 69–78.

Visintin, S., Ramos, L., Batista, N., Dolci, P., Schwan, F., & Cocolin, L. (2017). Impact of *Saccharomyces cerevisiae* and *Torulaspora delbrueckii* starter cultures on cocoa beans fermentation. *International Journal of Food Microbiology, 257,* 31–40.

Walter, J., Britton, R. A., & Roos, S. (2011). Host-microbial symbiosis in the vertebrate gastrointestinal tract and the *Lactobacillus reuteri* paradigm. *Proceedings of the National Academy of Sciences of the United States of America, 108*(Suppl. 1), 4645–4652.

Waters, D. M., Arendt, E. K., & Moroni, A. V. (2017). Overview on the mechanisms of coffee germination and fermentation and their significance for coffee and coffee beverage quality. *Critical Reviews in Food Science and Nutrition, 57,* 259–274.

Wie, G. A., Cho, Y. A., Kang, H. H., Ryu, K. A., Yoo, M. K., Kim, J., et al. (2017). Identification of major dietary patterns in Korean adults and their association with cancer risk in the Cancer Screening Examination Cohort. *European Journal of Clinical Nutrition, 71,* 1223–1229.

Wigle, D. T., Turner, M. C., Gomes, J., & Parent, M. E. (2008). Role of hormonal and other factors in human prostate cancer. *Journal of Toxicology and Environmental Health. Part B, Critical Reviews, 11,* 242–259.

Wilfrid Padonou, S., Nielsen, D. S., Hounhouigan, J. D., Thorsen, L., Nago, M. C., & Jakobsen, M. (2009). The microbiota of Lafun, an African traditional cassava food product. *International Journal of Food Microbiology, 133,* 22–30.

Xie, C., Cui, L., Zhu, J., Wang, K., Sun, N., & Sun, C. (2018). Coffee consumption and risk of hypertension: A systematic review and dose-response meta-analysis of cohort studies. *Journal of Human Hypertension, 32,* 83–93.

Xiong, T., Guan, Q., Song, S., Hao, M., & Xie, M. (2012). Dynamic changes of lactic acid bacteria flora during Chinese sauerkraut fermentation. *Food Control, 26,* 178–181. https://doi.org/10.1016/j.foodcont.2012.01.027.

Yang, Y., Deng, Y., Jin, Y., Liu, Y., Xia, B., & Sun, Q. (2017). Dynamics of microbial community during the extremely long-term fermentation process of a traditional soy sauce. *Journal of the Science of Food and Agriculture, 97,* 3220–3227.

Yu, J., Gao, W., Qing, M., Sun, Z., Wang, W., Liu, W., et al. (2012). Identification and characterization of lactic acid bacteria isolated from traditional pickles in Sichuan, China. *The Journal of General and Applied Microbiology, 58,* 163–172.

Yu, Z., Zhang, X., Li, S., Li, C., Li, D., & Yang, Z. (2013). Evaluation of probiotic properties of *Lactobacillus plantarum* strains isolated from Chinese sauerkraut. *World Journal of Microbiology and Biotechnology, 29,* 489–498.

Zago, M., Lanza, B., Rossetti, L., Muzzalupo, I., Carminati, D., & Giraffa, G. (2013). Selection of *Lactobacillus plantarum* strains to use as starters in fermented table olives: Oleuropeinase activity and phage sensitivity. *Food Microbiology, 34,* 81–87.

Zaragoza, J., Bendiks, Z., Tyler, C., Kable, M. E., Williams, T. R., Luchkovska, Y., et al. (2017). Effects of exogenous yeast and bacteria on the microbial population dynamics and outcomes of olive fermentations. *mSphere, 2*(1), e00315–e00316. https://doi.org/10.1128/mSphere.00315-16. eCollection 2017 Jan-Feb.

Zeidan, A. A., Poulsen, V. K., Janzen, T., Buldo, P., Derkx, P. M. F., Oregaard, G., et al. (2017). Polysaccharide production by lactic acid bacteria: From genes to industrial applications. *FEMS Microbiology Reviews, 41*, S168–S200.

Zhang, C., Derrien, M., Levenez, F., Brazeilles, R., Ballal, S. A., Kim, J., et al. (2016). Ecological robustness of the gut microbiota in response to ingestion of transient foodborne microbes. *The ISME Journal, 10*, 2235–2245.

Zheng, H., Yde, C. C., Clausen, M. R., Kristensen, M., Lorenzen, J., Astrup, A., et al. (2015). Metabolomics investigation to shed light on cheese as a possible piece in the French paradox puzzle. *Journal of Agricultural and Food Chemistry, 63*, 2830–2839.

Zhu, Y., Wang, H., Hollis, J. H., & Jacques, P. F. (2015). The associations between yogurt consumption, diet quality, and metabolic profiles in children in the USA. *European Journal of Nutrition, 54*, 543–550.

Zhu, Y., Xiao, L., Shen, D., & Hao, Y. (2010). Competition between yogurt probiotics and periodontal pathogens in vitro. *Acta Odontologica Scandinavica, 68*, 261–268.

Zvrko, E., Gledovic, Z., & Ljaljevic, A. (2008). Risk factors for laryngeal cancer in Montenegro. *Arhiv za Higijenu Rada i Toksikologiju, 59*, 11–18.

CHAPTER FOUR

Marine Waste Utilization as a Source of Functional and Health Compounds

**Amin Shavandi*, Yakun Hou*, Alan Carne†, Michelle McConnell‡,
Alaa El-din A. Bekhit*,1**
*Department of Food Science, University of Otago, Dunedin, New Zealand
†Department of Biochemistry, University of Otago, Dunedin, New Zealand
‡Department of Microbiology, University of Otago, Dunedin, New Zealand
1Corresponding author: e-mail address: aladin.bekhit@otago.ac.nz

Contents

Advances in Food and Nutrition Research, Volume 87
ISSN 1043-4526
https://doi.org/10.1016/bs.afnr.2018.08.001

Abstract

Consumer demand for convenience has led to large quantities of seafood being value-added processed before marketing, resulting in large amounts of marine by-products being generated by processing industries. Several bioconversion processes have been proposed to transform some of these by-products. In addition to their relatively low value conventional use as animal feed and fertilizers, several investigations have been reported that have demonstrated the potential to add value to viscera, heads, skins, fins, trimmings, and crab and shrimp shells by extraction of lipids, bioactive peptides, enzymes, and other functional proteins and chitin that can be used in food and pharmaceutical applications. This chapter is focused on reviewing the opportunities for utilization of these marine by-products. The chapter discusses the various products and bioactive compounds that can be obtained from seafood waste and describes various methods that can be used to produce these products with the aim of highlighting opportunities to add value to these marine waste streams.

1. INTRODUCTION

Recent data from the FAO (2016) showed that there has been more than an eightfold increase in total fisheries and aquaculture production in the period between 1954 and 2014 and a corresponding increase in *per capita* fish supply from <4 kg/capita in 1954 to 20.1 kg/capita in 2014. This massive increase in production has been driven by advances in fishing technologies (e.g., large fishing fleets equipped with advanced technologies to trace and capture fish and capability to sustain long fishing trips) and rapid developments in aquaculture that have resulted in a significant contribution to world fish production (estimated to be 44% of world production in 2014) (FAO, 2016). More than 70% of the total fish caught undergo further processing before going to market (Hou et al., 2016). Processing of fish is a requirement for large-scale fish processing companies to reduce the cost associated with transportation of inedible parts of the fish and to improve the stability and quality of products by removing parts that may harbor bacteria and enzymes (e.g., viscera) and become a risk during further processing and storage of fish. The mass harvesting of fish in the open water by large fishing vessels and the seasonal harvest of aquacultured species has resulted in the generation of large amounts of solid fish waste and wastewater. Depending on the processed species, the solid waste can comprise low-quality whole fish, fish skeleton including head and tail, viscera, skin, and filet trimmings, and shell material in the case of shellfish and crustaceans. Ghaly, Ramakrishnan, Brooks, Budge, and Dave (2013) estimated the average composition of fish waste to be head (21%), liver (5%), roe (4%), backbone (14%), fins, gut, and gills (17%), but this average is expected to vary widely depending on fish species, size, gender, and season. In terms of shellfish, the waste can be up to 60% of the catch weight (Hou et al., 2016). The traditional management of these wastes is either disposal in a landfill, processing into animal feed, or used as a fertilizer, with the exception of fish roe that may be consumed as a delicacy in some countries. The disposal of these waste materials in landfills can cause many environmental problems (e.g., generation of off-odors, air and soil pollution, and damage to the marine ecosystem) due to their high organic contents (protein, lipids, enzymes) and minerals, strict regulation is imposed on its management. The organic components in this waste are the main driving force toward maximizing the use of this available resource. For example, the recovery of protein and potential generation of bioactive peptides from proteins present in fish trimmings, skin, and other organs; the production of collagen and gelatin from skins; the recovery of enzymes from intestines; oil from fish frames, head, gut, liver, and roe; and in addition calcium, glucosamine, and chitosan are recognized to be valuable materials that provide significant opportunities for development of value-added products. This chapter discusses recent developments in the utilization of marine by-products and highlights opportunities to add value and produce novel health-promoting products.

2. FISHBONE

Filleting fish is a common processing step that is used in the preparation of the muscle for production of other products (e.g., fish fingers, fish steaks, marination, canning) or with the aim of improving the value of a unit of the final product and reducing transportation cost by removing inedible material. Large amounts of fishbone or frames are generated from the filleting step. Fishbone is a mainly composed of minerals such as calcium phosphate and proteins. The organic phase of the fishbone is largely composed of collagen, a valuable product, which accounts for 30% of the fishbone (Kim, 2014). Next to fish skin, fishbone is regarded as the richest source of collagen and gelatin (Herpandi, Huda, & Adzitey, 2011). The inorganic phase of the fishbone is composed of about 70% calcium phosphate mainly in the form of hydroxyapatite, which has diverse technical and biomedical applications such as air/water filtration and bone tissue engineering (Kim & Mendis, 2006; Shavandi, Bekhit, Sun, & Ali, 2015).

Fishbone peptides and hydrolysates with strong antioxidant activity have been suggested as food additives, and feed ingredients. Similar to fishbone, fish scale is also a rich source of proteins and minerals such as collagen and hydroxyapatite.

2.1 Fishbone as a Calcium Supplement

Although fishbone is potentially a rich source of calcium, an important and an essential mineral for human health and wellbeing, the bioavailability of calcium derived from fish bone has not been extensively investigated to evaluate the possibility to use fishbone calcium as a food supplement (Barrow & Shahidi, 2007). The required calcium for the body is normally supplied through the diet, however, the daily calcium intake in many countries is below the required level in the diet (Barrow & Shahidi, 2007). In this regard, the use of calcium supplements has become a growing market to provide a convenient way to obtain required calcium. Calcium in fishbone appears to have good bioavailability and the consumption of whole small fish has been known to be beneficial for human bone health. Larsen, Thilsted, Kongsbak, and Hansen (2000) fed rats with a diet containing small fish and concluded that Ca from the whole small fish was bioavailable and acted as a good source of calcium. The authors suggested small fish bone as a good source of calcium in the human diet. In another study (Hansen et al., 1998), Ca (397 mg) from small Bengali fish was added to the diet of 19 healthy individuals and the Ca adsorption determined and reported to be comparable to that of skim milk. The potential bioavailability of calcium from large fish bones has been less studied. In a double-blind randomized study, bones obtained from Atlantic salmon and cod were found to be a suitable source of calcium in human diets. The calcium from the enzymatically treated fish bones was well absorbed when added to the diet of 10 young healthy men (Malde, Bügel, et al., 2010). In a similar study, the biosorption of fish bone calcium in pigs was found to be as effective as calcium carbonate tablets (Malde, Graff, et al., 2010).

Bone structure is different in various fish species; the bone of *Teleosts*, for example, is acellular and has no enclosed osteocytes, whereas *Salmonoids* have a cellular bone structure. Despite this known difference in bone structure, information on the functionality and biaxiality of acellular and cellular bones is lacking (Horton & Summers, 2009). The acellular bone has lower mineral content and a smaller dimensioned crystalline structure and therefore has a higher surface area than the cellular bones. This higher surface area to volume ratio of the acellular bone may enhance the biosorption and availability of Ca as compared to cellular bone (Malde, Bügel, et al., 2010). Peptides isolated from fish bone were also found to be effective in increasing the bioavailability of the calcium in the ovariectomized rat. The retention of calcium was increased and also the loss of mineral from bone reduced when the diet of the rats was supplemented with fishbone peptide (Jung, Lee, & Kim, 2006). Fortified foods and supplements containing calcium obtained from fish bone appear to be bioavailable. In this regard, the possibility of sourcing calcium from various commercial species of fish needs to be further investigated and compared to determine their potential commercial use.

2.2 Fishbone for Biomedical Applications

Naturally derived biomaterials play an important role in the production of various biomedical devices for tissue engineering applications. Natural biomaterials are biocompatible and normally have a specific structure and design that allows cells to interact and proliferate. In this regard, for bone tissue engineering, biomaterials with similar properties to bone have been explored and investigated. Bone is composed of organic and inorganic phases. The organic phase is largely composed of collagen type 1 while the calcium phosphates, and in particular hydroxyapatite, form the inorganic phase of the bone. Therefore, biomaterials consisting of hydroxyapatite and collagen, which can mimic the natural composition and properties of the bone, have always attracted researchers in the field of bone tissue engineering. Lin et al. (2010) isolated collagen from the scales of Tilapia after decellularization and decalcification for use in fabrication of a collagen scaffold for a corneal tissue engineering application. The collagen three-dimensional structure remained intact after the isolation process and the obtained collagen was hydrophilic, biocompatible with SIRC [Statens Seruminstitut Rabbit Cornea] rabbit corneal cells and had a positive effect on cell proliferation. In addition, the scaffold exhibited a gas permeability property, a required characteristic for the production of a corneal substitute (Shusterov, Bragin, Bykanov, & Eliseeva, 2001). Fish scale collagen was also found to have a denaturation temperature (T_d) of 36.5°C, which is close to the T_d of mammalian collagen and therefore is a promising alternative for biomedical applications such as bone and skin grafting (Pati, Adhikari, & Dhara, 2010).

Hydroxyapatite composites from mussel shells have been reported to be biocompatible and bioactive, have osteoconductive properties and can interact with other tissues (Shavandi, Bekhit, Sun, Ali, & Gould, 2015). Due to the similar physicochemical properties, hydroxyapatite has been widely studied in the design and fabrication of a range of biomedical products such as dentine and bone applications (Matsuura et al., 2009). One major approach is the fabrication of composites containing hydroxyapatite and type I collagen (Hoyer et al., 2012).

Nano-sized crystals of hydroxyapatite have been synthesized from a range of natural and synthetic materials. Various synthetic approaches, including sol-gel, precipitation, sonochemical, emulsion combustion, and hydrothermal have been evaluated, however, additional chemicals are used in all of these synthetic methods. Hydroxyapatite synthesized from biogenic sources was reported to have better acceptability by the living organism due to having closer physicochemical characteristics to that of human bone and also not requiring additional chemicals for the synthesis (Sadat-Shojai, Khorasani, Dinpanah-Khoshdargi, & Jamshidi, 2013). Hydroxyapatite has also been extracted from fish bone and scales, including cod and catfish bone (Sadat-Shojai et al., 2013; Venkatesan et al., 2015). Marine sources of hydroxyapatite have several advantages over those isolated from land-based animals, such as porcine and bovine, since the latter may be associated with the risk of disease transmission and depending on culture may have religious restrictions, whereas hydroxyapatite from fish has fewer issues in these regards. In addition, abundant sources of fish and shellfish by-products are available and utilizing these by-products would also be beneficial in reducing environmental pollution (Gómez-Guillén, Giménez, López-Caballero, & Montero, 2011; Hoyer et al., 2012). A number of attempts have been made to extract hydroxyapatite from fishbone and use it as an alternative source for chemically synthesized hydroxyapatite in biomedical applications.

Nano-sized hydroxyapatite has been isolated from fishbones using a technique such as alkaline hydrolysis (Cahyanto, Kosasih, Aripin, & Hasratiningsih, 2017). The isolated hydroxyapatite was demonstrated to be beneficial for bone remineralization processes, in having positive effects on bone osteoblast cell proliferation. Hence, hydroxyapatite derived from fish bone has been suggested to be an excellent source of hydroxyapatite for biomaterial and bone tissue engineering application.

3. FISH SKIN

3.1 Fish Skin as a Rich Source of Gelatine and Collagen

Gelatin and collagen are derived from the same macromolecules and have diverse applications in cosmetic, food, and biomedical products. Gelatin is a popular derivative of collagen due to its clarity, bland flavor, and thermo-reversible gel properties with the ability to be used over a diverse range of pH. In the food industry, gelatin serves as an emulsifying agent, providing desired texture and rheological properties to food. In the pharmaceutical industry, gelatin is used to make capsules, tablets, encapsulation of nutrients, and vitamins and also in the fabrication of wound healing pads. Currently, gelatin is mainly obtained from porcine and bovine tendons. The outbreak of bovine spongiform encephalopathy, known as mad cow disease, highlighted the health risks of using bovine-derived products such as gelatin and finding alternative sources for gelatin and collagen and has been a research priority. In addition, in some cultures, religious sentiments limit the application of collagen and gelatin derived from bovine and porcine. For example, Islam Judaism prohibits the consumption of porcine-derived products while in Hinduism the consumption of cow-derived products is forbidden. Knowing that fish and its products are generally considered as halal, fishery waste may be an ideal source for production of gelatin and collagen for halal and Hindu-accepted products (Herpandi et al., 2011). Kosher gelatin will have stricter requirements based on the kashrut (Jewish food law). Marine-derived collagen and gelatin are generally known to be safe and free from diseases.

Collagen is composed of macromolecules forming a triple helical structure, with each building block unit having an average molecular weight of 300 kDa that is composed of three α-chain polypeptides that are folded around each other. Collagen and gelatin both have similar molecular structure. The alpha chains in the collagen are closely packed together due to the presence of a glycine as every third amino acid with centrally oriented hydrogen side chains. The main part of collagen is characterized by a high content of proline, glycine, and hydroxyproline (Shoulders & Raines, 2009). More than 80% of the collagen and gelatin amino acid composition is composed of non-polar amino acids such as glycine, valine, proline, and alanine (Kim & Mendis, 2006) that is different to the fish muscle protein.

The techniques used for extraction of collagen and gelatin can alter the structure of the molecule. Therefore, to preserve the triple helix of collagen, techniques such as acid treatment and solubilization have been used. Methods such as thermal treatment may break the protein covalent and hydrogen bonding by increasing the kinetic energy and causes the molecules to vibrate so rapidly that the bonds stabilizing the triple helical structure are disrupted and result in the conversion of the structure from a helical to a coiled conformation, which is known as gelatin.

Hot water extraction is the most common method for isolation and extraction of gelatin from fish skin. The final properties of the extracted gelatin, however, depend on the extraction parameters used, such as the extraction medium pH and the temperature of the process. Depending on the type of fish collagen, the extraction parameters need to be optimized (Kim & Mendis, 2006).

3.2 Extraction and Characterization

The isolation of collagen from fish skin provides an opportunity for seafood industries to convert this waste into a valuable product for which there is a global demand. Generally, collagen extraction from fish skin requires either acid or alkaline treatment followed by extraction and purification. Collagen yields of 2–64% have been obtained from different species of fish (Aewsiri, Benjakul, Visessanguan, & Tanaka, 2008; Chen et al., 2016; Tan & Chang, 2018).

Gelatin with suitable physicochemical properties such as good gelling ability, has been extracted from the skin of various fish species, including snapper (7%) (Benjakul, Oungbho, Visessanguan, Thiansilakul, & Roytrakul, 2009), hoki (Mohtar, Perera, & Quek, 2010), cuttlefish bream (2%) (Aewsiri et al., 2008), giant catfish (20.1 g/100 g skin) (Jongjareonrak et al., 2010), striped catfish (12.8%) (Singh, Benjakul, Maqsood, & Kishimura, 2011), channel catfish (Tan & Chang, 2018), catfish and leatherjacket (4.2%) (Ahmad, Benjakul, & Nalinanon, 2010). Fish skin gelatin extract is reported to have a high protein content (89.1 g/100 g) (Jongjareonrak et al., 2010) with a different amino acid profile (Jongjareonrak et al., 2010) and content (Benjakul et al., 2009) than bovine gelatin. Gelatin extracted from bigeye snapper had high contents of glycine (>245 mg/g), protein (>96 mg/g), and hydroxyproline (>87 mg/g). The presence of a high percentage of proline and hydroxyproline has been reported to have a positive effect on the structural stability of the extracted gelatin (Benjakul et al., 2009). Similarly, high proline and hydroxyproline content were reported for the gelatin extracted from giant catfish (Jongjareonrak et al., 2010). Glycine was reported as the predominant amino acid of collagen (359 residues/1000 residue) isolated from the skin of giant catfish (Jongjareonrak et al., 2010), leather jacket skin (321 residues/1000 residue) (Ahmad et al., 2010), bigeye snapper (246 residues/1000 residue), and striped catfish (309 residues/1000 residue) (Singh et al., 2011), while proline was the dominant amino acid in tuna fin (288 residues/1000 residue) followed by glycine (222 residues/1000 residue)

(Aewsiri et al., 2008). Various parameters such as the type of solvent used for the extraction, temperature, and extraction time may affect the yield of gelatine extraction from fish skin (Jongjareonrak et al., 2010).

3.3 Gel Strength

The strength of the gelatin gel (bloom strength) obtained from fish skin is reported as being lower than the gelatin derived from bovine (Benjakul et al., 2009; Jongjareonrak et al., 2010). Nevertheless, the gelatin obtained from some fish species demonstrated a higher gel strength compared to others. Gel from the skin of bigeye snapper showed a bloom strength of 254 g, which was higher than gelatin obtained from the skin of other fish species such as giant catfish (153 g) and hoki (197 g) (Mohtar et al., 2010). However, it was lower than the 293 g reported for bovine (Benjakul et al., 2009) and porcine gelatin (307 g) (Mohtar et al., 2010). Viscosity, emulsifying ability, turbidity, solubility, foam stability, capacity, and the color of the gelatin are other important physicochemical properties that have been evaluated for gelatin extracted from fish skin (Aewsiri et al., 2008; Benjakul et al., 2009; Jongjareonrak et al., 2010), which were again dependent on the type of fish and the extraction procedure. The gelatin obtained from giant catfish, for example, had better stability and viscosity compared to the gelatin obtained from calf skin (Jongjareonrak et al., 2010). Arnesen and Gildberg (2007) compared the gelatin from salmon and cod skin and reported a higher gel strength for the salmon gelatin. The higher amino acid content of salmon gelatin is suggested as the reason for the observed higher gel strength of salmon gelatin compared to cod. A considerable gel strength was observed for Atlantic cod gelatin during cold storage, which may be associated with the high content of hydroxylated amino acid in this sample compared to other fish-derived gelatins (Badii & Howell, 2006; Gómez-Guillén et al., 2002).

3.4 Solubility and Turbidity

The pH and concentration of NaCl have an important effect on the solubility and turbidity of the isolated gelatin (Fig. 1). Gelatin from the skin of striped catfish demonstrated the highest solubility at pH 2 (Singh et al., 2011), while gelatin from bigeye snapper skin solubilized well in a wide pH range of 1–10 (Benjakul et al., 2009). The solubility of gelatin showed a decreasing pattern by increasing the pH and the isoelectric point of the fish skin gelatin extracted by acid pretreatment varied from 6.5 to 9 (Johnston-Banks, 1990). During the process of fish skin acid treatment

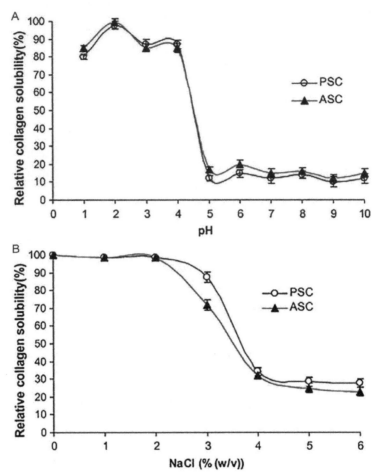

Fig. 1 Relative solubility (%) of acid soluble collagen (ASC) and pepsin soluble collagen (PSC) from the skin of striped catfish as affected by different pH (A) and NaCl (B) concentrations (Singh et al., 2011).

for gelatin isolation, amino acids such as glutamine and asparagine may be deamidated to glutamic and aspartic acids (Jamilah & Harvinder, 2002). At the isoelectric point (PI), the electrostatic repulsion force between protein molecules is compromised and, consequently, the protein molecules are precipitated (Sikorski, 2001). The gelatin from tuna fin and bigeye snapper skin had the lowest solubility at pH 6 and 8, respectively, while gelatin from bovine had the lowest solubility at pH 5. The difference in the molecular weight, amino acid sequence, and the content of polar and non-polar amino acids might explain the observed difference in solubility of gelatin from different species of fish compared to bovine gelatin. At their PI, protein molecules are electrically neutral and the least soluble and tend to aggregate and precipitate from solution, resulting in less interaction between protein molecules and water and therefore increasing the turbidity. With this regard, the gelatin solution turbidity increased by increasing the pH and the gelatin produced from tuna fin had the highest turbidity at pH 9 (Aewsiri et al., 2008). Fish gelatins with a diverse range of pH solubility are potential candidates for application in food industries (Benjakul et al., 2009).

Acid treatment has been normally utilized for extraction of gelatin from fish skin. In this method, the skin is first pre-treated with different NaOH or NaCl solutions to remove non–collagenous materials and then subjected to different concentrations of acids such as hydrochloric acid, acetic acid, lactic and citric acid (Tan & Chang, 2018), for different times and a number of extraction cycles. However, 0.5 M acetic acid has been commonly used for the extraction of fish collagen following the method developed by Nagai and Suzuki (2000). Fish skin may not dissolve completely during acid treatment (Tamilmozhi, Veeruraj, & Arumugam, 2013), due to cross-linked covalent bonding and intermolecular cross-linked structure of the collagen. Therefore, the insoluble residue in some studies (Singh et al., 2011; Tamilmozhi et al., 2013) has been further subjected to proteolysis using pepsin to isolate soluble collagen fractions, that does not affect the triple helix structure of the isolated collagen (Tamilmozhi et al., 2013).

3.5 Molecular Weight Distribution of Gelatin Isolated From Fish Skin

Collagen products isolated from fish skin have all been found to have two distinct chains of $\alpha1$ and $\alpha2$ with a ratio of 2–1 and the cross-linked chains of β (dimer) and γ (trimer) (Ahmad et al., 2010; Mohtar et al., 2010). The α-chain proteins were generally reported to have a molecular weight in the range of 100–140 kDa, while a higher molecular weight of 200 and above 200 kDa has been documented for β- and γ-chains, respectively (Aewsiri et al., 2008; Chen et al., 2016; Minh Thuy, Okazaki, & Osako, 2014; Tamilmozhi et al., 2013) (Fig. 2).

The protein isolated from fish skin and scale is similar to type I collagen obtained from calf skin and therefore, fish collagen has been recommended as a primary source of type I collagen (Ahmad et al., 2010; Singh et al., 2011; Tamilmozhi et al., 2013). Identical SDS–PAGE gel electrophoresis patterns of collagen have been reported for collagen samples isolated from the skin of other fish species such as Tilapia (Chen et al., 2016), catfish

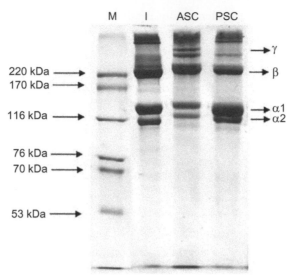

Fig. 2 SDS–PAGE of acid soluble collagen (ASC) and pepsin soluble collagen (PSC) from the skin of striped catfish. M, high molecular weight markers; I: type I calf skin collagen (Singh et al., 2011).

(Jongjareonrak et al., 2010), and sailfish (Tamilmozhi et al., 2013). The observed chains of α1 and α2 are also identical to standard type I collagen, which is found in the human placenta (Tamilmozhi et al., 2013).

3.6 Viscosity, Foam Forming Ability, and Stability of Fish Gelatin

Gelatin from bovine and porcine sources is traditionally known to have a superior viscosity, gel strength, and thermal stability properties compared to marine-derived gelatin (Cho, Gu, & Kim, 2005). Viscosity is an important characteristic of collagen, which is mainly due to the well-structured system of collagen. Due to the presence of strong repulsion force between chains of collagen molecules, even a low concentration of collagen results in high viscosity (Zhang, Liu, & Li, 2009).

Viscosity is a critical parameter for commercial applications of gelatin. Catfish skin-derived gelatin with a viscosity of 112.5 cP was found to be almost four times more viscous than calfskin gelatin with a viscosity of 31.3 cP (Jongjareonrak et al., 2010). Several factors affect the viscosity of the isolated gelatin including molecular weight, molecular size distribution, the source of the collagen, temperature, and pH (Arnesen & Gildberg, 2007; Jongjareonrak et al., 2010). The viscosity values of gelatin solutions obtained from hoki, bovine, and porcine skin were compared at a fixed concentration of 6.67% and the hoki gelatin with a viscosity of 10.8 cP was comparable to the bovine sample (9.8 cP) but was significantly higher than porcine gelatin (5 cP) (Mohtar et al., 2010). The viscosity of collagen is a function of temperature and normally decreases with increasing temperature (Fig. 3). This inverse relationship is due to the denaturating effect of the temperature, which converts the triple helix structure into a random coil conformation through depolymerization (Gurdak, Booth, Roberts, Rouxhet, & Dupont-Gillain, 2006). Data from several studies identified a denaturation temperature of 25–31°C for fish skin collagen (Ahmad et al., 2010; Arnesen & Gildberg, 2007). Given that the denaturation temperature of collagen from cold water fish species is lower than that of warm water fish, the variation in the viscosity of the collagen from different fish species may be partly due to the environmental conditions of the fish habitat (Nagai & Suzuki, 2000).

The T_d values of the fish collagen were lower compared to that of porcine collagen (37°C) (Nagai & Suzuki, 2000). Treatment at higher temperature can break and dissociate the intermolecular hydrogen bonding, resulting in destabilization of the collagen structure (Ahmad et al., 2010; Arnesen & Gildberg, 2007).

A protein with good foam forming properties can transport and migrate to an air–water interface and is able to unfold and rearrange at the interface.

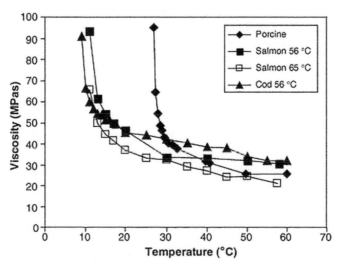

Fig. 3 Viscosities of solutions of gelatin extracted from porcine, Atlantic salmon, and Atlantic cod skin as functions of temperature (Arnesen & Gildberg, 2007).

The presence of hydrophobic residues can positively affect the foaming characteristics of the protein, making it more flexible and increasing the capability of the protein to reduce surface tension (Mutilangi, Panyam, & Kilara, 1996). A protein with hydrophobic residues can also better migrate to an air–water interface. With this regard, giant catfish skin collagen with a high content of non-polar amino acids such as alanine, proline, and valine has been demonstrated to have better foaming capability compared to calf-skin gelatin (Jongjareonrak et al., 2010). However, the foam expansion and stability of tuna fin gelatin was reported to be lower than porcine skin gelatin (Aewsiri et al., 2008).

3.7 Thermal Stability of Fish Skin Collagen

The successful application of gelatin in different food and pharmaceutical applications is partly related to its melting and gelling temperature. Gelatin with higher melting and gelation points has normally a wider range of applications. In this regard, porcine and bovine gelatins are known to have higher melting and gelling points compared to most fish-derived gelatins (Gudmundsson, 2002). Different melting temperatures ranging from 24.3 to 36.5°C have been reported for different types of fish-derived gelatin (Gilsenan & Ross-Murphy, 2000), however calfskin gelatin with a melting point of 40.8°C is higher than most of the reported melting point values for gelatin derived from fish skin (Zhang et al., 2009).

Farmed salmon is a major fish source in Europe and in other countries including Canada, Chile, and New Zealand. Due to selling the fish with skin on, salmon skin is not normally considered as a by-product of the industry, nevertheless the demand for skinless product has risen and salmon skin which accounts for 5% of the fish weight has attracted attention as a good source for collagen isolation (Arnesen & Gildberg, 2007; Kołodziejska, Skierka, Sadowska, Kołodziejski, & Niecikowska, 2008). Generally, gelatin can be successfully isolated from fish skin, however, this fish–derived gelatin usually has inferior properties compared to mammalian gelatin such as lower melting point and molecular weight. More studies are deemed necessary to improve the properties of gelatin isolated from fish skin.

3.8 Applications of Collagen and Gelatin

Collagen and gelatin have a wide range of applications in food, cosmetic and pharmaceutical industries. They also have been explored for production of protein hydrolysate bioactive compounds with biomedical and food applications. Protein hydrolysates of collagen and gelatin have been demonstrated to have good absorption and affect lipid adsorption and bone metabolism when tested in animal models (Gómez-Guillén et al., 2011). The rheological behavior of collagen and gelatin is important in certain food and pharmaceutical applications. In addition, taste, color, transparency, and solubility are also considered as important parameters that affect the suitability of these proteins for their applications. Some of these properties such as the viscoelastic and thermal behavior of the gelatin are directly related to the protein molecular weight and the amino acid content and composition. With this regard, the gel strength is influenced by the content of proline and hydroxyproline. It is also suggested that the pyrrolidine amino acid content of the protein extract can affect the gel stability (Hayes & McKeon, 2014). In this section, the use of gelatine in cosmetic, pharmaceutical, and adhesive applications is presented.

Gelatin isolated from cold water fish skin such as Alaska pollock has low-temperature gelling properties and normally gels below 11.9°C (Zhou, Mulvaney, & Regenstein, 2006). In comparison, gelatin from fish skin of catfish that lives in warm waters has a gelling point temperature of 22°C (Liu, Li, & Guo, 2008). With this regard, mammalian gelatin has a higher gelling temperature, which is 27°C for pork skin and 24.7°C for cattle hide (Gudmundsson, 2002; Liu et al., 2008). Gelatin from fish such as Tilapia or tuna has a gelling temperature of 25–27°C, which closely resembles mammalian gelatin and therefore can be used as alternatives (Karim & Bhat, 2009). Gelatin from cold water fish has potential application in the production of frozen or refrigerated products, which normally need to be consumed shortly after removal from fridge. Cold water fish skin collagen can also be used successfully to prevent texturization of food products (Karim & Bhat, 2009). Gelatin from cold water fish has also been largely used for the microencapsulation of vitamins, colorants, and supplement products such as azoxanthine (Gaonkar, Vasisht, Khare, & Sobel, 2014). Fish gelatin can be used in producing products such as low-fat spreads and dairy products.

3.9 Gelatin in Food and Industrial Application

Gelatin is in high demand in the food industry for applications such as gel formation, foam formation, and to improve the stability of food products. Gelatin is the main ingredient in products such as table jellies and gummy products. Gelatin, however, can interact with the sugar present in the formulation and precipitate or show weaker gelling properties, or even result in hardening of the product. The gelatin sugar interaction normally occurs in products formulated with low water content as a result of competition between sugar and gelatin to access water in the food system (Shimizu & Matubayasi, 2014). The foaming ability of gelatin has been used in marshmallow production, while in yogurt manufacturing gelatin has been used to suppress the floury mouth feeling of added starch. In ice cream or other desserts, gelatin prevents ice crystal formation in the product's structure and provides a favorable creamy smooth texture that melts at body temperature (Karim & Bhat, 2009; Soukoulis & Fisk, 2016). The result of a sensory study by Choi and Regenstein (2000) suggested that desserts containing fish gelatin had a better ability to release flavor and aroma than commercial non-fish gelatin. The sensory property of the gelatin can be related to the low gel melting properties of the fish gelatin (Choi & Regenstein, 2000). In another study, gel desserts made from different gelatin sources were compared using puncture test and texture profile analysis (TPA). Desserts prepared using Alaska pollock gelatin have been reported to be more resistant to mechanical stress and deformation compared with desserts made from pork and tilapia gelatin (Zhou & Regenstein, 2007).

The gelling and foaming characteristics of gelatins are variable and can change depending on parameters such as the type of fish, extraction method, and seasonal variation. Therefore, gelatin from fish needs to be carefully selected and characterized before use in a particular food application. Gelatin has also been widely used as a clarification agent in apple juice production and winemaking. Gelatin can precipitate floating particulates by coagulating them through charged ion interactions (Benitez & Lozano, 2007).

3.10 Nutritional Supplement, Cosmetic, and Biomedical Applications

Gelatin is a rich source of certain amino acids such as glycine, with a beneficial impact on nail and hair growth, as well as reducing the symptoms associated with arthritis (Karim & Bhat, 2009). In addition, gelatin has a relatively low calorific value of 3.5 kcal/g and has good digestibility (Johnston-Banks, 1990). Therefore, gelatin is a promising candidate for the preparation of low-fat products.

Gelatin and collagen are also important natural polymers with large application in the field of biomaterial and tissue engineering implants for tissues such as cartilage, bone, and skin. Collagen is biocompatible, safe, and does not cause mutagenic immunologic reactions. Collagen forms layers in the bone and provides a high-density filament network. The number and distribution of collagen fibrils (average diameter = 100 nm) in tissue usually determines its mineralization status (Hing, 2004). The superiority of collagen-based polymers could also be due to their availability, their ability to interact with the host tissue and low toxicity behavior. The high water holding capacity of collagen is also exploited in the production of different creams for the treatment of dry skin and also various hydrogels for tissue engineering applications (Salgado, Coutinho, & Reis, 2004).

Collagen has successfully been used in the production of tissue glues for various surgical and wound healing applications. The collagen-based glues were found to be less toxic than synthetic glues composed of cyanoacrylates and aldehydes (Taguchi et al., 2006). Fish gelatin can also have other applications in the food industry when combined with other thickener or gelling agents such as pectin. Cheng, Lim, Chow, Chong, and Chang (2008) have formulated low-fat spreads using different ratios of gelatin to pectin. The physicochemical characteristics of the spread such as bulk density, compressibility, elasticity, and firmness improved using a low gelatin to pectin ratio of 1:2 (Cheng et al., 2008). Cross-linking is another approach to extend the fish gelatin application. In one study, gelatin from the skin of Baltic cod (*Gadus morhua*) that was cross-linked with transglutaminase was stable for 30 min of heating in a boiling water bath. Fish gels with high thermal stability, therefore, can be good candidates for use in sterilized products (Kołodziejska, Kaczorowski, Piotrowska, & Sadowska, 2004).

4. FISH MUSCLE AND INTERNAL ORGAN PROTEINS

The fish processing industry produces a lot of bone frames and trimmings during the removal of the muscles. Being rich in various nutrients, these muscle proteins are a valuable source for production of different functional protein hydrolysates. Traditionally, fish protein extracts have been added to various sauces and fish flavor enhancers in several culinary applications. As food supplements, fish protein hydrolysates also have an important place in the modern food industry.

The fish filleting process is a major source of considerable quantities of fish frames that can be converted to valuable functional hydrolysate products containing bioactive peptides. A number of different techniques such as autolysis and thermal or enzymatic hydrolysis have been used for the production of fish by-product hydrolysates. Nevertheless, enzymatic hydrolysis is recognized as the most economical approach for recovery of proteins from fish processing waste and the protein hydrolysate from this method is a peptide mixture with diverse health-promoting characteristics such as antioxidant capacity and angiotensin-I-converting enzyme (ACE) inhibitory activity (Halim, Yusof, & Sarbon, 2016). Therefore, a lot of attention has been paid to proteases used in the production of fish waste protein hydrolysates (Gajanan, Elavarasan, & Shamasundar, 2016). Various proteases including pepsin, Alcalase, trypsin, bromelain, proteases N and A, and thermolysin have been used for this process (Chalamaiah, Dinesh Kumar, Hemalatha, & Jyothirmayi, 2012). The antioxidant properties and angiotensin-I-converting enzyme (ACE) inhibitory activity (antihypertensive properties) are reported as the major health-promoting properties of these fish-derived peptides (Halim et al., 2016).

During hydrolysis, proteins are converted into smaller peptides, some of which exhibit different health-promoting effects on physiological functions in the body. In addition, fish-derived hydrolysates have strong free radical scavenging activity, anti-inflammatory, antibacterial, and immune-modulatory activities (Chakrabarti, Jahandideh, & Wu, 2014). Fish muscle, skin, viscera, roe, head, fins, and trimmings are a major source for production of fish protein hydrolysates. In addition to fish muscle by-products generated during removal of the muscle from the bone frame, the dark flesh is another source for production of fish hydrolysates. Dark muscle is commercially of low value. The conversion of these by-products leads to the production of high-value products (Benhabiles et al., 2012).

There is a large number of studies dealing with the conversion of fish muscle to hydrolysates. Some recent examples include threadfin breams (*Nemipterus japonicus*) (Gajanan et al., 2016), sardine (*Sardina pilchardus*) (Vieira, Pinho, & Ferreira, 2017), and small-spotted catshark (*Scyliorhinus canicula*) (Vázquez, Blanco, Massa, Amado, & Pérez-Martín, 2017).

Fish skin is not only a good source for isolation of collagen and gelatin but also a good source for production of protein hydrolysates. In some studies, collagen (Wang et al., 2013) or gelatin (Razali, Amin, & Sarbon, 2015) was isolated from the skin and then converted to hydrolysates. There are also studies on direct hydrolysis of fish skin (Sampath Kumar, Nazeer, & Jaiganesh, 2012). In addition, the potential of fish internal organs or viscera was also explored for production of hydrolysates. In research on skipjack tuna (*Katsuwonus pelamis*) viscera, Alcalase was applied for the hydrolysis and the hydrolysate obtained contained more than 80% protein with a high content of essential amino acids (43.13%). Glutamic acid, methionine, and aspartic acid were reported as the major amino acids

(Klomklao & Benjakul, 2017). High-quality hydrolysates have also been obtained from carp (Bhaskar & Mahendrakar, 2008) and beluga internal organs (Molla & Hovannisyan, 2011), viscera and roe obtained from rohu (Chalamaiah et al., 2014) and channa or Labeo (Galla, Pamidighantam, Akula, & Karakala, 2012).

4.1 The Production of Fish Protein Hydrolysates

Enzymatic and thermal hydrolysis along with autolysis and bacterial fermentation are methods that have been used for the production of fish protein hydrolysates (FPH) (Halim et al., 2016). In an autolysis process, the endogenous proteases existing in muscle and internal organs of the fish are used for breaking down of proteins into smaller peptides and amino acids (Prabha, Narikimelli, Sajini, & Vincent, 2013). In a recent study, hydrolyzed collagen was extracted from croaker fish scales *via* thermal hydrolysis at temperatures ranging from 60 to 100°C. When the temperature was increased from 70 to 100°C the yield significantly increased from 6% to 30%, respectively. The process which was a zero order reaction at 70°C, became a first-order reaction when the temperature was increased to 90°C. Given that no other chemicals or enzymes were used in the process, thermal hydrolysis is an attractive approach for the production of FPH, however, the effect of temperature on physicochemical properties of the final product such as amino acid composition needs to be evaluated (Olatunji & Denloye, 2017). In another study (Wang et al., 2013), thermal and enzymatic hydrolyses were compared in the production of Tilapia skin hydrolysate. In the thermal process, the fish skin was added to buffers with pH values of 4, 6.5, and 9 and then autoclaved at 121°C for 3 h. Three different proteases alkaline protease, Alcalase, and papain were used in combination for the enzymatic process and it was found that compared to the thermal process, less processing time, and more gentle conditions were required, that resulted in hydrolysate products with low molecular weight peptides.

Jemil et al. (2014) investigated two proteolytic bacteria, *Bacillus subtilis* A26 (FSPH-A26) and *Bacillus amyloliquefaciens* An6 (FSPH-An6) in producing sardinelle fish hydrolysate *via* fermentation. A fish protein hydrolysate (FPH) with a high protein content of up to 81% was achieved, which had high DPPH radical scavenging activity (75% at 6 mg/mL).

Enzymatic processing is a relatively fast process and preserves the product nutritional content, hence there is a large body of literature on fish protein hydrolysis using proteases. In a normal process of enzymatic hydrolysis, the homogenized mixture is first heated up to 90°C to inactivate any endogenous enzymes to make it possible to evaluate the effect of exogenous proteases added later in the process (Halim et al., 2016; Villamil, Váquiro, & Solanilla, 2017). A number of parameters such as concentration and type of protease, the pH value of the mixture and the processing time and temperature must be controlled for the hydrolysis process (Villamil et al., 2017). Fig. 4 shows processes involved in the recovery of the proteins from fish viscera.

Under optimum enzymatic hydrolysis conditions, enzymes have been used in concentrations of up to 5% of the sample. However, a concentration range of 0.1–1.5% has been widely used in the literature (Salwanee, Wan Aida, Mamot, Maskat, & Ibrahim, 2013; Shirahigue et al., 2016; Villamil et al., 2017). In terms of optimum pH, values ranging from acidic (2.5) (Jai ganesh, Nazeer, & Sampath Kumar, 2011) to alkaline (11) (Liu, Wang, Peng, & Wang, 2013) have been tested, while neutral (7) to slightly basic pH of 8 have been commonly used (Villamil et al., 2017). Time of the reaction has also been varied widely from a few minutes to more than 20 h (Halim et al., 2016), while 1–2 h have been commonly suggested as the optimum processing time. A temperature around 50°C has also been demonstrated to be the optimum processing temperature (Halim et al., 2016; Villamil et al., 2017). The peptides become smaller in size and more peptide bonds are hydrolyzed by increasing the protease concentration and the temperature. This trend continues until the protease thermally denatures (Jamil, Halim, & Sarbon, 2016). After the completion of the hydrolysis, the proteases are deactivated by elevating the temperature to 95°C for 5–20 min (Intarasirisawat, Benjakul, Visessanguan, & Wu, 2014).

4.2 Characteristics of Fish Protein Hydrolysates

Fish protein hydrolysate (FPH) normally has a high content of protein primarily due to the solubilization of the protein during the hydrolysis reaction and removal of non-protein compounds such as fat and other insoluble materials. In some cases, the FPH has also been subjected to a drying step (Chalamaiah et al., 2012; Dong et al., 2008). Fish hydrolysates contain all of the essential amino acids, making them a good nutritional product. The amino acid composition of the FPH product can affect its bioactive and functional properties. The amino acid composition of the FPH is affected by the enzymatic reaction parameters such as the type of enzyme,

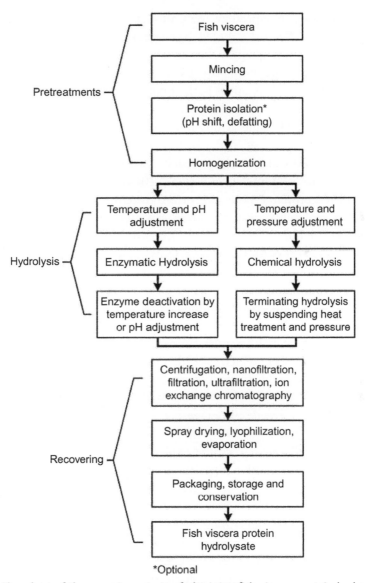

Fig. 4 Flowchart of the generic process of obtaining fish viscera protein hydrolysates (FVPH) (Villamil et al., 2017).

pH value, and time (Bhaskar, Benila, Radha, & Lalitha, 2008). The enzymatic hydrolysis influences the protein functional properties by changing the peptides and amino acid composition of the proteins. Therefore, a controlled enzymatic process can result in hydrolysate products with improved functional properties such as improved pH functionality compared to the native proteins (de Castro, Bagagli, & Sato, 2015). In particular, protease affects the molecular weight and hydrophobic properties of proteins and consequently the protein functional properties (Villamil et al., 2017). Therefore, with proper selection of the enzyme and substrates, it is possible to achieve peptides in the hydrolysate with a desired molecular weight range. Fish protein hydrolysates demonstrate and have good solubility over a wide pH range at different ionic strengths and has good heat stability (Li, Luo, Shen, & You, 2012). A hydrolysate with good emulsifying, foaming, and water holding capacity also can positively contribute to the food system properties (Jemil et al., 2014). The functional properties of the isolated protein can be categorized in terms of hydrodynamics, such as viscosity, solubility, water holding capacity, as well as foaming, film forming, and emulsifying which are known as surface-active properties. Among various properties of the proteins, solubility is recognized as one of the most influential characteristics which can significantly affect other properties (de Castro & Sato, 2014).

Hydrolysates with greater solubility can be obtained by increasing the time of the hydrolysis reaction resulting in smaller peptides with lower molecular weight. Small peptides have consequently more ionizable polar groups on their surface, which become more able to form hydrogen bonds with water molecules (de Castro & Sato, 2014; He, Franco, & Zhang, 2013). These low molecular weight peptides with more polar residues are also more soluble over a wide pH range compared to native proteins (Taheri, Anvar, Ahari, & Fogliano, 2013). The low solubility of large protein molecules such as collagen can also be explained by the low content of polar groups resulting in the solvent having limited possibility to form hydrogen bonds (Chi et al., 2014). In this regard, higher solubility has been reported for FHP compared to native collagen proteins (Liu et al., 2014). The solubility of surimi protein containing by-product increased from 10% in native form to above 60% after proteolysis with Protamex and Alcalase (Liu et al., 2014). When sardinella viscera samples were hydrolyzed with Alcalase the hydrolysate was completely soluble, indicating the potential of FPH for use in diverse food and biomedical applications (Souissi, Bougatef, Triki-Ellouz, & Nasri, 2007). Hydrolysis of proteins can result in the exposure of hydrophobic groups, with the only increase in ionizable groups due to the generation of amino and carboxyl termini on the peptides (Kristinsson & Rasco, 2000; Jemil et al., 2014).

The degree of hydrolysis influences the water and fat holding capacities. Water holding capacity is another important property of proteins, which affects the texture and mouthfeel characteristics of the food system (Kristinsson & Rasco, 2000). A direct relationship has been observed between water holding capacity and the degree of protein hydrolysis, due to the increase in polar carboxyl and amine groups as a result of increased hydrolysis (Balti et al., 2010). A water holding capacity of up to 7.7 g/g has been documented in the literature for fish-derived hydrolysates (Jemil et al., 2014).

One of the important physicochemical properties of proteins used in food applications is the ability to absorb and retain oil/fat. This property is directly related to the bulk density of the hydrolysate as well as degree of hydrolysis and in the reported literature it normally ranges from 1 to more than 10 mL/g for fish protein hydrolysates (dos Santos, Martins, Salas-Mellado, & Prentice, 2011; Halim et al., 2016; Pires & Batista, 2013; Taheri et al., 2013). A similar oil hold capacity has been observed for the hydrolysate produced from cuttlefish viscera (Balti et al., 2010). In food manufacturing, peptide ability to hold oil affects the product taste, which is particularly important in the meat and bakery industry (Taheri et al., 2013; Tanuja, Viji, Zynudheen, & Joshy, 2012).

Given their amphipathic structure, proteins can absorb at the interface of oil and water, and behave as an emulsifier by reducing interfacial tension between hydrophobic and hydrolytic components. The emulsifying property of a hydrolysate is influenced by the size and molecular weights of the peptides in the hydrolysate and their surface properties, the type of protease used to generate the hydrolysate and the composition of amino acids (Liu et al., 2014; Pires & Batista, 2013), the concentration of protein, solution pH, the volume fraction of oil and the instrument used for the preparation of emulsion (Šližyte, Daukšas, Falch, Storrø, & Rustad, 2005).

The pH of the solution, in particular, can affect the hydrophobicity of the protein surface and therefore change the protective layer covering the globules of lipid and consequently change the solubility (Taheri et al., 2013). An emulsion with optimum stabilization can be obtained in the pH range of 8–11. The better emulsifying properties of the hydrolysate at alkaline pH may be explained by the unfolding of the protein chain in the alkaline conditions. Due to this unfolding, the hydrophilic residues of the proteins are exposed and consequently improve the interaction at the oil–water interface (Taheri et al., 2013).

The stability of the emulsion reduces as the size of the protein decreases, which could be due to weaker interfacial films that surround the droplet of the emulsion (Galla et al., 2012). The hydroxylated amino acids, hydroxyproline and hydroxylysine, present in the hydrolysate of rainbow trout by-products were found to contribute to the emulsifying properties of the sample (Taheri et al., 2013). Good emulsifying properties have been reported for peptides with a minimum of 20 residues (Lee, Shimizu, Kaminogawa, & Yamauchi, 1987).

It was found that as the hydrolysis progresses, an increasing number of smaller peptides are produced and although the ability to achieve emulsification is initially enhanced, subsequently, the extensively hydrolyzed product exhibits a reduced ability to form an emulsion (Balti et al., 2010). In spite of this, fish-derived hydrolysate was shown to still have better emulsifying properties compared to some commercial emulsifiers that are used in the food industry (He et al., 2013). However, in spite of extensive laboratory results supporting good functional properties of fish protein hydrolysates, there is limited information on their commercial application.

4.3 Bioactive Compounds of Fish Protein Hydrolysate

Bioactive or biologically active peptides are beneficial for human health due to their positive effects on physiological activity such as that of antihypertension and anti-cancer activity (Wu et al., 2015). Marine-derived peptides with strong antioxidant activity are able to scavenge free radicals and prevent lipid oxidation. In contrast to peptide hydrolysates that are biologically active, the native or unhydrolyzed proteins are biologically inactive, requiring an *in vivo* or *in vitro* hydrolysis process to generate peptides exhibiting bioactivities such as anti-inflammation, antibacterial, or antioxidant activity (Chakrabarti et al., 2014).

In the food industry, lipid peroxidation and its undesirable outcomes such as off-flavors and potential toxicity issues have attracted a lot of attention among researchers. To prevent or reduce the negative effect of oxidation, a number of antioxidants have been developed and are utilized commercially in industry. Examples are *tert*-butylhydroquinone (TBHQ), butylated hydroxyanisole (BHA), and butylated hydroxytoluene (BHT). However, in spite of the efficiency of these synthetic antioxidants, their toxicity and potential health hazards limit their utilization (Centenaro, Mellado, & Prentice-Hernández, 2011). Studies have also been focused on the development of naturally synthesized alternative oxidants from fish hydrolysate (Centenaro et al., 2011).

Hypertension is another concern in relation to blood pressure health status. Current drugs for the treatment of hypertension, such as captopril or lisinopril, are synthetic medications with potential side effects

(Ghassem, Arihara, Babji, Said, & Ibrahim, 2011). Natural alternative medications, therefore, have been explored and fish-derived products with good ACE inhibitory activity have been reported by various studies (Halim et al., 2016; Wijesekara & Kim, 2010).

The anti-cancer properties of compounds derived from fish by-products have also highlighted their potential use for producing products with biomedical applications and thus converting by-products to high-value products (Suarez-Jimenez, Burgos-Hernandez, & Ezquerra-Brauer, 2012).

Fish hydrolysates have shown desirable antioxidant activities (Ketnawa & Liceaga, 2017; Lassoued, Mora, Nasri, Aydi, et al., 2015). The antioxidant capacity of the FPH can be determined using current analysis systems such as the 1,1-diphenyl-2-picryl hydrazyl (DPPH) method that is the most used assay for measurement of *in vitro* antioxidant activity (Alam, Bristi, & Rafiquzzaman, 2013). There are however other commonly used assays such as the ferric reducing antioxidant power (FRAP), thiobarbituric acid reactive substance (TBARS), and reactive oxygen species (ROS) scavenging (Chalamaiah et al., 2012).

The amino acid composition, hydrophobicity, and amino acid sequences of the peptides determine the antioxidant activity level of the hydrolysate (Lassoued, Mora, Nasri, Aydi, et al., 2015). Peptides with antioxidant activity are normally between 5 and 16 amino acids in length and composed of hydrophobic and aromatic residues such as valine, proline, histidine, and tyrosine. Tyrosine, in particular, has a substantial role in the free radical scavenging activity of a peptide which might be due to its phenolic chains that behave as electron donors and therefore terminate the chain reaction of the radicals (Picot et al., 2010). Typically it is understood that molecular weight also plays an important role in determining the activity of peptides (Chi, Hu, Wang, Li, & Luo, 2015). Previous studies have indicated that fish-derived peptides exhibiting antioxidant activities are typically of low molecular weight of up to 1.5 kDa (Centenaro et al., 2011; Lassoued, Mora, Nasri, Jridi, et al., 2015), that can act as electron donors which react with free radicals and consequently stop chain reactions (Chi et al., 2014). Antioxidative peptides of approximately 1.3 kDa obtained from ornate threadfin bream muscle have some of the highest reported ABTS radical scavenging activity (Nalinanon, Benjakul, Kishimura, & Shahidi, 2011). In another study, a hydrolysate of horse mackerel viscera was found to be better than α-tocopherol in the prevention of lipid oxidation (Sampath Kumar, Nazeer, & Jaiganesh, 2011). Hydrolysates of black pomfret and golden gray mullet fish waste were able to protect DNA from damage induced by hydroxyl radicals, and these hydrolysates were also found to be safe toward human erythrocytes (Jai ganesh et al., 2011). All of these outcomes suggest the potential of fish protein hydrolysate as functional additives for food and biomedical applications.

4.3.1 Antihypertension Activity

Angiotensinogen is a protein that is synthesized and released by the liver. This protein is then converted to angiotensin I by the action of renin, an enzyme produced in the kidney. Angiotensin I is not an active biological compound, however, an enzyme called angiotensin-converting enzyme (ACE) converts it to the reactive form "angiotensin II." Angiotensin II has numerous receptors in most human tissues and consequently results in blood vessel constriction and increased blood pressure. This normally occurs due to low blood sodium level and as a result, low blood pressure which can trigger increased renin production by the kidney (Yim & Yoo, 2008). Inhibition of ACE is typically a suitable hypertension treatment approach (Edgar et al., 2011) and there are chemically synthesized drugs available, but the side effects of these synthetic drugs have motivated the search for natural alternatives. Bioactive peptides with ACE inhibitory activity have been reported to be present in many fish by-product hydrolysates, including the freshwater fish *Cirrhinus mrigala* (Elavarasan, Shamasundar, Badii, & Howell, 2016), lizardfish (*Saurida elongata*) (Sun et al., 2017), and thornback ray (Lassoued, Mora, Nasri, Jridi, et al., 2015).

The ACE inhibitor activity of the protein hydrolysate also depends on the method used for the isolation and separation of bioactive compounds. Membrane size filtration and various chromatography techniques such as gel permeation, ion exchange, and HPLC have been evaluated and utilized to enrich for bioactive peptides (Hyoung Lee, Ho Kim, Sik Park, Jun Choi, & Soo Lee, 2004; Pihlanto, Virtanen, & Korhonen, 2010; Sun et al., 2017; Wu, Aluko, & Muir, 2009).

In a recent study, Sun et al. (2017) isolated an ACE inhibitory peptide with an IC50 value of 52 μM from *Saurida elongata* protein hydrolysate using nickel ion immobilized metal affinity chromatography (IMAC-Ni^{2+}). A peptide with the amino acid sequence Arg-Tyr-Arg-Pro was determined to have ACE inhibitory activity (Sun et al., 2017). Amino acid sequences containing tryptophan, tyrosine, proline, or phenylalanine can bind to ACE and act as a competitive inhibitor to block its activity (Cushman & Cheung, 1971). Although peptides present in fish by-product hydrolysates exhibit antihypertensive activity, they have not to date been as popular as antihypertensive peptides derived from plants and milk. In this regard, more research is required to achieve a clearer picture of the health benefits of fish-derived hydrolysates to facilitate commercial application (Halim et al., 2016; Rustad & Hayes, 2012). In addition, there is a shortage of information regarding the function of these fish-derived peptides in food systems, the production costs, and the sensory properties of these proteins (Rustad & Hayes, 2012).

4.3.2 Anti-Cancer Activity

Despite all of the advancement in research and investment in novel drugs, cancer remains a substantial health risk. Chemotherapy and radio-therapy approaches are widely used treatments, however, these treatments cause severe side effects resulting from chemical and radiation toxicity. Therefore, there is an urgent need for the development of accurate preventive/diagnostic and nontoxic therapeutic methods for cancer treatment (Jain, El-Sayed, & El-Sayed, 2007; Tietze et al., 2015).

Fish protein hydrolysates have been reported to have cellular antiproliferative activities, which may offer new approaches for cancer treatment (Suarez-Jimenez et al., 2012). Examples of hydrolysates that have been studied are that derived from the solitary tunicate *Styela clava* (Jumeri & Kim, 2011), which exhibited antiproliferative activity against AGS, DLD-1, HeLa cells, and hydrolysates of the flying fish *Exocoetus volitans* (Naqash & Nazeer, 2010) which was cytotoxic toward Vero cells, but exhibited an antiproliferative effect on human hepatocellular liver carcinoma Hep G2 cell lines. The anti-cancer properties of the fish hydrolysates indicate their potential for cancer treatment, that has also been proposed to be a suitable molecular model for drug delivery studies (Suarez-Jimenez et al., 2012). However, all of these *in vitro* studies of the hydrolysates are just indicative and animal trials should be used for confirmation of toxicity and curative effects.

Enzymatic hydrolysis of fish by-products is the commonly used technique for the production of peptides with anti-cancer characteristics, as the alkaline processing at high pH may damage the amino acid structure, and in addition, acid processing is not a desirable technique due to the complete destruction of tryptophan residues and damage to other amino acids such as serine, threonine, glutamine, and asparagine (Neklyudov, Ivankin, & Berdutina, 2000; Walker & Sweeney, 2002).

Oxidative damage to cell proteins and DNA due to free radicals can cause cancer (Oberley, 2002). Therefore, the anti-cancer properties of fish hydrolysates may be directly related to the antioxidant activity of the hydrolysates. Despite some amino acid sequences rich in Glu, Gly, Pro, Ala, Thr, Val, and Met that are reported to account for the anti-cancer properties of the hydrolysate *in vitro* (Hsu, Li-Chan, & Jao, 2011), there are no conclusive studies regarding the amino acids that contribute the most to the antiproliferative activity of the hydrolysate (Halim et al., 2016). Hence, more research is required to identify amino acid sequences that contribute the highest antiproliferative bioactivity.

5. FISH OIL

Fish contain from 2% to 30% oil depending on the type of fish, seasonal variation plus environmental and geographical differences (Kim & Mendis, 2006). Fishing of most species just for the production of oil is not a sensible approach, but by using fish waste it is possible to produce 20–80 kg of oil per ton of waste (Karadeniz & Kim, 2014). The annual worldwide production of fish oil is around 1 million tons which is mainly used in the aquaculture industry and only 5% of this amount is used for the production of omega-3 fatty acids (De Meester, Watson, & Zibadi, 2012).

Fish oil is largely composed of polyunsaturated fatty acids (PUFA), of which eicosapentaenoic acid (EPA) and docosahexaenoic acid (DHA), are known as omega-3 fatty acids. In this regard, fish from cold waters have been reported to contain a higher amount of unsaturated fat. PUFA (polyunsaturated fatty acids) in fish oil, compared to other oils, are better digested and are suggested to have positive bioactivities (Kim & Mendis, 2006).

Eicosapentaenoic acid (EPA) and docosahexaenoic acid (DHA) have proven to be beneficial for human health through imparting a positive effect on cardiovascular health, reducing the risk of heart attack, diabetes, and cancer, improving the immune system and the development of the neural system (Ivanovs & Blumberga, 2017).

The quality of fish oil primarily depends on the isolation process and condition. Methods such as supercritical fluid extraction, centrifugation, solvent extraction, microwave-assisted extraction (MAE), and enzymatic hydrolysis have been utilized for extraction of fish oil (Ivanovs & Blumberga, 2017).

Traditional methods such as pressing, heat treatment, and solvent extraction have some drawbacks such as resulting in lowering the quality of the oil as a result of high-temperature processing which could damage the nutritional properties of the oil and result in the retention of toxic solvent residues. These methods are also not environmentally safe due to the use of a large amount of solvent (Adeoti & Hawboldt, 2014). The use of relatively newer extraction methods such as supercritical CO_2 ($_{sc}CO_2$) and microwave irradiation has therefore widely replaced the traditional methods for the extraction of fish oil.

In this regard, $_{sc}CO_2$ has been the most common technique for production of high-quality fish oil. $_{sc}CO_2$ extraction is free from toxic solvents, is time efficient, and the extraction process is performed at a lower temperature (Sahena et al., 2009). The extraction solvent used in this process, CO_2, is a good solvent for lipophilic compounds. As the presence of moisture in the sample can negatively affect the extraction process by reducing the contact time between CO_2 and the oil solute, pretreatment, and drying of the fish by-product is a required pretreatment step (Sarker et al., 2012). Pressure, time, solvent flow rate, and temperature are major parameters affecting the SFE extraction process (Sarker et al., 2012). The yield of oil obtained by SFE is either comparable or even higher as compared to the traditional processes.

$_{sc}CO_2$ under optimized conditions isolated 36.2% of the oil from Tuna (Ferdosh et al., 2015), 35.6% from tuna head (Ferdosh et al., 2016), and 67% from catfish (Sarker et al., 2012). In a study on tuna by-products, DHA was the major PUFA which was largely obtained from tuna head (17.0–19.9%) followed by the skin (15.7–17.3%) and the viscera (14.3–16.1%) (Ferdosh et al., 2015).

Temperatures in the range of 40–80°C, with pressures of 25–40 MPa at 40–80°C, with a solvent flow rate of 2 mL/min are frequently reported in the literature as the optimum range of conditions for the extraction of oil from fish waste. The time of extraction, however, varied very widely from less than an hour up to several hours (Gedi, Bakar, & Mariod, 2015; Ivanovs & Blumberga, 2017; Rubio-Rodríguez et al., 2012; Sarker et al., 2012).

While conventional techniques are time-consuming, hard to control and involve possible chemical contamination, microwave irradiation can be used to reduce the time of reaction and provide better control over the process. In this method, microwave energy is used for stable and monotonous heating of the solvent to better extract the oil from the sample while minimizing heat shock of the materials, with a uniform heating process that will help achieve homogeneity of pressure and temperature in the reaction mixture (Adeoti & Hawboldt, 2014; Ramalhosa et al., 2012). Microwave heating is reproducible and requires less energy and time than conventional methods. Given the direct heating effect in the microwave process, the moisture content of the cells vaporizes generating a pressure inside the cell that breaks the cell membrane, resulting in the release of the cell contents, consequently achieving a better extraction of the oil (Mercer & Armenta, 2011). The microwave process, however, requires a strict control over the processing parameters of time and temperature as the generated heat may oxidize unsaturated fatty acids. In addition, the yield of oil from the microwave-assisted process is generally low when volatile solvents are used for the process (Wang & Weller, 2006). In a recent study, lipid from sardine fish waste was extracted using a microwave-assisted process and the results were compared with the Soxhlet extraction method. A 10-min microwave treatment time resulted in the highest lipid yield of 80.5 mg/g fish waste material when water was used as a solvent, whereas the extracted lipid yield was 46.6 mg/g fish waste material was extracted using the Soxhlet extraction method after 4 h of processing. When water was replaced with a mixture of hexane and isopropanol, the yield was 61 mg/g after 4 min of microwave heating, and a longer extraction time resulted in solvent evaporation. The high dielectric constant of water has been reported to be beneficial for the internal generation of heat and consequently, extraction of the lipid (Rahimi et al., 2017). In another study on catfish, a short microwave pretreatment (60 s) resulted in a 10% increase in the oil yield from an enzymatic extraction procedure (Chimsook & Wannalangka, 2015).

Given that the final oil product is required to be food grade, the enzyme needs to be food quality and not obtained from a pathogenic source. Alkaline and neutral proteases have been commonly used due to their better performance compared to acidic proteases. The yield and quality of the oil product depend on the type and concentration of enzyme, reaction temperature, time, pH, and particle size of the sample (Gbogouri, Linder, Fanni, & Parmentier, 2006; Hathwar, Bijinu, Rai, & Narayan, 2011). In the enzymatic process, the homogenized mixture is first heated up to 90°C to inactive endogenous enzymes to make it possible to evaluate the effect of the added protease.

5.1 Fish Oil: Pharmaceutical and Food Applications

Fish oil as a rich source of omega-3 fatty acids has been known to have a positive effect on human health and to be beneficial in the prevention and treatment of various diseases. In particular, the omega-3 fatty acid content of blood was found to be inversely related to the development of cardiovascular diseases (Kromhout, Yasuda, Geleijnse, & Shimokawa, 2012).

It has been suggested that omega-3 fatty acids can prevent the growth and occurrence of lipid-rich atherosclerotic plaques (von Schacky, 2000). Daily consumption of 4 g fish oil is reported to be effective in reducing the occlusion of aortocoronary venous bypass grafts and the addition of these fatty acids into a Mediterranean diet reduced cardiovascular mortality by 70% (von Schacky, 2000). In addition, various studies have indicated a beneficial effect of n-3 fatty acids in the prevention of cancer, mental illnesses, and improved infant development (Ginter & Simko, 2010; Riediger, Othman, Suh, & Moghadasian, 2009). Therefore, fish oil has become a popular dietary supplement with consumption linked to prevention of a wide range of diseases. Given the strong evidence for health benefits of PUFA, food and pharmaceutical companies have marketed fish oil in a variety of products as tablets or enriched food systems. Foods such as yoghurt, juices, milk, and soy beverages have been enriched with DHA and other omega-3s. The effectiveness of these omega-3 enriched products has not been verified. Major brands of fish oil supplements have been tested and only 3 of the 32 supplements contained the concentrations of fatty acids listed on the label and the rest had about a third lower omega-3 fatty acids than their claims (Albert et al., 2016, 2017). Given that fish oil is prone to oxidation, the processing conditions, exposure to air and delivery time between production and market can have a considerable effect on the product stability and composition.

In their study, supplementation of the diet of rats with unoxidized n-3 PUFA oils during pregnancy prevented the development of impaired insulin sensitivity in male adult offspring (Albert et al., 2017), whereas oxidized fish oil was found to cause high newborn mortality and increased maternal insulin resistance (Albert et al., 2016). This raises the question that there are regulatory issues that need to be put in place to provide consumers with greater reassurance about the status of marketed fish oils.

6. SEASHELL AND CRUSTACEAN BY-PRODUCT WASTE

6.1 Pigment: Carotenoids

The carotenoid pigments are ubiquitous in nature and more than 600 different carotenoids have been identified and characterized (Mercadante, Egeland, Britton, Liaaen-Jensen, & Pfander, 2004). They are responsible for the red and yellow color of crustaceans (Wade, Gabaudan, & Glencross, 2017) and the pigmentation in many plants, animals, and microorganisms. Carotenoids are a family of lipophilic compounds that can be classified into two groups, which comprise (i) hydrocarbons that are only composed of carbon and hydrogen atoms, such as carotene and the xantho-phylls, and (ii) oxygenated derivatives that contain at least one oxygen func-tional group, such as satacene, astaxanthin, cryptoxanthin, canthaxanthin, and lutein (de Quirós & Costa, 2006). Some of the carotenoids such as β-carotene possess pro-vitamin A activity. Carotenoids also possess remark-able antioxidant activities, acting as chain-breaking antioxidants and thus can protect cells and organisms against photo-oxidation (Edge, McGarvey, & Truscott, 1997). Crustacean wastes generated from lobsters, shrimps, crabs, crayfish, and krill are abundant sources of carotene and astaxanthin, an oxi-dized form of carotene (Sachindra, Bhaskar, Siddegowda, Sathisha, & Suresh, 2007). Astaxanthin is responsible for the pink-to-red pigmentation of crustaceans and wild salmonids (Chakrabarti, 2002).

6.1.1 The Process for Carotenoid (Astaxanthin) Recovery

The process of extraction of carotenoids from crustacean shell waste consists of an initial treatment to detach the pigments from the bound proteins and minerals (deproteinization and demineralization), followed by extrac-tion by organic solvents (Fig. 5). However, the conventional treatment of

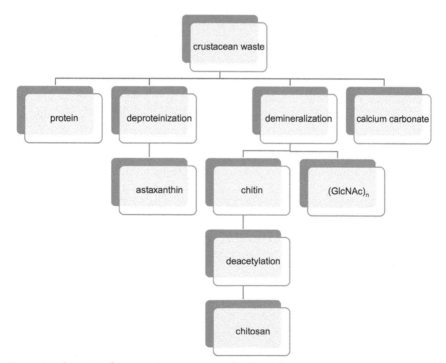

Fig. 5 A schematic diagram showing the utilization of crustacean waste.

crustacean waste by chemicals can damage the structure of biomolecules such as astaxanthin and destroy their bioactivities (Gortari & Hours, 2013). The astaxanthin recovery method based on solvent extraction is relatively expensive, time-consuming, and is considered to be hazardous to both workers and the environment (Han, Ma, Wang, & Xue, 2012). Moreover, the stability of astaxanthin is decreased and its structure is affected by the organic solvents (Han et al., 2012). Increasing concerns about environmental hazards and the stability of astaxanthin have prompted the search for a simple and environmental alternative method for extracting astaxanthin from crustacean waste.

Enzymatic hydrolysis using commercial, crude, or endogenous enzymes has been explored as an alternative preparation method to extract astaxanthin from crustacean shells (Table 1). Traditional organic solvent extraction achieves a good recovery rate by reducing both ash and chitin levels. The use of enzymes to release astaxanthin from carotenoproteins and recovery by an aqueous extraction method has been reported (Jafari et al., 2012; Sowmya et al., 2014). In order to avoid the growth of undesirable microflora, bioconversion of shrimp waste for the production of bioactive material has been carried out by addition of minerals or different acids or by fermentation with lactic acid bacteria (Armenta-López et al., 2002; Gimeno et al., 2007; Oh et al., 2007).

Fermentation can be used for stabilizing and extracting the pigments. As carotenoids are prone to degradation by acids, mild treatments such as fermentation may have beneficial effects on the stability of carotenoids (Prameela et al., 2017). Application of a microbial fermentation method has many advantages over chemical extraction methods (using organic solvents such as acetone, petroleum ether, benzene, hexane, diethyl ether, chloroform, ethanol, and methanol) as it is more eco-friendly, in being directed to the reduction of chemical usage, and can enhance the recovery of carotenoids along with chitin and chitosan that can be utilized as value-added products (Gimeno et al., 2007).

Other extraction technologies such as supercritical fluid, ultrasound, and high-pressure have been used to extract carotenoids from crustacean waste. Usage of supercritical carbon dioxide ($_{sc}CO2$) is suitable for thermolabile pigment compounds and are safer since this is a generally recognized as safe (GRAS) solvent type, is economical and easy to separate from the extract (Reverchon & De Marco, 2006). Several reports on the supercritical fluid extraction of carotenoids from crustacean waste (Armenta-López et al., 2002; Sánchez-Camargo et al., 2011, 2012), have described a procedure for supercritical fluid extraction of astaxanthin using different solvents at

Table 1 Different Treatments for Carotenoids (Astaxanthin) Extraction

Pigments	Crustacean Waste Sources	Treatment	Enzyme/Microorganisms	Carotenoids Recovery (%)	References
Carotenoids	*Penaeus semisulcatus*	Enzymatic hydrolysis and alkaline treatment	Trypsin	NA	Jafari, Gharibi, Farjadmand, and Sadighara (2012)
Xanthophylls	*Parapenaeus longirostris*	Enzymatic hydrolysis	Barbel and bovine trypsins	80 μg/g	Sila, Nasri, and Bougatef (2012)
Carotenoids	*Penaeus indicus*	Enzymatic hydrolysis Alcalase	Alcalase	82.5	Sowmya, Ravikumar, Vivek, Rathinaraj, and Sachindra (2014)
Carotenoids	*Penaeus monodon*	Enzymatic hydrolysis	Trypsin	87.91 μg/mL	Klomklao, Benjakul, Kishimura, and Chaijan (2011)
Carotenoids	*Enzymatic hydrolysis*	Enzymatic hydrolysis	Alcalase	47.0 mg/g	De Holanda and Netto (2006)
Carotenoids	*Xiphopenaeus kroyeri*	Enzymatic hydrolysis	Pancreatin	57.0 mg/g	De Holanda and Netto (2006)
Carotenoids	*Penaeus monodon*	Autolysis	Endogenous enzymes	63.4	Sowmya, Rathinaraj, and Sachindra (2011)
Carotenoids	*Penaeus indicus*	Fermentation	*Lactobacillus plantarum B 4496*	31.3 mg/g	Armenta-López, Guerrero, and Huerta (2002)

Free astaxanthin	Litopenaeus vannamei, stylirostris, and setiferus	Fermentation	Lactobacillus plantarum	115 mg/g	Gimeno et al. (2007)
NA	Crab shell	Fermentation	P. aeruginosa F722	NA	Oh, Kim, Jung, and Park (2007)
Astaxanthin	Farfantapenaeus paulensis	Supercritical fluid extraction	CO_2, 43°C, 37 MPa	20.7 mg/g	Sánchez-Camargo, Martínez-Correa, Paviani, and Cabral (2011)
Astaxanthin	Farfantapenaeus paulensis	Supercritical fluid extraction	CO_2 and ethanol (5%), 50°C, 30 MPa	1325 mg/g	Sánchez-Camargo, Meireles, Ferreira, Saito, and Cabral (2012)
Astaxanthin.	Penaeus monodon	Supercritical fluid extraction	CO_2 and ethanol (60–100%), 50–60°C 20 MPa	2.70–53.17 mg/g	Armenta-López et al. (2002)
Astaxanthin	Litopenaeus vannamei	High-pressure processing	The pressure was 210 MPa, the liquid-to-solid ratio was 32 mL per 1 g, pressure holding time was 10 min	89.12	Du, He, Yu, Zhu, and Li (2013)

different concentrations. The recovery of astaxanthin was found to be relatively low when using pure $_{sc}CO_2$ compared with those extracted by traditional methods (Sánchez-Camargo et al., 2011) because pure $_{sc}CO_2$ is not suitable for extraction of astaxanthin due to low solute solubility. The lipid and astaxanthin extraction yields were found to increase considerably with increasing the proportion of ethanol in the ethanol/$scCO_2$ mixture (Armenta-López et al., 2002). Another study showed that olive oil as co-solvent could increase the extraction efficiency and it was comparable to that obtained using ethanol as co-solvent (Krichnavaruk, Shotipruk, Goto, & Pavasant, 2008). A simple, effective, and green eutectic solvent-based ultrasound-assisted method was proposed for the extraction of astaxanthin from shrimp by-products (Xi, He, & Yan, 2015). High-pressure processing (HPP) showed advantages with higher yield and shorter extraction time. The use of HPP to extract astaxanthin from shrimp discards is a potential alternative to conventional methods (Du et al., 2013). The effect of different treatments for carotenoid (astaxanthin) extraction by environmentally friendly processes is summarized in Table 1.

6.1.2 Bioactivities of Carotenoids

Due to the attractive red-yellow-orange color range displayed by carotenoids β-carotene and other natural carotenoids are often utilized as coloring agents in foods, beverages, confectionery and food supplements, as well as in the nutraceutical and pharmaceutical industries. In addition, carotenoids have been referred to as having considerable bioactivities such as antioxidant, anti-inflammatory, anti-cancer, and anti-diabetes effects. Interestingly, most of the biological effects referred to above are attributable to the well-known antioxidant activity of these compounds. Astaxanthin, as one of the most studied carotenoid compounds, is defined as a super antioxidant because of its strong antioxidant ability, which is 10 and 500 times greater than that of other carotenoids and of vitamin E, respectively (Mao, Guo, Sun, & Xue, 2017). Table 2 briefly describes the most common biological effects of carotenoids and their process methods and sources.

6.1.3 Food Industry and Pharmaceutical Application

Due to the considerable antioxidant activity of carotenoids, they are able to improve the shelf life of numerous foodstuffs by acting either as electron donors, singlet oxygen quenchers, or free radical scavengers (Maria, Graziano, & Nicolantonio, 2015). For example, because of its superior antioxidant activity, astaxanthin has been used as a natural antioxidant in

Table 2 Carotenoids (Astaxanthin) Bioactivities

Carotenoids	Sources	Method	Bioactivities	References
Astaxanthin	Shrimp waste	Autolysis	Anti-inflammatory properties	Santos et al. (2012)
Astaxanthin	*Astaxanthin*	NA	Anti-cancer	Gal, Andrei, Cernea, Taulescu, and Catoi (2012)
Astaxanthin	*Parapenaeus longirostris*	Organic solvent	Preventing diabetic complications	Sila et al. (2015)
Astaxanthin	*Parapenaeus longirostris*	Organic solvent	Antioxidant activity/ antiproliferative activity	Sila et al. (2014)

edible oil (Rao, Sarada, & Ravishankar, 2007). In addition, as it was mentioned before, carotenoids are also used as colorants because of their characteristic red to orange color. Actually, they are not just only food additives but, more interestingly, they are also considered as nutritional supplements, which act as precursors of vitamin A (Guerin, Huntley, & Olaizola, 2003; Maria et al., 2015; Toldra, 2017). Astaxanthin is used in the cosmetic industry because as in terms of its dermatological actions, it was found to suppress hyper-pigmentation (Yamashita, 1995) and inhibit melanin synthesis and photoaging (Tominaga, Hongo, Karato, & Yamashita, 2012). Astaxanthin derived from *Haematococcus pluvialis* was reported to improve the skin condition in both women and men (Tominaga et al., 2012).

Generally, carotenoids have shown great potential for promoting human health and potential use as pharmaceutical compounds. Astaxanthin showed some benefits in diabetes, where it was more effective at both preventing and treating hepatic insulin resistance and non-alcoholic steatohepatitis (NASH) compared with vitamin E in mice (Ni et al., 2015). Furthermore, astaxanthin improved hepatic steatosis and tended to ameliorate the progression of NASH in biopsy-proven human subjects (Ni et al., 2015). It also improves the cardiovascular health and capillary circulation, as experimental investigations in a range of species using a cardiac ischaemia-reperfusion model demonstrated cardiac muscle preservation when astaxanthin is administered either orally or intravenously prior to the induction of ischaemia (Fassett & Coombes, 2011, 2012). Another study demonstrated that treatment with astaxanthin attenuates oxidative stress induced by decreased skeletal muscle use and that this attenuation prevents the associated capillary regression (Kanazashi et al., 2013). In addition, astaxanthin also benefits liver health and metabolic processes, as it was shown that the natural astaxanthin product, AstaREAL, had the effect of improving blood lipids and increased adiponectin, preventing fatty liver disease, reducing the risk for atherosclerotic plaque and lowering hypertension by improving vascular tone (Kindlund & BioReal, 2011). Astaxanthin is also reported to boost the immune system due to its immunomodulating activity (Park, Chyun, Kim, Line, & Chew, 2010).

6.2 Polyhydroxylated Naphthoquinone (PHNQ) Pigments

Sea urchin species exhibit many different colors, ranging from purple to red, brown, olive green, and black. Color of sea urchin shells is influenced by different amounts of the pigments, called spinochromes, which are polyhydroxyl naphthoquinone (PHNQ) pigments usually substituted with ethyl, acetyl, methoxyl, and amino groups. In addition to shell and spine, the pigments are also present in the eggs, ovaries, and coelomic fluid of the animals.

6.2.1 Isolation, Fractionation, and Characterization of PHNQ Pigments

As PHNQ pigments in sea urchin shell and spine are associated with minerals and proteins, it is necessary to use chemical reagents to extract the pigments. The general traditional scheme of PHNQ pigment extraction is to treat sea urchin powdered shell and spine with acidic solution to dissolve the mineral structure, followed by extraction of the pigments from the acidic solution with organic solvents, followed by further fractionation of the crude pigment mixture (Anderson, Mathieson, & Thomson, 1969; Kuwahara, Hatate, Chikami, Murata, & Kijidani, 2010; Shikov et al., 2011; Soleimani, Yousefzadi, Moein, Rezadoost, & Bioki, 2016). It is worth mentioning that Zhou and co-workers (Li et al., 2013; Zhou et al., 2011) investigated the optimization of the extraction process of PHNQs from *Strongylocentrotus nudus*, *Glyptocidaris crenularis*, and *Staphylococcus intermedius* using macroporous resins. As traditional extraction of PHNQs with organic solvents expose operators and the environment to toxic compounds, the use of macroporous resins was proposed as a more environmentally friendly and safe option for the extraction and purification of PHNQs (Zhou et al., 2011).

PHNQ pigments extracted from sea urchin shell and spine are a mixture of several compounds that can be separated by column chromatography on acid-washed silica-gel (Anderson et al., 1969; Chang & Moore, 1971;

Nishibori, 1959), gel-chromatography on Sephadex LH-20 (Amarowicz, Synowiecki, & Shahidi, 1994), or reversed-phase chromatography on Toyopearl HW-40 (Mischenko et al., 2005). Crude PHNQ pigment mixtures have been subjected to solid phase extraction using GIGA C18E units (Powell, Hughes, Kelly, Conner, & McDougall, 2014), which speed up the selection of phenolic material from the initial acid extracts generating a more reproducible PHNQ preparation (Powell et al., 2014), however the authors noted a considerable loss of pigments because of low desorption rate.

Although to date more than 30 different PHNQs have been identified from sea urchin shell. Six PHNQs (echinochrome A, spinochrome A, spinochrome B, spinochrome C, spinochrome D, and spinochrome E) are the most commonly found in different species. The minor identified PHNQs include amino PHNQs and PHNQ dimers. Most of the previous studies have combined physical characterization using melting point, ultra-violet (UV) spectrum (Vasileva, Mishchenko, & Fedorcyev, 2017; Zhou et al., 2011), mass (MS) spectrum (Li et al., 2013; Powell et al., 2014; Zhou et al., 2011), infrared radiation (IR) characterization, and nuclear magnetic resonance (NMR) data (Becher, Djerassi, Moore, Singh, & Scheuer, 1966; Moore, Singh, & Scheuer, 1966; Singh, Moore, Chang, Ogata, & Scheuer, 1968) to identify PHNQs.

6.2.2 Biomedical and Pharmaceutical Applications of PHNQs

Quinones are among the most studied natural products in the literature because of their considerable biological potential (Ferreira, Nicoletti, Ferreira, Futuro, & da Silva, 2016). Quinones constitute a structurally diverse class of phenolic compounds with a wide range of pharmacological properties, which are the basis for different applications in the broad fields of pharmacy and medicine (Martínez & Benito, 2005). Traditional medicine utilizes natural product materials rich in quinones for treatment of a variety of diseases (Martínez & Benito, 2005). The unique structure of quinones distinguishes the PHNQ pigments of sea urchin from other well-known antioxidants and makes them a promising base for the creation of new treatments for a variety of pathologies. Among the spinochromes, echinochrome A was the most well-studied of the compounds (Vasileva & Mishchenko, 2016). The commercial antioxidant pharmaceutical histochrome was formulated based on the PHNQ pigment echinochrome A (Mishchenko, Fedoreev, & Bagirova, 2003). This histochrome product was developed at the Laboratory for the Chemistry of Natural Quinoid Compounds, Pacific Ocean Institute of Bioorganic Chemistry, Far Eastern Branch, Russian Academy of Sciences

(Mishchenko et al., 2003). Histochrome is a water-soluble sodium salt of echinochrome A—a pentahydroxyethylnaphthoquinone dissolved in 0.9% sodium chloride solution (Mishchenko et al., 2003), that has been demonstrated to be an effective drug for the treatment of proliferative processes, degeneracies, and ophthalmic hemorrhages of various geneses (Agafonova, Kotel'nikov, Mischenko, & Kolosova, 2011; Egorov et al., 1998; Mishchenko et al., 2003). It can be widely used as antioxidant, cardioprotector, and antiarrhythmic medicine (Mishchenko et al., 2003). It was registered as a medicinal product authorized for use in the Russian Federation in the regulated pharmacotherapeutic group: antioxidant agent; registration number P N002363/01-2003, July 23, 2008 (authorization certificate No. P N02363/01-2008) (Kareva, Tikhonov, Mishchenko, Fedoreev, & Shimanovskii, 2014). However, the effect of this medicine is very difficult to explain only in terms of antioxidant properties. The results of experimental investigation of histochrome both *in vivo* and *in vitro* showed that it demonstrates remarkable pharmacologic effects. Table 3 shows PHNQs pigments reported in literature and their sources.

6.3 Chitin and Chitosan

Chitin (poly(β-(1 → 4)-*N*-acetyl-D-glucosamine)) is the second most abundant polysaccharide in nature. Depending on its source, chitin occurs as α- and β-isomorph forms (Blackwell, 1973). The α-chitin isomorph is the most abundant and it occurs in fungal and yeast cell walls, in krill, lobster, and crab tendons and in shrimp shells, as well as in insect cuticle (Blackwell, 1973). The less abundant β-chitin is found in association with proteins in squid pens and in the tubes synthesized by pogonophoran and vestimentiferan worms (Blackwell, 1973). The worldwide annual production of chitin as a natural product in animals has been estimated to be approximately 10^{11} tons (Hoell, Vaaje-Kolstad, & Eijsink, 2010). The production of aquatic animals from aquaculture in 2014 amounted to 73.8 million tons, with crustacean over 6.9 million tons (wet weight) (Moffitt & Cajas-Cano, 2014). Crustacean shells are the principal source for commercial chitin (Gortari & Hours, 2013). After deacetylation (DA), chitin can be transformed into chitosan, which comprises D-glucosamine chains (Yeul & Rayalu, 2013). Chitin, chitosan, and their derivatives have gained much attention due to their non-toxicity and variety of bioactivities.

6.3.1 Extraction of Chitin From Crustacean Shell

The primary sources of raw material for the production of chitin are the cuticles of various crustaceans, principally crab and shrimps. In crustaceans or

Table 3 Different PHNQs Reported in Literature and Their Sources

Compound	Source	Biological Activities	References
PHNQs	*S. droebachiensis*	Antiallergic: Histamine-induced contraction of isolated guinea pig ileum was inhibited by pigments	Pozharitskaya, Ivanova, Shikov, and Makarov (2013)
PHNQs	*S. droebachiensis*	Antidiabetic: The hypoglycemic activity	Kovaleva et al. (2013)
Echinochrome A	*S. mirabilis*	Antidiabetic: Reduce elevated blood glucose in a model of alloxan-induced diabetes in CBA mice	Popov and Krivoshapko (2013)
Echinochrome A	*Paracentrotus lividus*	Antidiabetic: Improvements in glucose, arginase, insulin, and glucose-6-phosphate dehydrogenase activities	Mohamed, Soliman, and Marie (2016)
Histochrome	Synthesized	Antihypertensive: Histochrome normalized the systolic and diastolic blood pressure in Wistar rats	Agafonova, Bogdanovich, and Kolosova (2015)
Histochrome	Synthesized	Anti-inflammatory: Effect of histochrome was similar to that of acetylsalicylic acid or diclofenac	Talalaeva et al. (2012)
Echinochrome A	Obtained from Pacific Institute of Bioorganic Chemistry (PIBOC)	The number of inflammatory aqueous cells and protein levels were lower in the groups treated with echinochrome A	Lennikov et al. (2014)
PHNQs	*Echinometra mathaei, Diadema savignyi, Tripneustes gratilla, Toxopneustes pileolus*	Pro-inflammatory activity	Brasseur et al. (2017)

Continued

Table 3 Different PHNQs Reported in Literature and Their Sources—cont'd

Compound	Source	Biological Activities	References
Echinochrome A (dissolved in solutions of proteins)	Obtained from Prof. R.H. Thomson, Department of Chemistry, University of Aberdeen	Antimicrobial: Against Gram-negative marine *Pseudomonas strain* 111, *Vibrio fischeri* and *Gram-positive Micrococcus* sp. and *Clyde isolate YP2*	Service and Wardlaw (1984)
Separate PHNQs	*T. pileolus*	Antimicrobial: Comparison of different compounds antimicrobial activities	Brasseur et al. (2017)
Echinochrome A	*S. mirabilis*	Cardioprotective	Jeong et al. (2014) and Kim et al. (2015)
Echinochrome A spinochrome A		Enzymes inhibition: Inhibited tyrosine hydroxylase/inhibition of dopamine-b-hydroxylase	Kuzuya, Ikuta, and Nagatsu (1973)
Echinochrome A	*S. mirabilis*	Effects on neurodegenerative disorders: Acetylcholinesterase (AChE) inhibitory activity	Lee et al. (2014)
Echinochrome A	*S. intermedius*	Hypocholesterolemic: Cholesterol synthesis was inhibited	Lakeev et al. (1992)
PHNQs	NA	Radical scavenging	Lebedev, Ivanova, and Krasnovid (1999) and Berdyshev, Glazunov, and Novikov (2007)

more specifically shellfish, chitin is found as a constituent of a complex network with proteins onto which calcium carbonate deposits to form the rigid shell. To isolate chitin from shell material, the removal of the protein and inorganic calcium carbonate (if calcium carbonate is the major constituent) is required by deproteinization and demineralization, together with some minor constituents such as pigments and lipids (Fig. 6). In some cases, decolorization has been applied to remove the pigment (Beaney, Lizardi-Mendoza, & Healy, 2005). Both the deproteinization and demineralization can be carried out using either chemical or biological treatments. On a commercial scale, chemical methods have been applied. An eco-friendly method including an enzymatic treatment and microbial fermentation to deproteinization has also reported (Kandra, Challa, & Jyothi, 2012).

6.3.2 Chemical Extraction of Chitin From Crustacean Shell

Chemical methods have been the first approach employed for deproteinization of crustacean shell, using a wide variety of chemicals such as either NaOH, Na_2CO_3, $NaHCO_3$, KOH, K_2CO_3, $Ca(OH)_2$, Na_2SO_3, $NaHSO_3$, $CaHSO_3$, Na_3PO_4 or Na_2S (Younes & Rinaudo, 2015). NaOH has been the preferred chemical, applied at concentrations ranging from 0.125 to 5.0 M, at a temperature ranging from room temperature up to 100°C for a few minutes up to several days. However, using NaOH results in partial deacetylation of chitin and hydrolysis of the biopolymer, reducing the molecular weight of the polymer.

Chemical demineralization has generally been carried out by acid treatment using either HCl, HNO_3, H_2SO_4, CH_3COOH or HCOOH (No & Hur, 1998; Percot, Viton, & Domard, 2003). HCl has been the preferred reagent because it converts the calcium carbonate into water-soluble calcium salts with the release of carbon dioxide. After that, the salt can be removed by washing the solid phase chitin with deionized water. Demineralization has been achieved by using dilute hydrochloric acid (up to 10%, w/v) at room temperature for various incubation times (0.5–28 h) (Younes & Rinaudo, 2015). However, the acid treatment conditions may cause depolymerization and deacetylation of the native chitin (Blackwell, 1973). Alternative mild acids such as ethylenediaminetetracetic acid (EDTA) (Austin, Brine, Castle, & Zikakis, 1981), acetic acid (Brine & Austin, 1981), sulfurous acid (Peniston & Johnson, 1978) have been evaluated, but those treatments result in chitin with a high ash content.

The use of these chemicals in both the deproteinization and demineralization steps can cause negative effects on the properties of the target

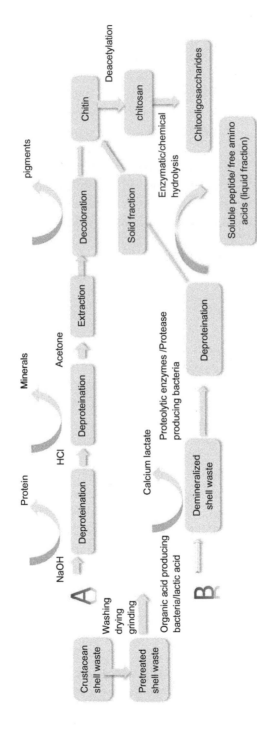

Fig. 6 Summary of processing of crustacean shell waste into chitin by different extraction methods.

product such as reducing the molecular weight of the polymer, affecting the viscosity, and altering the degree of acetylation (Younes & Rinaudo, 2015). Also, the use of chemicals such as HCl and NaOH on an industrial scale is hazardous, energy consuming, and potentially damaging to the environment. In addition, protein and other components that could have valuable biological activities are lost in the process. Therefore, there is a need to develop ways to process this shell waste material in a more environmentally friendly manner.

6.3.3 Biological Extraction of Chitin From Crustacean Shell

Biological extraction methods have been evaluated using different microorganisms that are capable of secreting hydrolyzing enzymes. These methods are relatively straightforward to use and can be more effective and environmentally friendly compared to chemical methods. Several comparative studies have been carried out and have shown that the biological method had advantages compared to the chemical one because it preserves the structure of chitin with retention of higher molecular weight polymers (Bustos & Healy, 1994; Khanafari, Marandi, & Sanati, 2008).

Chitin extraction usually requires the use of proteases, either purified or as crude extracts, for the deproteinization step. Crude proteases are not only cheaper but also more efficient due to the presence of other proteases in addition to the protease that is present in the commercially purified preparation (Fig. 6). The extraction of crude protease from seafood by-product waste could be utilized in the future, which may decrease the cost of the process (Younes & Rinaudo, 2015). Several reports have demonstrated the application of bacterial proteases for the deproteinization step (Manni, Ghorbel-Bellaaj, Jellouli, Younes, & Nasri, 2010; Younes et al., 2012). Microorganism-mediated deproteinization, along with demineralization generate a solubilized fraction rich in proteins, minerals, and carotenoids, especially astaxanthin, and a chitin-enriched solid fraction (Prameela, Murali, & Hemalatha, 2010; Rao & Stevens, 2005) (Fig. 6). However, with proteolytic deproteinization, the mineral component present in the cuticles may reduce the accessibility of the proteases and affect the efficiency of deproteinization of crab and shrimp shells. Therefore, the material should first be demineralized, and one way to achieve this is by combining a fermentation process with the deproteinization step.

Fermentation methods can be divided into two categories of lactic acid and non-lactic acid fermentation. Lactic acid fermentation can be combined with a microbial fermentation process, involving a protease-producing bacterial species such as *Serratia* or *Aspergillus* and an organic acid-producing bacterial species such as *Pseudomonas* or *Lactobacillus* to depolymerize chitin in shell waste (Hou et al., 2016). Lactic acid can react with calcium carbonate to form calcium lactate as precipitate, which could be separated from lighter demineralization shell waste. Following this, the chitin material can be deproteinized by proteases secreted by the added microorganisms, or by proteases present in the waste material itself. The efficiency of this fermentation depends on many factors such as the bacterial species used and inoculum size, carbon source and its concentration, initial pH and its change during fermentation, and the temperature and duration of the fermentation. Raw heads of African river prawns (*Macrobrachium vollenhovenii*) were fermented with *L. plantarum* inoculum at 5% (w/w) at 30°C using 15% (w/w) cane molasses as the carbohydrate source required for growth (Fagbenro, 1996). The fermentation of shrimp waste by *L. plantarum* 541 with and without pH control was carried out. Among four acids tested including acetic acid, citric acid, HCl, and lactic acid; acetic acid and citric acid were demonstrated to be the most effective to control pH both at the start and during the fermentation. In the presence of glucose, 75% deproteinization (DP) and 86% demineralization (DM) were obtained with the pH maintained at 6. However, without pH control, DP and DM were 68.1% and 64.1%, respectively (Rao, Munoz, & Stevens, 2000). In non-lactic acid fermentation, both bacteria and fungi were used for fermentation of crustacean shell. Ghorbel-Bellaaj, Younes, Maâlej, Hajji, and Nasri (2012) evaluated six proteolytic *Bacillus* strains for fermentation of shrimp waste: *Bacillus pumilus* A1, *B. mojavensis* A21, *B. licheniformis* RP1, *B. cereus* SV1, *B. amyloliquefaciens* An6, and *B. subtilis* A26. The highest deproteinization degree was obtained using *B. cereus* SV1.

6.3.4 Deacetylation and Hydrolysis of Chitin/Chitosan

Chitosan, which is prepared from chitin by deacetylation, is a polysaccharide comprising β-(1,4)-linked D-glucosamine. When the extent of deacetylation of chitin is more than 50%, it becomes soluble in acid solution. In general, chitin is treated with 40–50% alkali at 100–160°C for several hours in order to achieve 95% deacetylation (DA), resulting in the production of chitosan (Honarkar & Barikani, 2009). However, because of the energy consumption and potential hazard to the environment, the use of chitin deacetylase enzymes has become an alternative method to generate chitosan from chitin. There are still some disadvantages of the enzymatic DA process because of the crystallinity, insolubility, and high molecular weight of chitin (Pareek, Vivekanand, Saroj, Sharma, & Singh, 2012). In order to optimize the parameters of the enzymatic reaction, chitin was modified by physical or chemical modification methods (Beaney, Gan, Magee, Healy, & Lizardi-Mendoza, 2007). The most successful chitin modification method reported was to freeze dry a colloidal chitin suspension prior to deacetylation, which increased the yield of enzymatic generated DA by 20-fold (Beaney et al., 2007).

Chito-oligosaccharides and N-acetyl glucosamine can be generated as hydrolysis products of chitosan and chitin, respectively. These products have biological characteristics and therefore have potential application in food, agricultural, and medical industries. Hydrolysis of chitosan to produce chito-oligosaccharides can be achieved by different methods, including microwave processing (Xing et al., 2005), ultrasound (Wu, Zivanovic, Hayes, & Weiss, 2008; Xie, Hu, Wei, & Hong, 2009), hydrothermal methods (Sato, Saimoto, Morimoto, & Shigemasa, 2003), and chemically using acid. Although chemical methods using acid have some disadvantages such as being hard to control, resulting in low-quality production and environmental pollution, they were the most widely used methods traditionally. In recent years, a more gentle and controllable enzymatic hydrolysis method has been used as an alternative. Chitinase can hydrolyze β-(1,4) glycosidic bonds of the chitosan or chitin polymer down to the N-acetyl D-glucosamine dimer (Ashhurst, 2001). In addition, a wide variety of enzymes have been applied to produce chito-oligosaccharides such as cellulase, amylase, and other crude carbohydrate hydrolyzing enzymes (Chang et al., 2016; Nguyen-Thi & Doucet, 2016; Nidheesh, Kumar, & Suresh, 2015; Sinha, Chand, & Tripathi, 2014).

6.3.5 The Biological Activity of Chitin and Chitosan and Relation With Chemical Structure

Chitosan and its derivatives are reported to have antioxidant capability by scavenging free radicals such as hydroxyl, superoxide, and alkyl radicals, as well as chelating metals, having reducing power and prevention of lipid peroxidation (Aytekin, Morimura, & Kida, 2011; Cho, Kim, Ahn, & Je, 2011). It has also been reported that their antioxidant activities depend on the molecular weight (MW) and the degree of deacetylation. Park, Je, and Kim (2003) demonstrated that low MW chitosan (1–3 kDa) is more active than that with higher MW in terms of antioxidant activity. Another study showed that low MW chitosan can exhibit more than 80% of superoxide radical scavenging activity at 0.5 mg/mL concentration (Yin, Lin, Zhang, & Yang, 2002). The antioxidant activity of chitosan with different MWs (30, 90, and 120 kDa) was evaluated and results showed that all chitosan samples had antioxidant activity but the 30 kDa sample had the highest activity (Kim & Thomas, 2007).

Chitosan and its derivatives have been reported as good antimicrobial agents. Low MW chitosan could penetrate cell wall of bacteria, combine with DNA, and inhibit DNA transcription (Sudarshan, Hoover, & Knorr, 1992), while the high MW chitosan could alter cell permeability and block the transport of essential solutes into the cell (Choi et al., 2001; Eaton, Fernandes, Pereira, Pintado, & Malcata, 2008; Kato, Onishi, & Machida, 2005). Chitosan is more absorbable by bacterial cells at lower pH because of the increase of chitosan positive ionic charge. There is a direct relationship between the antibacterial activity of chitosan and their DA. Lower DA gives more antibacterial activity (Younes, Sellimi, Rinaudo, Jellouli, & Nasri, 2014). In addition, the antimicrobial activity increases with the increase of the MW of chitosan against *Staphylococcus aureus* (Gram-positive) while antibacterial activity increased with a decrease in MW against *Escherichia coli* (Gram-negative). The authors suggested that for *S. aureus*, a polymeric membrane was formed by chitosan on the surface of the cell, that prevented nutrients from entering into the cell. In the case of *E. coli*, chitosan with lower MW entered the cell through permeation (Zheng & Zhu, 2003).

Studies have shown that chitosan and its derivatives also exhibited anti-cancer and antitumor activities due to their immunostimulatory activity (Huang, Mendis, Rajapakse, & Kim, 2006). Other studies have demonstrated that chitosan exhibited a direct effect on tumor cells by inhibiting tumor cell proliferation by inducing apoptosis (Murata et al., 1989). The anti-cancer or antitumor activities of chitosan and its derivatives depend not only on the chitosan structural characteristics including the extent of DA and polymer MW but also on tumor species. Chitosan oligosaccharides with different MW were studied in relation to their antitumor activity, but it was found that only MW ranging from 1.5 to 5.5 kDa inhibited the growth of Sarcoma 180 solid (S180) or uterine cervix carcinoma No. 14 (U14) tumor in BALB/c mice 129. Another study showed that the chitosan samples with lower MW exhibited higher antitumor activity against Lewis lung carcinoma 220. However, there are some conflicting conclusions from other studies, because Huang, Khor, and Lim (2004) showed that decreasing the MW of chitosan from 213 to 10 kDa does not affect the cytotoxicity on human lung carcinoma cell line A 549 *in vitro*. In addition, chitosan samples with MWs ranging from 42 to 135 kDa had no cytotoxic effect on human bladder cancer cells and no effect of MW was observed (Younes, Frachet, Rinaudo, Jellouli, & Nasri, 2016).

6.3.6 Application of Chitosan and Its Derivatives

The characteristics of chitin and chitosan have been summarized and reviewed by Younes and Rinaudo (2015) and the authors reported these compounds to have biocompatibility, non-toxicity, non-allergenicity, and biodegradability in the body, in addition to being a renewable material. Because of these properties, chitin, chitosan, and their derivatives have found applications in several areas (Table 4).

The antimicrobial activity and film forming properties of chitosan make it potentially useful for preserving food or as a coating material (Ghaouth, Arul, Ponnampalam, & Boulet, 1991). In addition, chitin and chitosan

Table 4 Main Applications of Chitin and Chitosan in Pharmaceutical and Biomedical Domains

Form	Effect	Application	References
Chitin film and fiber		Medical and pharmaceutical applications as wound dressing material/controlled drug release	Kato, Onishi, and Machida (2003) and Yusof, Wee, Lim, and Khor (2003)
Spray, gel, and gauze	Support of medicaments or to control drug release	Nasal and oral delivery of polar drugs	Illum and Davis (2005) and Mi et al. (2003)
Chitosan	Bacteriostatic/fungistatic properties	Ingredient in tropical skin ointment (could healing agent)	Oh et al. (2007)
Hydroxyapatite–chitin–chitosan	Guided tissue regeneration	Bone filling material	Venkatesan, Vinodhini, Sudha, and Kim (2014)
Chitin	Immobilize enzymes and whole cells	Food industry to clarify the fruit juices and process milk	Krajewska (2004) and Muzzarelli et al. (2012)
Chitosan	Natural pseudo cationic polymer	Cosmetic applications, especially for hair care in relation to its electrostatic interactions	Senevirathne and Kim (2012)
Chitosan-based scaffolds		Tissue regeneration	Boucard et al. (2007)
Chitin		A source of dietary fiber in chicken feed	Venugopal (2011)

are widely used in the food industry to clarify fruit juices and in the processing of milk, due to the suitability of chitosan for the immobilization of enzymes (Krajewska, 2004). Chitosan could be used as an ingredient in tropical skin ointments as wound healing agent because of the bacteriostatic and fungistatic activities. The main development of chitin film and fiber in the medical and pharmaceutical application is as a wound dressing material and for encapsulated drug controlled release (Kato et al., 2003; Yusof et al., 2003). Chitosan is also a popular drug carrier (Illum & Davis, 2005; Mi et al., 2003), it was found to overcome the dissolution properties of poorly soluble drugs and to facilitate transdermal delivery of drugs. Chitosan-based scaffolds are useful for tissue regeneration (Muzzarelli et al., 2012). Chitin has been used in the animal feed industry as a source of dietary fiber to enhance the growth of bifidobacteria in the gut. Chitosan has also been used in cosmetic applications, especially for hair care in relation to its electrostatic interaction capability (Muzzarelli et al., 2012).

7. CONCLUSION

The considerable quantity of fish and shellfish waste generated worldwide by processing industries is currently an underutilized resource. A vast literature has demonstrated various approaches to harvesting the protein content of this waste and the generation of protein hydrolysates provides a key means to obtain the protein material in a soluble form with enhanced bioactivity that can be used to fortify processed food products. This fish waste also contains harvestable oils that are rich in health-promoting polyunsaturated fatty acids, that with appropriate extraction are a valuable nutraceutical resource. The chitin major component of crustacean shell can be converted to the soluble chitosan form that can be used in a wide variety of applications ranging from films and coatings to drug encapsulation and assembly of tissue regeneration implants.

However, in view of all that has been mentioned so far, it can be concluded that the experiments so far have mostly been performed in a laboratory setting and more work is required to apply the findings to a commercialization pathway. It needs to be recognized that there are challenges in the production of these materials in terms of investment, optimization of the recovery, and separation of the components and environmental impact that need to be closely evaluated.

REFERENCES

Adeoti, I. A., & Hawboldt, K. (2014). A review of lipid extraction from fish processing by-product for use as a biofuel. *Biomass and Bioenergy, 63*, 330–340.

Aewsiri, T., Benjakul, S., Visessanguan, W., & Tanaka, M. (2008). Chemical compositions and functional properties of gelatin from pre-cooked tuna fin. *International Journal of Food Science & Technology, 43*(4), 685–693.

Agafonova, I., Bogdanovich, R., & Kolosova, N. (2015). Assessment of nephroprotective potential of histochrome during induced arterial hypertension. *Bulletin of Experimental Biology and Medicine, 160*(2), 223–227.

Agafonova, I. G., Kotel'nikov, V. N., Mischenko, N. P., & Kolosova, N. G. (2011). Evaluation of effects of histochrome and mexidol on structural and functional characteristics of the brain in senescence-accelerated OXYS rats by magnetic resonance imaging. *Bulletin of Experimental Biology and Medicine, 150*(6), 739–743.

Ahmad, M., Benjakul, S., & Nalinanon, S. (2010). Compositional and physicochemical characteristics of acid solubilized collagen extracted from the skin of unicorn leatherjacket (*Aluterus monoceros*). *Food Hydrocolloids, 24*(6), 588–594.

Alam, M. N., Bristi, N. J., & Rafiquzzaman, M. (2013). Review on in vivo and in vitro methods evaluation of antioxidant activity. *Saudi Pharmaceutical Journal, 21*(2), 143–152.

Albert, B. B., Vickers, M. H., Gray, C., Reynolds, C. M., Segovia, S. A., Derraik, J. G. B., et al. (2016). Oxidized fish oil in rat pregnancy causes high newborn mortality and increases maternal insulin resistance. *American Journal of Physiology. Regulatory, Integrative and Comparative Physiology, 311*(3), R497–R504.

Albert, B. B., Vickers, M. H., Gray, C., Reynolds, C. M., Segovia, S. A., Derraik, J. G. B., et al. (2017). Fish oil supplementation to rats fed high-fat diet during pregnancy prevents development of impaired insulin sensitivity in male adult offspring. *Scientific Reports, 7*(1), 5595.

Amarowicz, R., Synowiecki, J., & Shahidi, F. (1994). Sephadex LH-20 separation of pigments from shells of red sea urchin (*Strongylocentrotus franciscanus*). *Food Chemistry, 51*(2), 227–229.

Anderson, H., Mathieson, J., & Thomson, R. (1969). Distribution of spinochrome pigments in echinoids. *Comparative Biochemistry and Physiology, 28*(1), 333–345.

Armenta-López, R., Guerrero, I., & Huerta, S. (2002). Astaxanthin extraction from shrimp waste by lactic fermentation and enzymatic hydrolysis of the carotenoprotein complex. *Journal of Food Science, 67*(3), 1002–1006.

Arnesen, J. A., & Gildberg, A. (2007). Extraction and characterisation of gelatine from Atlantic salmon (*Salmo salar*) skin. *Bioresource Technology, 98*(1), 53–57.

Ashhurst, D. (2001). *Chitin and chitinases*. Wiley Online Library.

Austin, P., Brine, C., Castle, J., & Zikakis, J. (1981). Chitin: New facets of research. *Science, 212*(4496), 749–753.

Aytekin, A. O., Morimura, S., & Kida, K. (2011). Synthesis of chitosan–caffeic acid derivatives and evaluation of their antioxidant activities. *Journal of Bioscience and Bioengineering, 111*(2), 212–216.

Badii, F., & Howell, N. K. (2006). Fish gelatin: Structure, gelling properties and interaction with egg albumen proteins. *Food Hydrocolloids, 20*(5), 630–640.

Balti, R., Bougatef, A., Ali, N. E. H., Zekri, D., Barkia, A., & Nasri, M. (2010). Influence of degree of hydrolysis on functional properties and angiotensin I-converting enzyme-inhibitory activity of protein hydrolysates from cuttlefish (*Sepia officinalis*) by-products. *Journal of the Science of Food and Agriculture, 90*(12), 2006–2014.

Barrow, C., & Shahidi, F. (2007). *Marine nutraceuticals and functional foods*. CRC Press.

Beaney, P. D., Gan, Q., Magee, T. R., Healy, M., & Lizardi-Mendoza, J. (2007). Modification of chitin properties for enzymatic deacetylation. *Journal of Chemical Technology and Biotechnology, 82*(2), 165–173.

Beaney, P., Lizardi-Mendoza, J., & Healy, M. (2005). Comparison of chitins produced by chemical and bioprocessing methods. *Journal of Chemical Technology and Biotechnology, 80*(2), 145–150.

Becher, D., Djerassi, C., Moore, R. E., Singh, H., & Scheuer, P. J. (1966). Mass spectrometry in structural and stereochemical problems. CXI. The mass spectrometric fragmentation of substituted naphthoquinones and its application to structural elucidation of echinoderm pigments. *The Journal of Organic Chemistry, 31*(11), 3650–3660.

Benhabiles, M. S., Abdi, N., Drouiche, N., Lounici, H., Pauss, A., Goosen, M. F. A., et al. (2012). Fish protein hydrolysate production from sardine solid waste by crude pepsin enzymatic hydrolysis in a bioreactor coupled to an ultrafiltration unit. *Materials Science and Engineering: C, 32*(4), 922–928.

Benitez, E. I. Y., & Lozano, J. E. (2007). Effect of gelatin on apple juice turbidity. *Latin American Applied Research, 37*(4), 261–266 [online].

Benjakul, S., Oungbho, K., Visessanguan, W., Thiansilakul, Y., & Roytrakul, S. (2009). Characteristics of gelatin from the skins of bigeye snapper, Priacanthus tayenus and *Priacanthus macracanthus. Food Chemistry, 116*(2), 445–451.

Berdyshev, D., Glazunov, V., & Novikov, V. (2007). 7-Ethyl-2,3,5,6,8-pentahydroxy-1,4-naphthoquinone (echinochrome A): A DFT study of the antioxidant mechanism. 1. Interaction of echinochrome A with hydroperoxyl radical. *Russian Chemical Bulletin, 56*(3), 413–429.

Bhaskar, N., Benila, T., Radha, C., & Lalitha, R. G. (2008). Optimization of enzymatic hydrolysis of visceral waste proteins of Catla (*Catla catla*) for preparing protein hydrolysate using a commercial protease. *Bioresource Technology, 99*(2), 335–343.

Bhaskar, N., & Mahendrakar, N. S. (2008). Protein hydrolysate from visceral waste proteins of Catla (*Catla catla*): optimization of hydrolysis conditions for a commercial neutral protease. *Bioresource Technology, 99*(10), 4105–4111.

Blackwell, J. (1973). *Chitin* (pp. 474–489). New York, NY, USA: Academic Press.

Boucard, N., Viton, C., Agay, D., Mari, E., Roger, T., Chancerelle, Y., et al. (2007). The use of physical hydrogels of chitosan for skin regeneration following third-degree burns. *Biomaterials, 28*(24), 3478–3488. https://doi.org/10.1016/j.biomaterials.2007.04.021.

Brasseur, L., Hennebert, E., Fievez, L., Caulier, G., Bureau, F., Tafforeau, L., et al. (2017). The roles of spinochromes in four shallow water tropical Sea urchins and their potential as bioactive pharmacological agents. *Marine Drugs, 15*(6), 179.

Brine, C. J., & Austin, P. R. (1981). Chitin variability with species and method of preparation. *Comparative Biochemistry and Physiology Part B: Comparative Biochemistry, 69*(2), 283–286.

Bustos, R. O., & Healy, M. G. (1994). In *Microbial deproteinisation of waste prawn shell. Paper presented at the proceedings of the 2nd international symposium on environmental biotechnology.*

Cahyanto, A., Kosasih, E., Aripin, D., & Hasratiningsih, Z. (2017). Fabrication of hydroxyapatite from fish bones waste using reflux method. *IOP Conference Series: Materials Science and Engineering, 172*(1), 012006.

Centenaro, G. S., Mellado, M. S., & Prentice-Hernández, C. (2011). Antioxidant activity of protein hydrolysates of fish and chicken bones. *Advance Journal of Food Science and Technology, 3*(4), 280–288.

Chakrabarti, R. (2002). Carotenoprotein from tropical brown shrimp shell waste by enzymatic process. *Food Biotechnology, 16*(1), 81–90.

Chakrabarti, S., Jahandideh, F., & Wu, J. (2014). Food-derived bioactive peptides on inflammation and oxidative stress. *BioMed Research International, 2014*, 608979.

Chakrabarti, S. K., Singh, B. P., Thakur, G., Tiwari, J. K., Kaushik, S. K., Sharma, S., et al. (2014). QTL analysis of late blight resistance in a diploid potato family of Solanum spegazzinii × S. chacoense. *Potato Research*, 57(1), 1–11.

Chalamaiah, M., Dinesh Kumar, B., Hemalatha, R., & Jyothirmayi, T. (2012). Fish protein hydrolysates: Proximate composition, amino acid composition, antioxidant activities and applications: A review. *Food Chemistry*, 135(4), 3020–3038.

Chalamaiah, M., Hemalatha, R., Jyothirmayi, T., Diwan, P. V., Uday Kumar, P., Nimgulkar, C., et al. (2014). Immunomodulatory effects of protein hydrolysates from rohu (*Labeo rohita*) egg (roe) in BALB/c mice. *Food Research International*, 62, 1054–1061.

Chang, C.-T., Lin, Y.-L., Lu, S.-W., Huang, C.-W., Wang, Y.-T., & Chung, Y.-C. (2016). Characterization of a chitosanase from jelly fig (Ficus awkeotsang Makino) latex and its application in the production of water-soluble low molecular weight chitosans. *PLoS One*, 11(3), e0150490.

Chang, C. W. J., & Moore, J. C. (1971). Pigments from some marine specimens. The isolation and spectral characterization of spinochromes. *Journal of Chemical Education*, 48(6), 408.

Chen, J., Li, L., Yi, R., Xu, N., Gao, R., & Hong, B. (2016). Extraction and characterization of acid-soluble collagen from scales and skin of tilapia (*Oreochromis niloticus*). *LWT—Food Science and Technology*, 66, 453–459.

Cheng, L. H., Lim, B. L., Chow, K. H., Chong, S. M., & Chang, Y. C. (2008). Using fish gelatin and pectin to make a low-fat spread. *Food Hydrocolloids*, 22(8), 1637–1640.

Chi, C. F., Cao, Z. H., Wang, B., Hu, F. Y., Li, Z. R., & Zhang, B. (2014). Antioxidant and functional properties of collagen hydrolysates from Spanish mackerel skin as influenced by average molecular weight. *Molecules*, 19(8), 11211–11230.

Chi, C.-F., Hu, F.-Y., Wang, B., Li, Z.-R., & Luo, H.-Y. (2015). Influence of amino acid compositions and peptide profiles on antioxidant capacities of two protein hydrolysates from skipjack tuna (*Katsuwonus pelamis*) dark muscle. *Marine Drugs*, 13(5), 2580–2601.

Chimsook, T., & Wannalangka, W. (2015). Effect of microwave pretreatment on extraction yield and quality of catfish oil in Northern Thailand. *MATEC Web of Conferences*, 35, 04001.

Cho, S. M., Gu, Y. S., & Kim, S. B. (2005). Extracting optimization and physical properties of yellowfin tuna (*Thunnus albacares*) skin gelatin compared to mammalian gelatins. *Food Hydrocolloids*, 19(2), 221–229.

Cho, Y.-S., Kim, S.-K., Ahn, C.-B., & Je, J.-Y. (2011). Preparation, characterization, and antioxidant properties of gallic acid-grafted-chitosans. *Carbohydrate Polymers*, 83(4), 1617–1622.

Choi, B.-K., Kim, K.-Y., Yoo, Y.-J., Oh, S.-J., Choi, J.-H., & Kim, C.-Y. (2001). In vitro antimicrobial activity of a chitooligosaccharide mixture against Actinobacillus actinomycetemcomitans and Streptococcus mutans. *International Journal of Antimicrobial Agents*, 18(6), 553–557.

Choi, S. S., & Regenstein, J. M. (2000). Physicochemical and sensory characteristics of fish gelatin. *Journal of Food Science*, 65(2), 194–199.

Cushman, D. W., & Cheung, H. S. (1971). Spectrophotometric assay and properties of the angiotensin-converting enzyme of rabbit lung. *Biochemical Pharmacology*, 20(7), 1637–1648.

de Castro, R. J. S., Bagagli, M. P., & Sato, H. H. (2015). Improving the functional properties of milk proteins: Focus on the specificities of proteolytic enzymes. *Current Opinion in Food Science*, 1(1), 64–69.

de Castro, R. J. S., & Sato, H. H. (2014). Comparison and synergistic effects of intact proteins and their hydrolysates on the functional properties and antioxidant activities in a simultaneous process of enzymatic hydrolysis. *Food and Bioproducts Processing*, 92(1), 80–88.

De Holanda, H. D., & Netto, F. M. (2006). Recovery of components from shrimp (Xiphopenaeus kroyeri) processing waste by enzymatic hydrolysis. *Journal of Food Science*, 71(5), C298–C303. https://doi.org/10.1111/j.1750-3841.2006.00040.x.

De Meester, F., Watson, R. R., & Zibadi, S. (2012). *Omega-6/3 fatty acids: Functions, sustainability strategies and perspectives*. Humana Press.

de Quirós, A. R.-B., & Costa, H. S. (2006). Analysis of carotenoids in vegetable and plasma samples: A review. *Journal of Food Composition and Analysis*, 19(2–3), 97–111.

Dong, S., Zeng, M., Wang, D., Liu, Z., Zhao, Y., & Yang, H. (2008). Antioxidant and biochemical properties of protein hydrolysates prepared from Silver carp (*Hypophthalmichthys molitrix*). *Food Chemistry*, 107(4), 1485–1493.

dos Santos, S. D., Martins, V. G., Salas-Mellado, M., & Prentice, C. (2011). Evaluation of functional properties in protein hydrolysates from bluewing searobin (*Prionotus punctatus*) obtained with different microbial enzymes. *Food and Bioprocess Technology*, 4(8), 1399–1406.

Du, J., He, J., Yu, Y., Zhu, S., & Li, J. (2013). In *Astaxanthin extracts from shrimp (Litopenaeus vannamei) discards assisted by high pressure processing. Paper presented at the 2013 Kansas City, Missouri, July 21–July 24, 2013*.

Eaton, P., Fernandes, J. C., Pereira, E., Pintado, M. E., & Malcata, F. X. (2008). Atomic force microscopy study of the antibacterial effects of chitosans on Escherichia coli and *Staphylococcus aureus*. *Ultramicroscopy*, 108(10), 1128–1134.

Edgar, L., Hogg, A., Scott, M., Timoney, M., Mc Elnay, J., Mairs, J., et al. (2011). ACE inhibitors for the treatment of hypertension drug selection by means of the SOJA method. *Reviews on Recent Clinical Trials*, 6(1), 69–93.

Edge, R., McGarvey, D., & Truscott, T. (1997). The carotenoids as anti-oxidants—A review. *Journal of Photochemistry and Photobiology B: Biology*, 41(3), 189–200.

Egorov, E., Alekhina, V., Volobueva, T., Fedoreev, S., Mishchenko, N., & Kol'tsova, E. (1998). Histochrome, a new antioxidant, in the treatment of ocular diseases. *Vestnik Oftalmologii*, 115(2), 34–35.

Elavarasan, K., Shamasundar, B. A., Badii, F., & Howell, N. (2016). Angiotensin I-converting enzyme (ACE) inhibitory activity and structural properties of oven- and freeze-dried protein hydrolysate from fresh water fish (*Cirrhinus mrigala*). *Food Chemistry*, 206, 210–216.

Fagbenro, O. A. (1996). Preparation, properties and preservation of lactic acid fermented shrimp heads. *Food Research International*, 29(7), 595–599.

FAO. (2016). *The state of world fisheries and aquaculture 2016 (SOFIA)*. (p. 2014). FAO.

Fassett, R. G., & Coombes, J. S. (2011). Astaxanthin: A potential therapeutic agent in cardiovascular disease. *Marine Drugs*, 9(3), 447–465.

Fassett, R. G., & Coombes, J. S. (2012). Astaxanthin in cardiovascular health and disease. *Molecules*, 17(2), 2030–2048.

Ferdosh, S., Sarker, M. Z. I., Norulaini Nik Ab Rahman, N., Haque Akanda, M. J., Ghafoor, K., & Kadir, M. O. A. (2016). Simultaneous extraction and fractionation of fish oil from tuna by-product using supercritical carbon dioxide (SC-CO2). *Journal of Aquatic Food Product Technology*, 25(2), 230–239.

Ferdosh, S., Sarker, Z. I., Norulaini, N., Oliveira, A., Yunus, K., Chowdury, A. J., et al. (2015). Quality of tuna fish oils extracted from processing the by-products of three species of neritic tuna using supercritical carbon dioxide. *Journal of Food Processing and Preservation*, 39(4), 432–441.

Ferreira, V. F., Nicoletti, C. D., Ferreira, P. G., Futuro, D. O., & da Silva, F. C. (2016). Strategies for increasing the solubility and bioavailability of anticancer compounds: ß-Lapachone and other naphthoquinones. *Current Pharmaceutical Design*, 22(39), 5899–5914.

Gajanan, P. G., Elavarasan, K., & Shamasundar, B. A. (2016). Bioactive and functional properties of protein hydrolysates from fish frame processing waste using plant proteases. *Environmental Science and Pollution Research, 23*(24), 24901–24911.

Gal, A. F., Andrei, S., Cernea, C., Taulescu, M., & Catoi, C. (2012). Effects of astaxanthin supplementation on chemically induced tumorigenesis in Wistar rats. *Acta Veterinaria Scandinavica, 54*(1), 50. https://doi.org/10.1186/1751-0147-54-50.

Galla, N. R., Pamidighantam, P. R., Akula, S., & Karakala, B. (2012). Functional properties and in vitro antioxidant activity of roe protein hydrolysates of Channa striatus and *Labeo rohita. Food Chemistry, 135*(3), 1479–1484.

Gaonkar, A. G., Vasisht, N., Khare, A. R., & Sobel, R. (2014). *Microencapsulation in the food industry: A practical implementation guide.* Elsevier Science.

Gbogouri, G. A., Linder, M., Fanni, J., & Parmentier, M. (2006). Analysis of lipids extracted from salmon (*Salmo salar*) heads by commercial proteolytic enzymes. *European Journal of Lipid Science and Technology, 108*(9), 766–775.

Gedi, M. A., Bakar, J., & Mariod, A. A. (2015). Optimization of supercritical carbon dioxide (CO2) extraction of sardine (*Sardinella lemuru* Bleeker) oil using response surface methodology (RSM). *Grasas y Aceites 2015, 66*(2), e074 (1–10).

Ghaly, A., Ramakrishnan, V., Brooks, M., Budge, S., & Dave, D. (2013). Fish processing wastes as a potential source of proteins, amino acids and oils: A critical review. *Journal of Microbial & Biochemical Technology, 5*, 107 129.

Ghaouth, A., Arul, J., Ponnampalam, R., & Boulet, M. (1991). Chitosan coating effect on storability and quality of fresh strawberries. *Journal of Food Science, 56*(6), 1618–1620.

Ghassem, M., Arihara, K., Babji, A. S., Said, M., & Ibrahim, S. (2011). Purification and identification of ACE inhibitory peptides from Haruan (Channa striatus) myofibrillar protein hydrolysate using HPLC-ESI-TOF MS/MS. *Food Chemistry, 129*(4), 1770–1777.

Ghorbel-Bellaaj, O., Younes, I., Maâlej, H., Hajji, S., & Nasri, M. (2012). Chitin extraction from shrimp shell waste using Bacillus bacteria. *International Journal of Biological Macromolecules, 51*(5), 1196–1201.

Gilsenan, P. M., & Ross-Murphy, S. B. (2000). Rheological characterisation of gelatins from mammalian and marine sources. *Food Hydrocolloids, 14*(3), 191–195.

Gimeno, M., Ramírez-Hernández, J. Y., Mártinez-Ibarra, C., Pacheco, N., García-Arrazola, R., Bárzana, E., et al. (2007). One-solvent extraction of astaxanthin from lactic acid fermented shrimp wastes. *Journal of Agricultural and Food Chemistry, 55*(25), 10345–10350.

Ginter, E., & Simko, V. (2010). Polyunsaturated fatty acids n-3: New data on heart disease, cancer, immune resistance and mental depression. *Bratislavské Lekárske Listy, 111*(12), 680–685.

Gómez-Guillén, M. C., Giménez, B., López-Caballero, M. E., & Montero, M. P. (2011). Functional and bioactive properties of collagen and gelatin from alternative sources: A review. *Food Hydrocolloids, 25*(8), 1813–1827.

Gómez-Guillén, M. C., Turnay, J., Fernández-Díaz, M. D., Ulmo, N., Lizarbe, M. A., & Montero, P. (2002). Structural and physical properties of gelatin extracted from different marine species: A comparative study. *Food Hydrocolloids, 16*(1), 25–34.

Gortari, M. C., & Hours, R. A. (2013). Biotechnological processes for chitin recovery out of crustacean waste: A mini-review. *Electronic Journal of Biotechnology, 16*(3), 14.

Gudmundsson, M. (2002). Rheological properties of fish gelatins. *Journal of Food Science, 67*(6), 2172–2176.

Guerin, M., Huntley, M. E., & Olaizola, M. (2003). Haematococcus astaxanthin: Applications for human health and nutrition. *Trends in Biotechnology, 21*(5), 210–216.

Gurdak, E., Booth, J., Roberts, C. J., Rouxhet, P. G., & Dupont-Gillain, C. C. (2006). Influence of collagen denaturation on the nanoscale organization of adsorbed layers. *Journal of Colloid and Interface Science, 302*(2), 475–484.

Halim, N. R. A., Yusof, H. M., & Sarbon, N. M. (2016). Functional and bioactive properties of fish protein hydolysates and peptides: A comprehensive review. *Trends in Food Science & Technology, 51*, 24–33.

Han, Y., Ma, Q., Wang, L., & Xue, C. (2012). Extraction of astaxanthin from Euphausia pacific using subcritical 1, 1, 1, 2-tetrafluoroethane. *Journal of Ocean University of China, 11*(4), 562–568.

Hansen, M., Thilsted, S. H., Sandstrom, B., Kongsbak, K., Larsen, T., Jensen, M., et al. (1998). Calcium absorption from small soft-boned fish. *Journal of Trace Elements in Medicine and Biology, 12*(3), 148–154.

Hathwar, S. C., Bijinu, B., Rai, A. K., & Narayan, B. (2011). Simultaneous recovery of lipids and proteins by enzymatic hydrolysis of fish industry waste using different commercial proteases. *Applied Biochemistry and Biotechnology, 164*(1), 115–124.

Hayes, M., & McKeon, K. (2014). Advances in the processing of marine discard and by-products. In S.-K. Kim (Ed.), *Seafood processing by-products: Trends and applications* (pp. 125–143). New York, NY: Springer New York.

He, S., Franco, C., & Zhang, W. (2013). Functions, applications and production of protein hydrolysates from fish processing co-products (FPCP). *Food Research International, 50*(1), 289–297.

Herpandi, H., Huda, N., & Adzitey, F. (2011). Fish bone and scale as a potential source of halal gelatin. *Journal of Fisheries and Aquatic Science, 6*(4), 379–389.

Hing, K. A. (2004). Bone repair in the twenty-first century: Biology, chemistry or engineering? *Philosophical Transactions of the Royal Society A, 362*(1825), 2821–2850.

Hoell, I. A., Vaaje-Kolstad, G., & Eijsink, V. G. H. (2010). Structure and function of enzymes acting on chitin and chitosan. *Biotechnology and Genetic Engineering Reviews, 27*(1), 331–366. https://doi.org/10.1080/02648725.2010.10648156.

Honarkar, H., & Barikani, M. (2009). Applications of biopolymers I: Chitosan. *Monatshefte für Chemie—Chemical Monthly, 140*(12), 1403.

Horton, J. M., & Summers, A. P. (2009). The material properties of acellular bone in a teleost fish. *The Journal of Experimental Biology, 212*(Pt. 9), 1413–1420.

Hou, Y., Shavandi, A., Carne, A., Bekhit, A. A., Ng, T. B., Cheung, R. C. F., et al. (2016). Marine shells: Potential opportunities for extraction of functional and health-promoting materials. *Critical Reviews in Environmental Science and Technology, 46*(11 − 12), 1047–1116.

Hoyer, B., Bernhardt, A., Heinemann, S., Stachel, I., Meyer, M., & Gelinsky, M. (2012). Biomimetically mineralized salmon collagen scaffolds for application in bone tissue engineering. *Biomacromolecules, 13*(4), 1059–1066.

Hsu, K. C., Li-Chan, E. C. Y., & Jao, C. L. (2011). Antiproliferative activity of peptides prepared from enzymatic hydrolysates of tuna dark muscle on human breast cancer cell line MCF-7. *Food Chemistry, 126*(2), 617–622.

Huang, M., Khor, E., & Lim, L.-Y. (2004). Uptake and cytotoxicity of chitosan molecules and nanoparticles: Effects of molecular weight and degree of deacetylation. *Pharmaceutical Research, 21*(2), 344–353.

Huang, R., Mendis, E., Rajapakse, N., & Kim, S.-K. (2006). Strong electronic charge as an important factor for anticancer activity of chitooligosaccharides (COS). *Life Sciences, 78*(20), 2399–2408.

Hyoung Lee, D., Ho Kim, J., Sik Park, J., Jun Choi, Y., & Soo Lee, J. (2004). Isolation and characterization of a novel angiotensin I-converting enzyme inhibitory peptide derived from the edible mushroom Tricholoma giganteum. *Peptides, 25*(4), 621–627.

Illum, L., & Davis, S. (2005). Chitosan as a delivery system for the transmucosal administration of drugs. In *Polysaccharides. Structural diversity and functional versatility* (pp. 643–660). CRC Press.

Intarasirisawat, R., Benjakul, S., Visessanguan, W., & Wu, J. (2014). Effects of skipjack roe protein hydrolysate on properties and oxidative stability of fish emulsion sausage. *LWT—Food Science and Technology, 58*(1), 280–286.

Ivanovs, K., & Blumberga, D. (2017). Extraction of fish oil using green extraction methods: A short review. *Energy Procedia, 128,* 477–483.

Jafari, A. M., Gharibi, S., Farjadmand, F., & Sadighara, P. (2012). Extraction of shrimp waste pigments by enzymatic and alkaline treatment: Evaluation by inhibition of lipid peroxidation. *Journal of Material Cycles and Waste Management, 14*(4), 411–413.

Jai ganesh, R., Nazeer, R. A., & Sampath Kumar, N. S. (2011). Purification and identification of antioxidant peptide from black pomfret, Parastromateus niger (Bloch, 1975) viscera protein hydrolysate. *Food Science and Biotechnology, 20*(4), 1087.

Jain, P. K., El-Sayed, I. H., & El-Sayed, M. A. (2007). Au nanoparticles target cancer. *Nano Today, 2*(1), 18–29.

Jamil, N. H., Halim, N. R. A., & Sarbon, N. M. (2016). Optimization of enzymatic hydrolysis condition and functional properties of eel (Monopterus sp.) protein using response surface methodology (RSM). *International Food Research Journal, 23*(1), 1–9.

Jamilah, B., & Harvinder, K. G. (2002). Properties of gelatins from skins of fish—Black tilapia (Oreochromis mossambicus) and red tilapia (Oreochromis nilotica). *Food Chemistry, 77*(1), 81–84.

Jemil, I., Jridi, M., Nasri, R., Ktari, N., Ben Slama-Ben Salem, R., Mehiri, M., et al. (2014). Functional, antioxidant and antibacterial properties of protein hydrolysates prepared from fish meat fermented by Bacillus subtilis A26. *Process Biochemistry, 49*(6), 963–972.

Jeong, S. H., Kim, H. K., Song, I.-S., Lee, S. J., Ko, K. S., Rhee, B. D., et al. (2014). Echinochrome A protects mitochondrial function in cardiomyocytes against cardiotoxic drugs. *Marine Drugs, 12*(5), 2922–2936.

Johnston-Banks, F. A. (1990). Gelatine. In P. Harris (Ed.), *Food gels* (pp. 233–289). Dordrecht: Springer Netherlands.

Jongjareonrak, A., Rawdkuen, S., Chaijan, M., Benjakul, S., Osako, K., & Tanaka, M. (2010). Chemical compositions and characterisation of skin gelatin from farmed giant catfish (Pangasianodon gigas). *LWT—Food Science and Technology, 43*(1), 161–165.

Jumeri, & Kim, S. M. (2011). Antioxidant and anticancer activities of enzymatic hydrolysates of solitary tunicate (Styela clava). *Food Science and Biotechnology, 20*(4), 1075.

Jung, W. K., Lee, B. J., & Kim, S. K. (2006). Fish-bone peptide increases calcium solubility and bioavailability in ovariectomised rats. *The British Journal of Nutrition, 95*(1), 124–128.

Kanazashi, M., Okumura, Y., Al-Nassan, S., Murakami, S., Kondo, H., Nagatomo, F., et al. (2013). Protective effects of astaxanthin on capillary regression in atrophied soleus muscle of rats. *Acta Physiologica, 207*(2), 405–415.

Kandra, P., Challa, M. M., & Jyothi, H. K. (2012). Efficient use of shrimp waste: Present and future trends. *Applied Microbiology and Biotechnology, 93*(1), 17–29. https://doi.org/10.1007/s00253-011-3651-2.

Karadeniz, F., & Kim, S.-K. (2014). Trends in the use of seafood processing by-products in Europe. In S.-K. Kim (Ed.), *Seafood processing by-products: Trends and applications* (pp. 11–19). New York, NY: Springer New York.

Kareva, E. N., Tikhonov, D. A., Mishchenko, N. P., Fedoreev, S. A., & Shimanovskii, N. L. (2014). Effects of histochrome on P53 expression in mouse red bone marrow cells in a model of chronic stress. *Pharmaceutical Chemistry Journal, 48*(3), 149–152.

Karim, A. A., & Bhat, R. (2009). Fish gelatin: Properties, challenges, and prospects as an alternative to mammalian gelatins. *Food Hydrocolloids, 23*(3), 563–576.

Kato, Y., Onishi, H., & Machida, Y. (2003). Application of chitin and chitosan derivatives in the pharmaceutical field. *Current Pharmaceutical Biotechnology*, *4*(5), 303–309.

Kato, Y., Onishi, H., & Machida, Y. (2005). Contribution of chitosan and its derivatives to cancer chemotherapy. *In Vivo (Athens, Greece)*, *19*(1), 301–310.

Ketnawa, S., & Liceaga, A. M. (2017). Effect of microwave treatments on antioxidant activity and antigenicity of fish frame protein hydrolysates. *Food and Bioprocess Technology*, *10*(3), 582–591.

Khanafari, A., Marandi, R., & Sanati, S. (2008). Recovery of chitin and chitosan from shrimp waste by chemical and microbial methods. *Journal of Environmental Health Science & Engineering*, *5*(1), 1–24.

Kim, S. K. (2014). *Seafood processing by-products: Trends and applications*. Springer New York.

Kim, S. K., & Mendis, E. (2006). Bioactive compounds from marine processing byproducts—A review. *Food Research International*, *39*(4), 383–393.

Kim, K. W., & Thomas, R. (2007). Antioxidative activity of chitosans with varying molecular weights. *Food Chemistry*, *101*(1), 308–313.

Kim, H. K., Youm, J. B., Jeong, S. H., Lee, S. R., Song, I.-S., Ko, T. H., et al. (2015). Echinochrome A regulates phosphorylation of phospholamban Ser16 and Thr17 suppressing cardiac SERCA2A Ca2+ reuptake. *Pflügers Archiv: European Journal of Physiology*, *467*(10), 2151–2163.

Kindlund, P., & BioReal, A. (2011). AstaREAL®, natural astaxanthin–nature's way to fight the metabolic syndrome. *Wellness Foods*, *1*, 8–13.

Klomklao, S., & Benjakul, S. (2017). Utilization of tuna processing byproducts: Protein hydrolysate from skipjack tuna (Katsuwonus pelamis) viscera. *Journal of Food Processing and Preservation*, *41*(3), e12970.

Klomklao, S., Benjakul, S., Kishimura, H., & Chaijan, M. (2011). 24 kDa trypsin: A predominant protease purified from the viscera of hybrid catfish (Clarias macrocephalus × Clarias gariepinus). *Food Chemistry*, *129*(3), 739–746.

Kołodziejska, I., Kaczorowski, K., Piotrowska, B., & Sadowska, M. (2004). Modification of the properties of gelatin from skins of Baltic cod (Gadus morhua) with transglutaminase. *Food Chemistry*, *86*(2), 203–209.

Kołodziejska, I., Skierka, E., Sadowska, M., Kołodziejski, W., & Niecikowska, C. (2008). Effect of extracting time and temperature on yield of gelatin from different fish offal. *Food Chemistry*, *107*(2), 700–706.

Kovaleva, M., Ivanova, S., Makarova, M., Pozharitskaia, O., Shikov, A., & Makarov, V. (2013). Effect of a complex preparation of sea urchin shells on blood glucose level and oxidative stress parameters in type II diabetes model. *Eksperimental'naia i Klinicheskaia Farmakologiia*, *76*(8), 27–30.

Krajewska, B. (2004). Application of chitin- and chitosan-based materials for enzyme immobilizations: A review. *Enzyme and Microbial Technology*, *35*(2–3), 126–139.

Krichnavaruk, S., Shotipruk, A., Goto, M., & Pavasant, P. (2008). Supercritical carbon dioxide extraction of astaxanthin from Haematococcus pluvialis with vegetable oils as co-solvent. *Bioresource Technology*, *99*(13), 5556–5560. https://doi.org/10.1016/j.biortech.2007.10.049.

Kristinsson, H. G., & Rasco, B. A. (2000). Fish protein hydrolysates: Production, biochemical, and functional properties. *Critical Reviews in Food Science and Nutrition*, *40*(1), 43–81.

Kromhout, D., Yasuda, S., Geleijnse, J. M., & Shimokawa, H. (2012). Fish oil and omega-3 fatty acids in cardiovascular disease: Do they really work? *European Heart Journal*, *33*, 436–443.

Kuwahara, R., Hatate, H., Chikami, A., Murata, H., & Kijidani, Y. (2010). Quantitative separation of antioxidant pigments in purple sea urchin shells using a reversed-phase high performance liquid chromatography. *LWT—Food Science and Technology*, *43*(8), 1185–1190.

Kuzuya, H., Ikuta, K., & Nagatsu, T. (1973). Inhibition of dopamine-β-hydroxylase by spinochrome A and echinochrome A, naphthoquinone pigments of echinoids. *Biochemical Pharmacology*, *22*(21), 2772–2774.

Lakeev, Y. V., Kosykh, V., Kosenkov, E., Novikov, V., Lebedev, A., & Repin, V. (1992). Effect of natural and synthetic antioxidants (polyhydroxynaphthaquinones) on cholesterol metabolism in cultured rabbit hepatocytes. *Bulletin of Experimental Biology and Medicine*, *114*(5), 1611–1614.

Larsen, T., Thilsted, S. H., Kongsbak, K., & Hansen, M. (2000). Whole small fish as a rich calcium source. *British Journal of Nutrition*, *83*(2), 191–196.

Lassoued, I., Mora, L., Nasri, R., Aydi, M., Toldrá, F., Aristoy, M.-C., et al. (2015). Characterization, antioxidative and ACE inhibitory properties of hydrolysates obtained from thornback ray (Raja clavata) muscle. *Journal of Proteomics*, *128*, 458–468.

Lassoued, I., Mora, L., Nasri, R., Jridi, M., Toldrá, F., Aristoy, M.-C., et al. (2015). Characterization and comparative assessment of antioxidant and ACE inhibitory activities of thornback ray gelatin hydrolysates. *Journal of Functional Foods*, *13*, 225–238.

Lebedev, A., Ivanova, M., & Krasnovid, N. (1999). Interaction of natural polyhydroxy-1, 4-naphthoquinones with superoxide anion-radical. *Biochemistry. Biokhimiia*, *64*(11), 1273–1278.

Lee, S. R., Pronto, J. R. D., Sarankhuu, B.-E., Ko, K. S., Rhee, B. D., Kim, N., et al. (2014). Acetylcholinesterase inhibitory activity of pigment echinochrome A from sea urchin Scaphechinus mirabilis. *Marine Drugs*, *12*(6), 3560–3573.

Lee, S. W., Shimizu, M., Kaminogawa, S., & Yamauchi, K. (1987). Emulsifying properties of peptides obtained from the hydrolyzates of β-Casein. *Agricultural and Biological Chemistry*, *51*(1), 161–166.

Lennikov, A., Kitaichi, N., Noda, K., Mizuuchi, K., Ando, R., Dong, Z., et al. (2014). Amelioration of endotoxin-induced uveitis treated with the sea urchin pigment echinochrome in rats. *Molecular Vision*, *20*, 171.

Li, X., Luo, Y., Shen, H., & You, J. (2012). Antioxidant activities and functional properties of grass carp (Ctenopharyngodon idellus) protein hydrolysates. *Journal of the Science of Food and Agriculture*, *92*(2), 292–298.

Li, D. M., Zhou, D. Y., Zhu, B.-W., Miao, L., Qin, L., Dong, X.-P., et al. (2013). Extraction, structural characterization and antioxidant activity of polyhydroxylated 1,4-naphthoquinone pigments from spines of sea urchin Glyptocidaris crenularis and Strongylocentrotus intermedius. *European Food Research and Technology*, *237*(3), 331–339.

Lin, C. C., Ritch, R., Lin, S. M., Ni, M. H., Chang, Y. C., Lu, Y. L., et al. (2010). A new fish scale-derived scaffold for corneal regeneration. *European Cells & Materials*, *19*, 50–57.

Liu, Y., Li, X., Chen, Z., Yu, J., Wang, F., & Wang, J. (2014). Characterization of structural and functional properties of fish protein hydrolysates from surimi processing by-products. *Food Chemistry*, *151*, 459–465.

Liu, H., Li, D., & Guo, S. (2008). Extraction and properties of gelatin from channel catfish (Ietalurus punetaus) skin. *LWT—Food Science and Technology*, *41*(3), 414–419.

Liu, L., Wang, Y., Peng, C., & Wang, J. (2013). Optimization of the preparation of fish protein anti-obesity hydrolysates using response surface methodology. *International Journal of Molecular Sciences*, *14*(2), 3124–3139.

Malde, M. K., Bügel, S., Kristensen, M., Malde, K., Graff, I. E., & Pedersen, J. I. (2010). Calcium from salmon and cod bone is well absorbed in young healthy men: A double-blinded randomised crossover design. *Nutrition & Metabolism*, *7*(1), 61.

Malde, M. K., Graff, I. E., Siljander-Rasi, H., Venalainen, E., Julshamn, K., Pedersen, J. I., et al. (2010). Fish bones—A highly available calcium source for growing pigs. *Journal of Animal Physiology and Animal Nutrition*, *94*(5), e66–e76.

Manni, L., Ghorbel-Bellaaj, O., Jellouli, K., Younes, I., & Nasri, M. (2010). Extraction and characterization of chitin, chitosan, and protein hydrolysates prepared from shrimp waste by treatment with crude protease from Bacillus cereus SV1. *Applied Biochemistry and Biotechnology, 162*(2), 345–357.

Mao, X., Guo, N., Sun, J., & Xue, C. (2017). Comprehensive utilization of shrimp waste based on biotechnological methods: A review. *Journal of Cleaner Production, 143*, 814–823. https://doi.org/10.1016/j.jclepro.2016.12.042.

Maria, A. G., Graziano, R., & Nicolantonio, D. O. (2015). Carotenoids: Potential allies of cardiovascular health? *Food & Nutrition Research, 59*(1), 26762.

Martínez, M. J. A., & Benito, P. B. (2005). Biological activity of quinones. *Studies in Natural Products Chemistry, 30*, 303–366.

Matsuura, A., Kubo, T., Doi, K., Hayashi, K., Morita, K., Yokota, R., et al. (2009). Bone formation ability of carbonate apatite-collagen scaffolds with different carbonate contents. *Dental Materials Journal, 28*(2), 234–242.

Mercadante, A., Egeland, E., Britton, G., Liaaen-Jensen, S., & Pfander, H. (2004). *Carotenoids handbook*. Birkhäuser Basel, Britton, G.

Mercer, P., & Armenta, R. E. (2011). Developments in oil extraction from microalgae. *European Journal of Lipid Science and Technology, 113*(5), 539–547.

Mi, F.-L., Shyu, S.-S., Lin, Y.-M., Wu, Y.-B., Peng, C.-K., & Tsai, Y.-H. (2003). Chitin/PLGA blend microspheres as a biodegradable drug delivery system: A new delivery system for protein. *Biomaterials, 24*(27), 5023–5036.

Minh Thuy, L. T., Okazaki, E., & Osako, K. (2014). Isolation and characterization of acid-soluble collagen from the scales of marine fishes from Japan and Vietnam. *Food Chemistry, 149*, 264–270.

Mischenko, N. P., Fedoreyev, S. A., Pokhilo, N. D., Anufriev, V. P., Denisenko, V. A., & Glazunov, V. P. (2005). Echinamines A and B, first animated hydroxynaphthazarins from the sea urchin Scaphechinus mirabilis. *Journal of Natural Products, 68*(9), 1390–1393.

Mishchenko, N. P., Fedoreev, S. A., & Bagirova, V. L. (2003). Histochrome: A new original domestic drug. *Pharmaceutical Chemistry Journal, 37*(1), 48–52.

Moffitt, C. M., & Cajas-Cano, L. (2014). Blue growth: The 2014 FAO state of world fisheries and aquaculture. *Fisheries, 39*(11), 552–553.

Mohamed, A. S., Soliman, A. M., & Marie, M. A. S. (2016). Mechanisms of echinochrome potency in modulating diabetic complications in liver. *Life Sciences, 151*, 41–49.

Mohtar, N. F., Perera, C., & Quek, S.-Y. (2010). Optimisation of gelatine extraction from hoki (Macruronus novaezelandiae) skins and measurement of gel strength and SDS–PAGE. *Food Chemistry, 122*(1), 307–313.

Molla, A. E., & Hovannisyan, H. G. (2011). Optimization of enzymatic hydrolysis of visceral waste proteins of beluga Huso huso using Protamex. *International Aquatic Research, 3*, 93–99.

Moore, R. E., Singh, H., & Scheuer, P. J. (1966). Isolation of eleven new spinochromes from echinoids of the genus Echinothrix. *The Journal of Organic Chemistry, 31*(11), 3645–3650.

Murata, J., Saiki, I., Nishimura, S. i., Nishi, N., Tokura, S., & Azuma, I. (1989). Inhibitory effect of chitin heparinoids on the lung metastasis of B16-BL6 melanoma. *Cancer Science, 80*(9), 866–872.

Mutilangi, W. A. M., Panyam, D., & Kilara, A. (1996). Functional properties of hydrolysates from proteolysis of heat-denatured whey protein isolate. *Journal of Food Science, 61*(2), 270–275.

Muzzarelli, R. A., Boudrant, J., Meyer, D., Manno, N., DeMarchis, M., & Paoletti, M. G. (2012). Current views on fungal chitin/chitosan, human chitinases, food preservation, glucans, pectins and inulin: A tribute to Henri Braconnot, precursor of the carbohydrate polymers science, on the chitin bicentennial. *Carbohydrate Polymers, 87*(2), 995–1012.

Nagai, T., & Suzuki, N. (2000). Isolation of collagen from fish waste material—Skin, bone and fins. *Food Chemistry, 68*(3), 277–281.

Nalinanon, S., Benjakul, S., Kishimura, H., & Shahidi, F. (2011). Functionalities and antioxidant properties of protein hydrolysates from the muscle of ornate threadfin bream treated with pepsin from skipjack tuna. *Food Chemistry, 124*(4), 1354–1362.

Naqash, S. Y., & Nazeer, R. A. (2010). Antioxidant activity of hydrolysates and peptide fractions of Nemipterus japonicus and Exocoetus volitans muscle. *Journal of Aquatic Food Product Technology, 19*(3–4), 180–192.

Neklyudov, A. D., Ivankin, A. N., & Berdutina, A. V. (2000). Properties and uses of protein hydrolysates (Review). *Applied Biochemistry and Microbiology, 36*(5), 452–459.

Nguyen-Thi, N., & Doucet, N. (2016). Combining chitinase C and N-acetylhexosaminidase from Streptomyces coelicolor A3 (2) provides an efficient way to synthesize N-acetylglucosamine from crystalline chitin. *Journal of Biotechnology, 220*, 25–32.

Ni, Y., Nagashimada, M., Zhuge, F., Zhan, L., Nagata, N., Tsutsui, A., et al. (2015). Astaxanthin prevents and reverses diet-induced insulin resistance and steatohepatitis in mice: A comparison with vitamin E. *Scientific Reports, 5*, 17192.

Nidheesh, T., Kumar, P. G., & Suresh, P. (2015). Enzymatic degradation of chitosan and production of d-glucosamine by solid substrate fermentation of exo-β-d-glucosaminidase (exochitosanase) by Penicillium decumbens CFRNT15. *International Biodeterioration & Biodegradation, 97*, 97–106.

Nishibori, K. (1959). Isolation of echinochrome A from the spines of the sea urchin, Diadema setosum (Leske). *Nature, 184*(4694), 1234.

No, H. K., & Hur, E. Y. (1998). Control of foam formation by antifoam during demineralization of crustacean shell in preparation of chitin. *Journal of Agricultural and Food Chemistry, 46*(9), 3844–3846.

Oberley, T. D. (2002). Oxidative damage and cancer. *The American Journal of Pathology, 160*, 403–408.

Oh, K.-T., Kim, Y.-J., Jung, W.-J., & Park, R.-D. (2007). Demineralization of crab shell waste by Pseudomonas aeruginosa F722. *Process Biochemistry, 42*(7), 1069–1074.

Olatunji, O., & Denloye, A. (2017). Temperature-dependent extraction kinetics of hydrolyzed collagen from scales of croaker fish using thermal extraction. *Food Science & Nutrition, 5*(5), 1015–1020.

Pareek, N., Vivekanand, V., Saroj, S., Sharma, A. K., & Singh, R. P. (2012). Purification and characterization of chitin deacetylase from Penicillium oxalicum SAEM-51. *Carbohydrate Polymers, 87*(2), 1091–1097.

Park, J. S., Chyun, J. H., Kim, Y. K., Line, L. L., & Chew, B. P. (2010). Astaxanthin decreased oxidative stress and inflammation and enhanced immune response in humans. *Nutrition & Metabolism, 7*(1), 18.

Park, P.-J., Je, J.-Y., & Kim, S.-K. (2003). Free radical scavenging activity of chitooligosaccharides by electron spin resonance spectrometry. *Journal of Agricultural and Food Chemistry, 51*(16), 4624–4627.

Pati, F., Adhikari, B., & Dhara, S. (2010). Isolation and characterization of fish scale collagen of higher thermal stability. *Bioresource Technology, 101*(10), 3737–3742.

Peniston, Q. P., & Johnson, E. L. (1978). Process for demineralization of crustacea shells: Google patents.

Percot, A., Viton, C., & Domard, A. (2003). Characterization of shrimp shell deproteinization. *Biomacromolecules, 4*(5), 1380–1385.

Picot, L., Ravallec, R., Fouchereau-Péron, M., Vandanjon, L., Jaouen, P., Chaplain-Derouiniot, M., et al. (2010). Impact of ultrafiltration and nanofiltration of an industrial fish protein hydrolysate on its bioactive properties. *Journal of the Science of Food and Agriculture, 90*(11), 1819–1826.

Pihlanto, A., Virtanen, T., & Korhonen, H. (2010). Angiotensin I converting enzyme (ACE) inhibitory activity and antihypertensive effect of fermented milk. *International Dairy Journal*, *20*(1), 3–10.

Pires, C., & Batista, I. (2013). *Utilization of fish waste*. CRC Press, pp. 59–75.

Popov, A. M., & Krivoshapko, O. N. (2013). Protective effects of polar lipids and redox-active compounds from marine organisms at modeling of hyperlipidemia and diabetes. *Journal of Biomedical Science and Engineering*, *6*, 543–550.

Powell, C., Hughes, A., Kelly, M., Conner, S., & McDougall, G. (2014). Extraction and identification of antioxidant polyhydroxynaphthoquinone pigments from the sea urchin, Psammechinus miliaris. *LWT—Food Science and Technology*, *59*(1), 455–460.

Pozharitskaya, O. N., Ivanova, S. A., Shikov, A. N., & Makarov, V. G. (2013). Evaluation of free radical-scavenging activity of sea urchin pigments using HPTLC with post-chromatographic derivatization. *Chromatographia*, *76*(19–20), 1353–1358.

Prabha, J., Narikimelli, A., Sajini, M. I., & Vincent, S. (2013). Optimization for autolysis assisted production of fish protein hydrolysate from underutilized fish Pellona ditchela. *International Journal of Scientific and Engineering Research*, *4*(12), 1863–1869.

Prameela, K., Murali, M., & Hemalatha, K. (2010). Extraction of pharmaceutically important chitin and carotenoids from shrimp biowaste by microbial fermentation method. *Journal of Pharmacy Research*, *3*, 2393–2395.

Prameela, K., Venkatesh, K., Immandi, S. B., Kasturi, A. P. K., Krishna, C. R., & Mohan, C. M. (2017). Next generation nutraceutical from shrimp waste: The convergence of applications with extraction methods. *Food Chemistry*, *237*, 121–132.

Rahimi, M. A., Omar, R., Ethaib, S., Mazlina, M. K. S., Biak, D. R. A., & Aisyah, R. N. (2017). Microwave-assisted extraction of lipid from fish waste. *IOP Conference Series: Materials Science and Engineering*, *206*(1), 012096.

Ramalhosa, M. J., Paíga, P., Morais, S., Rui Alves, M., Delerue-Matos, C., & Oliveira, M. B. P. P. (2012). Lipid content of frozen fish: Comparison of different extraction methods and variability during freezing storage. *Food Chemistry*, *131*(1), 328–336.

Rao, M., Munoz, J., & Stevens, W. (2000). Critical factors in chitin production by fermentation of shrimp biowaste. *Applied Microbiology and Biotechnology*, *54*(6), 808–813.

Rao, A. R., Sarada, R., & Ravishankar, G. A. (2007). Stabilization of astaxanthin in edible oils and its use as an antioxidant. *Journal of the Science of Food and Agriculture*, *87*(6), 957–965.

Rao, M., & Stevens, W. (2005). Quality parameters of chitosan derived from fermentation of shrimp biomaterial using a drum reaction. *The Journal of Chemical Technology & Biotechnology*, *80*, 1080–1087.

Razali, A. N., Amin, A. M., & Sarbon, N. M. (2015). Antioxidant activity and functional properties of fractionated cobia skin gelatin hydrolysate at different molecular weight. *International Food Research Journal*, *22*(2), 651–660.

Reverchon, E., & De Marco, I. (2006). Supercritical fluid extraction and fractionation of natural matter. *The Journal of Supercritical Fluids*, *38*(2), 146–166.

Riediger, N. D., Othman, R. A., Suh, M., & Moghadasian, M. H. (2009). A systemic review of the roles of n-3 fatty acids in health and disease. *Journal of the American Dietetic Association*, *109*(4), 668–679.

Rubio-Rodríguez, N., de Diego, S. M., Beltrán, S., Jaime, I., Sanz, M. T., & Rovira, J. (2012). Supercritical fluid extraction of fish oil from fish by-products: A comparison with other extraction methods. *Journal of Food Engineering*, *109*(2), 238–248.

Rustad, T., & Hayes, M. (2012). Marine bioactive peptides and protein hydrolysates: Generation, isolation procedures, and biological and chemical characterizations. In M. Hayes (Ed.), *Marine bioactive compounds: Sources, characterization and applications* (pp. 99–113). Boston, MA: Springer US.

Sachindra, N., Bhaskar, N., Siddegowda, G., Sathisha, A., & Suresh, P. (2007). Recovery of carotenoids from ensilaged shrimp waste. *Bioresource Technology, 98*(8), 1642–1646.

Sadat-Shojai, M., Khorasani, M. T., Dinpanah-Khoshdargi, E., & Jamshidi, A. (2013). Synthesis methods for nanosized hydroxyapatite with diverse structures. *Acta Biomaterialia, 9*(8), 7591–7621.

Sahena, F., Zaidul, I. S. M., Jinap, S., Karim, A. A., Abbas, K. A., Norulaini, N. A. N., et al. (2009). Application of supercritical CO_2 in lipid extraction—A review. *Journal of Food Engineering, 95*(2), 240–253.

Salgado, A. J., Coutinho, O. P., & Reis, R. L. (2004). Bone tissue engineering: State of the art and future trends. *Macromolecular Bioscience, 4*(8), 743–765.

Salwanee, S., Wan Aida, W. M., Mamot, S., Maskat, M. Y., & Ibrahim, S. (2013). Effects of enzyme concentration, temperature, pH and time on the degree of hydrolysis of protein extract from viscera of tuna (Euthynnus affinis) by using alcalase. *Sains Malaysiana, 42*(3), 279–287.

Sampath Kumar, N. S., Nazeer, R. A., & Jaiganesh, R. (2011). Purification and biochemical characterization of antioxidant peptide from horse mackerel (Magalaspis cordyla) viscera protein. *Peptides, 32*(7), 1496–1501.

Sampath Kumar, N. S., Nazeer, R. A., & Jaiganesh, R. (2012). Purification and identification of antioxidant peptides from the skin protein hydrolysate of two marine fishes, horse mackerel (Magalaspis cordyla) and croaker (Otolithes ruber). *Amino Acids, 42*(5), 1641–1649.

Sánchez-Camargo, A. P., Martinez-Correa, H. A., Paviani, L. C., & Cabral, F. A. (2011). Supercritical CO2 extraction of lipids and astaxanthin from Brazilian redspotted shrimp waste (Farfantepenaeus paulensis). *The Journal of Supercritical Fluids, 56*(2), 164–173.

Sánchez-Camargo, A. P., Meireles, M. Â. A., Ferreira, A. L., Saito, E., & Cabral, F. A. (2012). Extraction of ω-3 fatty acids and astaxanthin from Brazilian redspotted shrimp waste using supercritical CO2 + ethanol mixtures. *The Journal of Supercritical Fluids, 61*, 71–77.

Santos, S. D., Cahu, T. B., Firmino, G. O., de Castro, C. C., Carvalho, L. B., Jr., Bezerra, R. S., et al. (2012). Shrimp waste extract and astaxanthin: Rat alveolar macrophage, oxidative stress and inflammation. *Journal of Food Science, 77*(7), H141–H146. https://doi.org/10.1111/j.1750-3841.2012.02762.x.

Sarker, M. Z. I., Selamat, J., Habib, A. S. M. A., Ferdosh, S., Akanda, M. J. H., & Jaffri, J. M. (2012). Optimization of supercritical CO(2) extraction of fish oil from viscera of African catfish (Clarias gariepinus). *International Journal of Molecular Sciences, 13*(9), 11312–11322.

Sato, K., Saimoto, H., Morimoto, M., & Shigemasa, Y. (2003). Depolymerization of chitin and chitosan under hydrothermal conditions. *Sen-I Gakkaishi, 59*, 104–109.

Senevirathne, M., & Kim, S. K. (2012). Development of bioactive peptides from fish proteins and their health promoting ability. *Advances in Food and Nutrition Research, 65*, 235–248. https://doi.org/10.1016/b978-0-12-416003-3.00015-9.

Service, M., & Wardlaw, A. C. (1984). Echinochrome-A as a bactericidal substance in the coelomic fluid of Echinus esculentus (L.). *Comparative Biochemistry and Physiology Part B: Comparative Biochemistry, 79*(2), 161–165.

Shavandi, A., Bekhit, A. E.-D. A., Sun, Z. F., & Ali, A. (2015). A review of synthesis methods, properties and use of hydroxyapatite as a substitute of bone. *Journal of Biomimetics, Biomaterials and Biomedical Engineering, 25*, 98–117.

Shavandi, A., Bekhit, A. E.-D. A., Sun, Z., Ali, A., & Gould, M. (2015). A novel squid pen chitosan/hydroxyapatite/β-tricalcium phosphate composite for bone tissue engineering. *Materials Science and Engineering: C, 55*, 373–383.

Shikov, A. N., Ossipov, V. I., Martiskainen, O., Pozharitskaya, O. N., Ivanova, S. A., & Makarov, V. G. (2011). The offline combination of thin-layer chromatography and high-performance liquid chromatography with diode array detection and micrOTOF-Q mass spectrometry for the separation and identification of spinochromes from sea urchin (Strongylocentrotus droebachiensis) shells. *Journal of Chromatography A, 1218*(50), 9111–9114.

Shimizu, S., & Matubayasi, N. (2014). Gelation: The role of sugars and polyols on gelatin and agarose. *The Journal of Physical Chemistry B, 118*(46), 13210–13216. https://doi.org/10.1021/jp509099h.

Shirahigue, L. D., Silva, M. O., Camargo, A. C., Sucasas, L. F. d. A., Borghesi, R., Cabral, I. S. R., et al. (2016). The feasibility of increasing lipid extraction in Tilapia (Oreochromis niloticus) waste by proteolysis. *Journal of Aquatic Food Product Technology, 25*(2), 265–271.

Shoulders, M. D., & Raines, R. T. (2009). Collagen structure and stability. *Annual Review of Biochemistry, 78*, 929–958.

Shusterov, Y. A., Bragin, V. E., Bykanov, A. N., & Eliseeva, E. V. (2001). Refractive tunnel keratoplasty with synthetic implants modified by a gas-discharge plasma. *Artificial Organs, 25*(12), 983–993.

Sikorski, Z. E. (2001). *Chemical and functional properties of food proteins.* Taylor & Francis.

Sila, A., Kamoun, Z., Ghlissi, Z., Makni, M., Nasri, M., Sahnoun, Z., et al. (2015). Ability of natural astaxanthin from shrimp by-products to attenuate liver oxidative stress in diabetic rats. *Pharmacological Reports, 67*(2), 310–316. https://doi.org/10.1016/j.pharep.2014.09.012.

Sila, A., Nasri, M., & Bougatef, A. (2012). Isolation and characterisation of carotenoproteins from deep-water pink shrimp processing waste. *International Journal of Biological Macromolecules, 51*(5), 953–959. https://doi.org/10.1016/j.ijbiomac.2012.07.011.

Sila, A., Sayari, N., Balti, R., Martinez-Alvarez, O., Nedjar-Arroume, N., Moncef, N., et al. (2014). Biochemical and antioxidant properties of peptidic fraction of carotenoproteins generated from shrimp by-products by enzymatic hydrolysis. *Food Chemistry, 148*, 445–452. https://doi.org/10.1016/j.foodchem.2013.05.146.

Singh, P., Benjakul, S., Maqsood, S., & Kishimura, H. (2011). Isolation and characterisation of collagen extracted from the skin of striped catfish (Pangasianodon hypophthalmus). *Food Chemistry, 124*(1), 97–105.

Singh, I., Moore, R., Chang, C., Ogata, R., & Scheuer, P. (1968). Spinochrome synthesis. *Tetrahedron, 24*(7), 2969–2978.

Sinha, S., Chand, S., & Tripathi, P. (2014). Microbial degradation of chitin waste for production of chitosanase and food related bioactive compounds. *Applied Biochemistry and Microbiology, 50*(2), 125–133.

Šližyte, R., Daukšas, E., Falch, E., Storrø, I., & Rustad, T. (2005). Characteristics of protein fractions generated from hydrolysed cod (Gadus morhua) by-products. *Process Biochemistry, 40*(6), 2021–2033.

Soleimani, S., Yousefzadi, M., Moein, S., Rezadoost, H., & Bioki, N. A. (2016). Identification and antioxidant of polyhydroxylated naphthoquinone pigments from sea urchin pigments of Echinometra mathaei. *Medicinal Chemistry Research, 25*(7), 1476–1483.

Souissi, N., Bougatef, A., Triki-Ellouz, Y., & Nasri, M. (2007). Biochemical and functional properties of sardinella (Sardinetta aurita) by-product hydrolysates. *Food Technology and Biotechnology, 45*(2), 187–194.

Soukoulis, C., & Fisk, I. (2016). Innovative ingredients and emerging technologies for controlling ice recrystallization, texture, and structure stability in frozen dairy desserts: A review. *Critical Reviews in Food Science and Nutrition, 56*(15), 2543–2559.

Sowmya, R., Rathinaraj, K., & Sachindra, N. M. (2011). An autolytic process for recovery of antioxidant activity rich carotenoprotein from shrimp heads. *Marine Biotechnology (New York, NY)*, *13*(5), 918–927. https://doi.org/10.1007/s10126-010-9353-4.

Sowmya, R., Ravikumar, T., Vivek, R., Rathinaraj, K., & Sachindra, N. (2014). Optimization of enzymatic hydrolysis of shrimp waste for recovery of antioxidant activity rich protein isolate. *Journal of Food Science and Technology*, *51*(11), 3199–3207.

Suarez-Jimenez, G.-M., Burgos-Hernandez, A., & Ezquerra-Brauer, J.-M. (2012). Bioactive peptides and depsipeptides with anticancer potential: Sources from marine animals. *Marine Drugs*, *10*(5), 963–986.

Sudarshan, N., Hoover, D., & Knorr, D. (1992). Antibacterial action of chitosan. *Food Biotechnology*, *6*(3), 257–272.

Sun, L., Wu, S., Zhou, L., Wang, F., Lan, X., Sun, J., et al. (2017). Separation and characterization of angiotensin I converting enzyme (ACE) inhibitory peptides from Saurida elongata proteins hydrolysate by IMAC-Ni(2+). *Marine Drugs*, *15*(2), 29.

Taguchi, T., Saito, H., Aoki, H., Uchida, Y., Sakane, M., Kobayashi, H., et al. (2006). Biocompatible high-strength glue consisting of citric acid derivative and collagen. *Materials Science and Engineering: C*, *26*(1), 9–13.

Taheri, A., Anvar, S. A. A., Ahari, H., & Fogliano, V. (2013). Comparison the functional properties of protein hydrolysates from poultry byproducts and rainbow trout (Onchorhynchus mykiss) viscera. *Iranian Journal of Fisheries Sciences*, *12*(1), 154–169.

Talalaeva, O., Mishchenko, N., Bryukhanov, V., Zverev, Y., Fedoreyev, S., Lampatov, V., et al. (2012). The influence of histochrome on exudative and proliferative phases of the experimental inflammation. *Bull. Sib. Branch RAMS*, *32*, 28–31.

Tamilmozhi, S., Veeruraj, A., & Arumugam, M. (2013). Isolation and characterization of acid and pepsin-solubilized collagen from the skin of sailfish (Istiophorus platypterus). *Food Research International*, *54*(2), 1499–1505.

Tan, Y., & Chang, S. K. C. (2018). Isolation and characterization of collagen extracted from channel catfish (Ictalurus punctatus) skin. *Food Chemistry*, *242*, 147–155.

Tanuja, S., Viji, P., Zynudheen, A. A., & Joshy, C. G. (2012). Composition, functional properties and antioxidative activity of hydrolysates prepared from the frame meat of striped catfish (Pangasianodon hypophthalmus). *Egyptian Journal of Biology*, *14*, 27–35.

Tietze, R., Zaloga, J., Unterweger, H., Lyer, S., Friedrich, R. P., Janko, C., et al. (2015). Magnetic nanoparticle-based drug delivery for cancer therapy. *Biochemical and Biophysical Research Communications*, *468*(3), 463–470.

Toldra, F. (2017). *Advances in food and nutrition research* [Vol. 82]. Academic Press.

Tominaga, K., Hongo, N., Karato, M., & Yamashita, E. (2012). Cosmetic benefits of astaxanthin on humans subjects. *Acta Biochimica Polonica*, *59*(1).

Vasileva, E. A., & Mishchenko, N. P. (2016). Antioxidant quinonoid pigments from ceolomic fluid of Far Eastern sea urchins. *Journal of Pharmacognosy & Natural Products*, *1000*(483), 527.

Vasileva, E. A., Mishchenko, N. P., & Fedoreyev, S. A. (2017). Diversity of polyhydroxynaphthoquinone pigments in North Pacific sea urchins. *Chemistry & Biodiversity*.

Vázquez, J. A., Blanco, M., Massa, A. E., Amado, I. R., & Pérez-Martín, R. I. (2017). Production of fish protein hydrolysates from Scyliorhinus canicula discards with antihypertensive and antioxidant activities by enzymatic hydrolysis and mathematical optimization using response surface methodology. *Marine Drugs*, *15*(10), 306.

Venkatesan, J., Lowe, B., Manivasagan, P., Kang, K.-H., Chalisserry, P. E., Anil, S., et al. (2015). Isolation and characterization of nano-hydroxyapatite from salmon fish bone. *Materials (Basel, Switzerland)*, *8*(8), 5426–5439.

Venkatesan, J., Vinodhini, P. A., Sudha, P. N., & Kim, S. K. (2014). Chitin and chitosan composites for bone tissue regeneration. *Advances in Food and Nutrition Research*, *73*, 59–81. https://doi.org/10.1016/b978-0-12-800268-1.00005-6.

Venugopal, V. (2011). *Marine polysaccharides.* Boca Raton: CRC Press.

Vieira, E. F., Pinho, O., & Ferreira, I. M. (2017). Bio-functional properties of sardine protein hydrolysates obtained by brewer's spent yeast and commercial proteases. *Journal of the Science of Food and Agriculture, 97*(15), 5414–5422.

Villamil, O., Váquiro, H., & Solanilla, J. F. (2017). Fish viscera protein hydrolysates: Production, potential applications and functional and bioactive properties. *Food Chemistry, 224,* 160–171.

von Schacky, C. (2000). n-3 fatty acids and the prevention of coronary atherosclerosis. *The American Journal of Clinical Nutrition, 71*(1 Suppl), 224s–227s.

Wade, N. M., Gabaudan, J., & Glencross, B. D. (2017). A review of carotenoid utilisation and function in crustacean aquaculture. *Reviews in Aquaculture, 9*(2), 141–156. https://doi.org/10.1111/raq.12109.

Walker, J. M., & Sweeney, P. J. (2002). Production of protein hydrolysates using enzymes. In J. M. Walker (Ed.), *The protein protocols handbook* (pp. 563–566). Totowa, NJ: Humana Press.

Wang, W., Li, Z., Liu, J., Wang, Y., Liu, S., & Sun, M. (2013). Comparison between thermal hydrolysis and enzymatic proteolysis processes for the preparation of tilapia skin collagen hydrolysates. *Czech Journal of Food Sciences, 31*(1), 1–4.

Wang, L., & Weller, C. L. (2006). Recent advances in extraction of nutraceuticals from plants. *Trends in Food Science & Technology, 17*(6), 300–312.

Wijesekara, I., & Kim, S. K. (2010). Angiotensin-I-converting enzyme (ACE) inhibitors from marine resources: Prospects in the pharmaceutical industry. *Marine Drugs, 8*(4), 1080–1093.

Wu, J., Aluko, R. E., & Muir, A. D. (2009). Production of angiotensin I-converting enzyme inhibitory peptides from defatted canola meal. *Bioresource Technology, 100*(21), 5283–5287.

Wu, R. B., Wu, C. L., Liu, D., Yang, X. H., Huang, J. F., Zhang, J., et al. (2015). Overview of antioxidant peptides derived from marine resources: The sources, characteristic, purification, and evaluation methods. *Applied Biochemistry and Biotechnology, 176*(7), 1815–1833.

Wu, T., Zivanovic, S., Hayes, D. G., & Weiss, J. (2008). Efficient reduction of chitosan molecular weight by high-intensity ultrasound: Underlying mechanism and effect of process parameters. *Journal of Agricultural and Food Chemistry, 56*(13), 5112–5119.

Xi, J., He, L., & Yan, L. (2015). Kinetic modeling of pressure-assisted solvent extraction of polyphenols from green tea in comparison with the conventional extraction. *Food Chemistry, 166,* 287–291.

Xie, Y., Hu, J., Wei, Y., & Hong, X. (2009). Preparation of chitooligosaccharides by the enzymatic hydrolysis of chitosan. *Polymer Degradation and Stability, 94*(10), 1895–1899.

Xing, R., Liu, S., Yu, H., Guo, Z., Wang, P., Li, C., et al. (2005). Salt-assisted acid hydrolysis of chitosan to oligomers under microwave irradiation. *Carbohydrate Research, 340*(13), 2150–2153.

Yamashita, E. (1995). Suppression of post-UVB hyperpigmentation by topical astaxanthin from krill. *Flavour and Fragrance Journal, 14,* 180–185.

Yeul, V. S., & Rayalu, S. S. (2013). Unprecedented chitin and chitosan: A chemical overview. *Journal of Polymers and the Environment, 21*(2), 606–614.

Yim, H. E., & Yoo, K. H. (2008). Renin-angiotensin system—Considerations for hypertension and kidney. *Electrolyte & Blood Pressure, 6*(1), 42–50.

Yin, X.-Q., Lin, Q., Zhang, Q., & Yang, L.-C. (2002). Scavenging activity of chitosan and its metal complexes. *Chinese Journal of Applied Chemistry, 19*(4), 325–328.

Younes, I., Frachet, V., Rinaudo, M., Jellouli, K., & Nasri, M. (2016). Cytotoxicity of chitosans with different acetylation degrees and molecular weights on bladder carcinoma cells. *International Journal of Biological Macromolecules, 84,* 200–207.

Younes, I., Ghorbel-Bellaaj, O., Nasri, R., Chaabouni, M., Rinaudo, M., & Nasri, M. (2012). Chitin and chitosan preparation from shrimp shells using optimized enzymatic deproteinization. *Process Biochemistry*, *47*(12), 2032–2039.

Younes, I., & Rinaudo, M. (2015). Chitin and chitosan preparation from marine sources. Structure, properties and applications. *Marine Drugs*, *13*(3), 1133.

Younes, I., Sellimi, S., Rinaudo, M., Jellouli, K., & Nasri, M. (2014). Influence of acetylation degree and molecular weight of homogeneous chitosans on antibacterial and antifungal activities. *International Journal of Food Microbiology*, *185*, 57–63.

Yusof, N. L. B. M., Wee, A., Lim, L. Y., & Khor, E. (2003). Flexible chitin films as potential wound-dressing materials: Wound model studies. *Journal of Biomedical Materials Research Part A*, *66*(2), 224–232.

Zhang, M., Liu, W., & Li, G. (2009). Isolation and characterisation of collagens from the skin of largefin longbarbel catfish (Mystus macropterus). *Food Chemistry*, *115*(3), 826–831.

Zheng, L.-Y., & Zhu, J.-F. (2003). Study on antimicrobial activity of chitosan with different molecular weights. *Carbohydrate Polymers*, *54*(4), 527–530.

Zhou, P., Mulvaney, S. J., & Regenstein, J. M. (2006). Properties of Alaska pollock skin gelatin: A comparison with tilapia and pork skin gelatins. *Journal of Food Science*, *71*(6), C313–C321.

Zhou, D., Qin, L., Zhu, B.-W., Wang, X.-D., Tan, H., Yang, J.-F., et al. (2011). Extraction and antioxidant property of polyhydroxylated naphthoquinone pigments from spines of purple sea urchin Strongylocentrotus nudus. *Food Chemistry*, *129*(4), 1591–1597.

Zhou, P., & Regenstein, J. M. (2007). Comparison of water gel desserts from fish skin and pork gelatins using instrumental measurements. *Journal of Food Science*, *72*(4), C196–C201.

FURTHER READING

Abdullah, S., Mudalip, S. A., Shaarani, S. M., & Pi, N. C. (2010). Ultrasonic extraction of oil from *Monopterus albus*: Effects of different ultrasonic power, solvent volume and sonication time. *Journal of Applied Sciences*, *10*(21), 2713–2716.

Bremner, H. A. (2002). *Safety and quality issues in fish processing*. Elsevier.

Fallah, M., Bahram, S., & Javadian, S. R. (2015). Fish peptone development using enzymatic hydrolysis of silver carp by-products as a nitrogen source in *Staphylococcus aureus* media. *Food Science & Nutrition*, *3*(2), 153–157.

Horn, S. J., Aspmo, S. I., & Eijsink, V. G. H. (2005). Growth of Lactobacillus plantarum in media containing hydrolysates of fish viscera. *Journal of Applied Microbiology*, *99*(5), 1082–1089.

Jemil, I., Nasri, R., Abdelhedi, O., Aristoy, M.-C., Salem, R. B. S.-B., Kallel, C., et al. (2017). Beneficial effects of fermented sardinelle protein hydrolysates on hypercaloric diet induced hyperglycemia, oxidative stress and deterioration of kidney function in Wistar rats. *Journal of Food Science and Technology*, *54*(2), 313–325.

Pallela, R., Venkatesan, J., Janapala, V. R., & Kim, S.-K. (2012). Biophysicochemical evaluation of chitosan-hydroxyapatite-marine sponge collagen composite for bone tissue engineering. *Journal of Biomedical Materials Research. Part A*, *100A*(2), 486–495.

Priatni, S., Kosasih, W., Budiwati, T. A., & Ratnaningrum, D. (2017). Production of peptone from boso fish (Oxyeleotris marmorata) for bacterial growth medium. *IOP Conference Series: Earth and Environmental Science*, *60*(1), 012009.

Taravel, M. N., & Domard, A. (1996). Collagen and its interactions with chitosan: III. Some biological and mechanical properties. *Biomaterials*, *17*(4), 451–455.

Vázquez, J. A., González, M. P., & Murado, M. A. (2004). A new marine medium: Use of different fish peptones and comparative study of the growth of selected species of marine bacteria. *Enzyme and Microbial Technology, 35*(5), 385–392.

Wang, X., Wang, X., Tan, Y., Zhang, B., Gu, Z., & Li, X. (2009). Synthesis and evaluation of collagen–chitosan–hydroxyapatite nanocomposites for bone grafting. *Journal of Biomedical Materials Research. Part A, 89*(4), 1079–1087.

Zayas, J. F. (1997). Solubility of proteins. In J. F. Zayas (Ed.), *Functionality of proteins in food* (pp. 6–75). Berlin, Heidelberg: Springer Berlin Heidelberg.

CHAPTER FIVE

Advanced Analysis of Roots and Tubers by Hyperspectral Techniques

Wen-Hao Su, Da-Wen Sun[1],[*]

Food Refrigeration and Computerised Food Technology (FRCFT), School of Biosystems and Food
Engineering, Agriculture & Food Science Centre, University College Dublin (UCD), National
University of Ireland, Dublin, Ireland
[1]Corresponding author: e-mail address: dawen.sun@ucd.ie
[*]http://www.ucd.ie/refrig; http://www.ucd.ie/sun

Contents

Abstract

Hyperspectral techniques in terms of spectroscopy and hyperspectral imaging have
become reliable analytical tools to effectively describe quality attributes of roots and
tubers (such as potato, sweet potato, cassava, yam, taro, and sugar beet). In addition
to the ability for obtaining rapid information about food external or internal defects
including sprout, bruise, and hollow heart, and identifying different grades of food
quality, such techniques have also been implemented to determine physical properties (such as color, texture, and specific gravity) and chemical constituents (such as
protein, vitamins, and carotenoids) in root and tuber products with avoidance of extensive sample preparation. Developments of related quality evaluation systems based

Advances in Food and Nutrition Research, Volume 87
ISSN 1043-4526
https://doi.org/10.1016/bs.afnr.2018.07.003
255

on hyperspectral data that determine food quality parameters would bring about economic and technical values to the food industry. Consequently, a comprehensive review of hyperspectral literature is carried out in this chapter. The spectral data acquired, the multivariate statistical methods used, and the main breakthroughs of recent studies on quality determinations of root and tuber products are discussed and summarized. The conclusion elaborates the promise of how hyperspectral techniques can be applied for non-invasive and rapid evaluations of tuber quality properties.

1. INTRODUCTION

As a major category of plant foods worldwide, roots and tubers occupy an important place in our human diet and contribute significantly to sustainable development and food security (Scott, Rosegrant, & Ringler, 2000; Sharma, Njintang, Singhal, & Kaushal, 2016). Root and tuber crops belong to several species in terms of four tropical species (sweet potato, cassava, yam, and taro) and two temperate species (potato and sugar beet) (Aina, Falade, Akingbala, & Titus, 2012; Allemann, Laurie, Thiart, Vorster, & Bornman, 2004; Chiu, Peng, Tsai, & Lui, 2012; Lebot, 2012; Mhemdi, Bals, Grimi, & Vorobiev, 2014; Pedreschi, Granby, & Risum, 2010; Uchechukwu-Agua, Caleb, & Opara, 2015). The root is a part of plant body which bears no leaves below the surface of the soil (Bouwkamp, 2018). Edible roots mainly include sweet potato (*Ipomoea batatas* L.), cassava (*Manihot esculenta*), and sugar beet (*Beta vulgaris* L.) (El Sheikha & Ray, 2017; Fugate, Ribeiro, Lulai, Deckard, & Finger, 2016; Liu, Zainuddin, Vanderschuren, Doughty, & Beeching, 2017). Tuber is an underground stem but can generate new plants (Dasgupta, 2013). The typical tubers include potato (*Solanum* spp.), yam (*Dioscorea* spp.), and taro (*Colocasia esculenta*) (Imaizumi et al., 2014; Vandenbroucke et al., 2016). Compared to roots, tubers contain lower fiber content but have higher protein and dry matter. Furthermore, root crops such as sweet potato and cassava are of American origin, but sugar beet is native to the western and southern coast of Europe; potato tuber is originated from the southern Peru and northwestern Bolivia, while yam is of African or Asian origin, and taro is from the Indo Malayan region (Draycott, 2008; Najamuddeen, Sayaya, Ala, & Garba, 2012; Sharma & Kaushal, 2016). Sugar beet containing a high concentration of sucrose is usually grown for sugar production (Giaquinta, 1979). As the most important source of carbohydrates, potato, sweet potato, and cassava are among the top 10 crops being produced in the developing world and can be used in many aspects such as staple foods, animal feeds, and cash crops. These raw roots and tubers contain about 70–80% of moisture and 16–24% starch, and are also rich in protein, calcium, and vitamin C (Beninca et al., 2013; Flores, Walker, Guimarães, Bais, & Vivanco, 2003; Ugwu, 2009). Such crops are considered potential functional foods and nutrients that can manage many ailments and ensure general wellness (Chandrasekara, 2017). Specifically, it is known that potato is a very high-yield starchy crop which can produce more food calories per unit area and per unit time than wheat, maize, and rice (Oerke, 2006). The starch in raw potato is not easily digested by humans, thus most of potatoes are consumed as processed products such as chips, French fries, dehydrated, and mashed foods (Weurding, Veldman, Veen, van der Aar, & Verstegen, 2001). Sweet potatoes have a very distant relationship with potatoes although both belong to the family of solanales. The sweet potato can be regarded as a good source of natural health-promoting compounds containing high anthocyanins and β-carotene (Bovell-Benjamin, 2007). As the staple food for manufacture of starch, sago, and bioethanol, cassava is also one of the most drought-tolerant crops that can grow on poor soils still with reasonable yields (Chandrasekaran, 2012; El-Sharkawy, 1993).

Modern lifestyle expectations for high-quality food products are increasing in both homemade and fast food, which means that there is a great demand for rapid, accurate, and cost-effective devices to assure that the final product could meet the expected quality (Bahadır & Sezgintürk, 2015; Sun, 2009). The more importance given to the high quality of roots and tubers after harvest, the better the marketability of these products, and therefore more benefit for manufactures and consumers. Many quality parameters in root and tuber products may determine their end use and price. If certain parameters exceed the recommended thresholds, corresponding products will be banned. As described in Fig. 1, these parameters are, but not limited to, physical properties (such as color, texture, and specific gravity), chemical constituents (such as protein, vitamins, and carotenoids), and external or internal defects including sprout, bruise, and hollow heart (Su, He, & Sun, 2017). Additionally, accurate estimations of varieties and gradient information during root and tuber processing are critical as they strongly affect the product characters. Therefore, effective techniques should be designed for the quality evaluation of roots and tubers.

The common techniques, such as liquid chromatography (LC) (Liu et al., 2013; Soler & Pico, 2007), supercritical fluid chromatography (SFC) (Lee et al., 2012; Pereira & Meireles, 2010), gas chromatography-mass

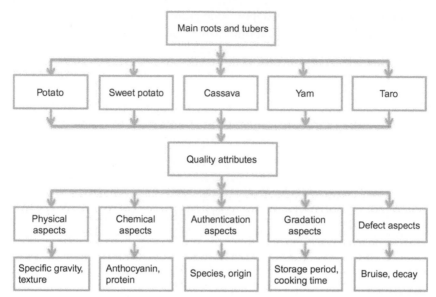

Fig. 1 Quality attributes of the main roots and tubers.

spectrometry (GC–MS) (Roessner, Wagner, Kopka, Trethewey, & Willmitzer, 2000; Sellami et al., 2012; Su, Sun, He, & Zhang, 2017), inductively coupled plasma mass spectrometry (ICP-MS) (Balcaen, Bolea-Fernandez, Resano, & Vanhaecke, 2015; Branch, Ebdon, & O'Neill, 1994), polymerase chain reaction (PCR) (Allmann, Candrian, Höfelein, & Lüthy, 1993; Debela et al., 2015; Severgnini, Cremonesi, Consolandi, De Bellis, & Castiglioni, 2011), and enzyme-linked immunosorbent assay (ELISA) (Alexandrakis, Downey, & Scannell, 2012; Asensio, González, García, & Martín, 2008; Petruccioli et al., 2016), have been employed to measure the main parameters of food products. However, these methods are time-consuming, complicated, destructive, and expensive. Such reasons have raised the interest of researchers and the food industry to develop high-technology systems to secure food quality. It was noticed that several rapid techniques based on the use of biosensors, electronic noses, computer vision, electronic tongue, vibrational spectroscopy, and nuclear magnetic resonance (NMR) have been widely employed to describe the quality attributes of food products (Arkhypova, Dzyadevych, Jaffrezic-Renault, Martelet, & Soldatkin, 2008; Biondi et al., 2014; Blanco, De la Fuente, Caballero, & Rodríguez-Méndez, 2015; Ding, Ni, & Kokot, 2015; Hansen et al., 2010; Pedreschi, Mery, Bunger, & Yanez, 2011;

Su & Sun, 2016d; Sun, 2016; Veloso, Dias, Rodrigues, Pereira, & Peres, 2016; Zhang, Saleh, & Shen, 2013). There have been quite a number of review studies regarding the application of these techniques for rapid evaluation of food quality attributes (Di Rosa, Leone, Cheli, & Chiofalo, 2017; Kirtil & Oztop, 2016; Lohumi, Lee, Lee, & Cho, 2015; Loutfi, Coradeschi, Mani, Shankar, & Rayappan, 2015; Narsaiah, Jha, Bhardwaj, Sharma, & Kumar, 2012; Rotariu, Lagarde, Jaffrezic-Renault, & Bala, 2016; Śliwińska, Wiśniewska, Dymerski, Wardencki, & Namieśnik, 2016). Particularly, hyperspectral techniques in terms of spectroscopy and hyperspectral imaging can acquire a spectral response with numerous contiguous wavelengths and are more promising to achieve rapid and accurate quality and safety analysis of food products (Cheng, Sun, Pu, & Zeng, 2014; Cubero, Aleixos, Moltó, Gómez-Sanchis, & Blasco, 2011; Lorente, Blasco, et al., 2013, Lorente et al., 2012; Menesatti et al., 2009; Sun, 2016; Wei, Liu, Qiu, Shao, & He, 2014; Wu & Sun, 2016; Yu et al., 2014; Zhu, Zhang, Shao, He, & Ngadi, 2014). With the advantages of low cost, simplicity, efficiency, and minimum sample preparation, hyperspectral techniques have shown greater potential in many fields including pharmaceutical (Brondi et al., 2014), environmental (e.g., powder flow) (Scheibelhofer, Koller, Kerschhaggl, & Khinast, 2012), geological (e.g., structural interpretation, regional mapping) (Kurz, Buckley, & Howell, 2013), medical (e.g., image-guided surgery, disease diagnosis) (Lu & Fei, 2014), and microbial (Gowen, Feng, Gaston, & Valdramidis, 2015; Leroux, Midahuen, Perrin, Pescatore, & Imbaud, 2015) aspects.

During the last several years, hyperspectral techniques in combination with chemometric approaches have resulted in the development of rapid methods to determine and monitor quality attributes of various foods including cereals, meats, fruits, and vegetables (Gómez-Sanchis et al., 2014; Liu, He, et al., 2016; Lorente et al., 2012; Naganathan et al., 2008; Ravikanth, Jayas, White, Fields, & Sun, 2017; Singh, Jayas, Paliwal, & White, 2010; Su & Sun, 2016b, 2018a). Although many reviews have been published on hyperspectral techniques, they have just focused on a handful of food products, such as muscle foods (Cheng, Nicolai, & Sun, 2017), chicken meat (Xiong, Xie, Sun, Zeng, & Liu, 2015), red meats (Xiong, Sun, Zeng, & Xie, 2014), fruits (Li, Sun, & Cheng, 2016), liquid foods (Wang, Sun, Pu, & Cheng, 2017), and powdery foods (Su & Sun, 2018a). As far as we know, no reviews on the applications of hyperspectral techniques for the analysis of different types of roots and tubers have been conducted. Therefore, it seems very important to present a comprehensive description of hyperspectral techniques in combination with chemometrics as used for assessing the quality parameters of roots and tubers (such as potato, sweet potato, cassava, yam, taro, and sugar beet), and also to pinpoint the research challenges and future trends of hyperspectral technique in root and tuber applications.

2. HYPERSPECTRAL TECHNIQUES

2.1 Basic Concepts of Visible/Infrared Spectroscopy

Visible/infrared (VIS/IR) spectroscopy is used to acquire spectral information over a wide spectral range where the molecular vibration of a particular frequency that matches the transition energy of a bond or group can be further analyzed (Abdel-Nour, Ngadi, Prasher, & Karimi, 2011; Antonucci et al., 2011; Griffiths & De Haseth, 2007; Klaypradit, Kerdpiboon, & Singh, 2011; Liu, He, Wang, & Sun, 2011; Magwaza et al., 2012). As a well-established non-invasive analytical technique, VIS/IR spectroscopy allows for the direct, reliable, and rapid determination of several quality attributes without the preprocessing of samples (Bao et al., 2014). As shown in Fig. 2, the VIS and IR regions have been highlighted on the electromagnetic spectrum. The VIS region ranges from 380 to 780 nm (26,316–12,821 cm^{-1}), indicating different color information. According to the distance from the spectra of VIS region, the IR spectra can be roughly divided into three parts in terms of the near-infrared (NIR) spectra in the region of 780–2500 nm (12,821–4000 cm^{-1}), the mid-infrared (MIR) spectra in the region of 2500–25,000 nm (4000–400 cm^{-1}), and the far-infrared (FIR) spectra in the region of 25,000–300,000 nm (400–33 cm^{-1}) (Su, He, et al., 2017). By extracting information from chemical constituents and generating reliable fingerprints, the NIR and MIR spectra with higher-energy were demonstrated to be more suitable than FIR spectra for the evaluation of food quality (Woodcock, Fagan, O'Donnell, & Downey, 2008). This is probably due to that MIR spectra can be more suitably used to study basic vibration and related rotational vibration structures, while the NIR spectra can more easily activate the overtone or harmonic vibrations (Cevoli et al., 2013). NIR spectroscopy is more meaningful for analysis of complex mixtures based on spectral stretching and bending of chemical bonds such as C–H, N–H, O–H, and S–H. When NIR radiation penetrates the sample, incoming radiation can be absorbed, transmitted, or reflected, and the beam of interest will be

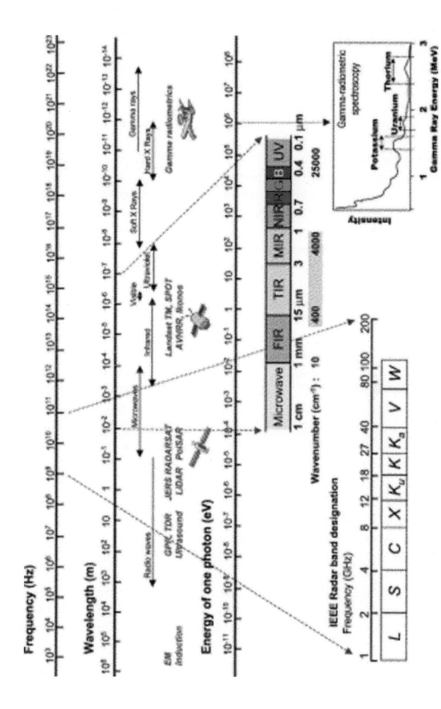

Fig. 2 The visible and infrared regions highlighted on the electromagnetic spectrum (McBratney, Santos, & Minasny, 2003).

collected and directed to the detector. Compared with NIR spectroscopy, MIR spectroscopy can provide more abundant spectral information related to chemical functional groups of a sample (Kačuráková & Wilson, 2001). Normally, the spectra in MIR region can be divided into four portions, including the fingerprint range ($1500-400\,cm^{-1}$), the double-bond range ($2000-1500\,cm^{-1}$), the triple-bond range ($2500-2000\,cm^{-1}$), and the $X-H$ stretching range ($4000-2500\,cm^{-1}$) (Stuart, 2005). For the chemical compounds which is inactive in the MIR region, FIR spectroscopy can be used in the transmission and attenuated total reflectance (ATR) modes for their characterizations as an efficient substitution.

It is of importance to understand the theory of the interaction between light and the surface of research object. Reflection is a process of the change in direction of an electromagnetic radiation on the surface of a sample so that the electromagnetic radiation can return into the same medium from which it is originated. Thus, reflectance spectroscopy is to study the light reflected or scattered from a sample. Normally, the light reflection can be divided into two main types in terms of specular reflection and diffuse reflection (Chambre & Schaaf, 2017). For the diffuse reflection, the electromagnetic radiation can be scattered at many different angles (Nocita et al., 2014). Most of the food materials reflect electromagnetic radiation in the diffusion mode. Reflectance is also the most commonly used pattern in food quality and safety analysis. The reflection pattern measurement is easy to carry out with no touching the material, and the level of light is relatively high in relation to the sample. By contrast, absorbance spectroscopy is the analytical approach based on measuring the amount of light retained in the sample at a given wavelength. Absorption spectroscopy has a stronger correlation with transmission spectroscopy which is based on collecting the light that has passed through the sample (Stuart, 2005). The resulting spectrum depends on a number of conditions, such as sample thickness or path length, the reflectivity of the sample, and the absorption coefficient of the sample. However, the transmission mode is not as popular as the reflection and absorption modes since it requires a stronger light source and a more sensitive detector. In addition, the absorbed electromagnetic radiation can possibly be re-emitted as longer wavelengths with lower energy, and this aspect of radiation is known as fluorescence. Fluorescence spectroscopy is based on the analysis of fluorescence from a sample. It usually involves using a beam of ultraviolet light to excite the electrons in molecules of some compounds and to cause them to emit visible light (Weiss, 1999).

2.2 Basic Concepts of Visible/Infrared Hyperspectral Imaging

Hyperspectral imaging which is also known as imaging spectroscopy integrates the main characteristics of imaging and spectroscopic techniques (Cen, Lu, Ariana, & Mendoza, 2014; Cheng & Sun, 2015; Cheng et al., 2014; Gómez-Sanchis et al., 2014; Su & Sun, 2016d). This technique is regarded as a very efficient detection tool to get over the increasing demand of the acquisition of both spatial and spectral information (Kamruzzaman, Makino, Oshita, & Liu, 2015; Liu, Sun, & Zeng, 2014; Lorente et al., 2012). In hyperspectral images, each pixel contains an almost continuous spectrum with hundreds of bands (Ravikanth et al., 2017; Su, Sun, et al., 2017; Tao & Peng, 2015). Based on the optical property of the sample, hyperspectral imaging can be normally conducted in the optical modes such as reflectance, transmittance, absorbance, or florescence mode. After the spectroscopies modalities (such as VIS, NIR, and MIR) are comprised in imaging technique, the data obtained will be three-dimensional (3-D) structures including two spatial and one spectral dimensions. There are three approaches for the generation of a hyperspectral image, which are area scanning (tunable filter or staredown), line (pushbroom) scanning, and point (whiskbroom) scanning (ElMasry & Nakauchi, 2016). The tunable filters such as filter wheels, acousto-optic filter, and liquid crystal optical filter have been widely used in staring focal plane array systems. Regarding assessments of food quality, the commonly used imaging spectroscopy systems (Figs. 3 and 4) are developed to record spectral information of VIS to MIR region. As the 3-D hypercube could be transformed into 2-D chemical images via multivariate modeling (Fig. 5), the hyperspectral imaging has been extensively used for the quantitative determination and visualization of various physicochemical properties of miscellaneous food products (ElMasry & Nakauchi, 2016). In principle, imaging spectroscopy is irrelevant for monitoring the quality properties of homogeneous samples. Multispectral technique is very similar to the hyperspectral technique in terms of acquiring VIS and IR spectral data from a test sample (Su & Sun, 2018b). Generally, multispectral system can obtain 3–20 wave bands or image-plans (such as 4 from VIS and 6 from IR). There is no continuum among different wavelengths. Multispectral technique that records information of several spaced spectral data can easily achieve the purpose of high-speed detection (Lu, 2004). Based on several selected feature wavelengths to indicate the specific features that characterize the objects of interest, multispectral technique can be the successor of hyperspectral technique with the merits of the hardware simplicity and light weight.

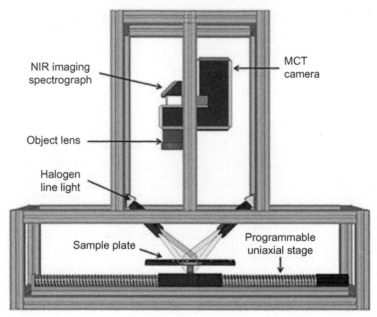

Fig. 3 The schematic diagram of the NIR imaging spectroscopy system (Lee et al., 2016).

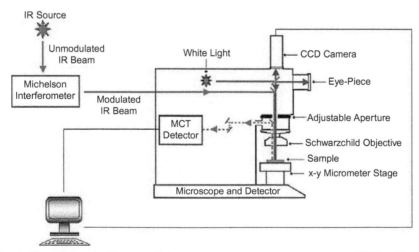

Fig. 4 The schematic diagram of the MIR imaging spectroscopy system (Türker-Kaya & Huck, 2017).

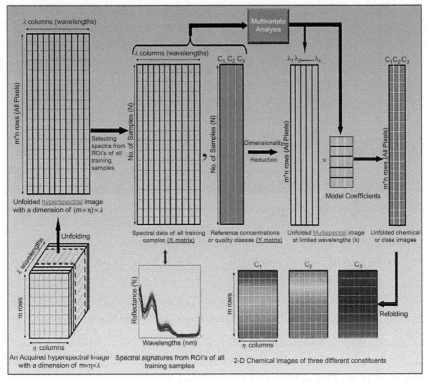

Fig. 5 Schematic representation for generating distribution maps of different food attributes (ElMasry & Nakauchi, 2016).

3. CHEMOMETRIC ANALYSES

There is a need of a multivariate data analysis method to interpret the obtained electromagnetic information (Clegg, Sklute, Dyar, Barefield, & Wiens, 2009). It should be capable of relating the spectral parameters with the quality information of the physical or chemical attributes. Based on suitable mathematical algorithms, this calculated relationship can then be utilized to predict the parameter values of unknown samples. The most commonly used chemometric methods for this process are partial least squares regression (PLSR), artificial neural networks (ANN), principal component regression (PCR), support vector machines (SVM), multiple linear regression (MLR), and least square support vector machine (LS–SVM)

(Lorente, Aleixos, Gómez-Sanchis, Cuberò, & Blasco, 2013; Nashat, Abdullah, Aramvith, & Abdullah, 2011; Pu, Sun, Ma, Liu, & Cheng, 2014; Shahin & Symons, 2011). Due to the complex nature and the large overlap of continuous spectral data, it is difficult to clearly find the positions of feature variables that represent the specific quality parameters in root and tuber products. The selection of important wavelengths is of great interest for the development of more robust and simplified calibration models (Pu, Kamruzzaman, & Sun, 2015). Several variable selection methods, including genetic algorithm (GA) (Jarvis & Goodacre, 2004), first-derivative and mean centering iteration algorithm (FMCIA) (Su & Sun, 2017b), successive projection algorithm (SPA) (Araújo et al., 2001), principal components analysis (PCA) (Shahin & Symons, 2011), competitive adaptive reweighted sampling (CARS) (Li, Liang, Xu, & Cao, 2009), and regression coefficient (RC) (He, Wu, & Sun, 2014), have been introduced to facilitate the reliability of multivariate models. In principle, a good model should obtain high accuracy values including correlation coefficients for prediction (R_P), determination coefficients for cross-validation (R_{CV}^2) and prediction (R_P^2), and low values of ratio of prediction to deviation (RPD), root mean square errors for cross-validation (RMSECV) and prediction (RMSEP).

4. QUALITY AND SAFETY ANALYSIS OF ROOTS AND TUBERS

During the past few years, the concepts of root and tuber quality and safety have drawn much attention, and many researchers have investigated the feasibility of hyperspectral technique for the assessment of quality-related parameters of roots and tubers. The following research briefly introduces the recent progresses of hyperspectral technique with appropriate chemometric approaches, and such latest applications in this research field are listed in Table 1.

4.1 Physical Properties
4.1.1 Color
Color is a particular combination of visible spectral compositions reflected or emitted from an object (Gordon & Morel, 2012). This indicates that color is the first quality property of food products evaluated by consumers and is an important indicator of food quality related to market acceptance (ElMasry, Sun, & Allen, 2012). Therefore, food quality control is required to conduct rapid and objective color assessment for the commercial grading of products.

Table 1 Application of Hyperspectral Techniques for Quality Determinations of Roots and Tubers

Parameter	Type of Sample	Spectral Range	Optimal Model	Maximum Accuracy	References
Hardness, resilience, springiness, cohesiveness, gumminess, chewiness	Potato, sweet potato	MIR	LW/PLSR	$R_p = 0.797, 0.881, 0.584, 0.574, 0.728,$ and 0.690 for hardness, resilience, springiness, cohesiveness, gumminess, and chewiness	Su, Bakalis, and Sun (2018)
Zebra chip disease	Potato	NIR	DA	98.35%	Liang et al. (2018)
MC, chromaticity	Potato	VIS–NIR	PLSR	$R_p^2 = 0.98$ for MC, $R_p^2 = 0.91$ for chromaticity	Amjad, Crichton, Munir, Hensel, and Sturm (2018)
Dry matter, starch	Potato, sweet potato	NIR	MLR, PLSR	$R_p^2 = 0.962$ for dry matter, $R_p^2 = 0.963$ for starch	Su and Sun (2017a)
Anthocyanin	Sweet potato	VIS–NIR	MLR	$R_p^2 = 0.866$	Liu, Sun, et al. (2017)
Bruise	Potato	VIS–NIR	GLCM	93.75%	Ye, Sun, Yang, Che, and Tan (2017)
MC, FWC	Sweet potato	VIS–NIR	MLR	$R_p^2 = 0.984$ for MC, $R_p^2 = 0.932$ for FWC	Sun et al. (2017)
Cultivar	Sweet potato	NIR	PLSDA	100%	Su and Sun (2017c)
MC, color	Potato	VIS–NIR	PLSR	$R_p^2 = 0.990$ for MC, $R_p^2 = 0.990$ for color	Moscetti, Sturm, Crichton, Amjad, and Massantini (2017)

Continued

Table 1 Application of Hyperspectral Techniques for Quality Determinations of Roots and Tubers—cont'd

Parameter	Type of Sample	Spectral Range	Optimal Model	Maximum Accuracy	References
Glycoalkaloids, chlorophyll	Potato	VIS–NIR	PLSR	$R_P^2 = 0.92$ for glycoalkaloids, $R_P^2 = 0.21$ for chlorophyll	Kjær et al. (2017)
Mechanical damage, greening, scab, soil deposits	Potato	VIS–NIR	SD	97.6%	Al Riza, Suzuki, Ogawa, and Kondo (2017)
VTC, TCD	Potato, sweet potato	NIR	TBPANN	$R_P^2 = 0.969$ for VTC, $R_P^2 = 0.983$ for TCD	Su and Sun (2016c)
Variety	Potato, sweet potato	NIR	PLSDA	$\geq 91.6\%$	Su and Sun (2016d)
WBC, SG	Potato, sweet potato	NIR	LWPCR	$R_P^2 = 0.966$ for WBC, $R_P^2 = 0.978$ for SG	Su and Sun (2016a)
MC	Potato, sweet potato	NIR	PLSR	$R_P^2 = 0.939$	Su and Sun (2016e)
Fat	Potato chip	NIR–MIR	PLSR	$R_P = 0.995$	Mazurek, Szostak, and Kita (2016)
Protein	Sweet potato	NIR	PLSR	$R_P^2 = 0.98$	Magwaza, Naidoo, Laurie, Laing, and Shimelis (2016)
Blackspot	Potato	VIS–NIR	PLSDA	98.56%	López-Maestresalas et al. (2016)

Analyte	Product	Spectral	Method	Results	Reference
Starch, glucose, asparagine	Potato	VIS–NIR	PLSR	$R^2_P=0.696$ for starch, $R^2_P=0.508$ for glucose, $R^2_P=0.703$ for asparagine	Kjær et al. (2016)
Crude protein, dry matter	Potato	NIR	MPLSR	$R^2_{CV}=0.936$ for crude protein, $R^2_{CV}=0.939$ for dry matter	Bernhard, Truberg, Friedt, Snowdon, and Wittkop (2016)
Chlorogenic acid, total phenolics, anthocyanin, sucrose, glucose, fructose, reducing sugar	Potato	MIR	PLSR	$R_P=0.92$ for chlorogenic acid, $R_P=0.91$ for total phenolics, $R_P=0.97$ for anthocyanin, $R_{CV}=0.93$ for sucrose, $R_{CV}=0.95$ for glucose, $R_{CV}=0.94$ for fructose, $R_{CV}=0.96$ for reducing sugar	Ayvaz et al. (2016)
Blackheart	Potato	VIS–NIR	PLSDA	97.11%	Zhou, Zeng, Li, and Zheng (2015)
Reducing sugar	Potato flour	NIR	PLSR	$R_P=0.981$	Sun and Dong (2015)
Mechanical damage, hole, scab, surface bruise, sprout, green skin	Potato	VIS–NIR	PCA	82.50%	Su, He, and Sun (2015)
Glucose, sucrose	Potato	NIR	PLSR	$R_P=0.97$ for glucose, $R_P=0.94$ for sucrose	Rady and Guyer (2015b)

Continued

Table 1 Application of Hyperspectral Techniques for Quality Determinations of Roots and Tubers—cont'd

Parameter	Type of Sample	Spectral Range	Optimal Model	Maximum Accuracy	References
Glucose, sucrose	Potato	VIS–NIR	PLSR	$R_P = 0.97$ for glucose, $R_P = 0.60$ for sucrose	Rady, Guyer, and Lu (2015)
Glucose, sucrose	Potato	VIS–NIR	PLSR	$R_P = 0.97$ for glucose, $R_P = 0.94$ for sucrose	Rady and Guyer (2015a)
Hollow heart	Potato	VIS–NIR	SVM	100%	Huang, Jin, and Ku (2015)
Dry matter	Potato	NIR	PLSR	$R_P^2 = 0.92$	Helgerud et al. (2015)
Total anthocyanin, total antioxidant activity, adulteration	Sweet potato	NIR	PLSR, KNN, LDA	$R_P = 0.988$ for total anthocyanin, $R_P = 0.982$ for total antioxidant activity, 100% for adulteration	Ding et al. (2015)
Sugars, free asparagine, glutamine	Potato	MIR	PLSR	$R_P > 0.95$	Ayvaz, Santos, Moyseenko, Kleinhenz, and Rodriguez-Saona (2015)
Acrylamide	Potato chip	NIR–MIR	PLSR	$R_P > 0.90$	Ayvaz and Rodriguez-Saona (2015)
Leaf counts, glucose, sucrose, soluble solids, specific gravity	Potato	VIS–NIR	PLSR	$R_P = 0.95$ for leaf counts, $R_P = 0.95$ for glucose, $R_P = 0.55$ for soluble solids, $R_P = 0.95$ for sucrose, $R_P = 0.61$ for specific gravity	Rady, Guyer, Kirk, and Donis-González (2014)
Total phenolic	Potato	NIR	PLSR	$R_P^2 = 0.74$	López et al. (2014)

Moisture, AIS, protein, starch, fiber	Sweet potato	NIR	MLR	$R^2_P=0.85$ for moisture, $R^2_P=0.94$ for AIS, $R^2_P=0.99$ for protein, $R^2_P=0.97$ for starch, $R^2_P=0.85$ for fiber	Diaz, Veal, and Chinn (2014)
Crude protein	Potato	NIR	PLSR	$R_P=0.88$	López et al. (2013)
Dry matter	Potato	NIR	PLSR	$P^2_P=0.95$	Helgerud et al. (2012)
Fat, moisture, acid value, peroxide	Potato chip	NIR	LS-SVM	$R^2_P=0.996$ for fat, $R^2_P=0.991$ for moisture, $R^2_P=0.990$ for acid value, $R^2_P=0.949$ for peroxide	Ni, Mei, and Kokot (2011)
Starch, sugars, proteins	Sweet potato	NIR	PLSR	$R^2_P=0.71$ for starch, $R^2_P=0.82$ for sugars, $R^2_P=0.87$ for proteins	Lebot, Ndiaye, and Malapa (2011)
Linoleic acid	Potato chip	MIR	PLSR	$R^2_P=0.96$	Kadamne, Castrodale, and Proctor (2011)
Dry matter, starch, reducing sugars	Potato	NIR	MLR	$R^2_P=0.99$ for dry matter, $R^2_P=0.96$ for starch, $R^2_P=0.43$ for reducing sugars	Haase (2011)
Sugar-end	Potato	NIR	PLSDA	9_.7%	Groinig, Burgstaller, and Pail (2011)
Cooking time	Potato	VIS–NIR	PLSDA	$R^2_P=0.956$	Do Trong, Tsuta, Nicolaï, De Baerdemaeker, and Saeys (2011)

Continued

Table 1 Application of Hyperspectral Techniques for Quality Determinations of Roots and Tubers—cont'd

Parameter	Type of Sample	Spectral Range	Optimal Model	Maximum Accuracy	References
Scab	Potato	NIR	SVM	97.1%	Dacal-Nieto, Formella, Carrión, Vazquez-Fernandez, and Fernández-Delgado (2011a)
Hollow heart	Potato	NIR	SVM	89.1%	Dacal-Nieto, Formella, Carrión, Vazquez-Fernandez, and Fernández-Delgado (2011b)
Dry matter, carotenoids	Cassava	VIS–NIR	PLSR	$R_P^2 = 0.95$ for dry matter, $R_P^2 = 0.99$ for carotenoids	Ikeogu et al. (2017)
Maleic acid	Cassava starch	NIR	OCPLS	Sensitivity = 0.954, specificity = 0.956	Fu et al. (2017)
Total carotenoids, total β-carotene, dry matter	Cassava	NIR	PLSR	$R_P^2 = 0.880$ for total carotenoids, $R_P^2 = 0.918$ for total β-carotene, $R_P^2 = 0.946$ for dry matter	Sánchez et al. (2014)
Acrylamide	French-fried potato	NIR	PLSR	$R_P^2 = 0.980$	Adedipe, Johanningsmeier, Truong, and Yencho (2016)

Analyte/property	Sample	Spectral region	Method	Results	Reference
Palm oil adulterated with lard	French-fried potato	MIR	PLSDA	$R_P^2 = 0.979$	Che Man, Marina, Rohman, Al-Kahtani, and Norazura (2014)
Trans fatty acid	French-fried potato	MIR	PLSR	$R_P^2 = 0.984$	Cho, Kim, Khurana, Li, and Jun (2011)
Moisture, fat content, color properties, maximum force	Taro chip	NIR	PLSR	$R_P^2 = 0.850-0.970$	Areekij et al. (2017)
Starch, sugars, proteins, minerals	Taro	NIR	PLSR	$R_P^2 = 0.76$ for starch, $R_P^2 = 0.74$ for sugars, $R_P^2 = 0.85$ for proteins, $R_P^2 = 0.85$ for minerals	Lebot, Malapa, and Bourrieau (2011)
Total sugar, polysaccharides, flavonoids	Yam	NIR–MIR	LS-SVM	$R_P = 0.989$ for total sugar, $R_P = 0.986$ for polysaccharides, $R_P = 0.986$ for flavonoids	Zhuang, Ni, and Kokot (2015)
Dioscin	Yam	MIR	PLSR	$R_P^2 = 0.721$	Kwon et al. (2015)
Starch, sugar, protein	Yam	NIR	PLSR	$R_P^2 = 0.84$ for starch, $R_P^2 = 0.86$ for sugar, $R_P^2 = 0.88$ for protein	Lebot and Malapa (2013)
Sucrose, soluble solids, moisture, mechanical properties	Sugar beet	VIS-NIR	PLSR	$R_P = 0.75-0.88$ for sucrose, soluble solids, and moisture; $R_P = 0.46-0.63$ for mechanical properties	Pan, Lu, Zhu, Tu, and Cen (2016)

Continued

Table 1 Application of Hyperspectral Techniques for Quality Determinations of Roots and Tubers—cont'd

Parameter	Type of Sample	Spectral Range	Optimal Model	Maximum Accuracy	References
Sucrose	Sugar beet	NIR	PLSR	$R_P^2 = 0.880$	Pan, Zhu, Lu, and McGrath (2015)
Moisture, soluble solids, sucrose, mechanical properties	Sugar beet	VIS–NIR	PLSR	$R_P = 0.89$–0.95 for moisture, soluble solids, and sucrose; $R_P = 0.31$–0.62 for mechanical properties	Pan, Lu, Zhu, McGrath, and Tu (2015)
Bioethanol yield	Sugar beet pulp	NIR	PLSR	$R_P^2 = 0.910$	Magaña et al. (2011)
Adulteration	Cassava flour	NIR	PLSR	$R_P^2 = 0.986$	Su and Sun (2017b)
Adulteration	Sweet potato starch, potato starch, cassava starch	NIR	PLSCM	100% for doping level of 4% or higher	Xu, Shi, Cai, Zhong, and Tu (2015)

LWPLSR, locally weighted partial least squares regression; DA, discriminant analysis; PLSR, partial least square fegression; MPLSR, modified partial least square fegression; GLCM, gray level co-occurrence matrix; SD, subspace discriminant; PLSDA, partial least square discriminant analysis; KNN: k-nearest neighbors, LDA, linear discriminant analysis; AIS, alcohol insoluble solids; OCPLS, one-class partial least squares; PLSCM, partial least squares class model; VTC, volatility of tuber compositions; TCD, tuber cooking degree; PCA, principal component analysis; LS-SVM, least-square support vector machine; SVM, support vector machines; MLR, multiple linear regression; TBPANN, three-layer back propagation artificial neural network; MC, moisture content; FWC, freezable water content; R_P, correlation coefficient for prediction; R_P^2, coefficient of determination for prediction.

Hyperspectral technique is a promising approach currently investigated for root and tuber color measurement. The potential of VIS/NIR (500–1010 nm) hyperspectral imaging was evaluated to monitor the color of potato slices of three thicknesses (5, 7, and 9 mm) subjected to air drying at 50 °C. The authors assessed the effectiveness of different feature selection methods such as interval partial least squares regression (iPLSR) for prediction of the color of potato slices. Excellent result was achieved based on the PLSR model using the iPLSR algorithm (Moscetti et al., 2017). Then, NIR (1100–2500 nm) spectroscopy combined with PLSR was used to determine the corresponding color properties of deep-fried taro chips, yielding R_P^2 as high as 0.97 (Areekij et al., 2017). In a recent study, the chromaticity of potato slices with 5, 7, and 9 mm thickness at three drying temperatures (50, 60, and 70 °C) was determined by the VIS/NIR hyperspectral imaging (Amjad et al., 2018). Based on the PLSR with several feature wavelengths, the highest accuracy reached R_P^2 of 0.91. These results demonstrated that VIS/IR hyperspectral techniques offered a rapid, accurate, and non-destructive approach for monitoring the color changes of tuber products during processing.

4.1.2 Textural Attributes

An acceptable food product for consumers should not only have a good appearance but also need to possess desirable sensory attributes. Texture of roots and tubers is a significant factor related to product quality at the time of consumption. Food texture is mainly determined by the sensory manifestation of the mechanical or structural properties of that product (Chung & McClements, 2015). Conventional methods for detection of food structural properties based on touch and taste require a considerable amount of time, which is low efficiency (Davies & Dixon, 1976). The textural parameters of roots and tubers mainly involve springiness, cohesiveness, firmness/hardness, gumminess, resilience, chewiness, and so on (Su et al., 2018). To eliminate the influence from human factors, the three-point bending test and the texture profile analysis have been used to measure food texture, but these methods can result in strong destructiveness for the tuber structure (Fagan et al., 2007).

Hyperspectral techniques have been used to evaluate the textural properties of root and tuber products. Pan, Lu, et al. (2015) investigated the feasibility of measurement of mechanical properties of sugar beet based on VIS/NIR (400–1100 nm) and NIR (900–1600 nm) spectroscopy. These authors reported that both spectrometers presented poor performances of detecting various mechanical parameters including maximum force, area, and the slope for the force/displacement curve of both intact and sliced beets, with the R_P values from 0.31 to 0.62. Similar results were found in another study where R_P values between 0.46 and 0.63 were obtained using NIR hyperspectral imaging (500–1000 nm) and PLSR (Pan et al., 2016). To improve the detection accuracy, a new NIR spectrometer (1100–2500 nm) was conducted to analyze textural attributes of taro chips. Based on the PLSR model, maximum force of break of deep-fried taro chips was successfully determined with R_P^2 of 0.85–0.97 and the ratio of prediction to deviation (RPD) ranging from 2.0 to 4.9 (Areekij et al., 2017). Besides VIS-NIR spectroscopy, the textural parameters of root and tuber products could also be detected by MIR spectroscopy. The predictability of MIR spectra in the fingerprint region (1500–900 cm^{-1}) and full-wavenumber region (4000–600 cm^{-1}) was evaluated for rapid measuring textural properties of both potato and sweet potato during microwave baking (Su et al., 2018). Based on locally weighted partial least squares regression (LWPLSR) and PLSR, the spectra from both regions showed acceptable prediction accuracy. Similar or even better predictions were obtained by models with fingerprint spectra. The LWPLSR model, developed using several feature wavenumbers, achieved the optimal results for the determination of root and tuber textural parameters (such as hardness, cohesiveness, resilience, gumminess, springiness, and chewiness) with R_P value as high as 0.881. It was concluded in their study that FT-MIR imaging spectroscopy could be used reliably to assess textural property of microwave baked potato and sweet potato products.

4.1.3 Specific Gravity and Water-Binding Capacity

As another very important factor for assessing roots and tubers, specific gravity is significantly affected by climate, plant density, irrigation, soil property, fertilization, and so on (Motes & Greig, 1970; Stark & Love, 2003). Normally, higher specific gravity can produce more output of French fry, chip, and other dehydrated products. Although higher tuber specific gravity is desirable, the very high specific gravity of fresh roots and tubers may also bruise more easily (Baritelle & Hyde, 2003). Moreover, water-binding capacity is defined as the ability of food tissue to retain water even if external pressures (such as gravity, heating) are attached (Chen, Piva, & Labuza, 1984; Robertson et al., 2000). The water-binding capacity can affect the organoleptic properties of food products. Poor water-binding capacity can result in undesired food appearances. Therefore, both specific gravity and water-binding capacity can indirectly affect tuber quality.

Hyperspectral modes including VIS/NIR interactance spectroscopy, NIR transmittance spectroscopy, and VIS/NIR hyperspectral imaging were investigated to determine the feasibility for rapid electronic estimation of specific gravity of potato tuber (Rady et al., 2014). The optimal R_P value (0.61) for specific gravity of sliced potato samples was calculated by the PLSR under interactance mode. Based on absorbance spectra (A_S), power spectra (P_S), reflectance spectra (R_S), and generalized logarithm spectra (GL_S) derived from NIR hyperspectral images, PLSR and locally weighted principal component regression (LWPCR) models were developed to measure the specific gravity and water-binding capacity of both potatoes and sweet potatoes (Su & Sun, 2016a). In that study, several feature wavelengths were selected from full-spectral region using reverse variable algorithm (RVA), FMCIA, and GA, then the selected variables were further reduced by the method of RC. The simplified FMCIA-RC-GL_S-LWPCR model collected the highest accuracy with R_P^2 of 0.978 for specific gravity, while the GA-RC-P_S-LWPCR model obtained best performance (R_P^2 of 0.966) of measuring water-binding capacity (Su & Sun, 2016a). Their results demonstrated that effective variable selection approach can improve the detection efficiency of hyperspectral imaging system.

4.2 Chemical Constituents

4.2.1 Protein

The protein of root and tuber is a valuable alternative in human nutrition because of its adequate nutritional value and acceptable protein yield per hectare. Tumwegamire et al. (2011) reported that the protein content can vary from about 4% to 10% of tuber dry matter, which provided a great possibility for increasing the protein content in the commonly used root and tuber cultivars. With the big variability in genotypes, protein content of root and tuber should be an important parameter to be considered in the qualification of superior genotype for wide-area production.

The newly developed NIR spectral model for the determination of protein in fresh potato tubers is very useful for selecting potato cultivars that contain high or low tuber protein content. The potential of spectroscopy and chemometrics was investigated to measure the proteins in sweet potato, taro, yam, and potato. NIR technique is considered a reliable approach that can be applied to quantify root and tuber components during processing. For the determination of protein of enzymatically processed sweet potatoes, the MLR using the NIR spectra was examined and yielded 99% confidence for raw products and a quantitative prediction ($R^2 = 0.69$) in processed material (Diaz et al., 2014). Afterward, the PLSR model using second derivative pre-processed NIR spectra was survived for performance enchantment and presented optimal result for the determination of protein in sweet potato with R_P^2 of 0.98 (Magwaza et al., 2016). In another study, a total of 240 accessions of sweet potato were used and obtained the R_P^2 of 0.87 using the PLSR (Lebot, Ndiaye, et al., 2011). The good performance of PLSR model has also been assessed by analyzing proteins in other roots and tubers. When 135 lyophilized samples of potato were considered, the crude protein could be predicted by PLSR using the NIR (1100–2300 nm) spectra ($R_P = 0.88$). Based on 117 samples of dried potato flour, the modified partial least square regression (MPLSR) was used to build practicable NIR calibrations for crude protein content of potato with R_{CV}^2 as high as 0.936 (Bernhard et al., 2016). The reasonably high confidence of PLSR was also realized based on the measurement of proteins in taro and yam. To predict proteins in taro corms, the R_P^2 values of 0.85 with RPD of 3.78 were obtained (Lebot, Malapa, et al., 2011). The superiority of NIR and PLSR was further verified for the prediction of protein in the yam tuber with R_P^2 of 0.88 and the ratio of RPD of 3.641, which also allows good quantitative predictions to be made (Lebot & Malapa, 2013). These developed calibration models for protein on root and tuber samples are useful for tuber cultivar selection.

4.2.2 Dry Matter

The dry matter is one of the main quality parameters of raw and processed potatoes. Many studies have demonstrated that dry matter content could be an equivalent predictor of flavor in numerous plant foods (Bally, Johnson, & Kulkarni, 1999; Burdon et al., 2004; Harker et al., 2009). The dry matter comprises most non-volatile compounds, containing both soluble and insoluble carbohydrates (Gibson, 2012). Moreover, dry matter has been measured to assess plant food maturity and consumer preference (Gamble et al., 2010; Palmer, Harker, Tustin, & Johnston, 2010). Accordingly, the prediction of dry matter content in roots and tubers is a necessary operation in the processing industry.

The technique of reflectance spectroscopy was widely applied to measure the processing quality of potatoes. Calibration equations using the NIR spectra of 850–2500 nm were computed for predicting the dry matter of dehydrated potatoes and potato crisps, yielding the best R_P^2 of 0.99 (Haase, 2011). Then, the dry matter of unpeeled potato tubers was effectively determined based on the PLSR developed using the spectra in the VIS/NIR region (760–1040 nm) with R_P^2 of 0.95 (Helgerud et al., 2012). The highly satisfactory results ($R_P^2 = 0.939$–0.960) for dry matter detection were also obtained by other NIR devices (Bernhard et al., 2016; Sánchez et al., 2014). During on-line measurements, still stable prediction performance was achieved by the PLSR model, thus showing the possibility of using the NIR instrument for on-line measurements (Helgerud et al., 2015). In a recent study, three calibration models including LWPLSR, PLSR, and MLR were, respectively, developed based on different spectra collected from hyperspectral images for rapid measurement of dry matter in both potato and sweet potato tubers. The MLR model with several important wavelengths was regarded as the best approach and presented a consistent accuracy ($R_P^2 = 0.962$) (Su & Sun, 2017a). In addition, a portable VIS/NIR system showed equally a high correlation (0.94) for dry matter determination of fresh cassava roots (Ikeogu et al., 2017).

4.2.3 Polysaccharide: Starch and Cellulose

Polysaccharide is polymerized carbohydrate molecules consisting of long chains of monosaccharides combined by glycosidic linkages. Starch and cellulose are two important polysaccharides in roots and tubers. Both starch and cellulose are insoluble in water. As a storage polysaccharide, starch is a glucose polymer where hundreds of glucose units are connected by alpha-linkages (Coppin et al., 2005). Cellulose is structural polysaccharide made with numerous glucose units bonded using beta-linkages (Lawther, Sun, & Banks, 1995). Predictions of starch and cellulose in roots and tubers are of interest. For instance, starch can be used to evaluate crop potential and the efficiency of processing enzymes which are used to convert starch into more valuable products for industrial applications.

Hyperspectral technique was investigated to predict the cellulose and starch in sweet potato, taro, and yam. It was found that the cellulose of yam, taro, and sweet potato could not be predicted precisely based on NIR spectroscopy (Lebot & Malapa, 2013; Lebot, Malapa, et al., 2011; Lebot, Ndiaye, et al., 2011). However, PLSR models showed acceptable performances for the measurement of starch in sweet potato, taro, and yam, with R_P^2 of 0.71, 0.76, and 0.84, respectively (Lebot & Malapa, 2013; Lebot, Malapa, et al., 2011; Lebot, Ndiaye, et al., 2011). The medium prediction accuracy ($R_P^2 = 0.696$) for potato starch-related parameters was also obtained by NIR hyperspectral imaging and PLSR (Kjær et al., 2016). Based on R_S, A_S, and exponent spectra (E_S) in NIR region, both

PLSR and MLR models were developed to determine the starch content of both potato and sweet potato tubers. The optimal result ($R_P^2 = 0.963$) using hyperspectral imaging was acquired by the E_S-PLSR model with six feature variables (Su & Sun, 2017a). Compared with PLSR, MLR could also predict the starch of potato and sweet potato with R_P^2 as high as 0.97 (Diaz et al., 2014; Haase, 2011).

4.2.4 Sugars

Sugar is a general term for soluble carbohydrates including monosaccharides (such as glucose and fructose) and disaccharides (such as sucrose) (Wilson, Work, Bushway, & Bushway, 1981; Xie, Ye, Liu, & Ying, 2009). The technique of NIR spectroscopy (850–2500 nm) was tested to predict reducing sugars, total sugar, and sucrose of dehydrated potatoes and potato crisps. The best R_P^2 values within independent validation sets were about 0.508 for reducing sugars, 0.660 for total sugar, and 0.710 for sucrose (Haase, 2011; Kjær et al., 2016). NIR spectroscopy combined with PLSR was then used to investigate sugar contents in taro, sweet potato, and yam, showing higher explained variances in R_P^2 of 0.74, 0.82, and 0.86, respectively (Lebot & Malapa, 2013; Lebot, Malapa, et al., 2011; Lebot, Ndiaye, et al., 2011). Compared with PLSR, LS-SVM calibration models produced better results (R_P^2 of 0.978) in a later study for measurement of total sugar in yam (Zhuang et al., 2015). Moreover, two portable NIR spectrometers with the regions of 400–1100 and 900–1600 nm were, respectively, employed to evaluate compositions of sugar beet (Pan, Lu, et al., 2015; Pan, Zhu, et al., 2015). Based on correlation analysis, both systems showed excellent determinations for the sucrose content of beet slices with R_P of 0.88–0.95. Feature wavelengths in the VIS/NIR range (446–1125 nm) were then selected by iPLSR and GA for determining glucose and sucrose in potato whole tubers. Results showed that the R values for prediction models of glucose and sucrose reached 0.97 and 0.94, respectively (Rady & Guyer, 2015a). Based on portable MIR systems, the sugars could also be effectively measured. For instance, a portable MIR instrument was applied for the rapid quantitation of major sugars including glucose, fructose, and sucrose in raw potato tubers. In that study, PLSR model showed a very high R_P of over 0.95 (Ayvaz et al., 2015). It was noticed both NIR and MIR techniques performed well for the analysis of sugars in root and tuber products. The potential of hyperspectral imaging was investigated toward determinations of fructose, glucose, or sucrose in potato. Based on the variable selection approach of iPLSR, both k-nearest neighbor (KNN) and PLSR models obtained similar results to those of full-wavelength models, yielding R_P^2 of 0.884 for glucose and 0.360 for sucrose (Rady et al., 2015). For the measurement of sucrose in sugar beet, PLSR model associated with uninformative variable elimination (UVE) also showed a comparable accuracy (Pan et al., 2016). When the UVE was combined with SPA for feature variable optimization, UVE-SPA-PLSR model performed better for rapid detection of reducing sugar content in potato flours with R_P^2 as high as 0.962. Their results demonstrated that NIR spectroscopy with UVE and SPA has significant potential to quantitatively analyze sugars in potato products (Sun & Dong, 2015). Additionally, soluble solids content is considered as a composite index of sweetness. VIS/NIR spectroscopy coupled with PLSR showed excellent predictions for the soluble solids content of beet slices with R_P of 0.89–0.95 (Pan, Lu, et al., 2015). Based on VIS/NIR hyperspectral technique, it was demonstrated that PLSR was also the most popular approach for quantifying and visualizing the spatial distribution of SSC in sugar beet (Pan et al., 2016).

4.2.5 Moisture Content

As a very basic and significant quality indicator, moisture content (MC) has great impacts on the overall quality of root and tuber products. The VIS-NIR spectroscopy in the spectral region of 400–1600 nm is potentially useful for rapid determination of MC in sugar beet during postharvest handling and processing. The PLSR models presented the R_P of 0.950 for MC prediction in beet slices (Pan, Lu, et al., 2015; Pan et al., 2016). The MC parameters in taro and sweet potato products were also successfully measured by NIS/NIR calibration models using MLR or PLSR ($R^2 = 0.820$–0.970) (Areekij et al., 2017; Diaz et al., 2014; Sun et al., 2017). Furthermore, hyperspectral imaging systems (500–1700 nm) and PLSR were also evaluated for monitoring of MC of potato slices (Amjad et al., 2018; Moscetti et al., 2017; Su & Sun, 2016e). Feature selection strategies such as CARS, PCA, and iPLSR were used to select the important variables, and the MC was accurately determined with R_P^2 as high as 0.990. Based on the LS-SVM model, similar prediction accuracy ($R_P^2 = 0.991$) of MC in potato crisp was obtained by the study of Ni et al. (2011).

4.2.6 Impurities

Impurities are substances which are either naturally occurring or added during processing so that differ from the chemical composition of the product (Eiceman & Carpen, 1982; Liu, Cheng, & Hong, 2016; Protte, Weiss, &

Hinrichs, 2018). Determinations of impurities (such as *trans* fatty acids and acrylamide contents) in French fries or potato chips are currently necessary due to its potentially toxic attributes. The contents of *trans* fatty acids in potato French fries could be successfully determined using Fourier transform-infrared (FT-IR) spectroscopic technique with no need for the fatty acid extraction pre-treatment. In the study of Cho et al. (2011), the spectrum at 966 cm^{-1} was found to be linearly correlated with the *trans* fatty acids ($R_P^2 = 0.984$). Based on PLSR analysis, different concentrations of conjugated linoleic acid and *trans* conjugated linoleic acid in unknown potato chip samples were determined with R_P^2 as high as 0.960 (Kadamne et al., 2011). In a later study, FT-IR calibration model was established using PLSR to predict the presence of lard adulterated with palm oil in French fries, and yielded R_{CV}^2 of 0.979 with 0.5% (w/w) of detection limit (Che Man et al., 2014). NIR and MIR spectrometers can also be used as rapid alternatives for screening acrylamide in potato chips (Ayvaz & Rodriguez-Saona, 2015). For prediction of acrylamide levels from 169 to 2453 μg/kg, PLSR model was survived with very good performance ($R_P^2 > 0.90$). Based on NIR (400–2500 nm) spectroscopy, higher prediction accuracy ($R^2 = 0.98$) of acrylamide content (135 μg/kg) in French-fried potatoes was achieved by the PLSR model (Adedipe et al., 2016).

4.2.7 Other Chemical Parameters

The carotenoid, anthocyanin, hydrolytic activities, dioscin, minerals, ethanol yield, and chlorophyll in roots and tubers were successfully determined by hyperspectral techniques. Specifically, carotenoids are known for their roles in promoting the immune system and reducing the risk of degenerative diseases (Fraser & Bramley, 2004). Because of these advantages, both consumer and food industry are concerned for food products with high carotenoids (Stahl & Sies, 2005). VIS and NIR spectroscopy presented great potential for rapid quantification of carotenoid content in cassava roots ($R_P^2 > 0.88$) (Ikeogu et al., 2017; Sánchez et al., 2014). Based on NIR and MIR hyperspectral techniques, quantitative analyses of anthocyanin in potato and sweet potato samples were conducted using chemometrics such as MLR and PLSR ($R_P > 0.866$) (Ayvaz et al., 2016; Ding et al., 2015; Liu, Sun, et al., 2017). The hydrolytic activities of different pectinases on sugar beet substrates were evaluated using FT-MIR spectroscopy (Adina, Fetea, Matei, & Socaciu, 2011). In their study, the specific peak areas for carbohydrates (650–4000 cm^{-1}) and galacturonic acid (1500–700 cm^{-1}) were calculated to indicate the concentrations of these compounds in the sugar beet. After their release by enzymatic treatment, an optimized protocol of action for these two enzymes could be established, and the pectinase activities including optimum concentration and hydrolysis timing were effectively determined. Compared with chromatographic methods, FT–MIR spectroscopy showed advantages to evaluate the enzymatic activity of the different enzyme types. Sugar beet is a raw material for the production of sugar and ethanol. The decision on which end product to pursue can be facilitated based on fast and reliable means of predicting the potential ethanol yield from the beets. In the study of Magaña et al. (2011), NIR spectroscopy and modified partial least squares regression (MPLSR) model were tested for the direct prediction of the potential ethanol yield from sugar beets, yielding high R^2_{CV} of 0.91 and low RMSECV of 0.51. FT–IR combined with multivariate analysis could also be used for taxonomic and metabolic discrimination of African yam lines. Based on the PLSR, the content of dioscin in yam tuber was determined with R^2_P of 0.721 (Kwon et al., 2015). NIR spectroscopy to predict minerals in taro and sweet potato showed higher explained variances in R^2_{CV} (0.90) for taro minerals and R^2_{CV} (0.74) for sweet potato minerals (Lebot, Malapa, et al., 2011; Lebot, Ndiaye, et al., 2011). Recently, hyperspectral imaging was explored to measure the concentrations of chlorophyll and glycoalkaloid in potatoes, and obtained a relatively high R^2_P of 0.920 for chlorophyll (Kjær et al., 2017).

PLSR models have shown great potential in evaluation of other parameters such as total antioxidant activity, asparagine, phenolics, flavonoids, acid, and peroxide values. Specifically, Ding et al. (2015) demonstrated that PLSR was the optimal model in the prediction of total antioxidant activity parameters in sweet potato ($R_P = 0.982$). However, the PLSR model only showed medium potential ($R^2_P = 0.703$) for prediction of asparagine by using VIS–NIR hyperspectral imaging (Kjær et al., 2016). The applicability of portable MIR spectral instruments for the rapid detection of phenolics and asparagine levels in raw potato tubers has also been assessed (Ayvaz et al., 2016, 2015). Based on the developed PLSR calibration models, excellent linear correlations ($R_P > 0.91$) between MIR predicted and reference values were realized. For the analysis of flavonoids in Chinese yams, the LS–SVM model with NIR spectra produced better prediction outcomes ($R_P = 0.986$) than the PLSR model using MIR spectra (Zhuang et al., 2015). PLSR calibration models were also qualified for fat measurement in potato chips using MIR and NIR spectra, yielding RMSEP in the range of 1.0–1.9% for both calibration and validation data sets (Mazurek et al., 2016). Acid and peroxide values of the extracted oil from potato chips were predicted using NIR spectroscopy. Compared with PLSR and kernel partial least squares regression (KPLSR), LS-SVM method produced better data linearity and non-linearity, with R_P^2 of 0.990 and 0.949 for acid and peroxide values (Ni et al., 2011). Based on one-class partial least squares regression (OCPLSR) and LS-SVM model developed with NIR spectra, the maleic acid in cassava starch was quantitatively analyzed with very high accuracy (sensitivity = 0.954, specificity = 0.956, RMSECP = 0.192) (Fu et al., 2017). Overall, hyperspectral technique is a prospective approach for real-time quality determinations of root and tuber products.

4.3 Varietal Authentication

Variety identification plays a key role in breeding and production of roots and tubers. Discrimination of the origin of different yams from four geographical regions was conducted by NIR and MIR spectroscopic methods. The NIR data performed better identification ability than the MIR based on PCA (Zhuang et al., 2015). Hyperspectral imaging in combination with the PLSDA has been successfully used to classify organic potatoes from non-organic tubers and obtained an accuracy of over 91.6% for grading tuber moisture levels (Su & Sun, 2016d). In another study, the PLSDA model together with variable selection method of PCA achieved 100% classification accuracy for the rapid identification of sweet potato cultivars (Su & Sun, 2017c). The potential of hyperspectral imaging to identify powdered root and tuber contaminants has also been investigated. There are various approaches of adulterations and new adulterants keep appearing, thus the usefulness of the traditional analysis methods is limited. Xu et al. (2015) explored the usability of NIR spectroscopy to estimate root and tuber starch adulterants such as sweet potato, potato, and cassava starches in kudzu starch. Partial least squares class model (PLSCM) showed high accuracy and could effectively detect 4% or more of the root and tuber starches. To differentiate powdered, pure, and adulterated sweet potato, Ding et al. (2015) utilized chemometrics such as PCA and radial basis function-partial least square regression (RBF-PLSR) to group the NIR reflectance spectra from many different types of tuber sample. Their results showed that the purple and white sweet potato varieties could be accurately distinguished from each other as well as from the different adulterated purple sweet potato samples. In addition, the NIR hyperspectral imaging showed great potential to quantitatively detect cassava flour adulterated with specific wheat flour in the range of 3–75% (w/w). The optimal detection result was achieved by the

PLSR model, with R_P^2 of 0.986 and RMSEP of 0.026 (Su & Sun, 2017b). Based on the methods used, it should be possible to identify the presence of other exotic flours.

4.4 Gradation Aspects

Hyperspectral techniques were exploited for determining gradation parameters such as component levels, cooking time, and cooking degree of root and tuber products. Both VIS/NIR (446–1125 nm) and NIR (900–1685 nm) spectroscopic systems have been successfully applied on classification of potatoes based on sugar levels that are crucial for the frying industry. Linear discriminant analysis such as KNN, PLSDA, and classifier fusion showed that classification errors of glucose (18%) were lower than that sucrose (26%) (Rady & Guyer, 2015a, 2015b). Hyperspectral imaging was exploited for determination of potato cooking properties. When the potatoes heat up from the exterior to the interior during the cooking process, the cooking front will move from the outside to the inside and finally arrive at the center of the potato, which indicates that this cooking front is the interface of uncooked and fully cooked parts. Based on the inspection of the potato cooking front over time, hyperspectral imaging (400–1000 nm) together with PLSDA and image processing techniques allowed to monitor the optimal cooking time of potatoes with <10% relative error (Do Trong et al., 2011). Su and Sun (2016c) reported that hyperspectral imaging (900–1700 nm) in combination with the three-layer back propagation artificial neural network (TBPANN) model could yield higher overall accuracy than that of the PLSR model for detection of the volatility of tuber compositions (VTC) and prediction of the tuber cooking degree (TCD). Based on eight feature variables selected by FMCIA, TBPANN model achieved higher R_P^2 of 0.969 for VTC and 0.983 for TCD.

4.5 Defect Aspects

4.5.1 External Defects

The presence of surface defects is another important factor affecting tuber price and consumer purchasing behavior. Before packaging, the removal of the defective tuber is beneficial to ensure that only good quality products can reach the market. The potato scab is a common skin disease that decreases the quality of the product. Based on hyperspectral system, the common scab on potatoes could be identified with a 97.1% of accuracy using the SVM classifier (Dacal-Nieto et al., 2011a). Then, six potato defects including mechanical damage, hole, scab, surface bruise, sprout, and green skin were detected using hyperspectral imaging in the VIS/NIR region (400–1000 nm). The overall rate for correct recognition of potato defects reached 82.50% by combining PCA and image processing methods such as threshold segmentation, corrosion, expansion, and connectivity analysis (Su et al., 2015). To further analyze potato bruise, the histogram equalization and gray level co-occurrence matrix were used to enhance the features of bruised region in principal component images generated by PCA, and the recognition rate reached 93.75% (Ye et al., 2017). Based on diffuse reflectance in the VIS–NIR region, the subspace discriminant (SD) classification model was applied to characterize the optical properties of various surface defects (such as mechanical damage, greening, common scab, and soil deposits), yielding the highest classification accuracy of 97.6% (Al Riza et al., 2017). Results of above studies demonstrated that such a sorting system could be effectively used to identify tubers with external defects.

4.5.2 Internal Disorders

As a physiological disorder occurred inside the tuber, potato hollow heart is not easy to be detected by traditional visual techniques. During the transport or storage period, another internal defect that potato is very easy to suffer is the blackheart which is associated with high carbon dioxide conditions or rapid change of environment temperature. These disorders can cause considerable economic losses since the symptoms are internal and cannot be distinguished visually without cutting them into halves.

Based on hyperspectral imaging in the reflection mode, the presence of the hollow heart in potato tubers was non-destructively detected using the SVM model, with an accuracy of 89.1% of correct classification (Dacal-Nieto et al., 2011b). Then, a semi-transmission hyperspectral image acquisition system (390–1040 nm) which can present clearer image of the internal quality information of agricultural products was investigated to inspect potato hollow heart instead of using the reflection hyperspectral image system. The accurate recognition rate of hollow heart was improved to 100% based on parameter optimization algorithms such as artificial fish swarm algorithm (AFSA) to optimize the SVM model (Huang et al., 2015). After, Zhou et al. (2015) investigated the possibility of VIS/NIR (513–850 nm) transmission spectroscopic technique coupled with chemometric methods to classify blackheart potato tubers. It was concluded in their study that the height corrected transmittance showed the best prediction performance (96.53%) in PCA-PLSDA model using six wavelengths (711, 817, 741, 839,

678, and 698 nm). Then, two hyperspectral imaging systems in the regions of VIS–NIR (400–1000 nm) and NIR (1000–2500 nm) were, respectively, used to inspect blackspot in potatoes. More accurate identification rate (98.56) of bruised tubers was achieved by the PLSDA model using the latter system (López-Maestresalas et al., 2016). Overall, hyperspectral technique achieved very high performance for internal disorder detection.

5. SUMMARY

In general, the feasibility of hyperspectral techniques for the quality detection of various roots and tubers has been illustrated by empirical studies. However, the frequency of investigated quality parameters in root and tuber products differs greatly. Fig. 6 shows the number of published studies on hyperspectral techniques for evaluations of different tuber quality parameters during the period of 2011–2018. It can be seen that chemical constituents have been the most frequently studied parameters, followed by physical properties and defect aspects which are two almost equally important research hotspots on roots and tubers. By contrast, gradation aspects have been studied less frequently. Additionally, it can be seen that the total number of studies over the past 4 years has been doubled compared with those from 2011 and 2014, which reveals that there is an upward trend in this field of research. The number of published studies for determinations of various chemical parameters is shown in Fig. 7. It was noticed that sugar content has been the most frequently studied single chemical component, followed by

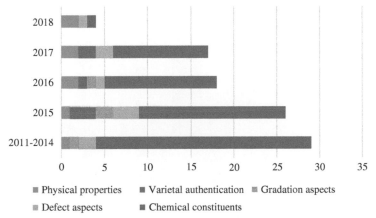

Fig. 6 Frequency of using hyperspectral techniques for evaluations of different quality parameters in roots and tubers from 2011 to 2018.

MC. Polysaccharide, dry matter, and protein are other three equally studied hotspots. For the spectral ranges used, there have been considerable differences in their frequency from 2011 to 2018. As shown in Fig. 8, NIR is the most remarkable and frequently used mode, which is followed in the distance by VIS and MIR modes. In other words, the NIR mode has attracted more attention for the quality evaluation of root and tuber products than others from 2011 to 2018. Furthermore, as reported in Fig. 9, PLSR is always the most popular modeling approach. The frequency of this linear PLSR method accounts for about 60% of all the optimal models used in

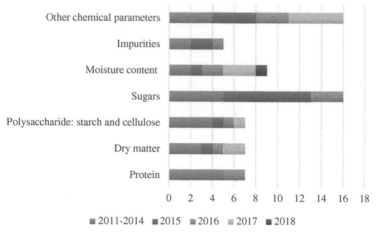

Fig. 7 Frequency of using hyperspectral techniques for evaluations of chemical constituents in roots and tubers from 2011 to 2018.

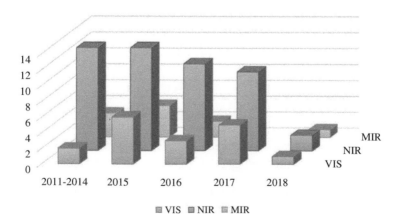

Fig. 8 Frequency of using different spectral ranges for quality evaluations of roots and tubers from 2011 to 2018.

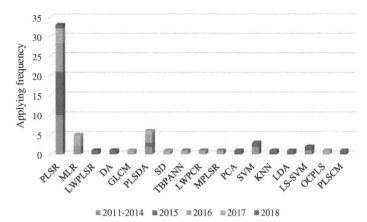

Fig. 9 Frequency of using different modeling methods for quality determinations of roots and tubers from 2011 to 2018.

studies between the years 2011 and 2018. The PLSDA is the second most frequently used model, followed by MLR and SVM models. Overall, the applicability of hyperspectral techniques for tuber quality detection has been demonstrated in numerous studies, and such techniques provide more comprehensive assessments for the future, together with different multivariate analysis methods.

6. CHALLENGES AND FUTURE TRENDS OF HYPERSPECTRAL TECHNIQUE IN ROOT AND TUBER APPLICATIONS

The main advantages of hyperspectral technique over reference methods are its precision, rapidness, ease-of-use, no sample preparation, no destructiveness, and no contaminant nature. Compared with conventional chemical analysis methods, hyperspectral technique is more economical as it reduces additional costs for reagents and labor. Due to its robust components and simplified mechanics, hyperspectral sensors are more suitable for on-line and in-line analysis. However, there are some problems common to both spectroscopy and hyperspectral imaging. It is difficult to collect the deep-seated spectral data of tuber samples as the light penetration depth of this spectrograph is limited to different varieties of samples and spectral ranges. Other difficulties are derived from the presentation of samples with high-water content when working with fresh products as there are strong absorption bands in certain spectral regions. Moreover, hyperspectral data contain hundreds of spectral variables. Such huge size of data set requires significant amount of storage space and increases the calculation burden as well as the costs of instruments, which prevents this technique for real-time application. Based on chemometric methods to extract spectral information, robust models can be developed after using enough number of samples with large variations, which means hyperspectral techniques require prior knowledge of the specific parameters that have been previously determined by a reference method. In addition, modeling and data processing procedures are time consuming since hyperspectral systems are easily influenced by various external factors such as noises in instrumentation.

In the future, more research on hyperspectral techniques should be conducted to achieve the high quality of roots and tubers in a more efficient and rational way, and to ensure the transformation of technological achievements from the laboratory setting to online detection systems in the industry. The suggestions for the future are listed below:

1. It is of importance to obtain precise reference values of target quality attributes from a larger sample set to improve the correction and the adaptation of developed models.
2. The optimal models with high accuracy should be re-optimized to guarantee its effectiveness as much as possible.
3. The fingerprint data in the MIR region with very complex information need to be further investigated to develop effective analysis systems.
4. More research should be carried out for reduction of data processing time and optimization of spectral resolution in real-time computing environments.
5. High-speed hardware platforms need to be assembled to generate high-quality spectral data. Based on such platforms, feature wavelengths associated with food quality parameters should be automatically qualified and rapidly applied for online monitoring systems.
6. The final well-developed system should be quick response, high efficiency, and easy operation, which is powerful enough to complete a given work.

7. CONCLUSIONS

This review article covers the current research status and prospects of hyperspectral techniques for the rapid and non-invasive quality assessment of roots and tubers. The review, for the first time, manages to summarize the research progress of hyperspectral techniques that so far has been applied on various root and tuber foods including potato, sweet potato, cassava, yam, taro, and sugar beet. Moreover, this review emphasizes how recent developments in hyperspectral techniques with chemometric methods have enhanced overall capabilities to evaluate physical, chemical, varietal, grading, and defect parameters of root and tuber products. The different spectral ranges (such as VIS, NIR, and MIR) and modeling methods (such as PLSR, SVM, and MLR) can be a basis for the further development of effective spectral sensors.

ACKNOWLEDGMENTS

The authors acknowledge the UCD-CSC Scholarship Scheme supported by the University College Dublin, Ireland, (UCD) and the China Scholarship Council (CSC).

REFERENCES

Abdel-Nour, N., Ngadi, M., Prasher, S., & Karimi, Y. (2011). Prediction of egg freshness and albumen quality using visible/near infrared spectroscopy. *Food and Bioprocess Technology*, 4(5), 731–736.

Adedipe, O. E., Johanningsmeier, S. D., Truong, V.-D., & Yencho, G. C. (2016). Development and validation of a near-infrared spectroscopy method for the prediction of acrylamide content in french-fried potato. *Journal of Agricultural and Food Chemistry*, 64(8), 1850–1860.

Adina, C., Fetea, F., Matei, H., & Socaciu, C. (2011). Evaluation of hydrolytic activity of different pectinases on sugar beet (Beta vulgaris) substrate using FT-MIR spectroscopy. *Notulae Botanicae Horti Agrobotanici Cluj-Napoca*, 39(2), 99.

Aina, A. J., Falade, K. O., Akingbala, J. O., & Titus, P. (2012). Physicochemical properties of Caribbean sweet potato (Ipomoea batatas (L) Lam) starches. *Food and Bioprocess Technology*, 5(2), 576–583.

Alexandrakis, D., Downey, G., & Scannell, A. G. (2012). Rapid non-destructive detection of spoilage of intact chicken breast muscle using near-infrared and Fourier transform mid-infrared spectroscopy and multivariate statistics. *Food and Bioprocess Technology*, 5(1), 338–347.

Al Riza, D. F., Suzuki, T., Ogawa, Y., & Kondo, N. (2017). Diffuse reflectance characteristic of potato surface for external defects discrimination. *Postharvest Biology and Technology*, 133, 12–19.

Allemann, J., Laurie, S., Thiart, S., Vorster, H., & Bornman, C. (2004). Sustainable production of root and tuber crops (potato, sweet potato, indigenous potato, cassava) in southern Africa. *South African Journal of Botany*, 70(1), 60–66.

Allmann, M., Candrian, U., Höfelein, C., & Lüthy, J. (1993). Polymerase chain reaction (PCR): A possible alternative to immunochemical methods assuring safety and quality of food detection of wheat contamination in non-wheat food products. *Zeitschrift für Lebensmitteluntersuchung und-forschung A*, 196(3), 248–251.

Amjad, W., Crichton, S. O., Munir, A., Hensel, O., & Sturm, B. (2018). Hyperspectral imaging for the determination of potato slice moisture content and chromaticity during the convective hot air drying process. *Biosystems Engineering*, 166, 170–183.

Antonucci, F., Pallottino, F., Paglia, G., Palma, A., D'Aquino, S., & Menesatti, P. (2011). Non-destructive estimation of mandarin maturity status through portable VIS-NIR spectrophotometer. *Food and Bioprocess Technology*, 4(5), 809–813.

Araújo, M. C. U., Saldanha, T. C. B., Galvao, R. K. H., Yoneyama, T., Chame, H. C., & Visani, V. (2001). The successive projections algorithm for variable selection in spectroscopic multicomponent analysis. *Chemometrics and Intelligent Laboratory Systems*, *57*(2), 65–73.

Areekij, S., Ritthiruangdej, P., Kasemsumran, S., Therdthai, N., Haruthaithanasan, V., & Ozaki, Y. (2017). Rapid and nondestructive analysis of deep-fried taro chip qualities using near infrared spectroscopy. *Journal of Near Infrared Spectroscopy*, *25*(2), 127–137.

Arkhypova, V., Dzyadevych, S., Jaffrezic-Renault, N., Martelet, C., & Soldatkin, A. (2008). Biosensors for assay of glycoalkaloids in potato tubers. *Applied Biochemistry and Microbiology*, *44*(3), 314–318.

Asensio, L., González, I., García, T., & Martín, R. (2008). Determination of food authenticity by enzyme-linked immunosorbent assay (ELISA). *Food Control*, *19*(1), 1–8.

Ayvaz, H., Bozdogan, A., Giusti, M. M., Mortas, M., Gomez, R., & Rodriguez-Saona, L. E. (2016). Improving the screening of potato breeding lines for specific nutritional traits using portable mid-infrared spectroscopy and multivariate analysis. *Food Chemistry*, *211*, 374–382.

Ayvaz, H., & Rodriguez-Saona, L. E. (2015). Application of handheld and portable spectrometers for screening acrylamide content in commercial potato chips. *Food Chemistry*, *174*, 154–162.

Ayvaz, H., Santos, A. M., Moyseenko, J., Kleinhenz, M., & Rodriguez-Saona, L. E. (2015). Application of a portable infrared instrument for simultaneous analysis of sugars, asparagine and glutamine levels in raw potato tubers. *Plant Foods for Human Nutrition*, *70*(2), 215–220.

Bahadır, E. B., & Sezgintürk, M. K. (2015). Applications of commercial biosensors in clinical, food, environmental, and biothreat/biowarfare analyses. *Analytical Biochemistry*, *478*, 107–120.

Balcaen, L., Bolea-Fernandez, E., Resano, M., & Vanhaecke, F. (2015). Inductively coupled plasma–tandem mass spectrometry (ICP-MS/MS): A powerful and universal tool for the interference-free determination of (ultra) trace elements—A tutorial review. *Analytica Chimica Acta*, *894*, 7–19.

Bally, I. S., Johnson, P., & Kulkarni, V. (1999). *Mango production in Australia VI international symposium on Mango Vol. 509*, (pp. 59–68).

Bao, Y., Liu, F., Kong, W., Sun, D.-W., He, Y., & Qiu, Z. (2014). Measurement of soluble solid contents and pH of white vinegars using VIS/NIR spectroscopy and least squares support vector machine. *Food and Bioprocess Technology*, *7*(1), 54–61.

Baritelle, A., & Hyde, G. (2003). Specific gravity and cultivar effects on potato tuber impact sensitivity. *Postharvest Biology and Technology*, *29*(3), 279–286.

Beninca, C., Colman, T. A. D., Lacerda, L. G., da Silva Carvalho Filho, M. A., Demiate, I. M., Bannach, G., et al. (2013). Thermal, rheological, and structural behaviors of natural and modified cassava starch granules, with sodium hypochlorite solutions. *Journal of Thermal Analysis and Calorimetry*, *111*(3), 2217–2222.

Bernhard, T., Truberg, B., Friedt, W., Snowdon, R., & Wittkop, B. (2016). Development of near-infrared reflection spectroscopy calibrations for crude protein and dry matter content in fresh and dried potato tuber samples. *Potato Research*, *59*(2), 149–165.

Biondi, E., Blasioli, S., Galeone, A., Spinelli, F., Cellini, A., Lucchese, C., et al. (2014). Detection of potato brown rot and ring rot by electronic nose: From laboratory to real scale. *Talanta*, *129*, 422–430.

Blanco, C. A., De la Fuente, R., Caballero, I., & Rodríguez-Méndez, M. L. (2015). Beer discrimination using a portable electronic tongue based on screen-printed electrodes. *Journal of Food Engineering*, *157*, 57–62.

Bouwkamp, J. C. (2018). *Sweet potato products: A natural resource for the tropics*. CRC Press.

Bovell-Benjamin, A. C. (2007). Sweet potato: A review of its past, present, and future role in human nutrition. *Advances in Food and Nutrition Research, 52,* 1–59.

Branch, S., Ebdon, L., & O'Neill, P. (1994). Determination of arsenic species in fish by directly coupled high-performance liquid chromatography–inductively coupled plasma mass spectrometry. *Journal of Analytical Atomic Spectrometry, 9*(1), 33–37.

Brondi, A. M., Terra, L. A., Sabin, G. P., Garcia, J. S., Poppi, R. J., & Trevisan, M. G. (2014). Mapping the polymorphic forms of fexofenadine in pharmaceutical tablets using near infrared chemical imaging. *Journal of Near Infrared Spectroscopy, 22*(3), 211–220.

Burdon, J., McLeod, D., Lallu, N., Gamble, J., Petley, M., & Gunson, A. (2004). Consumer evaluation of "Hayward" kiwifruit of different at-harvest dry matter contents. *Postharvest Biology and Technology, 34*(3), 245–255.

Cen, H., Lu, R., Ariana, D. P., & Mendoza, F. (2014). Hyperspectral imaging-based classification and wavebands selection for internal defect detection of pickling cucumbers. *Food and Bioprocess Technology, 7*(6), 1689–1700.

Cevoli, C., Gori, A., Nocetti, M., Cuibus, L., Caboni, M. F., & Fabbri, A. (2013). FT-NIR and FT-MIR spectroscopy to discriminate competitors, non compliance and compliance grated Parmigiano Reggiano cheese. *Food Research International, 52*(1), 214–220.

Chambre, P. A., & Schaaf, S. A. (2017). *Flow of rarefied gases.* Princeton University Press.

Chandrasekara, A. (2017). Roots and tubers as functional foods. In *Bioactive molecules in food* (pp. 1–29). Cham: Springer.

Chandrasekaran, M. (2012). *Valorization of food processing by-products.* CRC Press.

Che Man, Y., Marina, A., Rohman, A., Al-Kahtani, H., & Norazura, O. (2014). A Fourier transform infrared spectroscopy method for analysis of palm oil adulterated with lard in pre-fried French fries. *International Journal of Food Properties, 17*(2), 354–362.

Chen, J., Piva, M., & Labuza, T. (1984). Evaluation of water binding capacity (WBC) of food fiber sources. *Journal of Food Science, 49*(1), 59–63.

Cheng, J.-H., Nicolai, B., & Sun, D.-W. (2017). Hyperspectral imaging with multivariate analysis for technological parameters prediction and classification of muscle foods: A review. *Meat Science, 123,* 182–191.

Cheng, J.-H., & Sun, D.-W. (2015). Rapid quantification analysis and visualization of Escherichia coli loads in grass carp fish flesh by hyperspectral imaging method. *Food and Bioprocess Technology, 8*(5), 951–959.

Cheng, J.-H., Sun, D.-W., Pu, H., & Zeng, X.-A. (2014). Comparison of visible and long-wave near-infrared hyperspectral imaging for colour measurement of grass carp (Ctenopharyngodon idella). *Food and Bioprocess Technology, 7*(11), 3109–3120.

Chiu, H. W., Peng, J. C., Tsai, S. J., & Lui, W. B. (2012). Effect of extrusion processing on antioxidant activities of corn extrudates fortified with various Chinese yams (Dioscorea sp.). *Food and Bioprocess Technology, 5*(6), 2462–2473.

Cho, I. K., Kim, S., Khurana, H. K., Li, Q. X., & Jun, S. (2011). Quantification of trans fatty acid content in French fries of local food service retailers using attenuated total reflection–Fourier transform infrared spectroscopy. *Food Chemistry, 125*(3), 1121–1125.

Chung, C., & McClements, D. (2015). Structure and texture development of food-emulsion products. In *Modifying food texture* (pp. 133–155). Elsevier.

Clegg, S. M., Sklute, E., Dyar, M. D., Barefield, J. E., & Wiens, R. C. (2009). Multivariate analysis of remote laser-induced breakdown spectroscopy spectra using partial least squares, principal component analysis, and related techniques. *Spectrochimica Acta Part B: Atomic Spectroscopy, 64*(1), 79–88.

Coppin, A., Varre, J.-S., Lienard, L., Dauvillee, D., Guerardel, Y., Soyer-Gobillard, M.-O., et al. (2005). Evolution of plant-like crystalline storage polysaccharide in the protozoan parasite Toxoplasma gondii argues for a red alga ancestry. *Journal of Molecular Evolution, 60*(2), 257–267.

Cubero, S., Aleixos, N., Moltó, E., Gómez-Sanchis, J., & Blasco, J. (2011). Advances in machine vision applications for automatic inspection and quality evaluation of fruits and vegetables. *Food and Bioprocess Technology, 4*(4), 487–504.

Dacal-Nieto, A., Formella, A., Carrión, P., Vazquez-Fernandez, E., & Fernández-Delgado, M. (2011a). In *Common scab detection on potatoes using an infrared hyperspectral imaging system. International conference on image analysis and processing* (pp. 303–312): Springer.

Dacal-Nieto, A., Formella, A., Carrión, P., Vazquez-Fernandez, E., & Fernández-Delgado, M. (2011b). Non-destructive detection of hollow heart in potatoes using hyperspectral imaging. In *Computer analysis of images and patterns* (pp. 180–187). Springer.

Dasgupta, R. (2013). Tuber crop growth and pareto model. In *Advances in growth curve models* (pp. 185–197): Springer.

Davies, H., & Dixon, N. (1976). Evaluation of potato texture by taste and by appearance. *American Journal of Potato Research, 53*(6), 205–210.

Debela, A. M., Ortiz, M., Beni, V., Thorimbert, S., Lesage, D., Cole, R. B., et al. (2015). Biofunctionalization of polyoxometalates with DNA primers, their use in the polymerase chain reaction (PCR) and electrochemical detection of PCR products. *Chemistry—A European Journal, 21*(49), 17721–17727.

Di Rosa, A. R., Leone, F., Cheli, F., & Chiofalo, V. (2017). Fusion of electronic nose, electronic tongue and computer vision for animal source food authentication and quality assessment—A review. *Journal of Food Engineering, 210*, 62–75.

Diaz, J. T., Veal, M. W., & Chinn, M. S. (2014). Development of NIRS models to predict composition of enzymatically processed sweetpotato. *Industrial Crops and Products, 59*, 119–124.

Ding, X., Ni, Y., & Kokot, S. (2015). NIR spectroscopy and chemometrics for the discrimination of pure, powdered, purple sweet potatoes and their samples adulterated with the white sweet potato flour. *Chemometrics and Intelligent Laboratory Systems, 144*, 17–23.

Do Trong, N. N., Tsuta, M., Nicolaï, B., De Baerdemaeker, J., & Saeys, W. (2011). Prediction of optimal cooking time for boiled potatoes by hyperspectral imaging. *Journal of Food Engineering, 105*(4), 617–624.

Draycott, A. P. (2008). *Sugar beet.* John Wiley & Sons.

Eiceman, G., & Carpen, M. (1982). Determination of volatile organic compounds as impurities in polystyrene food containers and polystyrene cups. *Analytical Letters, 15*(14), 1169–1177.

El Sheikha, A. F., & Ray, R. C. (2017). Potential impacts of bioprocessing of sweet potato. *Critical Reviews in Food Science and Nutrition, 57*(3), 455–471.

ElMasry, G. M., & Nakauchi, S. (2016). Image analysis operations applied to hyperspectral images for non-invasive sensing of food quality—A comprehensive review. *Biosystems Engineering, 142*, 53–82.

ElMasry, G., Sun, D.-W., & Allen, P. (2012). Near-infrared hyperspectral imaging for predicting colour, pH and tenderness of fresh beef. *Journal of Food Engineering, 110*(1), 127–140.

El-Sharkawy, M. A. (1993). Drought-tolerant cassava for Africa, Asia, and Latin America. *Bioscience, 43*(7), 441–451.

Fagan, C. C., Everard, C., O'Donnell, C., Downey, G., Sheehan, E., Delahunty, C., et al. (2007). Evaluating mid-infrared spectroscopy as a new technique for predicting sensory texture attributes of processed cheese. *Journal of Dairy Science, 90*(3), 1122–1132.

Flores, H. E., Walker, T. S., Guimarães, R. L., Bais, H. P., & Vivanco, J. M. (2003). Andean root and tuber crops: Underground rainbows. *Hortscience, 38*(2), 161–167.

Fraser, P. D., & Bramley, P. M. (2004). The biosynthesis and nutritional uses of carotenoids. *Progress in Lipid Research*, *43*(3), 228–265.

Fu, H.-Y., Li, H.-D., Xu, L., Yin, Q.-B., Yang, T.-M., Ni, C., et al. (2017). Detection of unexpected frauds: Screening and quantification of maleic acid in cassava starch by Fourier transform near-infrared spectroscopy. *Food Chemistry*, *227*, 322–328.

Fugate, K. K., Ribeiro, W. S., Lulai, E. C., Deckard, E. L., & Finger, F. L. (2016). Cold temperature delays wound healing in postharvest sugarbeet roots. *Frontiers in Plant Science*, *7*, 499.

Gamble, J., Harker, F. R., Jaeger, S. R., White, A., Bava, C., Beresford, M., et al. (2010). The impact of dry matter, ripeness and internal defects on consumer perceptions of avocado quality and intentions to purchase. *Postharvest Biology and Technology*, *57*(1), 35–43.

Giaquinta, R. T. (1979). Sucrose translocation and storage in the sugar beet. *Plant Physiology*, *63*(5), 828–832.

Gibson, L. J. (2012). The hierarchical structure and mechanics of plant materials. *Journal of the Royal Society Interface*, *9*, 2749–2766. rsif20120341.

Gómez-Sanchis, J., Lorente, D., Soria-Olivas, E., Aleixos, N., Cubero, S., & Blasco, J. (2014). Development of a hyperspectral computer vision system based on two liquid crystal tuneable filters for fruit inspection. Application to detect citrus fruits decay. *Food and Bioprocess Technology*, *7*(4), 1047–1056.

Gordon, H. R., & Morel, A. Y. (2012). *Remote assessment of ocean color for interpretation of satellite visible imagery: A review*. Springer Science & Business Media.

Gowen, A. A., Feng, Y., Gaston, E., & Valdramidis, V. (2015). Recent applications of hyperspectral imaging in microbiology. *Talanta*, *137*, 43 54.

Griffiths, P. R., & De Haseth, J. A. (2007). *Fourier transform infrared spectrometry*. John Wiley & Sons.

Groinig, M., Burgstaller, M., & Pail, M. (2011). In *Industrial application of a new camera system based on hyperspectral imaging for inline quality control of potatoes*. ÖAGM/AAPR workshop.

Haase, N. U. (2011). Prediction of potato processing quality by near infrared reflectance spectroscopy of ground raw tubers. *Journal of Near Infrared Spectroscopy*, *19*(1), 37–45.

Hansen, C. L., Thybo, A. K., Bertram, H. C., Viereck, N., van den Berg, F., & Engelsen, S. B. (2010). Determination of dry matter content in potato tubers by low-field nuclear magnetic resonance (LF-NMR). *Journal of Agricultural and Food Chemistry*, *58*(19), 10300–10304.

Harker, F., Carr, B., Lenjo, M., MacRae, E., Wismer, W., Marsh, K., et al. (2009). Consumer liking for kiwifruit flavour: A meta-analysis of five studies on fruit quality. *Food Quality and Preference*, *20*(1), 30–41.

He, H.-J., Wu, D., & Sun, D.-W. (2014). Potential of hyperspectral imaging combined with chemometric analysis for assessing and visualising tenderness distribution in raw farmed salmon fillets. *Journal of Food Engineering*, *126*, 156–164.

Helgerud, T., Segtnan, V. H., Wold, J. P., Ballance, S., Knutsen, S. H., Rukke, E. O., et al. (2012). Near-infrared spectroscopy for rapid estimation of dry matter content in whole unpeeled potato tubers. *Journal of Food Research*, *1*(4), 55.

Helgerud, T., Wold, J. P., Pedersen, M. B., Liland, K. H., Ballance, S., Knutsen, S. H., et al. (2015). Towards on-line prediction of dry matter content in whole unpeeled potatoes using near-infrared spectroscopy. *Talanta*, *143*, 138–144.

Huang, T., Jin, R., & Ku, J. (2015). Non-destructive detection research for hollow heart of potato based on semi-transmission hyperspectral imaging and SVM. *Spectroscopy and Spectral Analysis*, *35*(1), 198–202.

Ikeogu, U. N., Davrieux, F., Dufour, D., Ceballos, H., Egesi, C. N., & Jannink, J.-L. (2017). Rapid analyses of dry matter content and carotenoids in fresh cassava roots using a portable visible and near infrared spectrometer (Vis/NIRS). *PLoS One*, *12*(12), e0188918.

Imaizumi, T., Muramatsu, Y., Orikasa, T., Hamanaka, D., Tanaka, F., Uchino, T., et al. (2014). Effect of soaking treatment in hot water on thermal and electrical properties of tubers and Japanese radish. *Nippon Shokuhin Kagaku Kogaku Kaishi, 61*(6), 244–250.

Jarvis, R. M., & Goodacre, R. (2004). Genetic algorithm optimization for pre-processing and variable selection of spectroscopic data. *Bioinformatics, 21*(7), 860–868.

Kačuráková, M., & Wilson, R. (2001). Developments in mid-infrared FT-IR spectroscopy of selected carbohydrates. *Carbohydrate Polymers, 44*(4), 291–303.

Kadamne, J. V., Castrodale, C. L., & Proctor, A. (2011). Measurement of conjugated linoleic acid (CLA) in CLA-rich potato chips by ATR-FTIR spectroscopy. *Journal of Agricultural and Food Chemistry, 59*(6), 2190–2196.

Kamruzzaman, M., Makino, Y., Oshita, S., & Liu, S. (2015). Assessment of visible near-infrared hyperspectral imaging as a tool for detection of horsemeat adulteration in minced beef. *Food and Bioprocess Technology, 8*(5), 1054–1062.

Kirtil, E., & Oztop, M. H. (2016). [1]H nuclear magnetic resonance relaxometry and magnetic resonance imaging and applications in food science and processing. *Food Engineering Reviews, 8*(1), 1–22.

Kjær, A., Nielsen, G., Stærke, S., Clausen, M. R., Edelenbos, M., & Jørgensen, B. (2016). Prediction of starch, soluble sugars and amino acids in potatoes (Solanum tuberosum L.) using hyperspectral imaging, dielectric and LF-NMR methodologies. *Potato Research, 59*(4), 357–374.

Kjær, A., Nielsen, G., Stærke, S., Clausen, M. R., Edelenbos, M., & Jørgensen, B. (2017). Detection of glycoalkaloids and chlorophyll in potatoes (Solanum tuberosum L.) by hyperspectral imaging. *American Journal of Potato Research, 94*(6), 573–582.

Klaypradit, W., Kerdpiboon, S., & Singh, R. K. (2011). Application of artificial neural networks to predict the oxidation of menhaden fish oil obtained from Fourier transform infrared spectroscopy method. *Food and Bioprocess Technology, 4*(3), 475–480.

Kurz, T. H., Buckley, S. J., & Howell, J. A. (2013). Close-range hyperspectral imaging for geological field studies: Workflow and methods. *International Journal of Remote Sensing, 34*(5), 1798–1822.

Kwon, Y.-K., Jie, E. Y., Sartie, A., Kim, D. J., Liu, J. R., Min, B. W., et al. (2015). Rapid metabolic discrimination and prediction of dioscin content from African yam tubers using Fourier transform-infrared spectroscopy combined with multivariate analysis. *Food Chemistry, 166*, 389–396.

Lawther, J. M., Sun, R., & Banks, W. (1995). Extraction, fractionation, and characterization of structural polysaccharides from wheat straw. *Journal of Agricultural and Food Chemistry, 43*(3), 667–675.

Lebot, V. (2012). Near infrared spectroscopy for quality evaluation of root crops: Practical constraints, preliminary studies and future prospects. *Journal of Root Crops, 38*, 3–14.

Lebot, V., & Malapa, R. (2013). Application of near infrared reflectance spectroscopy for the evaluation of yam (Dioscorea alata) germplasm and breeding lines. *Journal of the Science of Food and Agriculture, 93*(7), 1788–1797.

Lebot, V., Malapa, R., & Bourrieau, M. (2011). Rapid estimation of taro (Colocasia esculenta) quality by near-infrared reflectance spectroscopy. *Journal of Agricultural and Food Chemistry, 59*(17), 9327–9334.

Lebot, V., Ndiaye, A., & Malapa, R. (2011). Phenotypic characterization of sweet potato [Ipomoea batatas (L.) Lam.] genotypes in relation to prediction of chemical quality constituents by NIRS equations. *Plant Breeding, 130*(4), 457–463.

Lee, H., Kim, M. S., Lim, H. S., Park, E., Lee, W. H., & Cho, B. K. (2016). Detection of cucumber green mottle mosaic virus-infected watermelon seeds using a near-infrared (NIR) hyperspectral imaging system: Application to seeds of the "Sambok Honey" cultivar. *Biosystems Engineering, 148*, 138–147.

Lee, J. W., Uchikata, T., Matsubara, A., Nakamura, T., Fukusaki, E., & Bamba, T. (2012). Application of supercritical fluid chromatography/mass spectrometry to lipid profiling of soybean. *Journal of Bioscience and Bioengineering, 113*(2), 262–268.

Leroux, D. F., Midahuen, R., Perrin, G., Pescatore, J., & Imbaud, P. (2015). In *Hyperspectral imaging applied to microbial categorization in an automated microbiology workflow European conference on biomedical optics* (p. 953726): Optical Society of America.

Li, H., Liang, Y., Xu, Q., & Cao, D. (2009). Key wavelengths screening using competitive adaptive reweighted sampling method for multivariate calibration. *Analytica Chimica Acta, 648*(1), 77–84.

Li, J. L., Sun, D.-W., & Cheng, J. H. (2016). Recent advances in nondestructive analytical techniques for determining the total soluble solids in fruits: A review. *Comprehensive Reviews in Food Science and Food Safety, 15*(5), 897–911.

Liang, P.-S., Haff, R. P., Hua, S.-S. T., Munyaneza, J. E., Mustafa, T., & Sarreal, S. B. L. (2018). Nondestructive detection of zebra chip disease in potatoes using near-infrared spectroscopy. *Biosystems Engineering, 166*, 161–169.

Liu, Z., Cheng, F., & Hong, H. (2016). Identification of impurities in fresh shrimp using improved majority scheme-based classifier. *Food Analytical Methods, 9*(11), 3133–3142.

Liu, G., He, J., Wang, S., Luo, Y., Wang, W., Wu, L., et al. (2016). Application of near-infrared hyperspectral imaging for detection of external insect infestations on jujube fruit. *International Journal of Food Properties, 19*(1), 41–52.

Liu, F., He, Y., Wang, L., & Sun, G. (2011). Detection of organic acids and pH of fruit vinegars using near-infrared spectroscopy and multivariate calibration. *Food and Bioprocess Technology, 4*(8), 1331–1340.

Liu, Y., Sun, Y., Xie, A., Yu, H., Yin, Y., Li, X., et al. (2017). Potential of hyperspectral imaging for rapid prediction of anthocyanin content of purple-fleshed sweet potato slices during drying process. *Food Analytical Methods, 10*(12), 3836–3846.

Liu, D., Sun, D.-W., & Zeng, X.-A. (2014). Recent advances in wavelength selection techniques for hyperspectral image processing in the food industry. *Food and Bioprocess Technology, 7*(2), 307–323.

Liu, S., Zainuddin, I. M., Vanderschuren, H., Doughty, J., & Beeching, J. R. (2017). RNAi inhibition of feruloyl CoA 6'-hydroxylase reduces scopoletin biosynthesis and post-harvest physiological deterioration in cassava (Manihot esculenta Crantz) storage roots. *Plant Molecular Biology, 94*(1–2), 185–195.

Liu, B., Zhou, P., Liu, X., Sun, X., Li, H., & Lin, M. (2013). Detection of pesticides in fruits by surface-enhanced Raman spectroscopy coupled with gold nanostructures. *Food and Bioprocess Technology, 6*(3), 710–718.

Lohumi, S., Lee, S., Lee, H., & Cho, B.-K. (2015). A review of vibrational spectroscopic techniques for the detection of food authenticity and adulteration. *Trends in Food Science & Technology, 46*(1), 85–98.

López, A., Arazuri, S., Jarén, C., Mangado, J., Arnal, P., de Galarreta, J. I. R., et al. (2013). Crude protein content determination of potatoes by NIRS technology. *Procedia Technology, 8*, 488–492.

López, A., Jarén, C., Arazuri, S., Mangado, J., Tierno, R., & Ruiz de Galarreta, J. (2014). In *Estimation of the total phenolic content in potatoes by NIRS. Proceedings of an international conference of agricultural engineering, Zurich, Switzerland.*

López-Maestresalas, A., Keresztes, J. C., Goodarzi, M., Arazuri, S., Jarén, C., & Saeys, W. (2016). Non-destructive detection of blackspot in potatoes by Vis-NIR and SWIR hyperspectral imaging. *Food Control, 70*, 229–241.

Lorente, D., Aleixos, N., Gómez-Sanchis, J., Cubero, S., & Blasco, J. (2013). Selection of optimal wavelength features for decay detection in citrus fruit using the ROC curve and neural networks. *Food and Bioprocess Technology, 6*(2), 530–541.

Lorente, D., Aleixos, N., Gómez-Sanchis, J., Cubero, S., García-Navarrete, O. L., & Blasco, J. (2012). Recent advances and applications of hyperspectral imaging for fruit and vegetable quality assessment. *Food and Bioprocess Technology, 5*(4), 1121–1142.

Lorente, D., Blasco, J., Serrano, A. J., Soria-Olivas, E., Aleixos, N., & Gómez-Sanchis, J. (2013). Comparison of ROC feature selection method for the detection of decay in citrus fruit using hyperspectral images. *Food and Bioprocess Technology, 6*(12), 3613–3619.

Loutfi, A., Coradeschi, S., Mani, G. K., Shankar, P., & Rayappan, J. B. B. (2015). Electronic noses for food quality: A review. *Journal of Food Engineering, 144*, 103–111.

Lu, R. (2004). Multispectral imaging for predicting firmness and soluble solids content of apple fruit. *Postharvest Biology and Technology, 31*(2), 147–157.

Lu, G., & Fei, B. (2014). Medical hyperspectral imaging: A review. *Journal of Biomedical Optics, 19*(1), 010901.

Magaña, C., Núñez-Sánchez, N., Fernández-Cabanás, V., García, P., Serrano, A., Pérez-Marín, D., et al. (2011). Direct prediction of bioethanol yield in sugar beet pulp using near infrared spectroscopy. *Bioresource Technology, 102*(20), 9542–9549.

Magwaza, L. S., Naidoo, S. I. M., Laurie, S. M., Laing, M. D., & Shimelis, H. (2016). Development of NIRS models for rapid quantification of protein content in sweetpotato [Ipomoea batatas (L.) LAM.]. *LWT-Food Science and Technology, 72*, 63–70.

Magwaza, L. S., Opara, U. L., Nieuwoudt, H., Cronje, P. J., Saeys, W., & Nicolaï, B. (2012). NIR spectroscopy applications for internal and external quality analysis of citrus fruit—A review. *Food and Bioprocess Technology, 5*(2), 425–444.

Mazurek, S., Szostak, R., & Kita, A. (2016). Application of infrared reflection and Raman spectroscopy for quantitative determination of fat in potato chips. *Journal of Molecular Structure, 1126*, 213–218.

McBratney, A. B., Santos, M. M., & Minasny, B. (2003). On digital soil mapping. *Geoderma, 117*(1–2), 3–52.

Menesatti, P., Zanella, A., D'Andrea, S., Costa, C., Paglia, G., & Pallottino, F. (2009). Supervised multivariate analysis of hyper-spectral NIR images to evaluate the starch index of apples. *Food and Bioprocess Technology, 2*(3), 308–314.

Mhemdi, H., Bals, O., Grimi, N., & Vorobiev, E. (2014). Alternative pressing/ultrafiltration process for sugar beet valorization: Impact of pulsed electric field and cossettes preheating on the qualitative characteristics of juices. *Food and Bioprocess Technology, 7*(3), 795–805.

Moscetti, R., Sturm, B., Crichton, S. O., Amjad, W., & Massantini, R. (2017). Postharvest monitoring of organic potato (cv. Anuschka) during hot-air drying using visible-NIR hyperspectral imaging. *Journal of the Science of Food and Agriculture, 98*(7), 2507–2517.

Motes, J., & Greig, J. (1970). Specific gravity, potato chip color and tuber mineral content as affected by soil moisture and harvest dates. *American Potato Journal, 47*(11), 413.

Naganathan, G. K., Grimes, L. M., Subbiah, J., Calkins, C. R., Samal, A., & Meyer, G. E. (2008). Visible/near-infrared hyperspectral imaging for beef tenderness prediction. *Computers and Electronics in Agriculture, 64*(2), 225–233.

Najamuddeen, G., Sayaya, M., Ala, A., & Garba, M. (2012). Economics of potatoes production in Katsina Metropolis, Katsina state, Nigeria. *Scientific Journal of Crop Production, 1*(3), 48–54.

Narsaiah, K., Jha, S. N., Bhardwaj, R., Sharma, R., & Kumar, R. (2012). Optical biosensors for food quality and safety assurance—A review. *Journal of Food Science and Technology, 49*(4), 383–406.

Nashat, S., Abdullah, A., Aramvith, S., & Abdullah, M. (2011). Support vector machine approach to real-time inspection of biscuits on moving conveyor belt. *Computers and Electronics in Agriculture, 75*(1), 147–158.

Ni, Y., Mei, M., & Kokot, S. (2011). Analysis of complex, processed substances with the use of NIR spectroscopy and chemometrics: Classification and prediction of properties—The potato crisps example. *Chemometrics and Intelligent Laboratory Systems, 105*(2), 147–156.

Nocita, M., Stevens, A., Toth, G., Panagos, P., van Wesemael, B., & Montanarella, L. (2014). Prediction of soil organic carbon content by diffuse reflectance spectroscopy using a local partial least square regression approach. *Soil Biology and Biochemistry, 68,* 337–347.

Oerke, E.-C. (2006). Crop losses to pests. *The Journal of Agricultural Science, 144*(1), 31–43.

Palmer, J. W., Harker, F. R., Tustin, D. S., & Johnston, J. (2010). Fruit dry matter concentration: A new quality metric for apples. *Journal of the Science of Food and Agriculture, 90*(15), 2586–2594.

Pan, L., Lu, R., Zhu, Q., McGrath, J. M., & Tu, K. (2015). Measurement of moisture, soluble solids, sucrose content and mechanical properties in sugar beet using portable visible and near-infrared spectroscopy. *Postharvest Biology and Technology, 102,* 42–50.

Pan, L., Lu, R., Zhu, Q., Tu, K., & Cen, H. (2016). Predict compositions and mechanical properties of sugar beet using hyperspectral scattering. *Food and Bioprocess Technology, 9*(7), 1177–1186.

Pan, L., Zhu, Q., Lu, R., & McGrath, J. M. (2015). Determination of sucrose content in sugar beet by portable visible and near-infrared spectroscopy. *Food Chemistry, 167,* 264–271.

Pedreschi, F., Granby, K., & Risum, J. (2010). Acrylamide mitigation in potato chips by using NaCl. *Food and Bioprocess Technology, 3*(6), 917–921.

Pedreschi, F., Mery, D., Bunger, A., & Yanez, V. (2011). Computer vision classification of potato chips by color. *Journal of Food Process Engineering, 34*(5), 1714–1728.

Pereira, C. G., & Meireles, M. A. A. (2010). Supercritical fluid extraction of bioactive compounds: Fundamentals, applications and economic perspectives. *Food and Bioprocess Technology, 3*(3), 340–372.

Petruccioli, E., Vanini, V., Chiacchio, T., Cirillo, D. M., Palmieri, F., Ippolito, G., et al. (2016). Modulation of interferon-gamma response to QuantiFERON-TB-plus detected by enzyme-linked immunosorbent assay in patients with active and latent tuberculosis infection. *International Journal of Mycobacteriology, 5,* S143–S144.

Protte, K., Weiss, J., & Hinrichs, J. (2018). Insignificance of lactose impurities on generation and structural characteristics of thermally stabilised whey protein-pectin complexes. *International Dairy Journal, 80,* 46–51.

Pu, H., Kamruzzaman, M., & Sun, D.-W. (2015). Selection of feature wavelengths for developing multispectral imaging systems for quality, safety and authenticity of muscle foods—A review. *Trends in Food Science & Technology, 45*(1), 86–104.

Pu, H., Sun, D.-W., Ma, J., Liu, D., & Cheng, J.-H. (2014). Using wavelet textural features of visible and near infrared hyperspectral image to differentiate between fresh and frozen–thawed pork. *Food and Bioprocess Technology, 7*(11), 3088–3099.

Rady, A., & Guyer, D. (2015a). Utilization of visible/near-infrared spectroscopic and wavelength selection methods in sugar prediction and potatoes classification. *Journal of Food Measurement and Characterization, 9*(1), 20–34.

Rady, A. M., & Guyer, D. E. (2015b). Evaluation of sugar content in potatoes using NIR reflectance and wavelength selection techniques. *Postharvest Biology and Technology, 103,* 17–26.

Rady, A. M., Guyer, D. E., Kirk, W., & Donis-González, I. R. (2014). The potential use of visible/near infrared spectroscopy and hyperspectral imaging to predict processing-related constituents of potatoes. *Journal of Food Engineering, 135,* 11–25.

Rady, A., Guyer, D., & Lu, R. (2015). Evaluation of sugar content of potatoes using hyperspectral imaging. *Food and Bioprocess Technology, 8*(5), 995–1010.

Ravikanth, L., Jayas, D. S., White, N. D., Fields, P. G., & Sun, D.-W. (2017). Extraction of spectral information from hyperspectral data and application of hyperspectral imaging for food and agricultural products. *Food and Bioprocess Technology, 10*(1), 1–33.

Robertson, J. A., de Monredon, F. D., Dysseler, P., Guillon, F., Amado, R., & Thibault, J.-F. (2000). Hydration properties of dietary fibre and resistant starch: A European collaborative study. *LWT-Food Science and Technology, 33*(2), 72–79.

Roessner, U., Wagner, C., Kopka, J., Trethewey, R. N., & Willmitzer, L. (2000). Simultaneous analysis of metabolites in potato tuber by gas chromatography–mass spectrometry. *The Plant Journal, 23*(1), 131–142.

Rotariu, L., Lagarde, F., Jaffrezic-Renault, N., & Bala, C. (2016). Electrochemical biosensors for fast detection of food contaminants–trends and perspective. *TrAC Trends in Analytical Chemistry, 79*, 80–87.

Sánchez, T., Ceballos, H., Dufour, D., Ortiz, D., Morante, N., Calle, F., et al. (2014). Prediction of carotenoids, cyanide and dry matter contents in fresh cassava root using NIRS and Hunter color techniques. *Food Chemistry, 151*, 444–451.

Scheibelhofer, O., Koller, D., Kerschhaggl, P., & Khinast, J. (2012). In *Continuous powder flow monitoring via near-infrared hyperspectral imaging. Instrumentation and measurement technology conference (I2MTC), 2012 IEEE international* (pp. 748–753): IEEE.

Scott, G. J., Rosegrant, M. W., & Ringler, C. (2000). *Roots and tubers for the 21st century: Trends, projections, and policy options.* International Food Policy Research Institute.

Sellami, I. H., Rebey, I. B., Sriti, J., Rahali, F. Z., Limam, F., & Marzouk, B. (2012). Drying sage (Salvia officinalis L.) plants and its effects on content, chemical composition, and radical scavenging activity of the essential oil. *Food and Bioprocess Technology, 5*(8), 2978–2989.

Severgnini, M., Cremonesi, P., Consolandi, C., De Bellis, G., & Castiglioni, B. (2011). Advances in DNA microarray technology for the detection of foodborne pathogens. *Food and Bioprocess Technology, 4*(6), 936–953.

Shahin, M. A., & Symons, S. J. (2011). Detection of Fusarium damaged kernels in Canada Western Red Spring wheat using visible/near-infrared hyperspectral imaging and principal component analysis. *Computers and Electronics in Agriculture, 75*(1), 107–112.

Sharma, H. K., & Kaushal, P. (2016). Introduction to tropical roots and tubers. In *Tropical roots and tubers: Production, processing and technology* (p. 1). John Wiley & Sons.

Sharma, H. K., Njintang, N. Y., Singhal, R. S., & Kaushal, P. (2016). *Tropical roots and tubers: Production, processing and technology.* John Wiley & Sons.

Singh, C. B., Jayas, D. S., Paliwal, J., & White, N. D. (2010). Identification of insect-damaged wheat kernels using short-wave near-infrared hyperspectral and digital colour imaging. *Computers and Electronics in Agriculture, 73*(2), 118–125.

Śliwińska, M., Wiśniewska, P., Dymerski, T., Wardencki, W., & Namieśnik, J. (2016). Advances in electronic noses and tongues for food authenticity testing. In *Advances in food authenticity testing* (pp. 201–225): Elsevier.

Soler, C., & Pico, Y. (2007). Recent trends in liquid chromatography-tandem mass spectrometry to determine pesticides and their metabolites in food. *TrAC Trends in Analytical Chemistry, 26*(2), 103–115.

Stahl, W., & Sies, H. (2005). Bioactivity and protective effects of natural carotenoids. *Biochimica et Biophysica Acta (BBA)-Molecular Basis of Disease, 1740*(2), 101–107.

Stark, J., & Love, S. (2003). *Potato production systems.* Idaho, USA: University of Idaho Extension.

Stuart, B. (2005). *Infrared spectroscopy.* Wiley Online Library.

Su, W.-H., Bakalis, S., & Sun, D.-W. (2018). Fourier transform mid-infrared-attenuated total reflectance (FTMIR-ATR) microspectroscopy for determining textural property of microwave baked tuber. *Journal of Food Engineering, 218*, 1–13.

Su, W.-H., He, H.-J., & Sun, D.-W. (2015). Application of hyperspectral imaging technique for measurement of external defects of potatoes. *Biosystems Engineering and Research Review, 20*, 9.

Su, W.-H., He, H.-J., & Sun, D.-W. (2017). Non-destructive and rapid evaluation of staple foods quality by using spectroscopic techniques: A review. *Critical Reviews in Food Science and Nutrition, 57*(5), 1039–1051.

Su, W.-H., & Sun, D.-W. (2016a). Comparative assessment of feature-wavelength eligibility for measurement of water binding capacity and specific gravity of tuber using diverse spectral indices stemmed from hyperspectral images. *Computers and Electronics in Agriculture, 130*, 69–82.

Su, W.-H., & Sun, D.-W. (2016b). Facilitated wavelength selection and model development for rapid determination of the purity of organic spelt (Triticum spelta L.) flour using spectral imaging. *Talanta, 155*, 347–357.

Su, W.-H., & Sun, D.-W. (2016c). Multivariate analysis of hyper/multi-spectra for determining volatile compounds and visualizing cooking degree during low-temperature baking of tubers. *Computers and Electronics in Agriculture, 127*, 561–571.

Su, W.-H., & Sun, D.-W. (2016d). Potential of hyperspectral imaging for visual authentication of sliced organic potatoes from potato and sweet potato tubers and rapid grading of the tubers according to moisture proportion. *Computers and Electronics in Agriculture, 125*, 113–124.

Su, W.-H., & Sun, D.-W. (2016e). In *Rapid visualization of moisture migration in tuber during dehydration using hyperspectral imaging. CIGR-AgEng conference (Jun. 26–29, 2016), Aarhus, Denmark.*

Su, W.-H., & Sun, D.-W. (2017a). Chemical imaging for measuring the time series variations of tuber dry matter and starch concentration. *Computers and Electronics in Agriculture, 140*, 361–373.

Su, W.-H., & Sun, D.-W. (2017b). Evaluation of spectral imaging for inspection of adulterants in terms of common wheat flour, cassava flour and corn flour in organic Avatar wheat (Triticum spp.) flour. *Journal of Food Engineering, 200*, 59–69.

Su, W.-H., & Sun, D.-W. (2017c). Hyperspectral imaging as non-destructive assessment tool for the recognition of sweet potato cultivars. *Biosystems Engineering and Research Review, 22*, 21.

Su, W. H., & Sun, D.-W. (2018a). Fourier transform infrared and raman and hyperspectral imaging techniques for quality determinations of powdery foods: A review. *Comprehensive Reviews in Food Science and Food Safety, 17*(1), 104–122.

Su, W. H., & Sun, D.-W. (2018b). Multispectral imaging for plant food quality analysis and visualization. *Comprehensive Reviews in Food Science and Food Safety, 17*(1), 220–239.

Su, W.-H., Sun, D.-W., He, J.-G., & Zhang, L.-B. (2017). Variation analysis in spectral indices of volatile chlorpyrifos and non-volatile imidacloprid in jujube (Ziziphus jujuba Mill.) using near-infrared hyperspectral imaging (NIR-HSI) and gas chromatograph-mass spectrometry (GC–MS). *Computers and Electronics in Agriculture, 139*, 41–55.

Sun, D.-W. (2009). *Infrared spectroscopy for food quality analysis and control.* Academic Press.

Sun, D.-W. (2016). *Computer vision technology for food quality evaluation.* Academic Press.

Sun, X., & Dong, X. (2015). Improved partial least squares regression for rapid determination of reducing sugar of potato flours by near infrared spectroscopy and variable selection method. *Journal of Food Measurement and Characterization, 9*(1), 95–103.

Sun, Y., Liu, Y., Yu, H., Xie, A., Li, X., Yin, Y., et al. (2017). Non-destructive prediction of moisture content and freezable water content of purple-fleshed sweet potato slices during drying process using hyperspectral imaging technique. *Food Analytical Methods, 10*(5), 1535–1546.

Tao, F., & Peng, Y. (2015). A nondestructive method for prediction of total viable count in pork meat by hyperspectral scattering imaging. *Food and Bioprocess Technology, 8*(1), 17–30.

Tumwegamire, S., Kapinga, R., Rubaihayo, P. R., LaBonte, D. R., Grüneberg, W. J., Burgos, G., et al. (2011). Evaluation of dry matter, protein, starch, sucrose, β-carotene, iron, zinc, calcium, and magnesium in East African sweetpotato [Ipomoea batatas (L.) Lam] germplasm. *Hortscience, 46*(3), 348–357.

Türker-Kaya, S., & Huck, C. W. (2017). A Review of mid-infrared and near-infrared imaging: Principles, concepts and applications in plant tissue analysis. *Molecules, 22*(1), 168.

Uchechukwu-Agua, A. D., Caleb, O. J., & Opara, U. L. (2015). Postharvest handling and storage of fresh cassava root and products: A review. *Food and Bioprocess Technology, 8*(4), 729–748.

Ugwu, F. (2009). The potentials of roots and tubers as weaning foods. *Pakistan Journal of Nutrition, 8*(10), 1701–1705.

Vandenbroucke, H., Mournet, P., Vignes, H., Malapa, R., Duval, M.-F., & Lebot, V. (2016). Somaclonal variants of taro (Colocasia esculenta Schott) and yam (Dioscorea alata L.) are incorporated into farmers' varietal portfolios in Vanuatu. *Genetic Resources and Crop Evolution, 63*(3), 495–511.

Veloso, A. C., Dias, L. G., Rodrigues, N., Pereira, J. A., & Peres, A. M. (2016). Sensory intensity assessment of olive oils using an electronic tongue. *Talanta, 146*, 585–593.

Wang, L., Sun, D.-W., Pu, H., & Cheng, J.-H. (2017). Quality analysis, classification, and authentication of liquid foods by near-infrared spectroscopy: A review of recent research developments. *Critical Reviews in Food Science and Nutrition, 57*(7), 1524–1538.

Wei, X., Liu, F., Qiu, Z., Shao, Y., & He, Y. (2014). Ripeness classification of astringent persimmon using hyperspectral imaging technique. *Food and Bioprocess Technology, 7*(5), 1371–1380.

Weiss, S. (1999). Fluorescence spectroscopy of single biomolecules. *Science, 283*(5408), 1676–1683.

Weurding, R. E., Veldman, A., Veen, W. A., van der Aar, P. J., & Verstegen, M. W. (2001). Starch digestion rate in the small intestine of broiler chickens differs among feedstuffs. *The Journal of Nutrition, 131*(9), 2329–2335.

Wilson, A., Work, T., Bushway, A., & Bushway, R. (1981). HPLC determination of fructose, glucose, and sucrose in potatoes. *Journal of Food Science, 46*(1), 300–301.

Woodcock, T., Fagan, C. C., O'Donnell, C. P., & Downey, G. (2008). Application of near and mid-infrared spectroscopy to determine cheese quality and authenticity. *Food and Bioprocess Technology, 1*(2), 117–129.

Wu, D., & Sun, D.-W. (2016). The use of hyperspectral techniques in evaluating quality and safety of meat and meat products. In *Emerging technologies in meat processing: Production, processing and technology* (pp. 345–374), John Wiley & Sons.

Xie, L., Ye, X., Liu, D., & Ying, Y. (2009). Quantification of glucose, fructose and sucrose in bayberry juice by NIR and PLS. *Food Chemistry, 114*(3), 1135–1140.

Xiong, Z., Sun, D.-W., Zeng, X.-A., & Xie, A. (2014). Recent developments of hyperspectral imaging systems and their applications in detecting quality attributes of red meats: A review. *Journal of Food Engineering, 132*, 1–13.

Xiong, Z., Xie, A., Sun, D.-W., Zeng, X.-A., & Liu, D. (2015). Applications of hyperspectral imaging in chicken meat safety and quality detection and evaluation: A review. *Critical Reviews in Food Science and Nutrition, 55*(9), 1287–1301.

Xu, L., Shi, W., Cai, C.-B., Zhong, W., & Tu, K. (2015). Rapid and nondestructive detection of multiple adulterants in kudzu starch by near infrared (NIR) spectroscopy and chemometrics. *LWT-Food Science and Technology, 61*(2), 590–595.

Ye, D., Sun, L., Yang, Z., Che, W., & Tan, W. (2017). In *Determination of bruised potatoes by GLCM based on hyperspectral imaging technique Service systems and service management (ICSSSM), 2017 international conference on. IEEE* (pp. 1–6).

Yu, K. Q., Zhao, Y. R., Liu, Z. Y., Li, X. L., Liu, F., & He, Y. (2014). Application of visible and near-infrared hyperspectral imaging for detection of defective features in loquat. *Food and Bioprocess Technology, 7*(11), 3077–3087.

Zhang, Q., Saleh, A. S., & Shen, Q. (2013). Discrimination of edible vegetable oil adulteration with used frying oil by low field nuclear magnetic resonance. *Food and Bioprocess Technology, 6*(9), 2562–2570.

Zhou, Z., Zeng, S., Li, X., & Zheng, J. (2015). Nondestructive detection of blackheart in potato by visible/near infrared transmittance spectroscopy. *Journal of Spectroscopy, 2015,* 786709.

Zhu, F., Zhang, H., Shao, Y., He, Y., & Ngadi, M. (2014). Mapping of fat and moisture distribution in Atlantic salmon using near-infrared hyperspectral imaging. *Food and Bioprocess Technology, 7*(4), 1208–1214.

Zhuang, H., Ni, Y., & Kokot, S. (2015). A comparison of near-and mid-infrared spectroscopic methods for the analysis of several nutritionally important chemical substances in the Chinese Yam (Dioscorea opposita): Total sugar, polysaccharides, and flavonoids. *Applied Spectroscopy, 69*(4), 488–495.

> CHAPTER SIX

Advances in Sheep and Goat Meat Products Research

Alfredo Teixeira*,[1], Severiano Silva[†], Sandra Rodrigues*

*Mountain Research Centre (CIMO), Escola Superior Agrária/Instituto Politécnico de Bragança, Bragança, Portugal
[†]Veterinary and Animal Research Centre (CECAV), Universidade Trás-os-Montes e Alto Douro, Vila Real, Portugal
[1]Corresponding author: e-mail address: teixeira@ipb.pt

Contents

Abstract

The main goal of this chapter was to review the state of the art in the recent advances in sheep and goat meat products research. Research and innovation have been playing an important role in sheep and goat meat production and meat processing as well as food safety. Special emphasis will be placed on the imaging and spectroscopic methods for predicting body composition, carcass and meat quality. The physicochemical and sensory quality as well as food safety will be referenced to the new sheep and goat meat products. Finally, the future trends in sheep and goat meat products research will be pointed out.

Advances in Food and Nutrition Research, Volume 87
ISSN 1043-4526
https://doi.org/10.1016/bs.afnr.2018.09.002

1. INTRODUCTION

The world population is expected to reach 8.6 billion in 2030 and 9.8 billion in 2050 (UN, 2017) and according to a recent report of the Food and Agriculture Organization of the United Nations (FAO) (2011) the meat consumption is projected to increase almost 73%. Trends in production and consumption levels per person, as well as patterns of consumption, are expected to grow substantially and particularly sheep and goat meat will be driven in developing countries where the protein supply is today above the minimum recommended level. Sheep and goat meat products have been growing interest and gaining popularity for several reasons and especially for their nutritional properties, traditional quality and physicochemical composition and sensory attributes. Sheep and goat meat are unique in flavor and palatability and particularly goat meat is leaner than other red meats and today has been preferred for meats with less fat. Furthermore, the consumption of sheep and goat meat is normally linked to certain ethnic groups and associated to some religious festivities. Some countries around the world with great tradition of sheep and goat meat consumption have the habit of eating some processed products of these meats as sheep and goat "mantas" (Oliveira et al., 2014; Teixeira, Pereira, & Rodrigues, 2011), the Spanish *cecina de castron* (Hierro, de la Hoza, Juan, & Ordóñez, 2004), the Italian *violoin di capra* (Fratianni, Sada, Orlando, & Nazzaro, 2008) or the Brazilian *charque* and *manta* (Madruga & Bressan, 2011). In southern Europe, particularly in Mediterranean countries, the meat consumption from younger animals as lamb and kid is very usual and appreciated and some of them are commercialized as quality brands with protected origin designation (PDO) or protected geographical indication (PGI).

In this context, there is no doubt that research and innovation will play a key role in advances in sheep and goat production, carcass evaluation and classification, in the control of processes and in the quality (physicochemical and sensory) of meat and meat products, as well as in the creation of new products and food safety. Recently some studies on the usage of sheep and goat meats to produce sausages, patties, pâtês have been reported (Leite et al., 2015; Teixeira, Fernandes, Pereira, Manuel, & Rodrigues, 2017). In terms of sheep and goat meat production systems, slaughter procedures, carcass evaluation and classification and body composition or meat quality, several new methodologies have recently been used like the use of ultrasounds (Silva, 2017; Teixeira, Joy, & Delfa, 2008), spectrophotometer methods (Prieto, Pawluczyk, Dugan, & Aalhus, 2017), hyperspectral imaging (Xu & Sun, 2017), the use of computed tomography (Bünger et al., 2011) or the use of computer vision (Ma et al., 2016). This chapter reviews the recent advances in sheep and goat meat products research. Special emphasis will be placed to the imaging and spectroscopic methods for predicting carcass and meat quality. The physicochemical and sensory quality as well as food safety will be referenced to the new sheep and goat meat products. Finally, the future trends in sheep and goat meat products research will be pointed out.

2. IMAGING AND SPECTROSCOPIC METHODS FOR PREDICTING MEAT QUALITY

In recent years, there have been important developments in non-invasive and non-destructive spectroscopic and image techniques to obtain objective data on carcass and meat quality of meat species. The following sections will present the most important advances of the use of these techniques in sheep and goat species.

2.1 Use of Real-Time Ultrasonography

The real-time ultrasonography (RTU) has become the most widespread technique for carcass composition and meat quality assessment on sheep and goat (Silva, 2017; Teixeira, 2008). For over 40 years, broad work has been carried out to obtain RTU information both in vivo or post-mortem about sheep and goat carcass traits to be used in breeding programs to improve carcass quality and also as a tool to monitoring carcass composition and meat quality of those two species (Hopkins, Stanley, & Ponnampalam, 2007; Leeds et al., 2008; Teixeira et al., 2008; Teixeira, Matos, Rodrigues, Delfa, & Cadavez, 2006). This section presents an overview of the RTU use for assessing carcass composition and meat quality in sheep and goat.

Along the years, the RTU technique has evolved continuously in both the equipment and the capability of image analysis and the ability to use and manage the generated information as a result of improvements in computer and ultrasound processing and image analysis capability (Whitsett, 2009). Furthermore, the ability for on-line image analysis and the decrease of equipment expenses (Li, 2010; Szabo, 2004) led to the extensive use of RTU in selection programs of carcass and meat traits for sheep and goat (Huisman, Brown, & Fogarty, 2016; McGregor, 2017; Tait, 2016).

The use of RTU needs accurate and precise ultrasound measurements. Usually, the studies conducted to evaluate the RTU in sheep and goat are orientated to two objectives. The first is the study of the correlation between RTU and the corresponding carcass measurements. The second is the use of RTU to predict carcass composition or meat quality traits. The next points will discuss both objectives.

The correlation between RTU and carcass measurements has been typically used as an indicator of accuracy in sheep and goat studies (Notter et al., 2014; Teixeira et al., 2008). Challenges are expected in comparing different works because of the diversity in the materials and methods. There are several factors which influence results (Silva, 2017). In this way, Table 1 includes information about species, breed, number of animals, equipment, frequency probe and reference point to help in the interpretation of the correlation values. In general, the works are focused on the 12–13th thoracic vertebrae and between the third and fourth lumbar vertebrae reference points to capture RTU images to do the measurements. The option for those anatomical points is supported in three major reasons: First, the thoracolumbar region is very easy to identify in a standing animal; second, this region has a simple anatomical feature, with a long muscle (*Longissimus thoracis* et *lumborum muscle*—LM) and the vertebrae apophyses, which help in a precise location of the reference points; third, and related to the other two reasons, the RTU images are straightforward interpretation which allows a high repeatability of the measurements (Glasbey, Abdalla, & Simm, 1996; Simm, 1987).

The correlations between RTU and the equivalent measurement in the carcass show significant values. Considering all correlations in Table 1, only 10 out of 81 are not significant ($P > 0.05$). The correlations observed for muscle measurements show a high variation, but in general, the LMA and LMD are more accurate than LMW. For LMA and LMD 15 out of 40 correlation coefficients are superior to 0.6, whereas for LMW that ratio is 8 out of 10. Also, the LMA and the LMD for goat and sheep show similar accuracy. The correlation variation observed for the LM measurements has different causes. For example, inadequate identification of the LM borders during RTU image analysis can influence the accuracy (Silva et al., 2006). The problem in distinguishing vertical interfaces such as the edges of the LM was also shown as a cause for the low correlation found with the LMW (McEwan, Clarke, Knowler, & Wheeler, 1989). As a consequence Hopkins et al. (1993) considered the LMW of little value as a trait to use sheep breeding programs.

Table 1 Correlations Between In Vivo Ultrasound Measurements and Corresponding Carcass Measurements in Sheep and Goat

Specie	Breed	n	Equipment	MHz	Reference Point	Muscle Measurements						Reference
						LMA	LMD	LMW	SF	GR	ST	
Sheep	Barki	15	PieMedical100	8	12/13TV	0.06^{ns};0.83	-0.18^{ns};0.67	0.21^{ns};0.42	0.34;0.62			Agamy, Moneim, Alla, Mageed, and Ashmawi (2015)
	Aragonesa	14	Toshiba SAL–32B	5	3/4VL			0.22^{ns}	0.73;0.87			Delfa, Teixeira, Blasco, and Rocher-Colomber (1991)
	Suffolk	163	Aloka 500	3.5	12/13TV;WT	0.66			0.78	0.73		Emenheiser, Greiner, Lewis, and Notter (2010)
	Mixed breeds	124	Vet 180 Plus	5	12/13TV;1/2LV	0.72;0.78	0.61^{*};0.88	0.69;0.70	0.32;0.60			Esquivelzeta, Casellas, Fina, and Piedra (2012)
	3 genotypes	60	Toshiba SAL–32B	5	12/13TV	0.88	0.56		0.74			Fernández, Gallego, and Quintanilla (1997)
	Manchego	10	Toshiba SAL–32B	5	12/13TV;3/4LV		0.13^{ns};0.76	0.40^{ns};0.83	-0.06^{ns};0.92			Fernández, García, Vergara, and Gallego (1998)
	6 genotypes	36	Mindray DP-6900	5	12/13TV		0.76		0.82			Grill, Ringdorfer, Baumung, and Fuerst-waltl (2015)
	Rambouillet × Finn	30	Technicare 210	5	12/13TV	0.80						Hamby, Stouffer, and Smith (1986)

Continued

Table 1 Correlations Between In Vivo Ultrasound Measurements and Corresponding Carcass Measurements in Sheep and Goat—cont'd

Specie	Breed	n	Equipment	MHz	Reference Point	Muscle Measurements						Reference
						LMA	LMD	LMW	SF	GR	ST	
	5 genotypes	147	Honda HS-1201	5	12/13TV		0.55		0.67			Hopkins et al. (2007)
		58	Aloka 500	3.5	12TV;GR	0.42	0.36	-0.15^{ns}	0.17^{ns}	0.60		Hopkins, Pirlot, Roberts, and Beattie (1993)
	F1 wether lambs	168	Aloka 500	3.5	12/13TV	0.75	0.71		0.81			Leeds et al. (2008)
	Omani	19	Microimager	7.5	6TV;12TV;2LV		0.27*;0.48					Mahgoub (1998)
	Mixed breeds	512	Aloka 500	3.5	12/13TV	0.65			0.69			Notter et al. (2014)
	Awassi	13	Dynamic	7.5	12/13TV	0.87;0.89	0.58*;0.60	$-0.17^{ns};0.48$	0.79;0.82			Orman et al. (2008)
	Awassi	30	Dynamic	7.5	12/13TV	0.88	0.77	0.58	0.93			Orman, Caliskan, and Dikmen (2010)
		99;147	Toshiba SAL-22A	5	GR;BWT					0.87		Ramsey, Kirton, Hogg, and Dobbie (1991)
	Tensina	114	Aloka 900	7	10/11TV;12/13TV 1/2LV;3/4LV		0.42*;0.59	$0.16^{ns};0.37$	0.70;0.74			Ripoll, Joy, and Sanz (2010)
	Akkaraman	40	PieMedical100	8	12/13TV	0.82	0.60		0.77			Sahin, Yardimci, Cetingul, Bayram, and Sengor (2008)
		15	Aloka 500	5	12/13TV	0.89			0.93			Stouffer (1991)
	Churra Bragançana	67	Aloka 500	5;7.5	12/13TV;3/4LV				0.31;0.42			Teixeira et al. (2006)

Species	Breed	n	Machine	Freq	Location							Reference
	2 genotypes	96	Ultrscan 50	3.5	12/13TV; GR;3/4LV				0.34*;0.42	0.78;0.82	0.83	Thériault, Pomar, and Castonguay (2009)
		162	Technicare 210		13TV	0.58				0.42;0.63		Turlington (1989)
	Torki	99	PieMedical100	8	12/13TV	0.80	0.77	0.54		0.70		Vardanjani, Ashtiani, Pakdel, and Moradi (2014)
		89	Dynamic	3.5	12TV	0.76	0.79					Ward, Purchas, and Abdullah (1992)
Goat	Boer	77	Aloka 500	5	12/13TV		0.51			-0.09^{ns}		Carr, Waldron, and Willingham (2002)
	Spanish	40	Aloka 500	5	12/13TV	0.75	0.71		0.52	0.49		Mesta et al., 2016
	Alpine	25	Keikei CS-3000	3.5	12/13TV	0.47;0.64	0.23*;0.62					Stanford, Clark, and Jones (1995)
	Blanca Celtiberica	56	Toshiba SAL-32B	5	1/2LV		0.81	0.69			0.94	Teixeira et al. (2008)
	Blanca Celtiberica	56	Toshiba SAL-32B	5	3/4LV		0.84	0.70				Teixeira et al. (2008)
	Blanca Celtiberica	56	Toshiba SAL-32B	5	5/6LV		0.47	0.74				Teixeira et al. (2008)

Abbreviations: LW, live weight; LMA, longissimus thoracis et lumborum muscle area; LMD, longissimus thoracis et lumborum muscle depth; LMW, longissimus thoracis et lumborum muscle width; SF, subcutaneous fat depth; GR, total depth of soft tissues over the 12th rib 11 cm from the dorsal midline; ST, fat thickness measurement in sternum region; BWT, measurement of body wall thickness between 12th and 13th ribs 12 cm from the dorsal midline; WT, body wall thickness between the 12th and 13th ribs including the lateral edge of the LM but not the spine; TV, thoracic vertebra; LV, lumbar vertebra. All correlation coefficients values are significant to $P < 0.01$ unless noted otherwise; *$P < 0.05$; $^{\text{ns}}P > 0.05$.

The correlation values for the SF show a considerable variation (r between -0.09, $P > 0.05$ to 0.93, $P < 0.01$). In spite of this substantial variation, the majority of the studies present values above 0.6. The GR and ST measurements show values of correlation that are comparable to the highest values reported for SF. The GR measurement is a well-known measurement included in the carcass grading systems (Hopkins, 1994; Jones, Robertson, Price, & Coupland, 1996; Kirton & Johnson, 1979). The GR can also be of attention for in vivo studies for selection decisions made in young lambs as this measurement shows a higher dimension than the SF measurement. The reduced depth observed with the SF measurement was reported by Hopkins et al. (1993) as an explanation of the low accuracy when compared with the GR ($r = 0.17$; $P > 0.05$ vs $r = 0.60$; $P < 0.01$, respectively). In goat, the reduced SF depth is also a problem since this species has less amount of that fat depot (Teixeira et al., 2008). One way to overcome this constraint is the use of a high-frequency probe (for example, 7.5 MHz). Higher frequency probes have a higher resolution at the surface (Silva et al., 2006). Currently, the RTU equipment works with multi-frequency probes which can be very helpful to set the probe with the SF depth. Conscious of this problem Ripoll et al. (2010) operated a high frequency (8–10 MHz) for SF measurements and a 7 MHz frequency for LMD. The prediction of body and carcass composition in meat species was reviewed in the past (Houghton & Turlington, 1992; Simm, 1987). However, little attention was directed to the goat and sheep species. Therefore, this section presents an overview of the prediction of body and carcass composition of goat and sheep using RTU measurements. The prediction models are typically obtained by multiple linear regression using RTU measurements along with live animal weight. The accuracy and the precision of the models are achieved by the using the coefficient of determination (R^2) which indicates accuracy and the residual mean square errors (RMSE), respectively, to assess the model fit (Hopkins, Ponnampalam, & Warner, 2008).

The coefficients of determination (R^2) and residual mean square errors (RMSE) values for the prediction of goat and sheep carcass or body fat, from multiple regressions with LW and RTU measurements, are presented in Table 2.

The LW was normally introduced in the body and carcass prediction models. Usually this variable is the most powerful predictor of carcass composition (Kempster et al., 1982; Silva et al., 2006). In general, the models using LW and RTU measurements accurately explain the variation of the fat components of goat and sheep (R^2 between 0.36 and 0.98). Most of

Table 2 Coefficients of Determination (R^2) and Residual Mean Square Errors (RSME) for the Prediction of Goat and Sheep Carcass or Body Fat, From Multiple Regressions With LW and RTU Measurements

Specie	n	Breed	LW	Device	MHz	Dependent Variable	Independent Variables LW	RTU	R^2	RSME	Reference
Goat	10	Blanca Celtiberica	22	Toshiba SAL–32B	5	BF (g)	●	4	0.98	0.04	Delfa, Teixeira, González, Torrano, and Valderrábano (1999)
Sheep	31	2 genotypes		Aloka 500	7.5	ChemBF (g/kg)	●	2	0.83	19.4	Silva, Gomes, Dias-da-Silva, Gil, and Azevedo (2005)
Sheep	31	2 genotypes		Aloka 500	7.5	ChemBF (kg)	●	2	0.95	0.6	Silva et al. (2005)
Sheep	15	Romney		Aloka	3	ChemCF (%)	●	1	0.52	2.9	McEwan et al. (1989)
Sheep				Toshiba SAL–22A	5	ChemCF (%)	●	1	0.66	2.44	Ramsey et al. (1991)
Goat	22	Blanca Celtiberica		Toshiba SAL–32B	5	CF (g)	●	2	0.92	0.22	Delfa (2004)
Goat	38	BoerxWhite	25	Pie Medical 100LC	8	CF (kg)	●	4	0.92	0.21	Stanisz, Gut, and Ślósarz (2004)
Sheep				Aloka 900	7;10	CF	●	1	0.51		Ripoll, Joy, Alvarez-Rodriguez, Sanz, and Teixeira (2009)
Sheep	67	Churra Bragançana	36	Aloka 500	5;7.5	CF	●	1	0.88		Teixeira et al. (2006)
Sheep	147	5 genotypes		Honda HS–1201	5	CF (%)	●	1	0.48	2.85	Hopkins et al. (2007)
Sheep	147			Toshiba Sal–22A	5	CF (%)	●	1	0.66	2.4	Ramsey et al. (1991)

Continued

Table 2 Coefficients of Determination (R^2) and Residual Mean Square Errors (RSME) for the Prediction of Goat and Sheep Carcass or Body Fat, From Multiple Regressions With LW and RTU Measurements—cont'd

Specie	n	Breed	LW	Device	MHz	Dependent Variable	Independent Variables		R^2	RSME	Reference
							LW	RTU			
Sheep	45	3 genotypes		Pie Medical Falco 100	8	CF (kg)	●		0.39		Agamy et al. (2015)
Sheep	14	Noire de Thibar		Falco Vet	3.5	CF (kg)	●	4	0.90	0.151	Hajji, Atti, and Hamouda (2015)
Sheep	40	Akkaraman	42	Pie Medical Falco 100	8	CF (kg)	●	1	0.84	0.24	Sahin et al. (2008)
Goat	20	Serrana		Aloka 500	7,5	CIF (g)	●	2	0.95	23.6	Cadavez, Rodrigues, and Teixeira (2007)
Sheep	114	Tensina	22	Aloka 900	7;10	CIF	●	2	0.84		Ripoll et al. (2010)
Sheep	67	Churra Bragançana	36	Aloka 500	5;7.5	CIF	●	1	0.84		Teixeira et al. (2006)
Sheep	67	Churra Bragançana	36	Aloka 500	5;7.5	CIF	●	1	0.85		Teixeira et al. (2006)
Sheep	46			Aloka 500		CIF (g/kg)	●	1	0.68	8.3	Silva et al. (2006)
Sheep	46			Aloka 500		CIF (kg)	●	1	0.92	0.3	Silva et al. (2006)
Goat	56	Blanca Celtiberica		Toshiba SAL–32B	5	CSF (g)	●	2	0.91	0.32	Teixeira et al. (2008)
Sheep	114	Tensina	22	Aloka 900	7;10	CSF (g)	●	2	0.75	214.9	Ripoll et al. (2010)

Species	n	Breed	Equipment		Measurement			R^2	RSD	Reference
Sheep	254	Several breeds	Danscanner	2.2	CSF (g/kg)	●	1		28.6	Kempster, Arnall, Alliston, and Barker (1982)
Sheep	46		Aloka 500		CSF (g/kg)	●	2	0.89	8.0	Silva et al. (2006)
Sheep	67	Churra Bragançana	Aloka 500	5;7.5	KPF	●	1	0.66		Teixeira et al. (2006)
Goat	56	Blanca Celtiberica	Toshiba SAL–32B	5	OF (g)	●	2	0.91	686B207	Teixeira et al. (2008)
Goat	10	Blanca Celtiberica	Toshiba SAL–32B	5	CSF + Internal fat (g)		5	0.98	58	Delfa et al. (1999)

Abbreviations: LW, live weight; CF, carcass fat; CSF, carcass subcutaneous fat; CIF, carcass intermuscular fat; ChemCF, chemical carcass fat; ChemBF, chemical body fat; BF, body fat; KPF, kidney and pelvic fat; OF, omental fat; R^2, coefficient or determination; RSD, residual standard deviation.

the models include in addition to the LW one or two RTU measurements, and only one model contains only RTU measurements (Delfa et al., 1999). All studies show that RTU measurements allow accurate prediction of carcass fat tissues ($R^2 > 0.8$ in 13 out of 19 models) as well as chemical fat (R^2 from 0.52 to 0.95). In goats and sheep, a good prediction of internal fat depots was observed when using RTU (Delfa et al., 1999; Teixeira et al., 2008, 2006). These results are very relevant when monitoring the body fat reserves along the productive cycle of these species.

The coefficient of determination (R^2) and residual mean square errors (RSME) for the prediction models of the goat and sheep protein and muscle carcass from multiple regressions with LW and RTU measurements are presented in Table 3.

The results for protein and muscle are in line with those previously discussed for fat components. In general, a very good prediction ability of the models was observed with LW and RTU measurements explaining well the muscle and protein variations (R^2 between 0.32 and 0.97). It is not possible to discriminate between species regarding the accuracy and precision. In general, the higher R^2 values are related to the amount of protein ($R^2 = 0.97$) or muscle ($R^2 > 0.8$ in 11 out of 14 models), rather than for the percentage or proportion of protein and muscle (R^2 from 0.46 to 0.87).

As was previously discussed for fat prediction, the LW usually explains the most significant part of the variation of protein and muscle. For example, Teixeira et al. (2006) reported that LW accounted for 96% of the muscle weight variation, whereas for the amount of protein the LW explains 97% (Silva et al., 2005). One aspect to note is that despite the LW value, the RTU measurements are relevant to the model accuracy.

2.2 Use of Computed Tomography

Despite its complexity and cost, the computed tomography (CT) has been recognized useful in animal science since the beginning. However, only in the last 10 years CT have gained importance for the knowledge of body and carcass composition and meat quality, especially in pigs and sheep. (Bünger et al., 2011; Kongsro, 2014; Scholz, Bünger, Kongsro, Baulain, & Mitchell, 2015). Much of this knowledge increase with projects such as Farm Animal Imaging—FAIM (Bunger et al., 2015).

For sheep, as well for other meat species, the dissection of an animal into its components represents the most accurate method of measuring carcass composition. However, this method is invasive, destructive, time-consuming, and

Table 3 Coefficients of Determination (R^2) and Residual Mean Square Errors (RSME) for the Prediction of Goat and Sheep Protein and Muscle Carcass, From Multiple Regressions With LW and RTU Measurements

Specie	n	Breed	LW	Device	MHz	Dependent Variable	LW	RTU	R^2	RSME	Reference
							Independent Variables				
Sheep	99			Toshiba Sal-22A	5	Protein (%)	●	1	0.51	0.87	Ramsey et al. (1991)
Sheep	31	2 genotypes		Aloka 500	7.5	Protein (g/kg)	●	1	0.54	5.35	Silva et al. (2005)
Sheep	31	2 genotypes		Aloka 500	7.5	Protein (kg)	●		0.97	0.18	Silva et al. (2005)
Goat	20	Serrana		Aloka 500	7,5	Musculo (g)	●	2	0.97	129.8	Cadavez et al. (2007)
Goat	56	Blanca Celtiberica		Toshiba SAL–32B	5	Musculo (g)	●	2	0.90	533	Teixeira et al. (2008)
Goat	38	BoerxWhite	25	Pie Medical 100LC	8	Musculo (kg)	●	4	0.85	1.81	Stanisz et al. (2004)
Sheep	114	Tensina	22	Aloka 900	7;10	Muscle	●	1	0.96	154.2	Ripoll et al. (2010)
Sheep	147	5 genotypes		Honda HS-1201	5	Muscle (%)	●	1	0.46	2.71	Hopkins et al. (2007)
Sheep				Aloka 900	7;10	Muscle (g)	●	1	0.59	144.46	Ripoll et al. (2009)
Sheep	67	Churra Bragançana	36	Aloka 500	5;7.5	Muscle (g)	●	1	0.96	214.6	Teixeira et al. (2006)
Sheep	46			Aloka 500	7.5	Muscle (g/kg)	●	2	0.87	12.7	Silva et al. (2006)
Sheep	45	3 genotypes		100 LC, Pie Medical	8	Muscle (kg)	●	1	0.82		Agamy et al. (2015)
Sheep	14	Noire de Thibar		Falco Vet	3.5	Muscle (kg)	●	3	0.88	0.238	Hajji et al. (2015)
Sheep	147	5 genotypes		Honda HS-1201	5	Muscle (kg)	●	1	0.86	1.62	Hopkins et al. (2007)
Sheep	40	Akkaraman	42	Pie Medical 100LC	8	Muscle (kg)	●	1	0.80	0.63	Sahin et al. (2008)
Sheep	46			Aloka 500	7.5	Muscle (kg)	●	2	0.99	0.48	Silva et al. (2006)
Sheep	162			Technicare 210		Muscle (kg)	●	1	0.89		Turlington (1989)
Sheep	76	Coopworth		Aloka 210	5	Muscle (kg)	●	1	0.32	0.723	Young and Deaker (1994)

Abbreviations: LW, live weight; R^2, coefficient or determination; RSD, residual standard deviation.

costly. Moreover, as this method needs the animal slaughter and carcass destruction it is inadequate for genetic selection programs (Bünger et al., 2011). For genetic purposes, Simm (1987) estimated that CT would improve by 50% the rates of response to the selection, due to a more accurate measurement of tissue dimensions when compared with ultrasound measurements. The CT has been successfully applied for the selection of breeders in breeding programs, such as those practiced in the UK sheep industry (Bunger et al., 2015; Bünger et al., 2011). In these programs, a two-stage approach is used in which a first stage supported by real-time ultrasound (RTU) is used for all potential candidates followed by CT scans in 10–15% of the animals that were identified as the best by RTU (Bünger, Moore, McLean, Kongsro, & Lambe, 2014). In New Zealand Jopson, Newman, and McEwan (2009) reported that economic returns were maximized when all lambs were ultrasonically scanned, and the best 13% were CT scanned. Although the expenses evolved in selection programs based on CT traits measurements are relatively high, the cost of these measurements needs to be considered in relation to its benefits (Bünger et al., 2014). The measure of spine traits in vivo, such as spine length and vertebrae number, is other CT use for sheep meat production. Several reports show that CT method is a non-invasive feasible approach to take information about spine and vertebrae length measurements traits which will be included in breeding programs aimed the loin yield increase (Donaldson, Lambe, Maltin, Knott, & Bunger, 2013; Donaldson, Lambe, Maltin, Knott, & Bünger, 2014).

The creation of a database with images, which allows permanent access to information related to the composition of animals, is also an important attribute of CT (Bünger et al., 2014; Scholz et al., 2015). In addition to the work related to selection, CT can also be used successfully in longitudinal studies. In fact, CT is a non-invasive tool, suitable for in vivo work to model protein and fat deposition during growth, providing valuable information to optimize sheep breeding, nutrition, feed efficiency and whole production systems (Bünger et al., 2014; Lambe et al., 2012). Also, CT can be used to modeling optimal slaughter times and weights, and therefore to produce final products which better meet industry and consumer demands (Bünger et al., 2014).

Table 4 summarizes the use of CT for prediction of carcass composition of sheep and goat in sheep and goat species. Most CT studies with sheep species are related to carcass composition and have been made in vivo (e.g., Rosenblatt et al., 2017) or with carcasses (e.g., Kongsro et al., 2008). For goat species little information is available. The only work with goat examines the carcass composition of light carcasses of goat kid (Silva et al., 2015).

Table 4 Summary of Applications of Computed Tomography Imaging for Prediction of Carcass Composition of Sheep and Goat

Specie	Target	Traits	n	CT image	Anatomical Landmarks	Data Analysis	R^2	RMSE	Reference
Sheep	In vivo	Muscle, kg	21	2D	TV7,LV2, LV5,FEM	OLS	0.94	0.508	Young, Nsoso, Logan, and Beatson (1996)
		Fat, kg					0.73	0.262	
		Bone, kg					0.93	0.406	
	In vivo Leg	Muscle, kg	47	2D	CAV3,CAV4,SV4	OLS	0.93		Kvame and Vangen (2006)
		Fat, kg					0.95		
		Bone, kg					0.83		
	Shoulder	Muscle, kg	32		TV6,CV7		0.93		
		Fat, kg					0.96		
		Bone, kg					0.72		
	Mid-Region	Muscle, kg	104		LV4,TV8		0.89		
		Fat, kg					0.98		
		Bone, kg					0.69		
	In vivo	Muscle, kg	160	2D	ISC, LV5,TV8		0.92	0.078	Macfarlane, Lewis, Emmans, Young, and Simm (2006)
		Fat, kg					0.98	0.097	
		Bone, kg					0.83	0.107	

Continued

Table 4 Summary of Applications of Computed Tomography Imaging for Prediction of Carcass Composition of Sheep and Goat—cont'd

Specie	Target	Traits	n	CT image	Anatomical Landmarks	Data Analysis	R^2	RMSE	Reference
	Carcass	Muscle, kg	120			PLSR	0.94	0.710	Kongsro, Røe, Aastveit, Kvaal, and Egelandsdal (2008)
		Fat, kg					0.92	0.600	
	In vivo	Body fat, g	22	SCTS (1 mm)		OLS	0.92		Rosenblatt et al. (2017)
		Visceral fat, g					0.94		
		Body fat, g		SCTS (5 mm)			0.90		
		Visceral fat, g					0.96		
Goat	Carcass	Muscle, kg	19	SCTS (5 mm)		OLS	0.95		Silva, Teixeira, Monteiro, Guedes, and Ginja (2015)
		Fat, kg					0.65		

Abbreviations: SCTS, spiral computed tomography scanning; 2D, two-dimensional cross-sectional scans; anatomical landmarks [5th lumbar vertebra (LV5), 2nd lumbar vertebra (LV2), 8th thoracic vertebra (TV8), 6th thoracic vertebra (TV6), 7th thoracic vertebra (TV7), mid-shaft of the femur (FEM), 3rd caudal vertebra (CAV3), 4th caudal vertebra (CAV4), 4th sacral vertebra (SV4), 7th cervical (CV7), ischium (ISC)]; OLS, ordinary least squares regression; PLSR, partial least squares regression; R^2, coefficient of determination; RMSE, root-mean-square error.

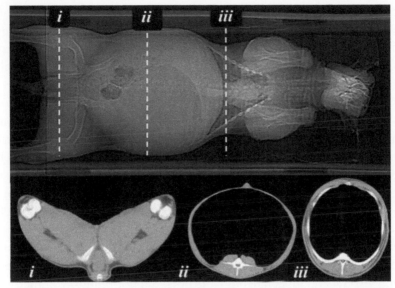

Fig. 1 Three anatomical landmarks that are considered in the cross-sectional CT scans which are obtained at the ischium, fifth lumbar vertebra and eighth thoracic vertebra.

Most of the CT work with sheep uses single slice scanning images resulting in two-dimensional images of the selected anatomical landmarks of the subject (Clelland et al., 2018). Typically, three anatomical landmarks are considered in the cross-sectional CT scans which are obtained at the ischium, 5th lumbar vertebra and 8th thoracic vertebra (Fig. 1). Those three anatomical landmarks are found as an optimum between the practicability of the scanning work and the accuracy. However, the advances in CT technology have led to the development of contiguous scanning procedures such as spiral CT scanning (SCTS). The SCTS is capable of producing a series of images in a single adjacent scan at intervals of as little as 0.6 mm apart (Daumas, Donkó, Maltin, & Bünger, 2015).

The thickness of the slice and the number of slices constitute an accuracy factor of CT. For example, Young et al. (1996) found the best accuracy when using five slices. For the cut thickness between 1 and 5 mm, it does not seem to have much influence when estimating the carcass composition. The advantage is that multiple images can be acquired faster, at reduced intervals, resulting in increased information acquisitions in less time (Clelland et al., 2018). To obtain genetic progress accurate information on the phenotypic characteristics to be chosen is required and CT represents a huge opportunity for this purpose (Bünger et al., 2011). In fact, several studies have shown that CT is a sufficiently precise tool to provide adequate information about the characteristics of interest of the carcass. Moreover, with the recent progress with the image analysis and with a greater diffusion of 3D reconstruction images resulting from SCTS procedures it is possible to increase CT capacity for in vivo studies (Bünger et al., 2014).

The more significant part of the work estimates the composition in carcass tissues since this is the most relevant information for the meat production and quality. There are however some studies that focus their objective on the internal fat depots (Lambe et al., 2003; Rosenblatt et al., 2017). These studies show that CT is competent in the determination of internal fat depots and therefore is a useful tool to understand how those reserves are depleted and repleted throughout an annual production cycle (Lambe et al., 2003).

For carcass composition, some reports in lambs also showed that CT could be used as a tool for virtual dissection (Ho et al., 2014; Kongsro et al., 2008). However, these studies did not specifically focus on differentiating visceral and subcutaneous fat depots, did not acquire volumetric CT data or did not use optimal statistical analyzes.

The prediction of spine length and muscularity is two other objectives that have been successfully addressed using CT (Jones, Lewis, Young, & Wolf, 2002; Lambe et al., 2015). The muscularity has increasingly been encouraged as being preferable to conformation as a measure of the shape of a lamb carcass because unlike conformation, muscularity is independent of carcass fatness and can be objectively obtained using muscle and bone measurements and weights (Hopkins, 1996; Purchas, Davies, & Abdullah, 1991). All that information can be accurately achieved in vivo using CT (Jones et al., 2002). In this work useful in vivo measurements of the width and depth of the LTL muscle and the length of the spine were obtained from CT scans. The ability to accurately obtain bone and muscle measurements places the CT as a suitable tool to include in vivo muscularity measures in sheep genetic selection programs to provide as a method to improve carcass conformation and leanness (Jones, Lewis, Young, & Simm, 2004; Navajas et al., 2007). Table 5 presents some data regarding muscularity prediction of sheep.

Table 6 presents a summary of the use of CT to predict IMF% and meat quality attributes. The results obtained in predicting IMF from CT measurements show that is possible but with moderate accuracy (R^2 between 0.36 and 0.70). However, this technique cannot accurately predict shear force or sensory traits of the meat (Clelland et al., 2018; Lambe et al., 2017). Despite the capacity shown by CT to predict the IMF%, its use in breeding programs is a challenge since this characteristic has an inverse relation with phenotypic CT lean% across all muscles of the carcass (Anderson, Pethick, & Gardner, 2014).

Table 5 Summary of Applications of Computed Tomography Imaging for Prediction of Muscularity Indices of Sheep

Traits	n	CT image	Anatomical Landmarks	Data Analysis	R^2	Reference
M3FL	160	2D	LV5,FEM,ISC	OLS	0.48–55.3	Jones et al. (2002)
HLMI	132	SCTS (8 mm)		OLS	0.26–0.79	Navajas et al. (2007)
LRMI	240				0.19–0.30	
CMI	240				0.30	

Abbreviations: M3FL, length of the femur and the combined weight of the three dissected muscles; HLMI, hind leg muscularity index; LRMI, lumbar muscularity index; CMI, carcass muscularity index; SCTS, spiral computed tomography scanning; 2D, two-dimensional cross-sectional scans; anatomical landmarks [5th lumbar vertebra (LV5), mid shaft of the femur (FEM), ischium (ISC)]; OLS, ordinary least squares regression; R^2, coefficient of determination.

Therefore, there is a balance that must be made to maintain an adequate level of IMF% for optimum eating quality (Anderson et al., 2014).

The scanning time and the CT image analysis are variable but increasingly faster with more modern equipment (Daumas et al., 2015). Also, for cuts, the possibility of doing a multi-object CT analysis in batches of three or more saves money and time (Lambe et al., 2017). The SCTS was found to be very useful for carcass composition prediction (Bunger et al., 2015) but for IMF prediction, the increased image analysis and processing currently required do not justify the increase in accuracy achieved when compared to current scan procedures (Clelland et al., 2018).

Prediction of sheep and goat carcass composition and meat quality using CT scanning has been investigated in several countries and it is expected that the confirmation about new CT phenotypic information in breeding programs will allow paying more attention to the characteristics of the carcass and the quality of the meat that are more valued by the market (Scholz et al., 2015). Finally the CT ability to have in vivo information about internal fat or pelvic dimensions as indicators for ease of lambing can have an impact on meat production systems and in the welfare of sheep (Lambe et al., 2003; Morgan-Davies et al., 2018; Scholz et al., 2015).

2.3 Use of Computer Vision

Many of the quality attributes affecting meat can be determined by visual inspection and image analysis. Computer vision has become an essential

Table 6 Summary of Applications of Computed Tomography Imaging for Prediction of Meat Quality Attributes of Sheep

Target	Traits	n	CT image	Anatomical Landmarks	Data Analysis	R^2	RMSE	Reference
In vivo	IMF, %	160	2D	ISC, LV5, LV2, TV8, TV6	OLS	0.57	0.608	Macfarlane, Young, Lewis, Emmans, and Simm (2005)
In vivo	IMF, %	370	2D	ISC, LV5, TV8	OLS	0.51–0.68	0.39–0.48	Clelland et al. (2014)
				LV5		0.51–0.65	0.40–0.48	
Loin	IMF, %	303	SCTS (8 mm)		OLS	0.36	0.620	Lambe et al. (2017)
	Shear force, kgF					0.03	−0.830	
	Texture					0.08	−0.530	
	Flavor					0.09	−0.370	
	Juiciness					0.06	−0.370	
	Liking					0.10	−0.390	
Loin	IMF, %	377	SCTS (8 mm)		OLS	0.51–0.70	0.48–0.38	Clelland et al. (2018)
	Shear force, kgF					0.02–0.06	0.16–0.16	
	IMF, %		SCTS+2D	ISC, TV8		0.50–0.71	0.47–0.37	
	Shear force, kgF					0.03–0.13	0.16–0.15	

Abbreviations: IMF, intramuscular fat; SCTS, spiral computed tomography scanning; 2D, two-dimensional cross-sectional scans; anatomical landmarks [5th lumbar vertebra (LV5), 2nd lumbar vertebra (LV2), 8th thoracic vertebra (TV8), 6th thoracic vertebra (TV6), ischium (ISC)]; OLS, ordinary least squares regression; R^2, coefficient of determination; RMSE, root-mean-square error.

technology for the quality control in the meat industry, which continually demands new and better applications (Ma et al., 2016).

Most studies that apply computer vision system (CVS) on carcass grading with small ruminants have been carried out on sheep (Hopkins, Safari, Thompson, & Smith, 2004; Ngo et al., 2016; Rius-Vilarrasa et al., 2010). In general, the CVS is used to the prediction of the lean meat yield (LMY%) and the saleable meat yield (SMY%). Typically, the predictions are supported in carcass weight and carcass measurements obtained from the CVS (Hopkins et al., 2004). Examples of CVS operating with lambs are the VIAScan, Cedar Creek Company, Australia (Hopkins et al., 2004), VSS 2000, E + V GmbH, Germany (Rius-Vilarrasa et al., 2010) and LVS, Research Management Systems—RMS, USA (Cunha et al., 2004). Fig. 2 illustrates lamb carcass images in the VSS 2000 system (Rius-Vilarrasa, Buenger, Maltin, Matthews, & Roehe, 2009).

Over the years, the CVS developed for sheep intent to be used in online in the abattoirs as an alternative to subjective grading systems. For example,

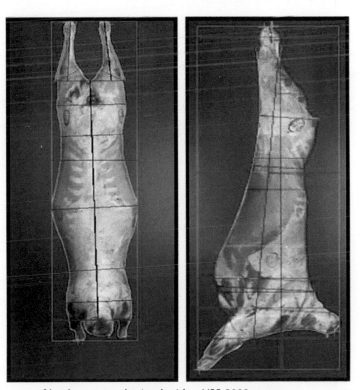

Fig. 2 Images of lamb carcass obtained with a VSS 2000 system.

in the study of Einarsson, Eythorsdottir, Smith, and Jonmundsson (2014), a CVS was applied for objectively evaluating the carcass as an alternative to the EUROP classification. Moreover, the carcass assessment can be performed without interrupting the normal flow of the slaughterhouse (Allen, 2007; Pabiou et al., 2011). Those CVS are costly but can offer a favorable cost/benefit when applied in a large number of carcasses (Craigie et al., 2012). The accuracy and speed of the assessment are key aspects to the success of a CVS and foreknowledge this technique as one to be elected for grading carcasses in the future (Hopkins, Gardner, & Toohey, 2015).

The prediction of LMY% and SMY% (Hopkins et al., 2004; Rius-Vilarrasa et al., 2010), assuming itself as a tool to discriminate the carcasses differently, aiming to achieve a value-based marketing system (Craigie et al., 2012). Table 7 summarizes the results of the estimated LMY% and SMY% of sheep carcasses by CVS.

The results obtained with CVS can be used commercially to accurately classify the carcasses which facilitate the development of a fair pricing system based on LMY% and not on carcass weight (Brady et al., 2003). When comparing the EUROP classification system and the CVS VIAscan, Einarsson et al. (2014) verified that the latter could be more effective in predicting LMY% in lambs. On the other hand, also using VIAscan in lamb carcasses, Hopkins et al. (2004) found reasonable values to predict lean meat yield ($R^2 = 0.52$). Concerning salable meat yield (SMY%), Stanford et al. (1998) reported that the VIAscan system showed the better ability for SMY% prediction when compared to the Canadian classification system. Using the CVS LVS technology, Cunha et al. (2004) found R^2 values of 0.68 for the SMY% when they included the warm carcass weight in the model, and when

Table 7 Examples of CVS to Predict Lean Meat Yield (LMY%) and Saleable Meat Yield (SMY%)

Carcass Trait	n	CVS	R^2	RMSE	Reference
LMY%	862	VIAscan[®]	0.36	2.57	Einarsson et al. (2014)
	360	VIAscan[®]	0.52	2.17	Hopkins et al. (2004)
SMY%	1211	VIAscan[®]	0.71		Stanford et al. (1998)
	149	LVS	0.68	0.02	Cunha et al. (2004)
	149	LVS	0.72	0.02	Cunha et al. (2004)
	246	LVS	0.60	0.03	Brady et al. (2003)

Abbreviations: RMSE, Residual mean square error; LVS, lamb vision system.

the fat percentage was involved in the prediction model, the accuracy was improved to 0.72. Similar results were found by Brady et al. (2003) who reported an R of 0.60 when evaluating the LVS system together with the hot carcass weight to predict SMY%.

With the use of CVS, it is possible to predict the weight and yield of the cuts in the carcass. The coefficients of determination and residual mean square errors for the prediction of carcass cut weight and yield of sheep obtained by CVS are presented in Table 8. Analyzing the prediction power

Table 8 Prediction of Carcass Cut Weight and Yield of Sheep Obtained by CVS

Cut	n	Equipment	R^2	RMSE	Reference
Tenderloin kg	792	LDG	0.60		Ngo et al. (2016)
Loin kg			0.62		
French rack kg			0.76		
Rump kg			0.75		
Leg kg			0.94		
Middle kg			0.94		
Shoulder kg			0.95		
Leg kg	443	VSS2000	0.97	0.16	Rius–Vilarrasa et al. (2009)
Chump kg			0.94	0.04	
Loin kg			0.89	0.20	
Middle kg			0.86	0.16	
Shoulder kg			0.96	0.22	
Total primal kg			0.99	0.25	
Shoulder kg	246	LVS	0.85	0.50	Brady et al. (2003)
Middle kg			0.72	0.35	
Loin kg			0.75	0.20	
Leg kg			0.85	0.45	
Subprimal yield%	149	LVS	0.66	0.02	Cunha et al. (2004)
Leg %	862	VIAscan®	0.60	1.04	Einarsson et al. (2014)
Loin %			0.31	0.72	
Shoulder %			0.47	1.16	

Subprimal yield%=percentage based on the cold carcass weight of the leg, loin, middle and shoulder.

of the carcass measurements obtained through a simple CVS (Lamb digital grading—LDG) on the weight of lamb carcass cuts, Ngo et al. (2016) found results ranging from moderate to high (R^2 ranging from 0.60 to 0.95). For the main cuts (leg and shoulder), they found R^2 with high accuracy capacity explaining 94% and 95% of the variation of their weight, respectively.

Similar results were reported by Rius-Vilarrasa et al. (2009), who compared the CVS technology with the EUROP classification system used in the United Kingdom, and observed that the CVS provided an excellent accuracy to predict leg and shoulder weight (R^2 of 0.97 and 0.96, respectively). On the other hand, Brady et al. (2003), to evaluate the potential of the CVS measurements plus hot carcass weight, show that it is possible to an explain of 85%, 72%, 75%, and 86% of the variation in weight of the shoulder, rib, loin and leg, respectively. According to these authors (Brady et al., 2003), the CVS technique has the potential to help in the grading process in the lamb meat industry. Regarding the prediction performance of the subprimal yield%, which represents the percentage based on the cold carcass weight of the leg, loin, middle and shoulder, Cunha et al. (2004) using a similar CVS approach verified that the method explained 66% of the variation of those cuts. Einarsson et al. (2014) showed that the CVS VIASscan system was able to explain 60%, 31% and 47% of the variation of the lean meat yield of the leg, loin and shoulder, respectively. According to these authors (Einarsson et al., 2014), the VIAscan is an automatic CVS able to predict the meat yield of cuts of carcass lambs. In addition to the cuts, Hopkins (1996) showed the feasibility of using a CVS to evaluate the muscularity of the carcass. On the other hand, the CVS is also suitable to predict carcass composition. For example, Lambe et al. (2009) using a CVS by combining different measurements, accurately predict the dissected carcass muscle weight (adjusted R^2 0.93 in Texel—TEX, and 0.88 in Scottish Black Sheep—SBF) and fat weight (adjusted R^2 0.84 in TEX, and 0.87 in SBF).

Unlike beef or pork meat, the use of CVS to assess intramuscular or marbling in sheep is less frequent. By combining different measurements, Lambe et al. (2009) achieved reasonable predictions of intramuscular fat (adjusted R^2 0.56 in TEX, and 0.48 in SBF). These authors also predicted the shear force using a CVS with low to moderate accuracy (adjusted $R^2 < 0.33$ across breeds and cuts). In a previous study, Lambe et al. (2008) have identified the potential of different in vivo measurement obtained with computed tomography and a CVS applied in live sheep, to predict intramuscular fat, shear force and ultimate pH of muscles. The results found are considered low for shear force and ultimate pH but for intramuscular fat, the predictions in combinations of CT and CVS taken in live lambs could be employed to genetic improvement of carcass quality traits.

2.4 Use of Near Infrared Spectroscopy

The near-infrared spectroscopy (NIRS) using the Fourier transform (FT) is a technology known by the end of 1960s when a computerized spectrophotometer NIR was developed, and its applicability to the analysis of meat was shown (Ben-Gera & Norris, 1968). NIRS measures the absorption of electromagnetic radiation from 750 to 2500 nm wavelengths, corresponding to overtones and combinations of vibrational modes of C—H, O—H and N—H chemical bonds. Recording the electromagnetic radiation absorbed from those molecular bonds in the NIR wavelengths produces spectra which are unique to a sample acting as a "fingerprint." The collected spectrum includes data related to the chemical and physical properties of organic molecules in the sample and, therefore, important information on sample composition (Prieto et al., 2017).

NIRS technology is currently a highly versatile tool used in diverse fields including the food industry and, particularly, in animal science to predict the chemical and physical composition of meat of different species (Weeranantanaphan, Downey, Allen, & Sun, 2011). The high versatility of the technology is being used for large-scale in meat quality evaluation, to predict chemical composition and identification and authentication of meat products. The use of technology was not so popular in lamb or goat meat because the information on the use of NIR spectroscopy in lamb or goat meat is relatively scant in comparison with other meats or meat products. The electromagnetic scanning was tested as objective means for assessing lamb carcass composition (Berg, Forrest, Thomas, Nusbaum, & Kauffman, 1994; Berg, Neary, Forrest, Thomas, & Kauffman, 1997). Kruggel, Field, Riley, and Horton (1981) in a study on ground raw lamb meat using NIRS suggested that the technology was more suitable for the fat and moisture determination than the protein. Viljoen, Hoffman, and Brand (2007) using NIRS concluded that the technology could be used as a rapid tool for predicting the proximal chemical composition and certain minerals in freeze-dried mutton. Andrés et al. (2007) investigated in lamb meat samples the association between chemical composition and meat quality traits scored by a trained sensory panel and absorbance data from NIR spectroscopy. The results suggested that the most important regions of spectra to estimate sensory characteristics of lamb meat are related to the absorbance of intramuscular fat and water parameters. Guy, Prache, Thomas, Bauchart, & Andueza (2011) comparing two lamb meat preparations (ground vs intact, non-ground meat samples) determined whether NIRS was feasible for accurate predictions of fatty acids profile. Results indicated that the prediction models are much better using ground than intact non-ground samples and the models obtained are satisfactory for fatty acids groups or individual fatty acids presented in medium to high concentrations as total saturated fat, *cis* and monounsaturated fat, but are lower for fatty acids presented at low or very low concentrations as polyunsaturated fatty acids. On the same way, Pullanagari, Yule, and Agnew (2015) evaluated the use of visible near infrared spectroscopy (Vis-NIRS) to quantify the fatty acid composition of intact lamb meat under commercial abattoir conditions. Those authors concluded that even though the prediction accuracies of individual fatty acids were low the Vis-NIRS could be used as a screening tool at abattoir.

The potential of NIRS to classify the geographical origin and predict the isotope of carbon $\delta^{13}C$ and nitrogen $\delta^{15}N$ of lamb meat combined with chemometrics particularly the application of partial least squares regression methodology (PLSR) was tested with promising results by Sun, Guo, Wei, and Fan (2012). However, in goats there are not many studies about the reliability and accuracy of the use of NIR spectroscopy to characterize the meat composition. A study by Teixeira et al. (2015) was the first approach to test the ability of NIRS to estimate the protein, moisture, connective tissue, ash and fat content in the LM muscle of goat meat. So, the NIRS technology combined with chemometrics would be a useful tool to know raw goat meat composition and select material for improving the quality of meat processing.

2.5 Use of Hyperspectral Imaging

Hyperspectral imaging is an emerging technology designed initially for remote satellite vigilance with military purposes but has been extended for use in astronomy and observing the territory (Goetz, 2009; Van der Meer et al., 2012). However, in the last years, it has begun to be applied as a rapid, reliable, non-destructive and non-invasive tool in the food industry (Gowen, O'Donnell, Cullen, Downey, & Frias, 2007; Liu, Pu, & Sun, 2017). Those characteristics are opening the possibility to apply hyperspectral imaging to the prediction of meat quality and meat classification (Cheng, Nicolai, & Sun, 2017; Xu & Sun, 2017).

In recent years, different image-based or spectroscopic techniques have been used to measure the various attributes of meat quality (Su, He, & Sun, 2017). Among these, the hyperspectral imaging (HIS) technique is recognized as a versatile technique for rapid quantitative, non-reagent and non-invasive applications in the red meat industry (Xu & Sun, 2017). In this point will be emphasized the use of HSI to sheep meat to evaluate sensorial, chemical, technological and classification attributes. For goat meat, like other techniques, there is scarce or just non available information.

The sensorial, chemical, technological, adulteration, authentication and discrimination attributes, which have been investigated using HIS for sheep meat and meat products, are presented in Table 9. In this section the quality attributes of red meats classification proposed by Xu and Sun (2017) are followed.

The sensory attributes have a significant influence on the evaluation of the meat by consumers. Color represents one of the most important sensory attributes since it is generally used as an indicator of the freshness of meat, and an attractive and stable color in the meat has a significant influence on the purchase decision made by the consumer (Grunert, Bredahl, & Brunsø, 2004). Several studies developed HSI as a non-contact measurement technique of color in meat (Table 1). It was found that HSI and multivariable models for prediction of color component L* (R^2 from 0.77 to 0.97) were good, yet prediction of the other color components b* and a* showed some discrepancies (R^2 of 0.48 and 0.84, respectively, and R^2 of 0.26 and 0.82, for b*and a*, respectively). The lower values were reported by Qiao, Ren, et al. (2015) who stated that HSI still proved to be a useful technique to predict complex quality traits such as color, even though prediction performance was low and requires further improvement.

The tenderness is an expression of meat texture and is the most important sensory quality attribute related to consumer satisfaction (Xu & Sun, 2017). Concerning sensory tenderness, only one reference associated with HSI and lamb meat was described (Kamruzzaman, ElMasry, et al., 2013). Four experienced panelists evaluated the sensory tenderness. The results are modest compared with those of Warner–Bratzler shear force (WBSF) ($R_{cv} = 0.84$ and 0.69 for WBSF and sensory tenderness, respectively). Nevertheless, Kamruzzaman, ElMasry, et al. (2013) argue that prediction of sensory characteristics may be improved if the tested samples are segregated into more specific sub-groups.

Efforts on using HSI for assessing lamb meat chemical composition have been investigated by several researchers (Kamruzzaman, ElMasry, et al., 2012a;

Table 9 Summary of Applications of HSI for Evaluating Quality Attributes of Sheep Meat

Quality Attributes		Wavelength Range (nm)	Multivariable Analysis	Accuracy	References
Sensory	a*	400–1000	PLSR	$R^2 = 0.84$	Kamruzzaman, Makino, and Oshita (2016a) and Kamruzzaman, Makino, and Oshita (2016b)
Sensory	a*	400–863	PCA. SVM	$R^2 = 0.48$	Qiao, Ren, et al., 2015
Sensory	b*	400–1000	PLSR	$R2 = 0.82$	Kamruzzaman et al. (2016a) and Kamruzzaman et al. (2016b)
Sensory	b*	400–863	PCA. SVM	$R^2 = 0.26$	Qiao, Ren, et al., 2015
Sensory	L*	900–1700	PLSR	$R^2 = 0.91$	Kamruzzaman, ElMasry, Sun, Allen (2012b)
Sensory	L*	400–1000	PLSR	$R^2 = 0.97$	Kamruzzaman et al. (2016a)
Sensory	L*	400–863	PCA. SVM	$R^2 = 0.77$	Qiao, Ren, et al., 2015
Sensory	Tenderness	900–1700	PLSR	$R_{cv} = 0.69$	Kamruzzaman, ElMasry, Sun, and Allen (2013)
Chemical	Protein	900–1700	PLSR	$R^2 = 0.85$	Kamruzzaman, ElMasry, Sun, and Allen (2012a)
Chemical	Protein	1021–1396	MLR	$R_c = 0.80$	Pu, Sun, Ma, Liu, and Kamruzzaman (2014)
Chemical	Water	900–1700	PLSR	$R^2 = 0.88$	Kamruzzaman, ElMasry, et al. (2012a)
Chemical	Water	1021–1396	MLR	$R_c = 0.91$	Pu et al. (2014)
Chemical	Fat	900–1700	PLSR	$R^2 = 0.91$	Kamruzzaman, ElMasry, et al. (2012a)
Chemical	Fat	1021–1396	MLR	$R_c = 0.95$	Pu et al. (2014)
Chemical	IMF%	550–1700	PLSR	$R^2_{cv} = 0.67$	Craigie et al. (2017)
Chemical	SFA	550–1700	PLSR	$R^2_{cv} = 0.68$	Craigie et al. (2017)
Chemical	MUFA	550–1700	PLSR	$R^2_{cv} = 0.70$	Craigie et al. (2017)
Chemical	FPUFA	550–1700	PLSR	$R^2_{cv} = 0.53$	Craigie et al. (2017)

Category	Parameter	Wavelength	Method	Value	Reference
Tecnological	pH	900–1700	PLSR	$R_{cv}^2 = 0.65$	Kamruzzaman, ElMasry, et al. (2012b)
Tecnological	pH	400–863	PCA. SVM	$R_{cv}^2 = 0.38$	Qiao, Ren, et al., 2015
Tecnological	pH	550–1700	PLSR	$R_{cv}^2 = 0.71$	Craigie et al. (2017)
Tecnological	MIRINZ SF	400–863	PCA. SVM	$R_{cv}^2 = 0.41$	Qiao, Ren, et al., 2015
Tecnological	WBSF	900–1700	PLSR. MLR. SPA	$R_{cv}^2 = 0.84$	Kamruzzaman, ElMasry, et al. (2013)
Tecnological	WBSF	400–1000	PLSR	$R_{cv}^2 = 0.89$	Wang et al. (2016)
Tecnological	WHC	900–1700	PLSR	$R_{cv}^2 = 0.77$	Kamruzzaman, ElMasry, et al. (2012b)
Tecnological	WHC	400–1000	PLSR. LS-SVM	$R^2 = 0.92$	Kamruzzaman et al. (2016b)
Adulteration	Minced lamb meat	900–1700	PCA. PLSR. MLR	$R_{cv}^2 = 0.98$	Kamruzzaman, Sun, ElMasry, and Allen (2013)
Adulteration	Red-meat products	548–1701	CNN	94.4%	Al-Sarayreh, Reis, Qi Yan, and Klette (2018)
Discrimination	Raw meat	900–1700	PCA. PLS-DA	98.7%	Kamruzzaman, Barbin, ElMasry, Sun, and Allen (2012)
Discrimination	LM, PM, ST, Sm	380–1028	PCA. LMS	96.7%	Sanz et al. (2016)
Discrimination	LM, PM, ST	900–1700	PCA. LDA	100.0%	Kamruzzaman, ElMasry, Sun, and Allen (2011)
Discrimination	Raw meat	1000–2500	LDA	100.0%	Qiao, Peng, Wei, and Li (2015)
Discrimination	Raw meat	1000–2500	LDA	87.5%	Qiao, Peng, Chao, and Qin (2016)

Abbreviations: LS-SVM, least square support vector machine; MLR, multiple linear regression; PLS, partial least square regression; PLS-DA, partial least square discrimination analysis; PCA, principal component analysis; R_c, correlation coefficient in the calibration set; R_c^2, determination coefficient in the calibration set; R_{cv}, correlation coefficient of cross validation; R_{cv}^2, determination coefficient of cross validation; Rp, correlation coefficient in the prediction set; Rp2, determination coefficient in the prediction set; SVM, support vector machine; CNN, convolution neural networks; WHC, water holding capability; IMF%, intramuscular fat percentage; SFA, saturated fatty acid; MUFA, monounsaturated fatty acid; PUFA, polyunsaturated fatty acid; WBSF, Warner-Bratzler shear force; MIRINZ SF, MIRINZ shear force; LM, longissimus thoracis et lumborum muscle; PM, Psoas major; Sm, semimembranosus; ST, semitendinosus.

Pu et al., 2014). The results show the feasibility of HSI to yield a good accuracy to predict protein, water and fat (R^2 range from 0.80 to 0.95). More recently, Kamruzzaman et al. (2016a, 2016b) using a HSI system (400–1000 nm) to predict water in lamb, but also in beef and pork meat in an industrial environment, showed that the best model explains 97% of the water variations in meat. This result confirms the HSI as a suitable technology for the prediction of water content in red meat.

When it comes to lamb, relatively few studies have been carried out using HSI for IMF%, and fatty acid content prediction. In a recent work, Craigie et al. (2017) used HSI (550–1700 nm) for simultaneous prediction of the IMF% and content of 34 fatty acids. Results demonstrated that HSI has the potential for predicting fatty acids content of lamb LM muscle samples, where cross-validated R^2 values ranging from 0.03 for lignoceric acid (C24:0) to 0.70 for oleic acid (C18:1c9). The prediction accuracy for IMF% ($R^2_{cv} = 0.67$), saturated fatty acid—SFA ($R^2_{cv} = 0.68$), monounsaturated fatty acid—MUFA ($R^2_{cv} = 0.70$), polyunsaturated fatty acid—PUFA ($R^2_{cv} = 0.53$) content in LM muscle. The encouraging results of this work allowed Craigie et al. (2017) to anticipate the possibility of improving the robustness of the technology for objective, rapid non-invasive assessment of lamb meat quality in a meat processing plant environment.

Besides sensory and chemical attributes, research on lamb meat quality using HSI has also been addressed to predict technological attributes such as pH, shear force and water holding capability (WHC). The pH has a significant influence on the storage and quality of red meat by affecting its WHC and color (Povše, Čandek-Potokar, Gispert, & Lebret, 2015). The models found for the pH showed reasonable prediction performance (R^2_{cv} between 0.38 and 0.71).

Warner–Bratzler shear force (WBSF) is the most common mechanical method to measure meat tenderness objectively. This trait is associated with juiciness and flavor of meat, which lead consumers to accept paying more for tender meat (Schulze-Ehlers & Anders, 2018). The HIS was utilized to predict instrumental (WBSF and MIRINZ SF) tenderness of lamb meat (Kamruzzaman, ElMasry, et al., 2013; Qiao, Ren, et al., 2015; Wang et al., 2016). Reasonable accuracy was obtained from HIS coupled with PLSR models to predict WBSF (R^2_{cv} of 0.84 and 0.89), but for MIRINZ SF the result was less predictable ($R^2 = 0.41$).

Regarding WHC, Kamruzzaman, ElMasry, et al. (2012b) applied a HIS using 237 wavelengths and a PLSR model which could explain 77% of the variation of that quality attribute. In more recent work,

Kamruzzaman et al. (2016a) tested effective wavelengths to be used in the design of a multispectral system for online monitoring of WHC in red meats including beef, lamb, and pork. With the approach in this work, a good accuracy was achieved ($R^2 = 0.92$).

Over the last few years, several non-destructive technologies have been developed to predict meat adulteration (Kamruzzaman, Makino, & Oshita, 2015). However, the non-destructive detection and quantification of adulteration in minced lamb remains a challenge. To solve this problem, HSI was used to detect the level of adulteration in minced lamb meat (Kamruzzaman, Sun, et al., 2013). In this work, the minced lamb meat was adulterated with minced pork, kidney, heart and lung around 2–40%, in increments of approximately 2%. Although it has not been possible to recognize the degree of adulteration in the different samples using visual evaluation in their RGB images, this adulteration is clearly distinguished in the final prediction maps with a linear color scale resulting from the HSI analysis (Fig. 3).

The HSI can also detect the adulteration of meat taking into account its state (fresh, frozen, thawed, and packing/unpacking). Using this, Al-Sarayreh et al. (2018) reported that the best model performance shows a 94.4% overall classification accuracy independent of the state of the products.

The HSI has also shown to be very accurate in discriminating different muscles and meats of different species. Using HSI and multivariate analysis, Sanz et al. (2016) and Kamruzzaman et al. (2011) showed that it is possible to discriminate correctly 96.7% and 100%, respectively, of four (LM, PM, ST, Sm) and three (LM, PM, ST) muscles of lamb carcasses. For the meat of

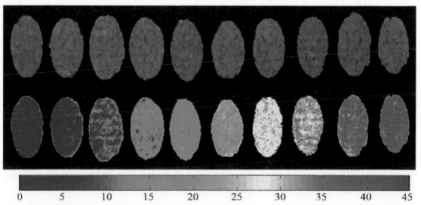

Fig. 3 Prediction maps with a linear color scale resulting from the HSI analysis for adulteration distinction.

different species (lamb, beef and pork) several works (Kamruzzaman, Barbin, et al., 2012; Qiao et al., 2016; Qiao, Peng, et al., 2015) clearly showed that the combination of HSI and multivariate analysis was accurate on the identification and authentication of red meat species. The accuracy yielded 87.5–100% of the overall classification.

2.6 Use of Raman Spectroscopy

Among the several spectroscopic technologies available, Raman spectroscopy has been in recent years the focus of particular interest to applications in meat quality assessment (Fowler, Schmidt, Scheier, & Hopkins, 2018; Kucha, Liu, & Ngadi, 2018). Raman spectroscopy is based on the inelastic scattering of light that occurs when a sample is exposed to a high energy monochromatic light beam (e.g., a laser), which interacts with the sample molecules (Motoyama, 2017; Qin, Kim, Chao, & Cho, 2017). The irradiation of molecular structures by laser light stimulates different molecules in a way that makes possible their measurement. In fact, the difference between the wavelength of the light source and the wavelength of diffuse light that is related to the presence of specific molecules and functional groups can be portrayed as spectral fingerprints (Herrero, 2008). Raman spectroscopy can, therefore, be used to measure the amount or concentration of a chemical constituent in a sample. This technique has been widely used as an analytical tool in many research fields ranging from archeology, forensics, biomedical, and food sciences (Fowler et al., 2018). In the latter area, several authors (Damez & Clerjon, 2008; Fowler et al., 2018) pointed out that Raman spectroscopy is of particular interest to applications in meat quality assessment. In this section, we will examine the Raman spectroscopy to predict meat quality traits of sheep and goat.

The Raman spectroscopy applied in the evaluation of meat characteristics has been mainly oriented to pork meat (Bauer, Scheier, Eberle, & Schmidt, 2016; Olsen, Rukke, Flåtten, & Isaksson, 2007; Scheier, Scheeder, & Schmidt, 2015; Scheier & Schmidt, 2013). However, there are also several studies in which Raman spectroscopy was applied to sheep (Table 10).

For goat, very little research has been conducted with the Raman spectroscopy, only one report related with determination of the origin of meat was published (Boyaci et al., 2014) and no studies have been conducted to investigate this technique to predict meat and meat eating quality on this species. For sheep most of the works aimed at the prediction of meat quality attributes such as shear force (related with tenderness), cooking losses, color and pH (Fowler et al., 2018).

Table 10 Summary of Applications of Raman Spectroscopy for Evaluating Quality Attributes of Lamb Meat

Quality Attributes	Unit	n	Muscle	Aging time (days)	Multivariable Analysis	R^2	R^2_{cv}	RMSE	RMSECV	Reference
Shear force	N	70	LM	1	PLSR	0.79		0.11	0.31	Schmidt, Scheier, and Hopkins (2013)
Shear force	N	70	LM	Day 1		0.86		0.10	0.26	Schmidt et al. (2013)
Cooking loss	%	70	LM	Day 1		0.79		3.20	0.09	Schmidt et al. (2013)
Cooking loss	%	70	LM	Day 1		0.83		0.03	0.08	Schmidt et al. (2013)
Shear force	N	80	LM	Day 1	PLSR		0.06	13.60		Fowler, Schmidt, van de Ven, Wynn, and Hopkins (2014a)
Shear force	N	80	LM	Day 5				10.00		Fowler et al. (2014a)
Shear force	N	80	SM	Day 1	PLSR		0.27		11.48	Fowler, Schmidt, van de Ven, Wynn, and Hopkins (2014b)
Shear force	N	81	SM	Day 5			0.17		12.20	Fowler et al. (2014b)
pH 24		80	SM	Day 1	PLSR		0.48		0.12	Fowler, Schmidt, van de Ven, Wynn, and Hopkins (2015)
pHu		80	SM	Day 1			0.59		0.07	Fowler, Schmidt, et al. (2015)
Purge loss	%	80	SM	Day 1			0.42		0.90	Fowler, Schmidt, et al. (2015)

Continued

Table 10 Summary of Applications of Raman Spectroscopy for Evaluating Quality Attributes of Lamb Meat—cont'd

Quality Attributes	Unit	n	Muscle	Aging time (days)	Multivariable Analysis	R^2	R^2_{cv}	RMSE	RMSECV	Reference
L*		80	SM	Day 1			0.32		1.96	Fowler, Schmidt, et al. (2015)
Purge loss	%	80	SM	Day 5			0.33		0.94	Fowler, Schmidt, et al. (2015)
L*		80	SM	Day 5			0.22		1.87	Fowler, Schmidt, et al. (2015)
PUFA	mg/100 g	80	LM	Day 1	PLSR	0.93	0.21		46.57	Fowler, Ponnampalam, Schmidt, Wynn, and Hopkins (2015)
MUFA	mg/100 g	80	LM	Day 1		0.54	0.16		400.30	Fowler, Ponnampalam, et al. (2015)
SFA	mg/100 g	80	LM	Day 1		0.08	0.01		358.72	Fowler, Ponnampalam, et al. (2015)
IMF	mg/100 g	80	LM	Day 1		0.08	0.02		1.12	Fowler, Ponnampalam, et al. (2015)
PUFA:SFA		80	LM	Day 1		0.21	0.13		0.06	Fowler, Ponnampalam, et al. (2015)

Abbreviations: N, Newton; LM, longissimus thoracis et lumborum muscle; SM, semimembranous muscle; PLSR, partial least squares regression; R^2, coefficient of determination; R^2_{cv}, coefficient of determination for cross validation; RMSE, root mean square error; RMSECV, root mean square error of validation; PUFA, polyunsaturated fatty acid; MUFA, monounsaturated fatty acid; SFA, saturated fatty acid; IMF, intramuscular fat.

The prediction of lamb meat quality attributes using Raman spectroscopy is generally significant, but there is some inconsistency in the results of the studies. For example, Fowler et al. (2014a) found that Raman spectroscopy has no ability to predict shear force in lamb LM ($R^2_{cv} = 0.06$), whereas Schmidt et al. (2013) obtained determination coefficients (R^2) of 0.79 and 0.86 to that trait from two measurement sites of the lamb LM muscle. In another study, Fowler et al. (2014b) found that the use of Raman spectra allowed a more accurate prediction of Semimembranosus (SM) muscle shear force (reduction in RMSE of 12.9% and 7.6%, for muscle aging during 1 and 5 days post-mortem, respectively) than for models using traditional predictors like cooking loss, sarcomere length, pHu and particle size.

Although all those studies used the same handheld Raman spectroscopy device and analyzed the LM and SM muscles of lambs with similar weight, the above-mentioned studies differ in some aspects of methodology that may explain the differences found. In fact, as in the work of Fowler et al. (2014a), Raman spectroscopy measurements were performed in fresh intact muscle samples while in the work of Schmidt et al. (2013) the measurements were taken after the muscle had been frozen and thawed. Some authors (Herrero, 2008; Li-Chan, 1996) stated that Raman spectroscopy is sensitive to the changes associated with freezing and thawing of meat, which may explain the differences. To address this problem, Fowler, Ponnampalam, et al. (2015) carried out a study in which they predicted meat quality traits in two experiments, one with fresh and other with freezing/thawing lamb SM muscle samples, and it was concluded the unability of Raman spectroscopy to predict shear force values in both experiments. However, the prediction was possible for other meat quality attributes (R^2 from 0.22 to 0.59).

In addition to the prediction of the meat quality attributes, the Raman spectroscopy is also able to classify samples of several species. For example, Beattie, Bell, Borggaard, Fearon, and Moss (2007) applied a combination of Raman spectroscopy with multivariate and neural network analytical methods, and reported an accuracy between 96.7% and 99.6% of classifying the fat from chicken, beef, lamb and pork species. This discrimination was also reported by Boyaci et al. (2014) working with Raman spectroscopy and using the principal component analysis (PCA). They successfully classified fat samples of seven different meat species (cattle, sheep, pig, fish, poultry, goat and buffalo). In addition to this fat classification attributes, the Raman spectroscopy was also suitable to predict the concentrations of the major fatty acids groups like PUFA, MUFA and SFA, as well IMF (Fowler, Ponnampalam, et al., 2015), without the drawbacks of traditionally methods which involve the extraction and purification of FAs which are costly, destructive, time consuming and requiring large amounts of chemicals and extensive sample preparation. The Raman spectroscopy was also considered valid to discriminate between tough and tender fresh lamb SM muscles using the intensity of spectral peaks that correspond to the tyrosine doublet at 826 and 853 cm^{-1} and α-helix at 930 cm^{-1} (Fowler et al., 2014b).

Most of the work performed with sheep used a handheld Raman spectroscopic device where its sensor head is in a robust waterproof casing (Fig. 4). This device was developed in Germany (Schmidt, Sowoidnich, Maiwald, Sumpf, & Kronfeldt, 2009) and was described as a better versatile approach for application in the meat industry than a benchtop instrument (Craigie et al., 2015; Fowler et al., 2018). Moreover, the fact that this technique requires almost no sample preparation is not influenced by variation in water content and is very fast to analyze a sample, makes Raman spectroscopy suitable for in line use in the meat industry (Fowler et al., 2018; Schmidt et al., 2013).

As for other spectroscopic techniques Raman spectra contain many dependent variables, so multivariate analysis techniques are required for prediction of meat quality attributes. The Partial least squares discriminant analysis performed well for classification (Beattie et al., 2007) and the PLSR has been the most used multivariate analysis method with this technique (Fowler et al., 2018).

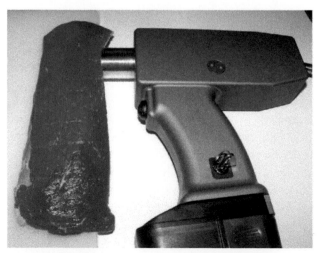

Fig. 4 Application of a handheld Raman spectroscopic device to a meat sample.

When compared with other spectroscopic techniques (NIRS or HSI), there has been relatively limited research with Raman spectroscopy for meat quality prediction and meat classification in sheep and goat species. The inconsistent results did not adequately demonstrate the ability of this technique to assess the meat quality of sheep and goat. However, before Raman spectroscopy can be adopted extensively in the industry, more research is needed to identify the best meat quality attributes to predict and identify factors that are contributing to the differences between predictive models. Nevertheless, as technology advances, it is expected that Raman spectroscopy is one of the techniques of choice for evaluating meat quality.

3. NEW SHEEP AND GOAT MEAT PRODUCTS

Small ruminants' meats are traditionally worldwide consumed and particularly in Mediterranean countries. Consumers prefer young or light sheep and goats animals' meat (Risvik, 1994; Rodrigues & Teixeira, 2010), characterized as tenderer when compared to older or heavier animals (Rodrigues & Teixeira, 2009). This type of meat is consumed as fresh meat, mainly roasted or grilled.

Meat from older and heavier animals has very low acceptability and market value, due to its hardness, poor structure and, normally, unpleasant taste and aroma. Occasionally, it is consumed in traditional dishes cooked for long time and very seasoned. Such meat is more suitable to process as drought, cured with salts or smoked meat products (Webb, Casey, & Simela, 2005), and new products can be produced. In the last years, there have been several studies concerning the incorporation of meat from culled sheep and goats in processed products, nuggets (Banerjee et al., 2012; Das, Anjaneyulu, Gadekar, Singh, & Pragati, 2008), dry-cured sheep and goat meat (Costa et al., 2011; de Andrade et al., 2017; Teixeira et al., 2011), *mantas* (Oliveira et al., 2014; Ortega, Chito, & Teixeira, 2016), fermented sausages (Cosenza, Williams, Johnson, Sims, & McGowan, 2003; Nassu, Aparecida, Gonçalves, & Beserra, 2002; Nassu, Gonçalves, & Beserra, 2002; Nassu, Gonçalves, Pereira da Silva, & Beserra, 2003; Stajić, Stanišić, Perunović, Živković, & Žujović, 2011), fresh sausages (Leite et al., 2015; Paulos et al., 2015), cured legs (Pugliese et al., 2009; Sañudo et al., 2016; Stojković et al., 2015; Teixeira et al., 2017; Tolentino, Estevinho, Pascoal, Rodrigues, & Teixeira, 2017; Villalobos-Delgado et al., 2014), pâtés or other patties (Amaral et al., 2015; Dalmás, Bezerra, Morgano,

Milani, & Madruga, 2011; Devatkal, Narsaiah, & Borah, 2010; Dutra et al., 2013; Villalobos-Delgado et al., 2015), or mortadella (Guerra et al., 2011). All those processed products gave added value to meats in view that they would not be well accepted by consumers as raw meat. In this section the physicochemical, sensory and microbiological analysis were made.

3.1 Physicochemical Quality

In the last two decades, several studies evaluating the physicochemical characteristics of those new sheep and goats' products were done. The main physicochemical attributes studied by the several authors are summarized in Table 11.

Studying the effects of ageing, salting and air-drying in goat meat, Teixeira et al. (2011), and comparing sheep and goat meat Ortega et al. (2016), both observed significant differences in meat color, referring that changes of a* and b* reflected the myoglobin oxidation during refrigeration and ageing processes. Also, salting and air-drying affected the C* and H* parameters and meat became darker. Very important for the final product preservation was the water activity (aw) reduction with salting and air-drying processes verified by the referred authors. The ageing process also promoted the toughness reduction, observed by a decrease in shear force. Salted goat and lamb meat obtained from heavier carcasses can have a leverage with the higher dressing yield (Costa et al., 2011). These products have high protein and a low balanced fat content (Costa et al., 2011; Oliveira et al., 2014). Oliveira et al. (2014) also reported the resistance to oxidative processes of both goat and ewe mantas.

Stajić et al. (2011) compared the use of beef vs goat meat in fermented sausages and observed no significant differences in terms of physicochemical parameters between the two variants at the end of production., except for a* value (11.72 beef and 14.15 goat). Similar results were registered by Lu et al. (2014).

Dutra et al. (2013) reported that increasing the substitution of pork meat by meat from adult sheep ham-type pâtés affected only the characteristics related to the increase in the concentration of total heme pigments. Increasing the amount of sheep meat in the formulation resulted in darker pâtés with a more intense red color.

A new sheep meat product (dry-cured lamb leg) was studied by Villalobos-Delgado et al. (2014) and tumbling and ripening time effects were considered. Conclusions from this study were that quality of dry-cured

Table 11 Summary of Principal Studies on Physicochemical Evaluation on the New Products From Sheep and Goat Meats

Attributes	Product	Effects	Reference
L*a*b* color, aw, pH, Water Holding Capacity, DO* (myoglobin), Shear Force	Goat *longissimus thoracis et lumborum, subscapular* and *semimembranosus* muscles	Aging, salting and air-drying	Teixeira et al. (2011)
Moisture, ashes, protein, lipids	Salted goat and lamb meat	Any (only descriptive)	Costa et al. (2011)
pH, aw, protein, moisture, ash, TBARS	Ewe and goat meat cured product: manta	Species	Oliveira et al. (2014)
pH, aw, Water Holding Capacity, texture, L*a*b* color	Ewe and goat meat cured product: manta	Species, aging, sampling area, salting	Ortega et al. (2016)
Aw, pH, ash, moisture, protein, total fat, fatty acids	Sheep and goats' fresh sausages with different levels of pork fat	Species, fat level	Leite et al. (2015)
Moisture, ash, protein, lipid, starch, chloride, emulsion stability, WHC, pH, aw, L*a*b* color	Goat mortadella with different levels of fat and goat meat from discarded animals	Lipid %	Guerra et al. (2011)
Moisture, NaCl, aw, protein, fat, ash, pH, free fatty acids, water soluble nitrogen, volatile compounds	Dry-cured lamb leg	Tumbling, time	Villalobos-Delgado et al. (2014)
Water, protein, lipids, ash contents, pH, aw, TBARS, heme pigments, L*a*b* color	Cooked ham-type pâté elaborated with sheep meat	% of adult sheep meat to replace pork meat	Dutra et al. (2013)
pH, aw, moisture, protein, lipids, L*a*b* color, nitrites	Lamb pâté prepared with "variety meat"	Storage time and packaging	Amaral et al. (2015)
pH, aw, moisture, NaCl, L*a*b* color, protein, collagen, fat, cholesterol, fatty acids, TBARS	Sheep and goat cured legs	Species, salting, ripening	Teixeira et al. (2017)

Continued

Table 11 Summary of Principal Studies on Physicochemical Evaluation on the New Products From Sheep and Goat Meats—cont'd

Attributes	Product	Effects	Reference
Moisture, ash, lipids, protein, carbohydrates, acidity, pH, aw, L*a*b* color	Lamb pâté prepared with "variety meat"	Different formulations (goat meat 20%)	Dalmás et al. (2011)
TBARS, proteincarbonyls, pH, L*a*b* color	Lean lamb patties	Addiction of hop (infusion or powder), storage period	Villalobos–Delgado et al. (2015)
Weight loss, pH, non-protein nitrogen content, basic chemical composition, instrumental color	Goat meat fermented sausage (Sucuk)	Species	Stajić et al. (2011)
pH, hardness, springiness, cohesiveness, adhesiveness, L*a*b* color	Sheep meat fermented sausage	Species	Lu, Young, and Brooks (2014)
Moisture, protein, fat, ash, pH, water activity, moisture/protein ratio, TBARS	Goat meat fermented sausage	Natural antioxidant percentage	Nassu et al. (2003)
pH, aw	Sheep and goat cured legs	Species, seasoning time	Tolentino et al. (2017)

lamb legs was mainly affected by processing time: moisture decreased, and protein, fat and ashes contents increased from day 1 to day 71. Also, pH, free fatty acids and water-soluble nitrogen increased with processing time. In a similar product studied by Teixeira et al. (2017) conclusions were that pH (5.7–5.8) and aw (0.87 and 0.83) values, which are comparable to those found by Tolentino et al. (2017), showed that goat and sheep cured legs can be a safe product with shelf life stability in relation to microbial growth. The high protein (46.2% and 38.4%) and low fat (5.3% and 8.7%) percentages show that the effect of salting and ripening processes makes the goat and sheep cured legs an interesting and nutritionally balanced meat product with a low cholesterol content and PUFA/SFA (P/S) and n-6/n-3 ratios within the guidelines recommended by the several world food organizations.

Leite et al. (2015) studying the effect of adding different pork fat levels to sheep and goats' meat fresh sausages found significant differences between species and fat levels addition in all physicochemical attributes. Pork backfat addition modified the fatty acid profile, prompting a significant reduction in the relative percentages of major fatty acids. With the addition of pork fat an increase in oleic and linoleic acids and PUFA/SFA ratio was observed. Authors pointed out that the ratio PUFA n6/n3 is impaired with the addition of pork backfat.

Effect on the quality of lamb pâté prepared with meat having different storage times and packaging was studied by Amaral et al. (2015). These authors observed a significant effect of the storage time, but not of the packing method, on the product stability. In this study, the lamb pâté also lead to a reduction in moisture content, protein, pH, aw, nitrite and redness intensity (a*), and a slight increase in luminosity (L*), yellow intensity (b*), TBARS and texture parameters.

To overcome the disadvantages of using synthetic antioxidants, several studies were made to study the effect of natural antioxidants in sheep and goats' new products. For example, in goat meat fermented sausages Nassu et al. (2003) studied the effect on lipids oxidation of two different levels of a natural antioxidant (Rosemary, 0.025% and 0.05%) and concluded that the formulations containing 0.05% rosemary showed the best characteristics in relation to oxidative stability, with the lowest initial values for TBARS. Antioxidant effects of fruit extracts were evaluated by Devatkal et al. (2010) on goat patties, who concluded that kinnow rind powder (KRP), pomegranate rind powder (PRP) and pomegranate seed powder (PSP) had potential to be used as natural anti-oxidants in meat products. The antioxidant potential of broccoli powder extract (BPE) was evaluated and validated in goat meat nuggets by Banerjee et al. (2012). Hop powder and hop infusion antioxidant activity were studied in lamb patties by Villalobos-Delgado et al. (2015). The addition of hop improved lipid and color stability of raw patties during refrigeration and frozen storage, although the improvement was stronger in refrigerated-stored patties. Similarly, the use of hop appeared to reduce lipid and protein oxidation of cooked patties during refrigerated storage. Hop powder showed higher antioxidant effect than hop infusion.

The overall results indicate that the use of meat from adult or discarded sheep and goats in the preparation of new products could be an alternative to the traditional use of these meats, generating new products with physico-chemical characteristics that comply with the legislative recommendations and exhibit quality and acceptance values similar to those of traditional prod-ucts, while conferring higher added value to the productive sector.

3.2 Sensory Quality

The final objective of a new product is to accomplish demands and expec-tations of consumers. Besides the physicochemical analysis of the new prod-ucts, their organoleptic characteristics and how consumers will accept them should be evaluated. Sensory analysis is the best way to study and predict food products acceptance by consumers. When studying the quality of new sheep and goats' products, several authors made their sensory evalua-tion, by stablishing a sensory profile by trained taste panels and/or evaluating likings or preferences by consumers.

Table 12 summarizes the main sensory studies performed on new meat processed products incorporating sheep and goat meat.

The possibility of the use of goat meat in the production of traditional sucuk (a traditional Turkish style dry-fermented sausage of ground meat, usually beef) was studied by Stajić et al. (2011), and the authors observed no significant differences in cut appearance, color and odor. However, in terms of appearance, texture and taste, evaluated in a 9 points scale from 1 (extremely unacceptable) to 9 (extremely acceptable) assessors gave smaller grades to goat than beef sucuk, but they also refer that those grades were higher than 5. The authors suggested the replacement of goat fat by beef fat, to appease the specific goat flavor, to make the product more acceptable to consumers that may not be used to such flavor.

Villalobos-Delgado et al. (2014) evaluated the effect of tumbling after dry-salting and processing time on the sensory characteristics of dry-cured lamb legs. They found no differences between Control, short tumbling,

Table 12 Summary of Principal Studies on Sensory Evaluation on New Products From Sheep and Goat Meat

Number of Panelists	Attributes	Product	Training	Scale	Reference
12	Flavor (saltiness, bitterness, lamb flavor, pungent flavor); Texture (harness, juiciness, pastiness)	Dry-cured lamb leg	Previous experience and 2 1-h sessions	1—lowest to 5—highest	Villalobos-Delgado et al. (2014)
9	Appearance, cut appearance, color, odor, texture and taste	Dry-fermented sausages (goat and beef)	Previous experience	1—extremely unacceptable, 9—extremely acceptable	Stajić et al. (2011)
9	Taste, flavor, texture and appearance	Sheep and goat dry-cured legs	Previous experience and specific training	10-cm continuous, left anchor—lowest intensity, right anchor—highest intensity	Tolentino et al. (2017)
10	Appearance (color, fat color, marbling and brightness), aroma (intensity, meat, acid, sweet, rancid, cured), taste (intensity, persistence, meat, rancid, salty, sweet, acid, cured) and texture (hardness, fibers feeling, and juiciness)	Sheep and goat cured legs	Previous experience and specific training	10-cm continuous, left anchor—lowest intensity, right anchor—highest intensity	Teixeira et al. (2017)
8	Appearance (fat yellowness, redness, marbling), texture (fat firmness, hardness, dryness, juiciness), flavor (smoke, garlic, saltiness, bitterness, acidity, mature, cured, metallic, rancid, and soapy), and aroma intensity	Dry-cured sheep hams	Trained	9-point structured scale, using quantitative–descriptive analysis	Stojković et al. (2015)

Continued

Table 12 Summary of Principal Studies on Sensory Evaluation on New Products From Sheep and Goat Meat—cont'd

Number of Panelists	Attributes	Product	Training	Scale	Reference
7	Oxidized flavor intensity (control vs hop infusion vs sodium ascorbate)	Lean lamb patties cooked (refrigerated 4°C 3 days)	Trained two 1-h sessions	Ranking test, 5-point descriptive scale, in which 1 denoted imperceptible oxidized flavor and 5 denoted extremely high oxidized flavor	Villalobos-Delgado et al. (2015)
6	Hop flavor intensity (control vs infusion vs powder)	Freshly cooked lamb patties	Trained one half-hour session	Ranking test	Villalobos-Delgado et al. (2015)
80 potential consumers	Appearance, color, odor, flavor, goat flavor and overall acceptance	Goat mortadella prepared with different levels of fat and goat meat from discarded animals	No training	Nine-point hedonic scale where 1 = dislike extremely, 5 = neither like or dislike, and 9 = like extremely	Guerra et al. (2011)
26 non-trained individuals two sessions	Taste, spicy taste, texture and overall acceptability	Sheep and goat fresh sausages with different levels of pork fat	No training	Unstructured 10cm scale with anchors at the extremities (0 "do not like" to 10 "like very much")	Leite et al. (2015)
30 non-trained individuals	Texture, taste, aroma, appearance, global acceptability	Goat meat fermented sausage with different levels of fat addition	No training	Hedonic nine-points scale from dislike extremely to like extremely	Nassu, Aparecida, et al. (2002)
50 non-trained panelists	Overall acceptance	Cooked ham-type pâté elaborated with sheep meat	No training	Nine-point hedonic scale (1 "less preferred than the reference," 5 "preferred equal to the reference" and 9 "preferred more than the reference")	Dutra et al. (2013)

Participants	Attribute	Product	Training	Scale	Reference
375 individuals	Maturation time—price, smoking, sodium reduction	Dry-cured sheep ham	No training	No scale	de Andrade et al. (2017)
375 individuals	Natural antioxidant, smoking, sodium reduction	Sheep meat coppa	No training	No scale	de Andrade et al. (2017)
320 individuals (Chinese, sub-Saharan, Andeans, Spanish)	Overall acceptability	"Cecina"—dry-cured ewe legs with different fatness levels	No training	Hedonic nine-points scale from dislike extremely to like extremely	Sañudo et al. (2016)
126 consumers (58 students and the staff members from the Catering School, 40 students from the Department of Food Hygiene and Technology, 28 staff members from the Instituto de Ganadería de Montaña research centre)	Flavor score/liking	Lean lamb patties	No training	Structured hedonic scale ranging from 1—extremely disliking and 9—extremely liking	Villalobos-Delgado et al. (2015)
60 consumers	Liking of flavor	Sheep meat fermented sausage	No training	9-point category scale ranging 1 (dislike extremely) to 9 (like extremely)	Lu et al. (2014)

ST and long tumbling, LT, except for pastiness, although the authors suggested the higher pastiness of tumbled legs would not be so strong as to be considered a defect of eating quality.

The production process and quality of two different dry-cured sheep hams from Western Balkan countries were studied by Stojković et al. (2015) and significant differences in volatile compounds content and sensory properties between sheep ham produced in different locations (Bosnia and Herzegovina—B&H, and Montenegro—MN) were found. B&H production differed from MN in: duration of smoking (14 vs 7 days), ripening time in air (7 vs 14 days) and additives; B&H hams were added garlic and peppercorn. Sheep ham from MN had a strong smoke flavor (from furans and phenols) and salty taste. B&H sheep hams were rich in sulfuric compounds due to added garlic, with less salty taste. The salting phase of the B&H ham seemed to involve fermentation and may be of critical importance for the safety of low salt dry cured ham production.

Villalobos-Delgado et al. (2015) studied the effect of hop as an antioxidant on lean lamb patties and tested the flavor and oxidized flavor intensity by a trained taste panel. The conclusions were that panelists found control (no antioxidant) patties as those with highest oxidized flavor intensity, showing that hop can be used as an ingredient in patty making to minimize the flavor deterioration through oxidation. Also, patties had higher flavor intensity when hop was used as a powder than when used as infusion or not used at all.

Sensory characteristics of sheep and goat cured legs were studied by Tolentino et al. (2017). Nineteen attributes of appearance, taste, aroma and texture were evaluated, and the conclusions were that significant differences among the treatments were found by the panelists. Goats legs were characterized as harder and less juicy than sheep legs and sheep meat with longer time of cure was the brightest and sheep meat with smaller time of cure the most succulent. Goat meat presented higher values of rancid and acid flavor, and sheep meat submitted to more seasoning time presented the most intense flavor and sheep in less time had the lowest intensity in all the attributes of taste. The sensory profile and the differences between sheep and goat cured legs were also evaluated later by Teixeira et al. (2017). These authors showed that the attribute that discriminates the best among sheep and goats cured legs are related to texture: hardness, followed by juiciness, adhesiveness and fibers feeling, and the least discriminating between one product and another is the acid taste, aroma intensity and marbling. Results from both studies indicate sheep and goat cured legs as tender, juicy, with median tastes and aromas, and with a dark color.

A sensory evaluation throughout acceptance and purchase intention of goat mortadella with different fat and goat meat percentages by 80 potential consumers was performed by Guerra et al. (2011). Their results showed that an acceptable value–added goat meat product can be produced, considering that all the formulations were accepted and sensory evaluated. The goat mortadella with less percentage of fat and more goat meat was the preferred choice of the consumers for all the studied sensory attributes, except for texture.

Lu et al. (2014) compared the sensory characteristics of fermented, cured sausages made from equivalent muscle groups of beef, pork, and sheep meat. They stated that the last had no commercial examples and represented an unexploited opportunity once have observed no significant differences between species in mean texture (hardness, springiness, adhesiveness, cohesiveness) following anaerobic fermentation (96 h, 30 °C), and only minor differences were observed in color. However, the same authors referred that although not consumer tested, it is argued that consumers would be able to pick a texture difference due to different fat melting point ranges, highest for sheep meat. Lu et al. (2014) also performed a sensory evaluation to understand if the peculiar sheep meat flavor could be covered or even eliminated to please consumers unused with this type of product. They simulated a very strong characteristic producing a mixed sheep meat and beef sausage, spicing it, or not, with 4-methyloctanoic, 4-methylnonanoic acid, and skatole (5.0, 0.35, and 0.08 mg/kg, respectively). They also, variably, added sodium nitrite (at 0.1 g/kg) and a garlic/rosemary flavor. Spiked sheep meat flavor caused an overall significant decrease from 5.83 to 5.35 in mean liking on a 1–9 scale using 60 consumers, but an increase from 5.18 to 6.00 was observed when garlic/rosemary was added (Lu et al., 2014). Nitrite had no effect on liking (5.61 vs 5.58). Conclusions suggested that "sheep meat flavor could be suppressed to appeal to unhabituated consumers. Commercial examples could thus be made for these consumers, but the mandatory use of the name 'mutton' in some markets would adversely affect prospects."

When writing about fermented products, particularly fermented sausages, the use of starter cultures should be considered. Everson, Danner, and Hammes (1970) had referred that the right physiologically active starter culture would improve the uniformity of fermented products in terms of flavor, appearance and texture. Sensory evaluation of fermented mutton sausage, using *Pediococcus acidilactici* H and *Lactobacillus plantarum* 27 as starter cultures, had shown acceptable scores after 60 days of storage at 4 °C (Wu, Rule, Busboom, Field, & Ray, 1991). The use of different starter cultures in the processing of goat meat fermented sausages by Nassu, Gonçalves, and Beserra (2002) produced average values between 5.5 and 5.9 for global sensory acceptability, using a 9 points hedonic scale. Global acceptability, aroma, taste and texture mean values presented no significant differences for all treatments, but appearance had the smallest value when the treatment with SPX (*Staphylococcus xylosus* and *Pd. pentosaceus*) culture was used. The authors referred the use of isolated observations from the judges as "rancid," "soap," which can be attributed to the products fat oxidation or even to the lipolytic action of microorganisms present in the cultures used. Using lactic starter cultures of *Lb. casei*, *Lb. plantarum* and *Pd. pentosaceus* (Mukherjee, Chowdhury, Chakraborty, & Chaudhuri, 2006) studied the effect of fermentation and drying temperature on the characteristics of goat meat (Black Bengal variety) dry sausages. Results were that the samples fermented at 30 °C, followed by drying at 10 °C, were the most acceptable samples regarding sensory characteristics as taste, flavor, texture and overall acceptability in a 5 points hedonic scale.

Consumers and processors are concerned about the safety of synthetic food additives, as some products like synthetic antioxidants used to mask or improve sensory characteristics can have health implications. So, a renewed interest in natural antioxidants and its research has increased. The use of additives in fermented sausages can improve sensory characteristics, as registered by Nassu et al. (2003). In a study on using goat meat in processing of fermented sausage, salami type, they observed that the incorporation of rosemary minimized oxidized goat aroma and flavor. Also, Paulos et al. (2015) observed that the use of paprika had an influence on the presence and intensity of flavor, spiciness, and off-odor in sausages made from heavy sheep and goat meat when studying their sensory characteristics. Sausages without paprika presented higher spicy intensity, flavor intensity, and off-flavor than sausages with paprika, which had higher odor intensity and sweetness. Paprika masks the less pleasant sensory characteristics of this type of meat. Related to species, these authors found that goat sausages were harder and more fibrous, while sheep sausages were juicier. Besides the effect of additives on sensory characteristics, results of Paulos et al. (2015) show that consumers generally accepted fresh sausages made of sheep and goat meat, with an average of 6 in a scale of 10, and no marked preferences were observed for sheep, goat or seasoning, used to mask some unpleasant characteristics like taste, odor or flavor.

The hedonic test performed by Villalobos-Delgado et al. (2015) on the effect of using hop as an antioxidant in lean lamb patties revealed significant differences in flavor acceptance between powder and both control and infused patties. Control and infused patties flavors were better scored than powder patties flavor. Hop is characterized by an intense smell and a bitter taste for which the numerous components of its essential oil and resin are responsible (Canbaş, Erten, & Özşahin, 2001; Steinhaus, Wilhelm, & Schieberle, 2006).

When processing meat products, fat addition can be an important matter. Verifying the addition of diverse fat contents (5%, 10% and 20%) effect on the sensory acceptance of a goat meat fermented sausage, Nassu, Aparecida, et al. (2002) observed no significant differences in the measured sensory attributes (appearance, aroma, taste, texture and global acceptability) using a 1–9 hedonic scale. Also, no significant differences were found by Dutra et al. (2013) in consumers preferences when studying the effect of meat replacement (pork by adult sheep meat) on cooked ham-type pâté elaborated with sheep meat. However, Leite et al. (2015) refer that overall acceptability was significantly affected by fat level and species of sheep and goat sausages. The goat sausages manufactured with higher fat content presented the highest scores of consumers' preference.

In a cross-cultural study of dry-cured sheep meat acceptability by native and immigrant consumers in Spain, Sañudo et al. (2016) concluded that "'cecina' from cull ewes was well accepted by the consumers independently from their cultural background and even when considering very low consumption rates of lamb. In general, the finishing level (fatness) of the animals was not a criterion to modify overall acceptability, although some groups of consumers prefer lean animals. Given its good overall acceptability and its economic advantages in comparison with the fresh product, it is evident that the production of 'cecina' would have a positive economic outcome for breeders".

Results from de Andrade et al. (2017) work showed a significant impact of process parameters on consumer choice for processed sheep meat dry-cured products. These authors reported that consumers have a positive attitude toward salt reduction and the use of natural antioxidants which can stress the development of improved meat processing techniques based on scientific knowledge offering potential benefits for both the meat industry and public health. Also, when understanding consumer perception for this type of products de Andrade et al. (2017) reinforce the need to include consumption contexts, as snack or as an appetizer with friends, which can lead to different results.

Effect of storage time and packaging on the quality of lamb pâté prepared with "variety meat" was studied by Amaral et al. (2015). Conclusions from this study were that "considering the trend toward the use of edible by-products, the preparation of pâtés is a viable alternative to add value to lamb 'variety meat', as well as providing greater profitability. As for stability, the lamb pâtés showed good microbiological quality being fit for human consumption, therefore met the requirements of the legislation, however, its shelf life was limited by decrease in sensory attributes texture and overall impression."

Generally, studies based on new sheep and goat meat products lead to the conclusion that such unique processed meat products, especially the ones with healthy claims, could represent opportunities for the sheep meat chain in general. Producers, industry and society would benefit from the development of better quality products that can satisfy modern market demands. Further research should be conducted to study how consumers perceive the sensory and hedonic characteristics of these products.

3.3 Food Safety

Food safety is a major concern for consumers and a major concern for the industries. When talking about food safety, microbiological safety is the most discussed, and to assess it several procedures and analysis are performed in order to achieve the safety and quality of food involved to safeguard public health and provide assurance on food safety (Centre for Food Safety, 2007), but microbiological analysis alone cannot guarantee the safety of food and microbiological criteria should be used to support good hygienic practice (GHP), good manufacturing practices (GMP), good agricultural practices (GAP) and implementation of food safety risk management systems such as hazard analysis and critical control point (HACCP) systems (Health Protection Agency, 2009; van Schothorst, Zwietering, Ross, Buchanan, & Cole, 2009). Microbiological safety plays an important role to be taken both by government and food industry for identifying, assessing and managing risks associated with the consumption of food and drink (Stringer, 2005). To accomplish their roles, the authorities can and should follow the recommended stepwise by International Commission on Microbiological Specifications for Foods (ICMSF, 1997), for the management of microbiological hazards in foods in international trade, applying existing Codex documents in a logical sequence (van Schothorst, 1998).

Fermentation and drying have been reported as the oldest methods for food preservation and consequently the consumption of these products by humans dates from immemorial times (Nassu et al., 2003). According to Bourdichon et al. (2012) fermentation plays different roles in food processing and the major roles related to food safety are (1) Preservation of food through formation of inhibitory metabolites such as organic acid (lactic acid, acetic acid, formic acid, propionic acid), ethanol, bacteriocins, etc., often in combination with decrease of water activity (by drying or use of salt) (Gaggia, Di Gioia, Baffoni, & Biavati, 2011; Paul Ross, Morgan, & Hill, 2002); (2) Improving food safety through inhibition of pathogens (Adams & Mitchell, 2002; Adams & Nicolaides, 1997) or removal of toxic compounds (Hammes & Tichaczek, 1994). Recent studies focused on fermented foods, including fermented meat products, proved that these products are an excellent source of microorganisms with probiotic characteristics (Marco et al., 2017). In a study on the effect of curing and fermentation on the microflora of meat of various animal species (cattle, wild boar, deer, goat and horse), the inhibitory effect of aw, pH and the produced lactic acid bacteria (LAB) on the pathogenic bacteria was observed during the fermentation process (Paleari, Bersani, Moretti Vittorio, & Beretta, 2002). In the raw materials, the referred authors noted the normal flora as the presence of *St. aureus* and coliforms in all the samples and none had *Salmonella* or *L. monocytogenes*. Nevertheless, at the final of the fermentation process, an increase of LAB that exerts an antagonistic action on contaminating flora was noted.

Dominant volatiles in MN sheep hams were smoke components (furans and phenols), acids, aldehydes, alcohols, and esters (Stojković et al., 2015). This fact suggests that smoking process could be better controlled in an industrial facility. Furans and derivates are classified by the International Agency for Research on Cancer (IRAC, 1995) as possible carcinogenic compounds. Increased attention is paid to their derivates (furan, 2-furfural, furfuryl alcohol, and penthylfuran), which are considered toxic for human and animal health (Perez-Palacios, Petisca, Pinho, & Ferreira, 2012).

The final content of 4.5% NaCl (or lower) in sheep ham is challenging in terms of unwanted bacterial growth, but preferable regarding health recommendations (Stojković et al., 2015). Salt content depends on production practices; for Spanish lamb ham Villalobos-Delgado et al. (2014) reported 7.96% NaCl and water activity (aw) of 0.88, for dry cured legs Teixeira et al. (2017) found values of 3.8% NaCl and aw of 0.83 on goats, and 4.7% and 0.87 on sheep, after salting and ripening processes. The pH and aw values found showed that processing could have an important role in controlling the meat spoiling promoting safety and shelf life stability of the products with respect to microbial growth. For Norwegian Fenalår (dry-cured lamb or sheep leg) 5–10% NaCl was registered (Håseth, Thorkelsson, Puolanne, & Sidhu, 2014). Skerpikjøt (air-dried and unsalted lamb meat product) with aw value 0.90 has no salt addition (Håseth et al., 2014), which was linked directly to incidences of botulism (Stojković et al., 2015).

4. CONCLUSIONS AND FUTURE TRENDS

The growing demand for sheep and goat meat products, in recent past, has been driven mainly by animal science and technology, especially regarding production systems, slaughter procedures and carcass fabrication and grading, meat processing, food quality and safety as well consumer preferences and satisfaction.

Technological developments will continue to contribute to a higher efficiency of procedures and better quality in sheep and goat meat products. Because of the importance of carcass composition, ultrasonic techniques, optical and spectroscopy methods or vision image analysis systems have been and will be devised to assess it from rapid and single measurements and to use more sophisticated procedures. For research purposes, some of these techniques have been used to assess body composition in both the live animal and the carcass in sheep and goats and will continue to be studied in the coming years with redoubled interest from the producer, abattoir, retailer and meat industry. Particularly the use of computed tomography and computer vision as alternative methods to carcass dissection should be further studied and its applicability confirmed and officially recognized by the competent livestock entities. The research on the use of non-invasive and non-destructive technologies like NIRS, Hyperspectral Imaging or Raman Spectroscopy to predict food composition, to monitoring composition and physicochemical food properties during processing will be increasingly important for a higher performance in processing sheep and goat meat products. The development of general, robust, and more reliable models to swiftly assess sheep and goat carcass and body compositions should be developed in the near future, as well as the implementation of a modern and objective on-line technique for carcass evaluation and marketing classification based in the electronic technologies associated to new statistical analysis methods. The use of these technologies will be more and more important once offers the possibility to enhance the quality of processing and to improve consumer safety. However, more research should be developed to optimize and promote the use of different technologies applications in sheep and goat meat processing, reducing costs, saving energy and being environmentally friendly and at the same time ensuring the quality and food safety.

The rediscovery of a new generation of meat processed products based in sheep or goat meat associated with a developing of functional foods will be an interesting food research field meeting the new deals of meat industry.

Eating quality is also an important factor in food choice especially if it is a new sheep and goat processed product. If it looks attractive, smells good and tastes good the consumer decision to purchase is stimulated. Consequently, sensory quality studies supported by trained and consumer panels associated with different statistical analysis methodologies will always be a priority. At the same time the studies on the nutritional value of sheep and goat products should be improved contributing to product differentiation in the meat market and to add value across the meat chain, giving better information to consumers and enhancing goat and sheep meat properties of wholesome food. Additionally, everything concerning food safety should be considered in future, particularly the increasing importance of traceability and the consumer request of detailed information (from the farm to the fork) corresponding to their expectations. Quality and meat safety will undoubtedly be important issues with increasing concerns in the next future and research should be awarded of this fact.

REFERENCES

Adams, M., & Mitchell, R. (2002). Fermentation and pathogen control: A risk assessment approach. *International Journal of Food Microbiology, 79*(1), 75–83.

Adams, M. R., & Nicolaides, L. (1997). Review of the sensitivity of different foodborne pathogens to fermentation. *Food Control, 8*(5), 227–239.

Agamy, R., Moneim, A. Y., Alla, M. S., Mageed, I. I., & Ashmawi, G. M. (2015). Use of ultrasound measurements to predict carcass characteristics of Egyptian ram-lambs. *Asian Journal of Animal and Veterinary Advances, 10*(5), 203–214.

Allen, P. (2007). New methods for grading beef and sheep carcasses. In C. Lazzaroni, S. Gigli, & D. Gabiña (Eds.), *Evaluation of carcass and meat quality in cattle and sheep* (pp. 39–48). The Netherlands: Wageningen Academic Publishers. EAAP publications no. 123.

Al-Sarayreh, M. M., Reis, M., Qi Yan, W., & Klette, R. (2018). Detection of red-meat adulteration by deep spectral–spatial features in hyperspectral images. *Journal of Imaging, 4*, 1–20.

Amaral, D. S., Silva, F. A. P., Bezerra, T. K. A., Arcanjo, N. M. O., Guerra, I. C. D., Dalmás, P. S., et al. (2015). Effect of storage time and packaging on the quality of lamb pâté prepared with 'variety meat'. *Food Packaging and Shelf Life, 3*, 39–46.

Anderson, F., Pethick, D. W., & Gardner, G. E. (2014). The correlation of intramuscular fat content between muscles of the lamb carcass and the use of computed tomography to predict intramuscular fat percentage in lambs. *Animal, 9*(7), 1239–1249.

Andrés, S., Murray, I., Navajas, E. A., Fisher, A. V., Lambe, N. R., & Bünger, L. (2007). Prediction of sensory characteristics of lamb meat samples by near infrared reflectance spectroscopy. *Meat Science, 76*, 509–516.

Banerjee, R., Verma, A. K., Das, A. K., Rajkumar, V., Shewalkar, A. A., & Narkhede, H. P. (2012). Antioxidant effects of broccoli powder extract in goat meat nuggets. *Meat Science,* *91*(2), 179–184.

Bauer, A., Scheier, R., Eberle, T., & Schmidt, H. (2016). Assessment of tenderness of aged bovine gluteus medius muscles using Raman spectroscopy. *Meat Science, 115,* 27–33.

Beattie, J. R., Bell, S. E. J., Borggaard, C., Fearon, A. M., & Moss, B. W. (2007). Classification of adipose tissue species using Raman spectroscopy. *Lipids, 42,* 679–685.

Ben-Gera, I., & Norris, K. H. (1968). Direct spectrophotometric determination of fat and moisture in meat products. *Journal of Food Science, 33,* 64–67.

Berg, E. P., Forrest, J. C., Thomas, D. L., Nusbaum, N., & Kauffman, R. G. (1994). Electromagnetic scanning to predict lamb carcass composition. *Journal of Animal Science, 72,* 1728–1736.

Berg, E. P., Neary, M. K., Forrest, J. C., Thomas, D. L., & Kauffman, R. G. (1997). Evaluation of electronic technology to assess lamb carcass composition. *Journal of Animal Science, 75*(9), 2433–2444.

Bourdichon, F., Casaregola, S., Farrokh, C., Frisvad, J. C., Gerds, M. L., Hammes, W. P., et al. (2012). Food fermentations: Microorganisms with technological beneficial use. *International Journal of Food Microbiology, 154*(3), 87–97.

Boyaci, İ. H., Uysal, R. S., Temiz, T., Shendi, E. G., Yadegari, R. J., Rishkan, M. M., et al. (2014). A rapid method for determination of the origin of meat and meat products based on the extracted fat spectra by using of Raman spectroscopy and chemometric method. *European Food Research and Technology, 238*(5), 845–852.

Brady, A. S., Belk, K. E., LeValley, S. B., Dalsted, N. L., Scanga, J. A., Tatum, J. D., et al. (2003). An evaluation of the lamb vision system as a predictor of lamb carcass red meat yield percentage. *Journal of Animal Science, 81,* 1488–1498.

Bünger, L., Macfarlane, J. M., Lambe, N. R., Conington, J., McLean, K. A., Moore, K., et al. (2011). Use of X-ray computed tomography (CT) in UK sheep production and breeding. In S. Karuppasamy (Ed.), *CT scanning—Techniques and applications* (pp. 329–348). Rijeka, Croatia: INTECH Open access Publisher.

Bunger, L., Menezes, A. M., McLean, K. A., Gordon, J., Yates, J., Moore, K., et al. (2015). Selecting terminal sire breed rams for lean meat percentage—Effects on their crossbred lambs. In C. Maltin, C. Craigie, & L. Bünger (Eds.), *Farm animal imaging* (pp. 56–61). Edinburgh: COST Action.

Bünger, L., Moore, K., McLean, K., Kongsro, J., & Lambe, N. (2014). Integrating computed tomography into commercial sheep breeding in the UK: Cost and value. In C. Maltin, C. Craigie, & L. Bünger (Eds.), *Farm animal imaging* (pp. 22–27). Copenhagen: COST Action.

Cadavez, V., Rodrigues, S., & Teixeira, A. (2007). The use of ultrasonography to predict carcass composition in kids. *Agriculture—Scientific and Professional Review, 13*(1), 213–217.

Canbaş, A., Erten, H., & Özşahin, F. (2001). The effects of storage temperature on the chemical composition of hop pellets. *Process Biochemistry, 36*(11), 1053–1058.

Carr, M. A., Waldron, D. F., & Willingham, T. D. (2002). Relationships among weights, ultrasound and carcass characteristics in Boer-cross goats. *Sheep and Goat, Wool and Mohair CPR,* 55–59.

Centre for Food Safety. (2007). *Microbiological guidelines for ready-to-eat food.* Revised Hong Kong: Food and Environmental Hygiene Department.

Cheng, J. H., Nicolai, B., & Sun, D. W. (2017). Hyperspectral imaging with multivariate analysis for technological parameters prediction and classification of muscle foods: A review. *Meat Science, 123,* 182–191.

Clelland, N., Bunger, L., McLean, K. A., Conington, J., Maltin, C., Knott, S., et al. (2014). Prediction of intramuscular fat levels in Texel lamb loins using X-ray computed tomography scanning. *Meat Science, 98*(2), 263–271.

Clelland, N., Bunger, L., McLean, K. A., Knott, S., Matthews, K. R., & Lambe, N. R. (2018). Prediction of intramuscular fat content and shear force in Texel lamb loins using combinations of different X-ray computed tomography (CT) scanning techniques. *Meat Science*, *140*, 78–85.

Cosenza, G. H., Williams, S. K., Johnson, D. D., Sims, C., & McGowan, C. H. (2003). Development and evaluation of a fermented cabrito snack stick product. *Meat Science*, *64*(1), 51–57.

Costa, R. G., de Medeiros, G. R., Duarte, T. F., Pedrosa, N. A., Voltolini, T. V., & Madruga, M. S. (2011). Salted goat and lamb meat: Typical regional product of the city of Petrolina, state of Pernambuco. *Small Ruminant Research*, *98*(1), 51–54.

Craigie, C. R., Fowler, S., Knight, M., Stuart, A., Hopkins, D., & Reis, M. M. (2015). Spectral imaging techniques for predicting meat quality—An Australasian perspective. In C. Maltin, C. Craigie, & L. Bünger (Eds.), *FAIM farm animal imaging* (pp. 75–79). Edinburgh: SRUC. Edinburgh 2015.

Craigie, C. R., Johnson, P. L., Shorten, P. R., Charteris, A., Maclennan, G., Tate, M. L., et al. (2017). Application of hyperspectral imaging to predict the pH, intramuscular fatty acid content and composition of lamb M. longissimus lumborum at 24 h post mortem. *Meat Science*, *132*, 19–28.

Craigie, C. R., Navajas, E. A., Purchas, R. W., Maltin, C. A., Buenger, L., Hoskin, S. O., et al. (2012). A review of the development and use of video image analysis (VIA) for beef carcass evaluation as an alternative to the current EUROP system and other subjective systems. *Meat Science*, *92*, 307–318.

Cunha, B. C. N., Belk, K. E., Scanga, J. A., LeValley, S. B., Tatum, J. D., & Smith, G. C. (2004). Development and validation of equations utilizing lamb vision system output to predict lamb carcass fabrication yields. *Journal of Animal Science*, *82*, 2069–2076.

Dalmás, P. S., Bezerra, T. K. A., Morgano, M. A., Milani, R. F., & Madruga, M. S. (2011). Development of goat pâté prepared with 'variety meat'. *Small Ruminant Research*, *98*(1), 46–50.

Damez, J.-L., & Clerjon, S. (2008). Meat quality assessment using biophysical methods related to meat structure. *Meat Science*, *80*, 132–149.

Das, A. K., Anjaneyulu, A. S. R., Gadekar, Y. P., Singh, R. P., & Pragati, H. (2008). Effect of full-fat soy paste and textured soy granules on quality and shelf-life of goat meat nuggets in frozen storage. *Meat Science*, *80*(3), 607–614.

Daumas, G., Donkó, T., Maltin, C., & Bünger, L. (2015). *Imaging facilities (CT & MRI) in EU for measuring body composition*. Edinburgh: SRUC.

de Andrade, J. C., Nalério, E. S., Giongo, C., de Barcellos, M. D., Ares, G., & Deliza, R. (2017). Consumer perception of dry-cured sheep meat products: Influence of process parameters under different evoked contexts. *Meat Science*, *130*, 30–37.

Delfa, R. (2004). *Los ultrasonidos como predictores del reparto del tejido adiposo y de la composición tisular de la canal en cabras adultas*. Tesis doctoral, Facultad de Veterinária. Universidad de Zaragoza.

Delfa, R., Teixeira, A., Blasco, I., & Rocher-Colomber, F. (1991). Ultrasonic estimates of fat thickness, C measurement and longissimus dorsi depth in rasa aragonesa ewes with same body condition score. *Options Méditerranéenes—Série Séminaires*, *13*, 25–30.

Delfa, R., Teixeira, A., González, C., Torrano, L., & Valderrábano, J. (1999). Utilización de ultrasonidos en cabritos vivos de raza Blanca Celtibérica, como predictores de la composición tisular de sus canales. *Archivos de Zootecnia*, *48*, 123–134.

Devatkal, S. K., Narsaiah, K., & Borah, A. (2010). Anti-oxidant effect of extracts of kinnow rind, pomegranate rind and seed powders in cooked goat meat patties. *Meat Science*, *85*(1), 155–159.

Donaldson, C. L., Lambe, N. R., Maltin, C. A., Knott, S., & Bunger, L. (2013). Between- and within-breed variations of spine characteristics in sheep. *Journal of Animal Science*, *91*(2), 995–1004.

Donaldson, C. L., Lambe, N. R., Maltin, C. A., Knott, S., & Bünger, L. (2014). Effect of the Texel muscling QTL (TM-QTL) on spine characteristics in purebred Texel lambs. *Small Ruminant Research, 117*(1), 34–40.

Dutra, M. P., Palhares, P. C., Silva, J. R. O., Ezequiel, I. P., Ramos, A. L. S., Perez, J. R. O., et al. (2013). Technological and quality characteristics of cooked ham-type pâté elaborated with sheep meat. *Small Ruminant Research, 115*(1), 56–61.

Einarsson, E., Eythorsdottir, E., Smith, C. R., & Jonmundsson, J. V. (2014). The ability of video image analysis to predict lean meat yield and EUROP score of lamb carcasses. *Animal, 8*, 1170–1177.

Emenheiser, J. C., Greiner, S. P., Lewis, R. M., & Notter, D. R. (2010). Validation of live animal ultrasonic measurements of body composition in market lambs. *Journal of Animal Science, 88*, 2932–2939.

Esquivelzeta, C., Casellas, J., Fina, M., & Piedra, J. (2012). Backfat thickness and longissimus dorsi real-time ultrasound measurements in light lambs. *Journal of Animal Science, 90*, 5047–5055.

Everson, C. W., Danner, W. E., & Hammes, P. A. (1970). Bacterial starter cultures in sausage products. *Journal of Agricultural and Food Chemistry, 18*(4), 570–571.

FAO. (2011). *World livestock 2011—Livestock in food security.* Rome: FAO.

Fernández, C., Gallego, L., & Quintanilla, A. (1997). Lamb fat thickness and longissimus muscle area measured by a computerized ultrasonic system. *Small Ruminant Research, 26*, 277–282.

Fernández, C., Garcia, A., Vergara, H., & Gallego, L. (1998). Using ultrasound to determine fat thickness and longissimus dorsi area on Manchego lambs of different live weight. *Small Ruminant Research, 27*, 159–165.

Fowler, S. M., Ponnampalam, E. N., Schmidt, H., Wynn, P., & Hopkins, D. L. (2015). Prediction of intramuscular fat content and major fatty acid groups of lamb M. longissimus lumborum using Raman spectroscopy. *Meat Science, 110*, 70–75.

Fowler, S. M., Schmidt, H., Scheier, R., & Hopkins, D. L. (2018). Raman spectroscopy for predicting meat quality traits. In F. Toldrá & L. M. L. Nollet (Eds.), *Advanced technologies for meat processing* (2nd ed., pp. 83–112). CRC Press.

Fowler, S. M., Schmidt, H., van de Ven, R., Wynn, P., & Hopkins, D. L. (2014a). Raman spectroscopy compared against traditional predictors of shear force in lamb m. longissimus lumborum. *Meat Science, 98*(4), 652–656.

Fowler, S. M., Schmidt, H., van de Ven, R., Wynn, P., & Hopkins, D. L. (2014b). Predicting tenderness of fresh ovine semimembranosus using Raman spectroscopy. *Meat Science, 97*(4), 597–601.

Fowler, S. M., Schmidt, H., van de Ven, R., Wynn, P., & Hopkins, D. L. (2015). Predicting meat quality traits of ovine m. semimembranosus, both fresh and following freezing and thawing, using a hand held Raman spectroscopic device. *Meat Science, 108*, 138–144.

Fratianni, F., Sada, A., Orlando, P., & Nazzaro, F. (2008). Micro-electrophoretic study of sarcoplasmic fraction in the dry-cured goat raw ham. *The Open Food Science Journal, 2*, 89–94.

Gaggia, F., Di Gioia, D., Baffoni, L., & Biavati, B. (2011). The role of protective and probiotic cultures in food and feed and their impact in food safety. *Trends in Food Science & Technology, 22*, S58–S66.

Glasbey, C., Abdalla, I., & Simm, G. (1996). Towards automatic interpretation of sheep ultrasound scans. *Animal Science, 62*, 309–315.

Goetz, A. F. (2009). Three decades of hyperspectral remote sensing of the Earth: A personal view. *Remote Sensing of Environment, 113*, S5–S16.

Gowen, A. A., O'Donnell, C., Cullen, P. J., Downey, G., & Frias, J. M. (2007). Hyperspectral imaging—An emerging process analytical tool for food quality and safety control. *Trends in Food Science & Technology, 18*(12), 590–598.

Grill, L., Ringdorfer, F., Baumung, R., & Fuerst-waltl, B. (2015). Evaluation of ultrasound scanning to predict carcass composition of Austrian meat sheep. *Small Ruminant Research*, *123*, 260–268.

Grunert, K. G., Bredahl, L., & Brunsø, K. (2004). Consumer perception of meat quality and implications for product development in the meat sector—A review. *Meat Science*, *66*(2), 259–272.

Guerra, I. C. D., Félex, S. S. S., Meireles, B. R. L. M., Dalmás, P. S., Moreira, R. T., Honório, V. G., et al. (2011). Evaluation of goat mortadella prepared with different levels of fat and goat meat from discarded animals. *Small Ruminant Research*, *98*(1), 59–63.

Guy, F., Prache, S., Thomas, A., Bauchart, D., & Andueza, D. (2011). Prediction of lamb meat fatty acid composition using near-infrared reflectance spectroscopy (NIRS). *Food Chemistry*, *127*, 1280–1286.

Hajji, H., Atti, N., & Hamouda, M. B. (2015). In vivo fat and muscle weight prediction for lambs from fat- and thin-tailed breeds by real-time ultrasonography. *Animal Science Papers and Reports*, *33*, 277–286.

Hamby, P. L., Stouffer, J. R., & Smith, S. B. (1986). Muscle metabolism and real-time ultrasound measurement of muscle and subcutaneous adipose tissue growth in lambs fed diets containing a beta-agonist. *Journal of Animal Science*, *63*, 1410–1417.

Hammes, W., & Tichaczek, P. S. (1994). The potential of lactic acid bacteria for the production of safe and wholesome food. *Zeitschrift fur Lebensmittel-Untersuchung und -Forschung*, *198*(3), 193–201.

Håseth, T. T., Thorkelsson, G., Puolanne, E., & Sidhu, M. S. (2014). Nordic products. In F. Toldrá (Ed.), *Handbook of fermented meat and poultry* (2nd ed., pp. 371–376). Wiley-Blackwell.

Health Protection Agency. (2009). *Guidelines for assessing the microbiological safety of ready-to-eat foods*. London: Health Protection Agency.

Herrero, A. M. (2008). Raman spectroscopy a promising technique for quality assessment of meat and fish: A review. *Food Chemistry*, *107*(4), 1642–1651.

Hierro, E., de la Hoza, L., Juan, A., & Ordóñez, J. A. (2004). Headspace volatile compounds from salted and occasionally smoked dried meats (cecinas) as affected by animal species. *Food Chemistry*, *85*(4), 649–657.

Ho, H., Patoir, A., Hunter, P., Quinn, K., Thomson, A., & Pearson, G. J. (2014). Image and model based virtual cutting of lamb carcasses. In C. Maltin, C. Craigie, & L. Bünger (Eds.), *Farm animal imaging* (pp. 98–101). Copenhagen: COST Action.

Hopkins, D. L. (1994). Predicting the weight of lean meat in lamb carcasses and suitability of this characteristic as a basis for valuing carcasses. *Meat Science*, *38*, 235–241.

Hopkins, D. L. (1996). The relationship between muscularity, muscle: Bone ratio and cut dimensions in male and female lamb carcasses and the measurement of muscularity using image analysis. *Meat Science*, *44*(4), 307–317.

Hopkins, D. L., Gardner, G. E., & Toohey, E. S. (2015). Australian view on lamb carcass and meat quality—The role of measurement technologies in the Australian sheep industry. In C. Maltin, C. Craigie, & L. Bünger (Eds.), *Farm animal imaging* (pp. 17–21). Edinburgh: COST Action.

Hopkins, D. L., Pirlot, K. L., Roberts, A. H. K., & Beattie, A. S. (1993). Changes in fat depths and muscle dimensions in growing lambs as measured by real-time ultrasound. *Australian Journal of Experimental Agriculture*, *33*, 707–712.

Hopkins, D. L., Ponnampalam, E. N., & Warner, R. D. (2008). Predicting the composition of lamb carcases using alternative fat and muscle depth measures. *Meat Science*, *78*, 400–405.

Hopkins, D. L., Safari, E., Thompson, J. M., & Smith, C. R. (2004). Video image analysis in the Australian meat industry—Precision and accuracy of predicting lean meat yield in lamb carcasses. *Meat Science*, *67*, 269–274.

Hopkins, D. L., Stanley, D. F., & Ponnampalam, E. N. (2007). Relationship between real-time ultrasound and carcass measures and composition in heavy sheep. *Australian Journal of Experimental Agriculture*, *47*, 1304–1308.

Houghton, P. L., & Turlington, L. M. (1992). Application of ultrasound for feeding and finishing animals: A review. *Journal of Animal Science*, *70*, 930–941.

Huisman, A. E., Brown, D. J., & Fogarty, N. M. (2016). Ability of sire breeding values to predict progeny bodyweight, fat and muscle using various transformations across environments in terminal sire sheep breeds. *Animal Production Science*, *56*, 95–101.

ICMSF (International Commission on Microbiological Specifications for Foods). (1997). Establishment of microbiological safety criteria for foods in international trade. *World Health Statistics Quarterly*, *50*, 119–123.

IRAC (International Agency for Research on Cancer). (1995). Dry cleaning, some chlorinated solvents and other industrial chemicals. In *IARC monographs on the evaluation of carcinogenic risks to humans*: *Vol. 63*. Lyon: World Health Organization (WHO Press).

Jones, H. E., Lewis, R. M., Young, M. J., & Simm, G. (2004). Genetic parameters for carcass composition and muscularity in sheep measured by X-ray computer tomography, ultrasound and dissection. *Livestock Production Science*, *90*(2), 167–179.

Jones, H. E., Lewis, R. M., Young, M. J., & Wolf, B. T. (2002). The use of X-ray computer tomography for measuring the muscularity of live sheep. *Animal Science*, *75*(3), 387–399.

Jones, S. D. M., Robertson, W. M., Price, M. A., & Coupland, T. (1996). The prediction of saleable meat yield in lamb carcasses. *Canadian Journal of Animal Science*, *76*, 49–53.

Jopson, N. B., Newman, S. A. N., & McEwan, J. C. (2009). Developments in the sheep meat industry: Genetic evaluation of meat yield. *Proceedings of the New Zealand Society of Animal Production*, *69*, 161–164.

Kamruzzaman, M., Barbin, D., ElMasry, G., Sun, D.-W., & Allen, P. (2012). Potential of hyperspectral imaging and pattern recognition for categorization and authentication of red meat. *Innovative Food Science & Emerging Technologies*, *16*, 316–325.

Kamruzzaman, M., ElMasry, G., Sun, D. W., & Allen, P. (2011). Application of NIR hyperspectral imaging for discrimination of lamb muscles. *Journal of Food Engineering*, *104*(3), 332–340.

Kamruzzaman, M., ElMasry, G., Sun, D.-W., & Allen, P. (2012a). Non-destructive prediction and visualization of chemical composition in lamb meat using NIR hyperspectral imaging and multivariate regression. *Innovative Food Science & Emerging Technologies*, *16*, 218–226.

Kamruzzaman, M., ElMasry, G., Sun, D.-W., & Allen, P. (2012b). Prediction of some quality attributes of lamb meat using near-infrared hyperspectral imaging and multivariate analysis. *Analytica Chimica Acta*, *714*, 57–67.

Kamruzzaman, M., ElMasry, G., Sun, D.-W., & Allen, P. (2013). Non-destructive assessment of instrumental and sensory tenderness of lamb meat using NIR hyperspectral imaging. *Food Chemistry*, *141*(1), 389–396.

Kamruzzaman, M., Makino, Y., & Oshita, S. (2015). Non-invasive analytical technology for the detection of contamination, adulteration, and authenticity of meat, poultry, and fish: A review. *Analytica Chimica Acta*, *853*, 19–29.

Kamruzzaman, M., Makino, Y., & Oshita, S. (2016a). Online monitoring of red meat color using hyperspectral imaging. *Meat Science*, *116*, 110–117.

Kamruzzaman, M., Makino, Y., & Oshita, S. (2016b). Hyperspectral imaging for real-time monitoring of water holding capacity in red meat. *LWT—Food Science and Technology*, *66*, 685–691.

Kamruzzaman, M., Sun, D.-W., ElMasry, G., & Allen, P. (2013). Fast detection and visualization of minced lamb meat adulteration using NIR hyperspectral imaging and multivariate image analysis. *Talanta*, *103*, 130–136.

Kempster, A. J., Arnall, D., Alliston, J. C., & Barker, J. D. (1982). An evaluation of two ultrasonic machines (Scanogram and Danscanner) for predicting the body composition of live sheep. *Animal Production, 34,* 249–255.

Kirton, A. H., & Johnson, D. L. (1979). Interrelationships between GR and other lamb carcass fatness measurements. *Proceedings New Zealand Society of Animal Production, 39,* 195–201.

Kongsro, J. (2014). Genetic gain on body composition in pigs by Computed Tomography (CT). Return on investment. In C. Maltin, C. Craigie, & L. Bünger (Eds.), *Farm animal imaging* (pp. 28–30). Copenhagen: COST Action.

Kongsro, J., Røe, M., Aastveit, A. H., Kvaal, K., & Egelandsdal, B. (2008). Virtual dissection of lamb carcasses using computer tomography (CT) and its correlation to manual dissection. *Journal of Food Engineering, 88,* 86–93.

Kruggel, W. G., Field, R. A., Riley, M. L., & Horton, K. M. (1981). Near-infrared reflectance determination of fat, protein, and moisture in fresh meat. *Journal—Association of Official Analytical Chemists, 64*(3), 692–696.

Kucha, C. T., Liu, L., & Ngadi, M. O. (2018). Non-destructive spectroscopic techniques and multivariate analysis for assessment of fat quality in pork and pork products: A review. *Sensors, 18,* 377.

Kvame, T., & Vangen, O. (2006). In-vivo composition of carcass regions in lambs of two genetic lines, and selection of CT positions for estimation of each region. *Small Ruminant Research, 66,* 201–208.

Lambe, N. R., Donaldson, C. L., McLean, K. A., Gordon, J., Menezes, A. M., Clelland, N., et al. (2015). Genetic control of CT-based spine traits in elite Texel rams. In C. Maltin, C. Craigie, & L. Bünger (Eds.), *Farm animal imaging* (pp. 52 55). Edinburgh: COST Action.

Lambe, N. R., McLean, K. A., Gordon, J., Evans, D., Clelland, N., & Bunger, L. (2017). Prediction of intramuscular fat content using CT scanning of packaged lamb cuts and relationships with meat eating quality. *Meat Science, 123,* 112–119.

Lambe, N. R., Navajas, E. A., Fisher, A. V., Simm, G., Roehe, R., & Buenger, L. (2009). Prediction of lamb meat eating quality in two divergent breeds using various live animal and carcass measurements. *Meat Science, 83,* 366–375.

Lambe, N. R., Navajas, E. A., Schofield, C. P., Fisher, A. V., Simm, G., Roehe, R., et al. (2008). The use of various live animal measurements to predict carcass and meat quality in two divergent lamb breeds. *Meat Science, 80*(4), 1138–1149.

Lambe, N., Wood, J. D., McLean, K. A., Walling, G. A., Whitney, H., Jagger, S., et al. (2012). Use of computed tomography (CT) in a longitudinal body composition study in pigs fed different diets. In *Farm animal imaging* (pp. 24–28). Dublin: COST Action.

Lambe, N. R., Young, M. J., Brotherstone, S., Kvame, T., Conington, J., Kolstad, K., et al. (2003). Body composition changes in Scottish blackface ewes during one annual production cycle. *Animal Science, 76,* 211–219.

Leeds, T. D., Mousel, M. R., Notter, D. R., Zerby, H. N., Moffet, C., & Lewis, G. S. (2008). B-mode, real-time ultrasound for estimating carcass measures in live sheep: Accuracy of ultrasound measures and their relationships with carcass yield and value. *Journal of Animal Science, 86,* 3203–3214.

Leite, A., Rodrigues, S., Pereira, E., Paulos, K., Oliveira, A., Lorenzo, J. M., et al. (2015). Physicochemical properties, fatty acid profile and sensory characteristics of sheep and goat meat sausages manufactured with different pork fat levels. *Meat Science, 105,* 114–120.

Li, C. (2010). A web service model for conducting research in image processing. *Journal of Computing Sciences in Colleges, 25*(5), 294–299.

Li-Chan, E. C. Y. (1996). The applications of Raman spectroscopy in food science. *Trends in Food Science and Technology, 7,* 361–370.

Liu, Y., Pu, H., & Sun, D. W. (2017). Hyperspectral imaging technique for evaluating food quality and safety during various processes: A review of recent applications. *Trends in Food Science & Technology*, *69*, 25–35.

Lu, Y., Young, O. A., & Brooks, J. D. (2014). Physicochemical and sensory characteristics of fermented sheepmeat sausage. *Food Science & Nutrition*, *2*(6), 669–675.

Ma, J., Sun, D. W., Qu, J. H., Liu, D., Pu, H., Gao, W. H., et al. (2016). Applications of computer vision for assessing quality of agri-food products: A review of recent research advances. *Critical Reviews in Food Science and Nutrition*, *56*(1), 113–127.

Macfarlane, J. M., Lewis, R. M., Emmans, G. C., Young, M. J., & Simm, G. (2006). Predicting carcass composition of terminal sire sheep using X-ray computed tomography. *Animal Science*, *82*(3), 289–300.

Macfarlane, J. M., Young, M. J., Lewis, R. M., Emmans, G. C., & Simm, G. (2005, June). Using X-ray computed tomography to predict intramuscular fat content in terminal sire sheep. In *Proceedings of the 56th annual meeting of the European association for animal production June, Uppsala, Sweden*, Vol. 264.

Madruga, M. S., & Bressan, M. C. (2011). Goat meats: Description, rational use, certification, certification, processing and technological developments. *Small Ruminant Research*, *98*, 39–45.

Mahgoub, O. (1998). Ultrasonic scanning measurements of the longissimus thoracis et lumborum muscle to predict carcass muscle content in sheep. *Meat Science*, *48*, 41–48.

Marco, M. L., Heeney, D., Binda, S., Cifelli, C. J., Cotter, P. D., Foligné, B., et al. (2017). Health benefits of fermented foods: Microbiota and beyond. *Current Opinion in Biotechnology*, *44*, 94–102.

McEwan, J. C., Clarke, J. N., Knowler, M. A., & Wheeler, M. (1989). Ultrasonic fat depths in Romney lambs and hoggets from lines selected for different production traits. *Proceedings of the New Zealand Society of Animal Production*, *49*, 113–119.

McGregor, B. A. (2017). Relationships between live weight, body condition, dimensional and ultrasound scanning measurements and carcass attributes in adult angora goats. *Small Ruminant Research*, *147*, 8–17.

Mesta, G. C., Will, P. A., & Gonzalez, J. M. (2016). The measurement of carcass characteristics of goats using the ultrasound method. *Texas Journal of Agriculture and Natural Resources*, *17*, 46–52.

Morgan-Davies, C., Lambe, N., Wishart, H., Waterhouse, T., Kenyon, F., McBean, D., et al. (2018). Impacts of using a precision livestock system targeted approach in mountain sheep flocks. *Livestock Science*, *208*, 67–76.

Motoyama, M. (2017). Raman spectroscopy for meat quality and safety assessment. In A. E. D. Bekhit (Ed.), *Advances in meat processing technology* (pp. 269–297). London: CRC Press.

Mukherjee, R. S., Chowdhury, B. R., Chakraborty, R., & Chaudhuri, U. R. (2006). Effect of fermentation and drying temperature on the characteristics of goat meat (Black Bengal variety) dry sausage. *African Journal of Biotechnology*, *5*(16), 1499–1504.

Nassu, R. T., Aparecida, L., Gonçalves, G., & Beserra, F. J. (2002). Efeito do teor de gordura nas características químicas e sensoriais de embutido fermentado de carne de caprinos— Effect of fat level in chemical and sensory characteristics of goat meat fermented sausage. *Pesquisa Agropecuária Brasileira*, *37*(8), 1169–1173.

Nassu, R. T., Gonçalves, L. A. G., & Beserra, F. J. (2002). Use of different starter cultures in processing of goat meat fermented sausages. *Ciência Rural Santa Maria*, *32*, 1051–1055.

Nassu, R. T., Gonçalves, L. A. G., Pereira da Silva, M. A. A., & Beserra, F. J. (2003). Oxidative stability of fermented goat meat sausage with different levels of natural antioxidant. *Meat Science*, *63*(1), 43–49.

Navajas, E. A., Lambe, N. R., McLean, K. A., Glasbey, C. A., Fisher, A. V., Charteris, A. J. L., et al. (2007). Accuracy of in vivo muscularity indices measured by computed tomography and their association with carcass quality in lambs. *Meat Science*, 75(3), 533–542.

Ngo, L., Ho, H., Hunter, P., Quinn, K., Thomson, A., & Pearson, G. (2016). Post-mortem prediction of primal and selected retail cut weights of New Zealand lamb from carcass and animal characteristics. *Meat Science*, 112, 39–45.

Notter, D. R., Mousel, M. R., Leeds, T. D., Zerby, H. N., Moeller, S. J., Lewis, G. S., et al. (2014). Evaluation of Columbia, USMARC composite, Suffolk, and Texel rams as terminal sires in an extensive rangeland production system: VII. Accuracy of ultrasound predictors and their association with carcass weight, yield, and value. *Journal of Animal Science*, 92, 2402–2414.

Oliveira, A. F., Rodrigues, S., Leite, A., Paulos, K., Pereira, E., & Teixeira, A. (2014). Quality of ewe and goat meat cured product mantas. An approach to provide value added to culled animals. *Canadian Journal of Animal Science*, 94(3), 459–462.

Olsen, E. F., Rukke, E.-O., Flåtten, A., & Isaksson, T. (2007). Quantitative determination of saturated-, monounsaturated- and polyunsaturated fatty acids in pork adipose tissue with non-destructive Raman spectroscopy. *Meat Science*, 76, 628–634.

Orman, A., Caliskan, G. U., & Dikmen, S. (2010). The assessment of carcass traits of Awassi lambs by real-time ultrasound at different body weights and sexes. *Journal of Animal Science*, 88, 3428–3438.

Orman, A., Calışkan, G. Ü., Dikmen, S., Ustüner, H., Ogan, M. M., & Calışkan, C. (2008). The assessment of carcass composition of Awassi male lambs by real-time ultrasound at two different live weights. *Meat Science*, 80, 1031–1036.

Ortega, A., Chito, D., & Teixeira, A. (2016). Comparative evaluation of physical parameters of salted goat and sheep meat blankets "mantas" from Northeastern Portugal. *Journal of Food Measurement and Characterization*, 10(3), 670–675.

Pabiou, T., Fikse, W. F., Cromie, A. R., Keane, M. G., Nasholm, A., & Berry, D. P. (2011). Use of digital images to predict carcass cut yields in cattle. *Livestock Science*, 137, 130–140.

Paleari, M. A., Bersani, C., Moretti Vittorio, M., & Beretta, G. (2002). Effect of curing and fermentation on the microflora of meat of various animal species. *Food Control*, 13(3), 195–197.

Paul Ross, R., Morgan, S., & Hill, C. (2002). Preservation and fermentation: Past, present and future. *International Journal of Food Microbiology*, 79(1), 3–16.

Paulos, K., Rodrigues, S., Oliveira, A. F., Leite, A., Pereira, E., & Teixeira, A. (2015). Sensory characterization and consumer preference mapping of fresh sausages manufactured with goat and sheep meat. *Journal of Food Science*, 80(7), S1568–S1573.

Perez-Palacios, M. T., Petisca, C., Pinho, O., & Ferreira, I. M. P. L. V. O. (2012). Headspace solid-phase microextraction of volatile and furanic compounds in coated fish sticks: Effect of the extraction temperature. *World Academy of Science, Engineering and Technology, International Journal of Nutrition and Food Engineering*, 6, 975–980.

Povše, M. P., Čandek-Potokar, M., Gispert, M., & Lebret, B. (2015). pH value and water-holding capacity. In M. Font-i-Furnols, M. Candek-Potokar, C. Maltin, & M. Prevolnik Povše (Eds.), *A handbook of reference methods for meat quality assessment* (pp. 22–32). Brussels, Belgium: European Cooperation in Science and Technology (COST).

Prieto, N., Pawluczyk, O., Dugan, M. E. R., & Aalhus, J.-L. (2017). A review of the principles and applications of near-infrared spectroscopy to characterize meat, fat, and meat products. *Applied Spectroscopy*, 71(7), 1403–1426, 2017.

Pu, H., Sun, D. W., Ma, J., Liu, D., & Kamruzzaman, M. (2014). Hierarchical variable selection for predicting chemical constituents in lamb meats using hyperspectral imaging. *Journal of Food Engineering*, 143, 44–52.

Pugliese, C., Sirtori, F., Ruiz, J., Martin, D., Parenti, S., & Franci, O. (2009). Effect of pasture on chestnut or acorn on fatty acid composition and aromatic profile of fat of Cinta Senese dry-cured ham. *Grasas y Aceites, 60*(3), 271–276.

Pullanagari, R., Yule, I., & Agnew, M. (2015). On-line prediction of lamb fatty acid composition by visible near infrared spectroscopy. *Meat Science, 100*, 156–163. https://doi.org/10.1016/j.meatsci.2014.10.008.

Purchas, R. W., Davies, A. S., & Abdullah, A. Y. (1991). An objective measure of muscularity: Changes with animal growth and differences between genetic lines of Southdown sheep. *Meat Science, 30*(1), 81–94.

Qiao, L., Peng, Y., Chao, K., & Qin, J. (2016). Rapid discrimination of main red meat species based on near-infrared hyperspectral imaging technology (2016). *Proceedings of SPIE—The International Society for Optical Engineering, 9864*, art. No. 98640U.

Qiao, L., Peng, Y., Wei, W., & Li, C. (2015). Identification of main meat species based on spectral characteristics. In *2015 ASABE Annual International Meeting*. American Society of Agricultural and Biological Engineers.

Qiao, T., Ren, J., Yang, Z., Qing, C., Zabalza, J., & Marshall, S. (2015). Visible hyperspectral imaging for lamb quality prediction. *Tm-Technisches Messen, 82*(12), 643–652.

Qin, J., Kim, M. S., Chao, K., & Cho, B. K. (2017). Raman chemical imaging technology for food and agricultural applications. *Journal of Biosystems Engineering, 42*, 170–189.

Ramsey, C. B., Kirton, A. H., Hogg, B., & Dobbie, J. L. (1991). Ultrasonic, needle, and carcass measurements for predicting chemical composition of lamb carcasses. *Journal of Animal Science, 69*, 3655–3664.

Ripoll, G., Joy, M., Alvarez-Rodriguez, J., Sanz, A., & Teixeira, A. (2009). Estimation of light lamb carcass composition by in vivo real-time ultrasonography at four anatomical locations. *Journal of Animal Science, 87*, 1455–1463.

Ripoll, G., Joy, M., & Sanz, A. (2010). Estimation of carcass composition by ultrasound measurements in 4 anatomical locations of 3 commercial categories of lamb. *Journal of Animal Science, 88*, 3409–3418.

Risvik, E. (1994). Sensory properties and preferences. *Meat Science, 36*(1), 67–77.

Rius-Vilarrasa, E., Buenger, L., Brotherstone, S., Macfarlane, J. M., Lambe, N. R., Matthews, K. R., et al. (2010). Genetic parameters for carcass dimensional measurements from video image analysis and their association with conformation and fat class scores. *Livestock Science, 128*, 92–100.

Rius-Vilarrasa, E., Buenger, L., Maltin, C., Matthews, K. R., & Roehe, R. (2009). Evaluation of video image analysis (VIA) technology to predict meat yield of sheep carcasses on-line under UK abattoir conditions. *Meat Science, 82*, 94–100.

Rodrigues, S., & Teixeira, A. (2009). Effect of sex and carcass weight on sensory quality of goat meat of Cabrito Transmontano. *Journal of Animal Science, 87*(2), 711–715.

Rodrigues, S., & Teixeira, A. (2010). Consumers' preferences for meat of Cabrito transmontano. Effects of sex and carcass weight. *Spanish Journal of Agricultural Research, 8*(4), 936–945.

Rosenblatt, A. J., Scrivani, P. V., Boisclair, Y. R., Reeves, A. P., Ramos-Nieves, J. M., Xie, Y., et al. (2017). Evaluation of a semi-automated computer algorithm for measuring total fat and visceral fat content in lambs undergoing in vivo whole body computed tomography. *The Veterinary Journal, 228*, 46–52.

Sahin, E. H., Yardimci, M., Cetingul, I. S., Bayram, I., & Sengor, E. (2008). The use of ultrasound to predict the carcass composition of live Akkaraman lambs. *Meat Science, 79*, 716–721.

Sañudo, C., Gomes, M. A. L., Velandia, V. M., Fugita, C. A., Monge, P., Guerrero, A., et al. (2016). Cross-cultural study of dry-cured sheep meat acceptability by native and immigrant consumers in Spain. *Journal of Sensory Studies, 31*(1), 12–21.

Sanz, J. A., Fernandes, A. M., Barrenechea, E., Silva, S., Santos, V., Gonçalves, N., et al. (2016). Lamb muscle discrimination using hyperspectral imaging: Comparison of various machine learning algorithms. *Journal of Food Engineering, 174*, 92–100.

Scheier, R., Scheeder, M., & Schmidt, H. (2015). Prediction of pork quality at the slaughter line using a portable Raman device. *Meat Science, 103*, 96–103.

Scheier, R., & Schmidt, H. (2013). Measurement of the pH value in pork meat early post-mortem by Raman spectroscopy. *Applied Physics B, 111*, 289–297.

Schmidt, H., Scheier, R., & Hopkins, D. L. (2013). Preliminary investigation on the relationship of Raman spectra of sheep meat with shear force and cooking loss. *Meat Science, 93*(1), 138–143.

Schmidt, H., Sowoidnich, K., Maiwald, M., Sumpf, B., & Kronfeldt, H. D. (2009). Handheld Raman sensor head for in-situ characterization of meat quality applying a mircosystem 671 nm diode laser. In T. Vo-Dinh, R. A. Lieberman, & G. Gauglitz (Eds.), *Proceedings of advanced environmental, chemical and bio sensing technologies VI* (pp. 1–8). Orlando, FL: United States International Society for Optics and Photonics.

Scholz, A. M., Bünger, L., Kongsro, J., Baulain, U., & Mitchell, A. D. (2015). Non-invasive methods for the determination of body and carcass composition in livestock: Dual-energy X-ray absorptiometry, computed tomography, magnetic resonance imaging and ultrasound: Invited review. *Animal, 9*, 1250–1264.

Schulze-Ehlers, B., & Anders, S. (2018). Towards consumer-driven meat supply chains: Opportunities and challenges for differentiation by taste. *Renewable Agriculture and Food Systems, 33*(1), 73–85.

Silva, S. R. (2017). Use of ultrasonographic examination for in vivo evaluation of body composition and for prediction of carcass quality of sheep. *Small Ruminant Research, 152*, 144–157.

Silva, S. R., Afonso, J. J., Santos, V. A., Monteiro, A., Guedes, C. M., Azevedo, J. M. T., et al. (2006). In vivo estimation of sheep carcass composition using real time ultrasound with two probes of 5 and 7.5 MHz and image analysis. *Journal of Animal Science, 84*, 3433–3439.

Silva, S. R., Gomes, M. J., Dias-da-Silva, A., Gil, L. F., & Azevedo, J. M. T. (2005). Estimation in vivo of the body and carcass chemical composition of growing lambs by real-time ultrasonography. *Journal of Animal Science, 85*, 350–357.

Silva, S. R., Teixeira, A., Monteiro, A., Guedes, C. M., & Ginja, M. (2015). Using computer tomography to predict composition of light carcass kid goats. In C. A. Maltin, C. R. Craigie, & L. Bunger (Eds.), *FAIM Edinburgh* (p. 133). United Kingdom, Edinburgh: SRUC.

Simm, G. (1987). Carcass evaluation in sheep breeding programmes. In I. Fayez, M. Marai, & J. B. Owen (Eds.), *New techniques in sheep production* (pp. 125–144). London: Butterworth.

Stajić, S., Stanišić, N., Perunović, M., Živković, D., & Žujović, M. (2011). Possibilities for the use of goat meat in the production of traditional sucuk. *Biotechnology in Animal Husbandry, 27*(4), 1489–1497.

Stanford, K., Clark, I., & Jones, S. D. M. (1995). Use of ultrasound in prediction of carcass characteristics in lambs. *Canadian Journal of Animal Science, 75*, 185–189.

Stanford, K., Richmond, R. J., Jones, S. D. M., Robertson, W. M., Price, M. A., & Gordon, A. J. (1998). Video image analysis for on-line classification of lamb carcasses. *Animal Science, 67*, 311–316.

Stanisz, M., Gut, A., & Ślósarz, P. (2004). The live ultrasound measurements to assess slaughter value of meat-type male kids. *Animal Science Papers and Reports, 22*(4), 687–693.

Steinhaus, M., Wilhelm, W., & Schieberle, P. (2006). Comparison of the most odour-active volatiles in different hop varieties by application of a comparative aroma extract dilution analysis. *European Food Research and Technology, 226*(1), 45.

Stojković, S., Grabež, V., Bjelanović, M., Mandić, S., Vučić, G., Martinović, A., et al. (2015). Production process and quality of two different dry-cured sheep hams from Western Balkan countries. *LWT-Food Science and Technology*, *64*(2), 1217–1224.

Stouffer, J. R. (1991). *Using ultrasound to objectively evaluate composition and quality of livestock. 21st century concepts important to meat-animal evaluation* (pp. 49–54). Madison: University of Wisconsin.

Stringer, M. (2005). Summary report: Food safety objectives—Role in microbiological food safety management. *Food Control*, *16*(9), 775–794.

Su, W. H., He, H. J., & Sun, D. W. (2017). Non-destructive and rapid evaluation of staple foods quality by using spectroscopic techniques: A review. *Critical Reviews in Food Science and Nutrition*, *57*(5), 1039–1051.

Sun, S., Guo, B., Wei, Y., & Fan, M. (2012). Classification of geographical origins and prediction of δ13C and δ15N values of lamb meat by nearinfrared reflectance spectroscopy. *Food Chemistry*, *135*, 508–514.

Szabo, T. L. (2004). *Diagnostic ultrasound imaging: Inside out. Academic press series in biomedical engineering*. Connecticut, USA: Academic Press/ Elsevier.

Tait, R. (2016). Ultrasound use for body composition and carcass quality assessment in cattle and lambs. *Veterinary Clinics: Food Animal Practice*, *32*, 207–218.

Teixeira, A. (2008). Avaliação "in vivo" da composição corporal e da carcaça de caprinos: uso de ultrasonografia. *Revista Brasileira de Zootecnia*, *37*, 191–196.

Teixeira, A., Fernandes, A., Pereira, E., Manuel, A., & Rodrigues, S. (2017). Effect of salting and ripening on the physicochemical and sensory quality of goat and sheep cured legs. *Meat Science*, *134*, 163–169.

Teixeira, A., Joy, M., & Delfa, R. (2008). In vivo estimation of goat carcass composition and body fat partition by real-time ultrasonography. *Journal of Animal Science*, *86*(9), 2369–2376.

Teixeira, A., Matos, S., Rodrigues, S., Delfa, R., & Cadavez, V. (2006). In vivo estimation of lamb carcass composition by real-time ultrasonography. *Meat Science*, *74*, 289–295.

Teixeira, A., Oliveira, A., Paulos, K., Leite, A., Marcia, A., Amorim, A., et al. (2015). An approach to predict chemical composition of goat longissimus thoracis et lumborum muscle by near infrared reflectance spectroscopy. *Small Ruminant Research*, *126*, 40–43.

Teixeira, A., Pereira, & Rodrigues, S. (2011). Goat meat quality. Effects of salting, air-drying and ageing processes. *Small Ruminant Research*, *98*, 55–58.

Thériault, M., Pomar, C., & Castonguay, F. W. (2009). Accuracy of real-time ultrasound measurements of total tissue, fat, and muscle depths at different measuring sites in lamb. *Journal of Animal Science*, *87*, 1801–1813.

Tolentino, G. S., Estevinho, L. M., Pascoal, A., Rodrigues, S. S., & Teixeira, A. J. (2017). Microbiological quality and sensory evaluation of new cured products obtained from sheep and goat meat. *Animal Production Science*, *57*(2), 391–400.

Turlington, L. M. (1989). *Live animal evaluation of carcass traits for swine and sheep using real-time ultrasound*. MSc Thesis. Kansas State University, pp. 1–76.

UN. (2017). Population Division (2017). In *World population prospects: The 2017 revision, data booklet*: United Nations, Department of Economic and Social Affairs. ST/ESA/SER.A/401.

Van der Meer, F. D., Van der Werff, H. M., Van Ruitenbeek, F. J., Hecker, C. A., Bakker, W. H., Noomen, M. F., et al. (2012). Multi-and hyperspectral geologic remote sensing: A review. *International Journal of Applied Earth Observation and Geoinformation*, *14*(1), 112–128.

van Schothorst, M. (1998). Principles for the establishment of microbiological food safety objectives and related control measures. *Food Control*, *9*(6), 379–384.

van Schothorst, M., Zwietering, M. H., Ross, T., Buchanan, R. L., & Cole, M. B. (2009). Relating microbiological criteria to food safety objectives and performance objectives. *Food Control*, *20*(11), 967–979.

Vardanjani, S. M. H., Ashtiani, S. R. M., Pakdel, A., & Moradi, H. (2014). Accuracy of real-time ultrasonography in assessing carcass traits in Torki-Ghashghaii sheep. *Journal of Agriculture Science Technology, 16*, 791–800.

Viljoen, M., Hoffman, L. C., & Brand, T. S. (2007). Prediction of the chemical composition of mutton with near infrared reflectance spectroscopy. *Small Ruminant Research, 69*, 88–94.

Villalobos-Delgado, L. H., Caro, I., Blanco, C., Bodas, R., Andrés, S., Giráldez, F. J., et al. (2015). Effect of the addition of hop (infusion or powder) on the oxidative stability of lean lamb patties during storage. *Small Ruminant Research, 125*, 73–80.

Villalobos-Delgado, L. H., Caro, I., Blanco, C., Morán, L., Prieto, N., Bodas, R., et al. (2014). Quality characteristics of a dry-cured lamb leg as affected by tumbling after dry-salting and processing time. *Meat Science, 97*(1), 115–122.

Wang, S.-L., Wu, L.-G., Kang, N.-B., Li, H.-Y., Wang, J.-Y., & He, X.-G. (2016). Study on tan-lamb mutton tenderness by using the fusion of hyperspectral spectrum and image information. *Guangdianzi Jiguang/Journal of Optoelectronics Laser, 27*(9), 987–995.

Ward, B., Purchas, R. W., & Abdullah, A. Y. (1992). The value of ultrasound in assessing the leg muscling of lambs. *Proceedings of the New Zealand Society of Animal Production, 52*, 33–36.

Webb, E. C., Casey, N. H., & Simela, L. (2005). Goat meat quality. *Small Ruminant Research, 60*(1), 153–166.

Weeranantanaphan, J., Downey, G., Allen, P., & Sun, D. W. (2011). A review of near infra-red spectroscopy in muscle food analysis: 2005–2010. *Journal of Near Infrared Spectroscopy, 19*, 61–104.

Whitsett, C. M. (2009). Ultrasound imaging and advances in system features. *Ultrasound Clininics, 4*, 391–401.

Wu, W. H., Rule, D. C., Busboom, J. R., Field, R. A., & Ray, B. (1991). Starter culture and time/temperature of storage influences on quality of fermented mutton sausage. *Journal of Food Science, 56*(4), 916–919.

Xu, J. L., & Sun, D. W. (2017). Hyperspectral imaging technique for online monitoring of meat quality and safety. In *Advanced technologies for meat processing* (2nd ed., pp. 17–82). CRC Press.

Young, M., & Deaker, J. M. (1994). Ultrasound measurements predict estimated adipose and muscle weights better than carcass measurements. *Proceedings of the New Zealand Society of Animal Production, 54*, 215–217.

Young, M. J., Nsoso, S. J., Logan, C. M., & Beatson, P. R. (1996). Prediction of carcass tissue weight in vivo using live weight, ultrasound or X-ray CT measurements. *Proceedings of the New Zealand Society of Animal Production, 56*, 205–211.

FURTHER READING

Anderson, F., Pethick, D. W., & Gardner, G. E. (2016). The impact of genetics on retail meat value in Australian lamb. *Meat Science, 117*, 147–157.

Brosnan, T., & Sun, D. W. (2004). Improving quality inspection of food products by computer vision—A review. *Journal of Food Engineering, 61*(1), 3–16.

Cuthberson, A. (1978). Carcass evaluation of cattle, sheep and pigs. *World Review of Nutrition and Dietetics, 28*, 210–235.

De Smet, S., & Vossen, E. (2016). Meat: The balance between nutrition and health. A review. *Meat Science, 120*, 145–156.

Elmasry, G., Barbin, D. F., Sun, D.-W., & Allen, P. (2012). Meat quality evaluation by hyperspectral imaging technique: An overview. *Critical Reviews in Food Science and Food Nutrition, 52*, 689–711.

Feng, C. H., Makino, Y., Oshita, S., & Martín, J. F. G. (2018). Hyperspectral imaging and multispectral imaging as the novel techniques for detecting defects in raw and processed meat products: Current state-of-the-art research advances. *Food Control, 84,* 165–176.

Shapiro, L., & Stockman, G. C. (2001). *Computer vision. 2001.* Prentice Hall.

Skjervold, H., Gronseth, K., Vangen, O., & Evensen, A. (1981). In vivo estimation of body composition by computerised tomography. *Zeitschrift für Tierzüchtung und Züchtungsbiologie, 98,* 77–79.

World Population. (2017). Population Division (2017). In *World population prospects: The 2017 revision, data booklet.* Wallchart United Nations, Department of Economic and Social Affairs. ST/ESA/SER.A/401.

Xiong, Z., Sun, D.-W., Zeng, X.-A., & Xie, A. (2014). Recent developments of hyperspectral imaging systems and their applications in detecting quality attributes of red meats: A review. *Journal of Food Engineering, 132,* 1–13.

> CHAPTER SEVEN

Particular Alimentations for Nutrition, Health and Pleasure

José Miguel Aguilera*,[1], Bum-Keun Kim[†], Dong June Park[†]
*Department of Chemical and Bioprocess Engineering, Pontificia Universidad Católica de Chile, Santiago, Chile
[†]Division of Strategic Food Research, Korea Food Research Institute, Seoul, South Korea
[1]Corresponding author: e-mail address: jmaguile@ing.puc.cl

Contents

Abstract

People around the world select their foods and meals according to particular choices based on physiological disorders and diseases, traditions, lifestyles, beliefs, etc. In this chapter, two of these particular alimentations are reviewed: those of the gourmet and the frail elderly. They take place in an environment where food is usually

Advances in Food and Nutrition Research, Volume 87
ISSN 1043-4526
https://doi.org/10.1016/bs.afnr.2018.07.005

synonymous of body health disregarding its effects on social, cultural and psychological aspects, including emotions. Based on an extensive literature review, it is proposed that the paradigm changes from food equals health to food means well-being, the latter encompassing physical and physiological aspects as well as psychological, emotional and social aspects at the individual and societal levels. The growing food and nutrition requirements of an aging population are reviewed and special nutritious and enjoyable products available for this group are discussed.

1. INTRODUCTION

Humans, being omnivorous, have thrived in almost every corner of this planet and derived their foods from what was available in each agro-ecological niche. In most cases, edible materials and diets were based on a few staple foods that provided the biological requirements for nutrients (Milton, 2000). The Inuit (Eskimos) have eaten a high-protein, high-fat diet consisting mainly of marine mammals such as seals and walrus, some meat (caribou), and lots of fish (Bang, Dyerberg, & Sinclair, 1980). East African pastoralists have diets centered on milk, meat and blood, the Nochmani of the Nicobar Islands get abundant protein from insects, whereas India's Jains are mostly vegetarians (Gibbons, 2014). The Inkas, dwelling in the highlands of the Andes, consumed only twice a day dishes prepared from tubers and corn (Bollinger, 1993). Our prehistoric ancestors ate mainly meat, fruits, and vegetables, but after agriculture was developed, the Western diet expanded to include a high number of cereal-grains, milk products, refined sugar and fat sources which now make up around 70% of the diet (Cordain et al., 2005). All these dietary patterns insinuate that humans have a large dietary flexibility and nutritional resilience. In fact, as demonstrated in dwellers of the Artic region, changes from these simple ancestral diets to a more varied and nutrient-rich Western diet have led to increased health problems, such as obesity, cardiovascular disease, and diabetes, and a decline in mental health (McGrath-Hanna, Greene, Tavernier, & Bult-Ito, 2003).

The individualization of the alimentary regimes (or diets) probably started in the Middle Ages with the belief that each person knew best what food was good or bad for his/her body (Vigarello, 2013). This trend has been exacerbated in recent times, predominantly among urban consumers around the world. People can now choose among many alimentation options because the food supply is ample, varied, ubiquitous and relatively cheap. In fact, the average proportion of household expenditures devoted to food in the United States continues to fall and it is now as low as 6.5%, and less than 20% in other 11 countries (Plumer, 2015). Particular alimentations are a concept introduced by the French anthropologist Claude Fischler to refer to this trend in which individuals embark on new diets or acquire food habits beyond the mere like or dislike of specific foods (Fischler, 2013). People may eat certain foods and not others for medical or health reasons, as is the case in long-term illnesses, the presence of food allergies and food intolerances, or when following a recommended weight reduction diet. There are also more subjective reasons for selecting a particular diet such as religious beliefs, ethical considerations (e.g., some expressions of vegetarianism and veganism) or cultural traditions. Other people such as athletes and those having a physically active lifestyle adopt specific diets and consume dietary supplements in a quest to improve their performance. Additionally, there are the fad diets that offer amazing health effects and attract many followers.

The word alimentation is seldom utilized in the English language. The Oxford Dictionary defines alimentation as "…the provision of nourishment or other necessities of life." In French and in many languages of Latin origin, alimentation means something similar: the process of procurement, preparation, consumption and enjoyment of foods. So, "alimentation" is a human process or activity while "food" is mostly a material that nourishes. We will come back to this difference later on in the text. The word "particular," in turn, denotes that we are all different when it comes to selecting, eating and digesting the food, and that there is an enormous variability in metabolic responses to diets that affect overall health (Ohlhorst et al., 2013). For example, identical (monozygotic) twins are neither genetically nor phenotypically "identical" and they differ in their preferences for food as well as in the physiological reactions to them, as is the case of the susceptibility to food allergies and prevalence of some diet-based diseases (Bruder et al., 2008; Fildes et al., 2014; Hong, Tsai, & Wang, 2009). People have also a dissimilar microbiota—bacteria, yeasts, fungi and virus in the gut—which plays a vital role in digestion, nutrition and health protection (Dicksved et al., 2008). It is not surprising that people around the world (and even within the same family) select their foods and meals according to their particular physiological disorders, lifestyles, beliefs, traditions, etc. Hence, "particular alimentation" is a much broader and multifaceted concept than "personalized nutrition" which proposes that optimal "health" may be achieved by tuning the person's genomics to the intake of nutrients (German, Zivkovic, Dallas, & Smilowitz, 2011; Ghosh, 2010). Table 1 presents a list of some of the most important particular alimentations in vogue.

Table 1 Examples of Conditions or Choices Leading to Particular Alimentations

Condition or Dietary Choice	Description and Main Diet Implications	Worldwide Impact or Specific Comment	References
Obesity	Excessive body fat accumulation (BMI ≥ 30) that severely impairs health.[a] Diet low in calories is enforced	Obesity has doubled since 1980. In 2014, more than 600 million adults aged 18 years and over were obese	WHO (2017b)
Overweight	Body fat accumulation (BMI ≥ 25) that may impair health.[a] Diet low in calories recommended	In 2014, more than 1.9 billion adults 18 years and older were overweight, and 41 million children under the age of 5 were overweight or obese	WHO (2017b)
Allergies	Adverse immune-mediated response on exposure to one or more of 170 foods. Abstinence from their consumption is required	Food allergies may be present in 4–7% of preschool children and 1–2% of the adult population	Turnbull, Adams, and Gorard (2015)
Food intolerance	Undesirable reactions to foods (e.g., lactose and gluten) not involving the immune system. Restricted intake of these foods	About 70% of the population suffer from lactose intolerance. Lactose intolerance and celiac disease affect approximately 1% of individuals	Mattar, de Campos Mazo, and Carrilho (2012)
Diabetes	High blood sugar levels over a prolonged periods leading to a series of NCDs. Restricted intake of sugar and starchy foods	People with diabetes almost quadrupled since 1980 (108 million to 422 million in 2014)	WHO (2016)
Hypertension	Persistent raised blood pressure, leading to heart attacks and strokes. Restriction in the amount of salt	Hypertension affects 1 billion people and kills 9 million people every year	WHO (2013)
Aging	Elderly that requires special nutrition (e.g., protein, fiber) or has mastication/swallowing problems (dysphagia)	Elderly are 3.8% of the world's population, projected to be 8.3% (700 million) in 2050	Aguilera and Park (2016) and United Nations (2015)

			References
Religious—Jewish	Kosher food is suitable for consumption. Kosher laws forbid eating some animals, fish without fins and scales, some shellfish, blood and combine meat and milk	In the United States alone more than 10 million people consume Kosher food. The market exceeds $24 billion/year and grows at a rate of 11.6%	Regenstein, Chaudry, and Regenstein (2003) and Persistence Market Research (2017)
Religious—Muslim	Halal dietary laws determine which foods are permitted for Muslims. E.g., meat (but not pork) and foods without alcohol	With over 1600 million Muslim population, the market for halal foods is estimated at more than $600 million/year	Regenstein et al. (2003) and Imarat Consultants (2015)
Vegetarian	Basically, vegetarians exclude any kind of flesh (e.g., red meat, fish, seafood). Some may include dairy products (lacto-vegetarian) and eggs (ovo-lacto-vegetarian)	Around 375 million strict vegetarians worldwide and 1–10% of the population in the United States and Europe	Figus (2014)
Vegan	Strict vegetarian who consumes no food that comes from animals (e.g., meat, eggs, dairy products, etc.)	In Britain, the number of vegans is now more than 3.5 times that of the past decade (around 1% of population)	Quinn (2016)
Eat organic foods	Likes food produced without using synthetic fertilizers and pesticides, antibiotics or growth hormones	Global retail market for organic foods was $90 billion in 2016. In the United States it is over 5% of total at-home food consumption. It grows at double-digit figures since 2011	Greene (2017)
Gourmet	Enjoys eating good food, derives pleasure from it and knows about gastronomy, cooking and food ingredients	Continuous increase in number of full-service restaurants, cooking TV shows, gastronomy books and blogs. Emergence of "gastronomic tourism"	Aguilera (2013)

[a]BMI is a person's weight in kilograms divided by the square of his/her height in meters.

In this chapter, the cases of the particular alimentations of the gourmet, who privileges the experience of eating, and of the frail elderly who has to restrain the food choices mainly due to physiological conditions are addressed. Also, there is an attempt to reconcile the notions of nutrition and pleasure of eating, often regarded as antagonistic. Because of our backgrounds, the emphasis will be on the food science and nutrition aspects but it has been unavoidable to make reference to sociological and psychological aspects. In passing, concepts that are rarely addressed in the context of food and nutrition such as commensality, gastronomy and wellness will be also addressed.

2. THE SCENARIO FOR PARTICULAR ALIMENTATIONS

2.1 Main Trends

Particular alimentations take place in a scenario with several contradictions. Per capita food supply in the last 30 years has steadily increased worldwide and approaches now 2900 kcal/person/day, which on average is enough to satisfy the caloric needs of every inhabitant of this planet. Yet, ca. 815 million people are still affected by hunger (WHO, 2017a). The extensive options of food choices and food availability beyond seasonality rather than leading to healthy habits have not changed poor diets or helped in curbing obesity. In fact, in 2016 over 340 million children and adolescents throughout the world were overweight or obese, and 650 million adults were obese (WHO, 2017b). The medical profession continues to find links between the excessive consumption of some food components and the presence of non-communicable diseases (NCDs); however, most policy efforts to lessen the intake of these foods have had modest effects (Hyseni et al., 2017). In spite of the demonstrated positive effect of physical activity as primary prevention against several chronic conditions and accelerated aging, sedentarism continues to increase among people of all ages (Booth, Roberts, & Laye, 2012). The mean life expectancy at birth continues to rise worldwide and is now 72 years, but most elderly people have a poor quality of life (Tesch-Roemer, 2010). Food consumed out of home is rapidly expanding in cities all around the planet, and in the United States it already accounts for over half of every dollar spent in food (Aguilera, 2018). Time devoted to cooking at home, that may be essential to promote healthier dietary habits, is steadily decreasing (Monsivais, Aggarwal, & Drewnowski, 2014).

The food industry, that in the near past contributed to the development and sustainability of modern urban lifestyles, is now being blamed for supplying "ultraprocessed" foods that presumably are directly conducive to an "obesogenic environment" (Monteiro, Moubarac, Cannon, Ng, & Popkin, 2013). All these trends and issues give a background to explain the emergence of particular alimentations in which individuals adopt eating habits and regimes in a quest to improve their physical and mental selves.

2.2 Nutritionism

Nutrition science got most of its reputation in the past century by proposing cures to diseases caused by the deficiency of single nutrients (e.g., some vitamins and minerals), and its contribution to alleviate child protein-calorie malnutrition in developing countries. The recent prominence of nutritionism or the exaggerated connection between particular nutrients and bodily health centers the main purpose of eating in a narrow concept of physical health, disregarding that food has been and will continue to be also a source of pleasure, community feelings, social identity, preservation of culture and traditions (Pollan, 2008; Scrinis, 2013). Nutritionism has exacerbated the idea that there are "good foods" and "bad foods," a divergence which started in the 1960s when nutrition scientists argued that there were "good fats" (polyunsaturated fats), "bad fats" (saturated fats) and that the total fat consumption should be limited. The end of the story is now well known: the bad turned out to be better than its replacement and the most recent recommendation is to move away from total fat reduction and engage in healthy diets and lifestyles (Mozaffarian & Ludwig, 2015). From the ethical viewpoint, foods are bad when they are produced at the expense of environmental damage or cause health problems to consumers. Any food may become "unhealthy" if consumed routinely, in large quantities and/or as a part of an unbalanced diet (Hirvonen, 2014). In the short term, what matters is that the overall nutrient mix in our meals does not contain excessive amounts of calories, saturated fat, salt and sugar, while in the long term, what counts are healthy alimentation habits. Calorie restriction accompanied with a varied diet that provides adequate nutrition remains as the most consistent intervention to assure a healthy life and reduce the risk of age-related diseases, as demonstrated in a variety of species (Anton & Leeuwenburgh, 2013; Heilbronn & Ravussin, 2003; Testa, Biasi, Poli, & Chiarpotto, 2014).

2.3 The Fear of Eating

As omnivores we experience the "omnivore paradox": selecting what we eat from a wide variety of foods but at the same time, feeling fear and distrust of them (Fischler, 1980). The dilemma has taken a simplistic new character in the minds of many people that "food is a source of risks" while "nutrition provides health," exacerbating the dichotomy between foods and nutrition (Buchler, Smith, & Lawrence, 2010; Pollan, 2008). This paradox is aggravated by the modest understanding among most consumers of the risks and hazards of what is on their plates. Risk and hazard are two terms that often mistakenly confused by people and the media (and by a few scientists as well). A food hazard refers to a biological, chemical or physical agent in foods with the potential to cause an adverse health effect. For example, there is the potential that compounds produced by ancient and widely practiced culinary techniques, such as frying, roasting and grilling, have carcinogenic activity when assessed under certain experimental conditions (Gibis, 2016; Rice, 2005). Risk is the probability of an adverse health effect that may occur at a given exposure dose (Barlow et al., 2015). For example, each time we swallow a food, there is the hazard (potential) that we get choked with a perfectly nutritious and safe piece of food. The risk, however, is quite low; in the United States about one in 4400 deaths are caused by choking with food and it usually occurs in children (Jagger, 2011). The risk of consuming most food compounds depends on the intake: it is high in deficiency, flat over a broad range of doses, and high again in excess (e.g., vitamin A and manganese). This is the so-called J-shape curve which means that the effect of a food component on health is not due to its presence (now detected up to the ppb level) but on its dose and the individual susceptibility to it (Mulholland & Benford, 2007). Ultraviolet radiation is a demonstrated physical hazard leading to skin lesions and carcinogenesis, including life-threatening melanomas, yet people continue to sunbath on beaches in spite of the high risk involved (Hansen & Bentzen, 2016; Krutmann, Morita, & Chung, 2012).

2.4 We Eat Meals Not Nutrients

The approach of blaming a disease or health condition on a single factor is still applied in nutrition research. As stated by Jacobs and Tapsell (2013), the single-nutrient approach has been in many ways counterproductive because it disregards the possible synergisms between different food components in a meal. Isolated components are processed by our bodies differently than when they are part of a complex food matrix. Digestion and absorption of nutrients is not 100% efficient and only a fraction of several nutrients or their metabolites reach the systemic circulation and become bioavailable (Parada & Aguilera, 2007). Moreover, it has been recognized that nutritional studies in relation to diseases have considerable methodological limitations. What we eat in our meals is subject to many endogenous and exogenous effects which are likely to induce non-linear relationships with a disease (Willet, 1987). Most conclusions of epidemiological studies in scientific articles linking food components and NCDs end up with a sentence like "…the available evidence supports a positive association (between a food component and a disease), but data are not conclusive" (see, for instance, Jakszyn & González, 2006 on nitrosamines, and Borek, 2017 on isolated antioxidant supplements and cancer risk or mortality). In summary, according to the best available scientific information (examined and enforced through regulation by national food authorities), foods and food components (e.g., additives) that are legally approved, sold and consumed in our meals are "healthy"; the individual's health effects depend on the amount and frequency in which they are consumed.

2.5 Do We Understand Enough About Foods?

According to Scopus, a database of peer-reviewed scientific literature, 370 journals were published in 2017 having "food science" and "nutrition and dietetics" as subject areas (accessed on June 1, 2018). A good number of articles published in such journals convey important information for consumers but they are inscrutable for lay people. Usually, the general public has access to the conclusions of these researches by the interpretation and sensationalistic versions of journalists knowing little about science (Halfon, 2018; Warner, 2017). Then, the long recognized psychological phenomenon of "confirmation bias" operates: a tendency to accept information that supports one's own dietary habits and beliefs, and to reject information that contradicts them. In other words, there is an "unconscious" selectivity in the acquisition and use of evidence that is favorable (Nickerson, 1998). Moreover, changes in governmental recommendations, periodical amends to food pyramids and nutritional guidelines as well as several nutrition "fiascos," i.e., the butter versus margarine issue, no eggs and eggs again, have confused the average consumer who knows little about nutrition about the real value of the information reaching them (McNamara, 2015; Rozin, 2005). As obvious as it may seem, only knowledgeable consumers can make informed food choices. Thus, in order to curb the rise in diet-related diseases it has been proposed that in addition to compulsory normative restrictions (e.g., front-of-pack nutrition warnings), a more positive "food literacy" be implemented to increase the knowledge, skills and tools of individuals, communities or nations for a healthy lifelong relationship with food (Truman, Lane, & Elliott, 2017; Vidgen & Gallegos, 2014).

It is also ignored (or purposely forgotten) that humans are extremely resilient when it comes to the temporal availability and consumption of foods and nutrients. There is widespread evidence that starvation and famines during evolution exerted a strong selection effect on the human genome. Amazingly, the time over which an individual can survive severe or total caloric restriction varies from over a month in thin individuals, to a year or more in obese ones (Prentice, 2005). Even today, there are two extreme cases where the regular consumption of real foods is suspended or vastly diminished, yet human life is maintained for extended periods (e.g., several days to even months). One is parenteral (intravenous) feeding where patients are given a fluid containing amino acids (both essential and non-essentials), electrolytes, vitamins and micronutrients, while the calorie supply is provided by lipids emulsified in an aqueous carbohydrate source, generally glucose (Dibb, Teubner, Theis, Shaffer, & Lal, 2013). Others are voluntary or involuntary drastic reductions in food intake which may last for up to a few months (e.g., hunger strikers and castaways, respectively).[a] Obviously, parental feeding and hunger strikes are exceptional situations that demand the adequate supervision by medical professionals and monitoring of their consequences (Redman & Ravussin, 2011).

3. GASTRONOMY AND THE PLEASURE OF EATING

Gastronomy or the art of selecting, preparing, serving and enjoying fine food highlights the eating-related pleasures, the social and cultural dimensions of eating as well as its esthetic experience. The etymology of the word gastronomy comes from Ancient Greek (*gastro* = stomach, and *nomos* = law). According to *Larousse Gastronomique*, the bible of gastronomists, the term "gastronomy" became popular in France after the publication in 1801 of *La Gastronomie ou l'Homme des Champs to Table* by J. Berchoux, and the

---

[a] One example is the 22 crew members of Shackleton's expedition to the Antarctica who survived 4 months in Elephant Island. Although by the early 1900s the life expectancy of men in England was 47 years, most of these survivors died well above this age, including one which died at 90 (https://en.wikipedia.org/wiki/Lionel_Greenstreet).

Academie Francaise made it an official term in 1835 (Aguilera, 2013). Recently, the subject of gastronomy and its relation with food science and technology has been given increased attention in several scientific journals. A peer-reviewed *International Journal of Gastronomy and Food Science* focusing on the interface of both disciplines was launched in 2012, and has in its editorial board reputed scientists and several chefs holding Michelin stars.

3.1 Pleasure and Health

Since antiquity the pleasure of eating has been considered among the major joys in life. In an obvious exaggeration, Archestratus of Gela, the author of one of Europe's oldest cookery books (dating back to the fourth century B.C.), promoted the pleasure of eating even at the risk of being immoral (Wilkins, 2015). Recently, in what may also be considered as an overstatement, a food anthropologist has affirmed that "…eating makes us feel good; it is more important than sex" (Fox, 2014). Some of the most sensible thoughts about the pleasures of the table come from Jean Anthelme Brillat-Savarin, the French writer and gourmet, considered as one of the most influential figures in the history of gastronomy. He wrote the book *Physiologie du Gout, ou Méditations de Gastronomie Transcendante* in 1826, and in one of his many aphorisms suggested that "… the pleasures of the table belong to all times and ages, to every country and every day; they go hand in hand with all our other pleasures, outlast them, and remain to console us for their loss" (Brillat-Savarin, 1826).

In more recent times and on the anecdotal side, the first NASA astronauts complained that their food was unappetizing although nutritionally sound and microbiologically safe. Their protests were noted and later in the 1970s, the Skylab astronauts taking part in 6–8 week-long missions had a refrigerator and a freezer as well as a menu of 72 dishes selected in terms of a "measure of pleasure," transforming them in the first "gastronauts" (Lesso & Kenyon, 1972). It is now recognized that foods in space must not only provide safety and adequate nutrition but also enhance the well-being of the astronauts under very hostile living conditions (Perchonok & Bourland, 2002). A similar "gourmetization" has experienced in the last 40 years the field rations of the US army; they are more delicious and varied than ever (National Geographic, 2014).

In 1948, the constitution of the World Health Organization of the United Nations defined health as "a state of complete physical, mental and social well-being and not merely the absence of disease or infirmity" (WHO, 1947). Eventually, the notion of health became dominated by the biomedical perspective of the absence of disease (or abnormalities) that has almost completely overshadowed the possible roles of mental and social aspects on health (Misselbrook, 2014). Thus, when it comes to the positive effects of foods on well-being, the pleasure of eating has been paid a low attention compared to that given to nutrition. A search in the database PubMed (accessed on May 5, 2018) using the combined words (food + nutrition + health) and (food + pleasure + health) revealed that articles in the latter group were less than 0.7% of those containing the word "nutrition" (49,645 vs 328 items). The reductionist approach that health has only to do with the body (but conveniently excluding the brain) has neglected the role of foods in so many social activities unique to humans such as family life, the practice of culinary traditions, solidarity in times of hunger, etc., not to mention the positive effects of some food components for good mental health, happiness and cognitive power (Gómez-Pinilla, 2008). Block et al. (2011) have proposed a change in paradigm from food = health to food = well-being, the latter defined as a positive connection of foods with psychological, physical, emotional, and social aspects at the individual and societal levels. This new paradigm has stronger connections with the rise in popularity of cooking, gastronomy and socio-cultural attitudes such as commensality, pleasure and empathy. In the same vein, the term "comfort food" has been used to refer to foods that provide consolation, improve the mood or give a feeling of well-being, in other words, foods that offer some sort of psychological or emotional reassurance (Spence, 2017).

3.2 Perceiving Pleasure During Eating

Pleasure perception as related to food procurement and intake involves three components, each one linked to different neurobiological mechanisms in the brain (Kringelbach, 2015): "wanting," also called incentive salience or motivation; "liking," a hedonic reaction to the actual pleasure; and "learning," that encompasses associations, representations and predictions about future rewards based on the eating experience (Berridge, 2009). The time course of the different stages in the food reward cycle (before, during and after eating) is shown in Fig. 1. Initially, there is an appetitive phase of expectation or wanting that is a behavioral response that may or may not coincide with hunger (a physiological response that we need to eat). This initial stage is followed during actual eating of a meal by the liking phase

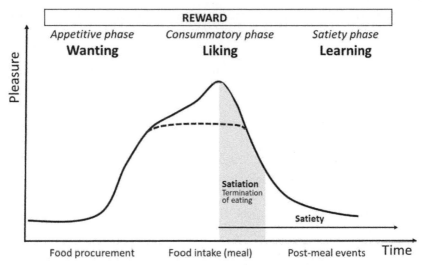

Fig. 1 Stages in the food reward cycle (before, during and after eating) and pleasure perception. *Adapted from Kringelbach, M. L. (2015). The pleasure of food: Underlying brain mechanisms of eating and other pleasures.* Flavor, 4, 20. doi:10.1186/s13411-014-0029-2.

of increased pleasure, that typically hits a plateau in the case of a normal meal or may exhibit a peak of pleasure (e.g., after the dessert of a tasty and enjoyable meal). After ingesting enough food there is a decrease in the pleasantness of the food leading to a physiological sensation of fullness called satiation (gray area in Fig. 1). The amount of food consumed and the point of satiation may vary among individuals, the occasion and whether people choose their food or not (Osdoba, Mann, Redden, & Vickers, 2015). To our misfortune, foods that are joyfully overeaten are usually highly palatable and energy dense. The cycle then enters a cognitive or learning phase where the expectations, manifestations and predictions for the food reward are updated and stored in the brain based on the eating experience. After completing eating, satiety or the fullness persisting afterward sets in until the physiological signals of hunger emanated from the hypothalamus emerge again. In the brain, wanting is associated with the neurotransmitter dopamine (as in addictions) while liking is related to the opioids that stimulate the receptors in the brain (Kuhar, 2012). Pleasure appears as neural activity in the orbitofrontal cortex of the brain that can be detected by neuroimaging techniques such as positron emission tomography (PET) and functional magnetic resonance imaging (f-MRI) (Berridge & Kringelbach, 2013). Identifying these brain regions is a first step toward understanding of the functional neuroanatomy of happiness (Kringelbach & Berridge, 2009).

There are two emerging fields in neurobiology related to foods and food consumption: neurogastronomy and nutritional psychiatry. Neurogastronomy deals with how the brain translates the signals resulting from the interactions of some food molecules with receptors in the mouth and nose, and converts them into patterns representing the sensations elicited by the food in our brains (Shepherd, 2012). Nutritional psychiatry, in turn, aims to link some foods or isolated food components (e.g., in the form of supplements) with psychiatric symptoms (e.g., depression) and their underlying neurobiology. This novel field nurtures from past evidence of the neurochemical properties of amino acids like tryptophan and other nutrients which are precursors for specific neurotransmitters in the brain such as serotonin (Zepf, Hood, & Guillemin, 2015). Interestingly, although there are multiple factors that contribute to the development of mental disorders, it is believed that the type of diet may be a modifiable factor for good mental well-being (Marx, Moseley, Berk, & Jacka, 2017).

3.3 Well-Being and Happiness

Well-being, together with pleasure, is frequently mentioned as values of a meal or diet. One definition of well-being describes it as an "equilibrium" between resources and challenges of psychological, physical and social situations (Dodge, Daly, Huyton, & Sanders, 2012). Well-being is also regarded as a "dynamic state," in which individuals are able to develop their potential, work productively and creatively, build strong and positive relationships with others, and contribute to their community (Anonymous, 2008). Psychologists have developed a well-being scale and used it in more than 500 published studies, that takes into account psychological dimensions such as self-acceptance, positive relations with others, purpose in life, personal growth, etc. (Ryff, 1989).

Happiness is often regarded as a feeling of satisfaction within an individual's long-term emotional state. Eating is an important source of personal pleasure (hedonics), hence, it may contribute to happiness (Berenbaum, 2002; Macht, Meininger, & Roth, 2005). However, in most cases the pursuit of happiness is not seen as something that individuals can achieve in solitude but as an experience of shared relationships (Uchida & Oishi, 2016). A few studies address the issue of the positive psychological implications of cooking, eating pleasurable foods, following "healthy" diets and sharing meals with others. They are generally based on the analysis of large data from national surveys, questionnaires involving target groups, or intervention studies (Table 2). These studies reinforce the dual role of food in

Table 2 Studies Reporting the Effect of Eating and Diets on Pleasure and Well-Being

Group	Type of Study	Outcomes/Conclusions	References
Young people	Semi-structured interviews with nine female and seven males (mostly university students)	People who enjoy eating were physiologically ready to do so, ate slowly, focused upon salient features of foods, and engaged in social activities during the meal	Macht et al. (2005)
	Interviews with 45 young adults in France and Germany	Although eating with others is usually pleasurable, it may also be stressful. Young adults may often choose to eat alone	Danesi (2012)
Adults	National survey of members of Australian households aged 15 years and older. Total: around 12,000 individuals	Happiness gains and well-being from increased consumption of fruit and vegetables may occur faster than any long-term improvement to people's health	Mujcic and Oswald (2016)
	Survey to adults and college students from different countries on diet–health links, worry about food, food as a positive force in life, etc.	Americans associate food mostly with health and least with pleasure, and are the least likely to classify themselves as healthy eaters. French and Belgians tend to occupy the pleasure extreme	Rozin, Fischler, Imada, Sarubin, and Wrzesniewski (1999)
Elderly	Interviews to find out the impact of cooking meals among retired women (some widows)	Most women had lost the meaning of cooking and felt loneliness at mealtimes. Widows, in particular, found joy in cooking for guests and the commensality during meals	Sidenvall, Nydahl, and Fjellström (2000)
	Mailed questionnaire to assess the effect of candy on men born in 1919–1934	Chocolate preference among elderly was associated with better health, optimism and better psychological well-being	Strandberg et al. (2008)
	Intervention based on nutrition education and cooking classes to 59 elderly (average 69 y/o)	The group improved their diet quality (e.g., intake of vitamin C and fiber). The intervention had a favorable effect on their psychological well-being	Jyväkorpi et al. (2014)

Continued

Table 2 Studies Reporting the Effect of Eating and Diets on Pleasure and Well-Being—cont'd

Group	Type of Study	Outcomes/Conclusions	References
	Olfactory and gustatory functions were measured on 239 healthy individuals (65–101 y/o) and good cognitive status	Chemosensory impairment detected was 41% for taste and 33% for olfaction, however, it was not related with eating pleasure and loss of appetite leading to malnutrition	Arganini and Sinesio (2015)
	Randomized control study of 626 seniors at Meals on Wheels programs	Home-delivered daily meals improve the well-being of older adults beyond nutrition by reducing feelings of loneliness	Thomas, Akobundu, and Dosa (2016)
	Mailed questionnaire responded by over 11,000 adults age 65 years or older on issues like life satisfaction and food patterns	Elderly persons in Norway with "healthier" food patterns had higher scores on life satisfaction and felt less depressed than those on "unhealthy" diets	André et al. (2017)
Large cohorts	Mailed questionnaire on aspects such as life satisfaction and happiness. Around 88,000 participants in Thailand	Being unhappy was associated with eating alone frequently and implied possible adverse psychological effects	Yiengprugsawan et al. (2015)

fulfilling basic bodily functions as well as a source of pleasure, positive emotions, social construction, and the support of personal identity (Hausman, 2005). Eating tasty foods provoke positive feelings such as warmth, contentment and relaxation (Hill, Magson, & Blundell, 1984). Commensality or sharing meals with others, that is so typical of humans and an event of daily social occurrence because of biological needs, has been associated with happiness (Blades, 2009; Fischler, 2011). Eating in company enhances the emotional experience of having a meal and is an opportunity to bond with friends (Brown, Edwards, & Hartwell, 2013). Special feasts or celebrations (birthdays, anniversaries, etc.) involving food have also been found to be occasions for happiness (Yiengprugsawan et al., 2015). Nevertheless, results from most of these intervention studies should be taken with caution as they are not longitudinal studies and may be subject to the Hawthorne effect: the tendency of some participants to respond better to the questions when they know are part of an experiment.

3.4 Eating for Pleasure

While food technology deals with bringing food into supermarkets and nutrition is mainly concerned in food that has been swallowed, gastronomy makes sure that food during eating has a value in terms of flavor, traditions, remembrances, habits and beliefs (Coveney & Santich, 1997). As pointed out previously, delicious dishes and meals cause great pleasure to many people, particularly when they are eaten in a social context. People who appreciate eating have the explicit intention to enjoy the food, eat it slowly and focus upon salient features of the meals and the environment (Macht et al., 2005). Although the typology of diners is far from being clear and consistently agreed on, several terms are used to classify diners in the scientific as well as popular literature (Aguilera, 2013):

- The gastronome, the referee of good taste in food, occupies the highest level on the scale. A gastronome is someone who enjoys cooking and eating food elaborately prepared, and values the finest dishes of the culinary art as well as the quality and taste of the ingredients.
- The gourmet is a connoisseur of good food and a person with a discriminating palate who is highly selective and refined in eating behaviors. The gourmet considers food selection and consumption as intellectual pleasures (Bruce, 1999). An epicure is someone who enjoys eating food that is of very good quality, especially unusual, novel or exotic food, and not in excess.

- The gourmand, one who is excessively fond of eating and drinking (sometimes too much), enjoys everything that is good food, but is one level below the gourmet because he/she eats with more voracity and less overall refinement (Mick, Burroughs, Hetzel, & Brannen, 2004).
- The term foodist is in vogue to describe a person who is knowledgeable about or keenly interested in foods.
- At the bottom of the hierarchy is the glutton, who represents someone who eats excessively without savoring the food, unable to control his tastes and inclinations (Bruce, 1999).

It has been suggested that there are the "gastro-nomers" who enjoy food per se and the "gastro-anomers," that think that eating is a necessity for good health (Fischler, 1980). This dichotomy has been exemplified by the alimentation habits of the French and the Americans. The French spend more time eating, eat less (probably because of smaller portion sizes) and snack less, do more physical activity and follow traditions of moderation, focus on quality, and emphasis on the joys of eating as a cultural and social activity (Rozin, 2005). Although the French eat a diet high in saturated fat, people have low rates of cardiovascular diseases (a phenomenon known as the "French paradox") and the incidence of overweight and obesity are much lower than in the United States. Thus, keeping a healthy weight should not be at the expense of the pleasure of eating, rather, it appears to be a matter of making more exercise and eating less (Hill & Peters, 1998; Rozin, 2005).

3.5 Restaurants and Fine Dining

The word restaurant dates back to the 1760s, when a Parisian merchant started to supply "restorative broths" or *restaurants*, and now it applies to an establishment where meals are served at certain times, either from a set menu or *a la carte*. A fine dining restaurant provides full service with a high quality menu in terms of raw materials, ingredients and creativity imparted by highly qualified chefs, as well as a unique ambiance in terms of a lavish decoration and table service (Mealey, 2018).

It is quite difficult to assess the number of fine dining restaurants around the world. Supposedly, there are 1 million restaurants located in the United States (650,000 according to other sources) (Meggiato, 2016), and fine dining restaurants may represent approximately 10% of the total (based on sales) (Ban, 2012). Another appraisal of the number of this type of restaurants may be inferred from the 2017 Michelin Red Guide, a hallmark of fine dining quality. Although limited mostly to 22 countries in Europe, its webpage lists 27,446 restaurants in 31 countries around the world with 4362 of them located in France (ViaMichelin, 2018).

Until the end of the 20th century French chefs dominated the gastronomic scene. Marie Antoine Carême (1784–1833) was the founder of the *haute cuisine* and Auguste Escoffier (1846–1935) established the principles of modern cuisine and revolutionized the way restaurant dining was conceived. In the 1970s, young chefs founded the *nouvelle cuisine*: a lighter and more delicate version of the French cuisine using local fresh produces and reduced cooking times. The end of the past century saw the ephemeral existence of a techno-emotional or molecular cooking using scientific techniques, powder ingredients and novel equipment (Liberman, 2014). The top restaurants in the world today aside from offering a flavorful cooking and a unique experience, use locally grown and produced ingredients, invite to discover exotic terroirs and ancestral cuisines, and offer evocative dishes (The World 50 Best, 2017).

3.6 The Role of Modern Chefs

Haute-cuisine chefs are becoming the most innovative actors on the food scene because they are compelled to update their menus to guard themselves against plagiarism (Albors-Garrigos, Barreto, García-Segovia, Martínez-Monzó, & Hervás-Oliver, 2013). In the search to create new dishes and gastronomic experiences, some of these chefs have adopted novel techniques and used refined ingredients to make the dishes of their tasting menus, which have become later a standard in the business (Hill, 2009). Modern top chefs assert that the pleasure of eating at their restaurants goes beyond the enjoyable sensorial perception in our mouths, and projects to the brain where it takes the form of emotions, evocations, and imagination (Arboleya, Lasa, Oliva, Vergara, & Aduriz, 2012).

As can be deduced from their gastronomic manifestos, several famous chefs are promoting sustainable agriculture, ancestral cooking, the use of locally grown or sourced products, reduced food wastes, and "healthier" diets (Schösler & de Boer, 2018). Their extensive communication networks and credibility facilitate the diffusion of these concepts into the society. Top chefs are becoming progressively concerned about the nutritional value of their dishes and menus, and many restaurants are increasing their "healthy" and nutritional food options and designing ad hoc menus (Navarro, Serrano, Lasa, Aduriz, & Ayo, 2012; Ozdemir & Caliskan, 2015).

Chefs in collaboration with nutritionists and food technologists may provide attractive nutritional messages to children and culinary advice to food service workers involved in feeding the young. Tasty, attractive and cost-effective menus in school lunch programs may improve the perception of some foods, such as vegetables (Cohen et al., 2012; Just, Wansink, & Hanks, 2014). Chefs may also help in endorsing the consumption of underutilized food sources (e.g., algae, small fish, mushrooms, etc.), foods grown locally and ancestral dishes.

4. THE NEW PARTICULAR ALIMENTATION: FEEDING THE OLD

The steadily worldwide increase in human lifespan is usually attributed to improvements in living conditions, sanitation and healthcare. The phenomenon has prolonged human aging with the biological consequence of the progressive accumulation of damage to cells and tissues during life. This biological deterioration is the primary risk factor for major diseases, including cancer, diabetes, cardiovascular disorders, and neurodegenerative conditions (López-Otín, Blasco, Partridge, Serrano, & Kroemer, 2013). Since aging takes place in a socio-cultural environment, it also has psychological and sociological implications which are often neglected as part of the quality of life of the elderly (Corner, Brittain, & Bond, 2006).

4.1 An Aging Population

Never in the history of mankind there have been so many old people alive. Individuals over 60 years old approach a billion while centenarians (individuals over 100 years of age) exceed 415,000 (Robine & Cubaynes, 2017). In fact, the group of older people is the fastest growing segment of the world population, representing a "silver tsunami" (Shlisky et al., 2017). By 2020 more than 700 million people will be over 65 years of age and by 2045 life expectancy at birth will have risen from the current 70 to 77 years. It is anticipated that around 2 billion people by 2050 will be aged 60 and over and in many countries (e.g., Japan and South Korea) ca. 15% of their population will be over 80 years of age, contributing to a worldwide total of around 400 million (Table 3). This sudden surge in the number of very old people occurs at a time when practically there is no information or past experience of what extreme human longevity means.

Many elderly people become disabled and dependent on others for acquiring, preparing and/or consuming their food. Thus, in order to understand the implications of this recent phenomenon in regards to alimentation

Table 3 Percentage of Population in Age Groups Over 60 and Over 80 Years Old, by Country

	2015		2050	
Country	60+	80+	60+	80+
Australia	20.4	3.9	28.3	8.3
Austria	24.2	5.1	37.1	12.9
Canada	22.3	4.2	32.4	10.6
Chile	15.7	2.7	32.9	10.3
China	15.2	1.6	36.5	8.9
France	25.2	6.1	31.8	11.1
Germany	27.6	5.7	39.3	14.4
Japan	33.1	7.8	42.5	15.1
Mexico	9.6	1.5	24.7	5.4
Netherlands	24.5	4.4	33.2	11.8
Rep. of Korea	18.5	2.8	41.5	15.9
United States	20.7	3.8	37.9	8.3
World	12.3	1.7	21.5	4.5

Source: Based on United Nations (2015). *World population prospects: The 2015 revision, key findings and advance tables*. Working Paper No. ESA/P/WP.241. Department of Economic and Social Affairs, Population Division.

and aging it is worthwhile to examine the cases of Japan and South Korea. Fig. 2 shows that the average age of death of the 30 oldest people in Japan increased for both sexes in the period from 1950 to 2005 but it increased faster than predicted by the historic trend line since 1973. This is the result of improved medical care programs that provide better and cheaper health care to those aged 70 or over (Ogawa, Mason, Chawla, & Matsukura, 2010). Two other trends in both countries are also relevant to the care of the elderly. The decrease in multi-generational households that use to be fairly common in Japan and South Korea, and the increase in the number of elderly people who die in hospitals rather than while being taken care of by their families (Lee, Yoon, & Kropf, 2007; Ogawa et al., 2010).

Food requirements for elderly people may be divided into those related to the oral experience (e.g., mastication, sensory enjoyment, safe swallowing) and those associated to other physiological changes of aging (e.g., changes in body composition, nutritional needs and related diseases).

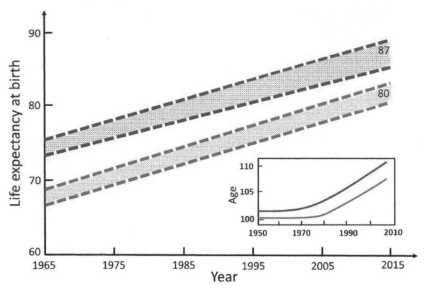

Fig. 2 Variation in life expectancy in nine countries (Japan, Canada, U.S.A., France, Germany, Iceland, Italy, Switzerland and U.K.; 1965–2015). Inset: Change in average age of death among 30 oldest persons by sex (Japan, 1950–2006). Lines and range: male (blue), female (red). Numbers indicate upper bound. *Sources: From http://www.mhlw.go.jp/english/ database/db-hw/dl/81-1a2en.pdf; Ogawa, N., Mason, A., Chawla, A., & Matsukura, R. (2010). Japan's unprecedented aging and changing intergenerational transfers. In T. Ito & A. Rose (Eds.),* The economic consequences of demographic change in East Asia *(pp. 131–160). Chicago: University of Chicago Press.*

A limitation of published data is that the majority of the studies on food intake are on seniors aged between 60 and 75 years, whereas elderly people aged 80 and above have been largely understudied (Doets & Kremer, 2016).

4.2 Food Consumption and Nutritional Needs During Aging

Many older people experience an impairment in the oral processing of foods, a gradual loss in appetite and flavor perception, weight loss and nutritional frailty due to changes in body physiology, psychological functioning and age-related diseases (Pilgrim, Robison, Sayer, & Roberts, 2015; Shlisky et al., 2017). The following sections are largely based on our recent review on the subject (Aguilera & Park, 2016).

4.2.1 Masticatory and Swallowing Dysfunctions

Aging individuals experience increasing difficulties in masticating and swallowing the oral contents due to anatomical and physiological alterations.

The difficulty in chewing caused by the loss of teeth (edentulism) and the impairment of jaw muscle activity are common among older adults. Dysphagia or the difficulty in pharyngeal swallowing during the transit of the oral contents to the stomach is estimated to affect around 590 million people globally (Cichero et al., 2017). Dysphagia increases dramatically during aging and can cause malnutrition and death by aspiration pneumonia, particularly in the elderly in nursing homes and rehabilitation hospitals (Khan, Carmona, & Traube, 2014). The ratio of aspirational pneumonia to total cases of pneumonia in patients hospitalized in Japan increases with advancing age and over 80% of the pneumonia patients diagnosed with aspiration pneumonia were elderly aged 70 years and older (Teramoto, 2014). In a recent review it was reported that the rate of aspiration of food was high for thin liquids and very low for most solid foods (Konishi et al., 2018). Thus, it has been recommended that foods for patients suffering from dysphagia should break down into soft and small particles (e.g., <1.5 mm) and that the swallowed bolus should be moist, cohesive and slippery (Cichero et al., 2013).

4.2.2 Loss of Chemosensory Perception and Appetite

A large number of elderly people exhibit a progressive loss of taste, smell and trigeminal stimuli, which has a negative effect on their food preferences, dietary habits and the enjoyment of meals (Doets & Kremer, 2016). In particular, olfaction and taste decline considerably after the seventh and fifth decade of age, respectively (Boyce & Shone, 2006). The cause has been attributed to neuroanatomical changes in the respective receptor cells and in the orbitofrontal cortex involved in processing the pleasant stimuli. This loss of chemosensory sensations often results in reduced appetite, poor meal appreciation, a decreased food intake, loss in body weight and eventually in nutritional deficiencies (Ross Watson, 2017). All this is quite unfortunate since for most elderly flavor perception is a strong determinant of their food choices. Therefore, their loss of flavor sensitivity has to be compensated by the addition of extra tastants and odorants to foods (Rolls, 1993).

4.2.3 Loss of Muscle and Bone Mass

Most people start to lose weight and feel weak around age 65 due to a progressive decline in skeletal muscle mass and function (sarcopenia) that is partly attributed to the atrophy and denervation of the muscle fibers (Beaudart, Zaaria, Pasleau, Reginster, & Bruyère, 2017). Thus, older adults require an average daily intake of protein at least in the range of 1.0–1.2 g per kilogram of body weight per day, to maintain and regain muscular mass

(Bauer et al., 2013). However, exercise is the most effective intervention to prevent and treat sarcopenia, and to extend an independent functional lifespan in the elderly (Montero-Fernández & Serra-Rexach, 2013). Osteoporosis is a prevalent skeletal disorder during senescence characterized by a loss of bone mass and bone strength leading to an increased risk of fracture. Older women are more exposed to fractures because their bone loss accelerates after menopause, hence, they suffer around 80% of hip fractures (Liberman & Cheung, 2015). A poor diet (e.g., one with a low intake of protein, vitamin D and calcium) and a sedentary life are closely associated with sarcopenia and osteoporosis.

4.2.4 Some Specific Nutritional Needs

Older people are particularly exposed to nutritional deficiencies due to a reduced appetite and a decrease in food intake. Several nutrients may be inadequately consumed among older adults, including protein, omega-3 fatty acids, dietary fiber, carotenoids (vitamin A precursors), calcium, magnesium, potassium, and vitamins B6, B12, D, and E (Shlisky et al., 2017). Energy requirements decrease with age and are different for males and females aged 85 and over. Total plasma and LDL cholesterol are known to increase with age and constitute risk factors associated with cardiovascular diseases (Liu & Li, 2015). Many old adults suffer from dementia and Alzheimer disease but there is not enough scientific evidence that ingestion of omega-3 fatty acids improves the cognitive function or quality of life of seniors without pathological conditions (Ubeda, Achón, & Varela-Moreiras, 2012).

Aside from a deterioration of the general physical health, aging is a risk factor for depression occasioned by psychological and social changes such as the loss of community contact, withdrawal from the workplace, independence from their children and a limited mobility. Several studies show that elderly who eat alone have increased risks of poor nutritional status and weight control, depression and ultimately death (Tani, Sasaki, Haseda, Kondo, & Kondo, 2015). In conclusion, physiological dysfunctions and specific nutritional needs developed during aging require a sourcing of special foods which in most cases have to be soft, easily and safely swallowed, nutritious and tasty. Moreover, the psycho-social aspects surrounding eating should not be underestimated.

5. FOOD FOR *GASTROSENIORS*

The enjoyment of eating lasts during the whole life. Old people with impaired physiological functions complain that their foods are tasteless, have a poor appearance and lack variety. These elderly expect to derive several positive values from their daily meals, including health and well-being, pleasure and happiness (Costa & Jongen, 2010). However, having meals in solitude and eating unappetizing foods only deepen their condition of frailty and disability (Table 2). Thus, the particular alimentations of the elderly demand not only nutrition but also attractive foods which are easy to handle and cook, and menu designs as well as enriching dining experiences that improve the quality of life of those who live alone and cannot cook by themselves or reside in long-term care facilities (Edfors & Westergren, 2012; Farmer, Poirier, & Tuan, 2017).

5.1 Foods for the Elderly

Approaches to supply adequate and attractive foods to elderly people vary in different countries. The Chinese draw on their unique medicinal cuisine by combining herbal ingredients with traditional culinary materials to produce tasty foods with health restoring properties (Fu, 2011). In South Korea the market value of the "senior-friendly" food industry in 2010 (year of last data available) was around US$ 4 million and it was growing at 11% per year (KHIDI, 2013). Since around 50% of elderly South Koreans have chewing difficulty (Kwon et al., 2017), techniques used in molecular cooking, particularly the gelation and spherification methods, have been used to produce soft foods with a very high preference among the elderly (Kim & Joo, 2015).

Because of convenience in preparation and serving as well as easy-to-open packaging, ready-to-eat meals are quite popular among older people, particularly the independent elderly (Hoffman, 2016). Special foods for the frail elderly can be broadly classified into complete meals delivered at home, modified foods for easy chewing, and thickened fluid foods for dysphagia patients. Social programs for home-delivery of prepared meals (e.g., as those supplied by Meal on Wheels) keep the elderly out of nursing homes while helping them to meet the basic food needs (Thomas et al., 2016). Modified foods use alterations of the texture to reduce the need to chew. They include foods that have been minced, mashed or pureed which are consumed by an estimated 15–20% of patients in long-term care facilities (Cichero, 2016; Stahlman, Garcia, Hakel, & Chambers, 2000). Some pureed items have been mixed with thickeners and molded to look as esthetically pleasing as the original food (Reilly, Frankel, & Edelstein, 2013). Within the category of modified foods are also those that have been softened by physical or enzymatic methods, retaining their original shape and color

(Aguilera & Park, 2016). New foods for the particular alimentations of the elderly include the design of novel ready-to-eat meal components and ready meals (Costa & Jongen, 2010). 3D printing appears as a promising technology to design personalized foods for the elderly with mastication and swallowing problems (Liu, Zhang, Bhandari, & Wang, 2017).

In the case of dysphagia, texture modification (TM) is one of the most common intervention approaches to cope with needs of patients. TM-foods and thickened drinks are classified into several categories based on properties such as hardness, cohesiveness, adhesiveness and viscosity. Examples are a guide developed by Swedish researchers (Wendin et al., 2010), and the standards developed by the Japanese Society of Dysphagia Rehabilitation (JSDR, 2013) and the International Dysphagia Diet Standardization Initiative (IDDSI, 2015). Additional categories of TM-foods are usually generated independently by hospitals. The IDDSI Framework, released in November of 2015, is summarized as a graphical continuum eight levels of textures ranging from regular and pureed foods to slightly thick and liquid foods, and identified by numbers, text labels and color codes. This classification is recommended for implementation throughout the world (Cichero et al., 2017). A list of some of the commercial thickeners used for dysphagia patients and recommendations for use and costs is presented in a bulletin prepared by National Health Service in England (NHS, 2015). Nevertheless, some studies have questioned the effectiveness of TM-foods on clinical outcomes such as aspiration pneumonia, nutrition, hydration, morbidity, and mortality for people with dementia and living in residential care facilities (Painter, Le Couteur, & Waite, 2017).

5.2 Foods for the Elderly in Japan

Japan is a super-aging society where ca. 8% of the population (equivalent to over 10 million people) is aged over 80 years (Table 3). A distinguishing negative feature of the current aging Japanese society is that as it gets older their happiness levels do not increase (they may even decrease), contrary to what occurs in several Western countries where well-being follows a U-shaped curve with age (The Cabinet Office of Japan, 2011). In response, the Japanese government has deployed several public services benefiting the elderly in the areas of pensions, health and long-term care (Muramatsu & Akiyama, 2011). Due to decrease in familial care for aging relatives, the country is facing a rapid increase in demand for aged care facilities and presently over 6% of Japanese aged 65 years or older live in this type of institutions (Annear, Otani, & Sun, 2016). Moreover, malnutrition among the elderly has become a serious problem. In a survey, more than 50% of the elderly in hospitals and institutions were found to be malnourished (Iisaka, Tadaka, & Sanada, 2008).

Several Japanese food companies offer a variety of soft-food products aimed at people presenting mastication problems, at risk of malnutrition and/or subject to aging diseases (Higashiguchi, 2015). Companies emphasize the "enjoyment of their meals" and their products are getting greater prominence on supermarket shelves (Food Makers, 2017). Commercial foods for people who have difficulty in chewing and swallowing are called "Engay" or easy-to-swallow foods (Penso, 2017). They are based on technologies that soften the original food structure (e.g., freeze-thawing with or without enzyme infusion, enzyme impregnation, high-pressure processing, pulsed electric fields and sonication) while preserving their color and flavor (Aguilera & Park, 2016). Other types of Engay products are pureed foods reshaped using a gelling agent and molded to simulate the original appearance of the food (Sugimoto, 2013).

In 2013, the Ministry of Agriculture, Forestry and Fisheries (MAFF) created three labels with the logo "Smile Care Foods" to be placed on packaged foods aimed at: (i) people without eating dysfunctions but needing supplementary nutrition (blue label); (ii) individuals having trouble with swallowing foods (red label); and (iii) people with chewing problems (yellow label) (MAFF, 2017). The types of products in "Smile Care Foods" include processed single-item foods, prepared dishes, and single-serving meals (e.g., for delivery service), but above all, they are required "to provide delicious taste and enjoyment of eating" (Mastication Friendly Foods, 2016).

6. CONCLUSIONS

The concept of particular alimentations responds to the empirical observation that people today have multiple ways to achieve a sensation of fulfillment for their bodies and minds through the foods they eat. The gourmet and the elderly belong to two particular alimentations that have received less attention relative to those based on medical or personal beliefs. Nevertheless, the worldwide "gourmetization" of taste together with the alimentation needs of a fast aging population with prolonged lifespans are phenomena that deserve a better attention. The present dominating view that foods contribute to health and well-being only through nutrition should make room for their impacts on psychological, cultural and social aspects of life. Modern nutrition research should encompass the effects and synergies of nutrients from whole meals, the role of microbiota in health, and the multiple relations between the gut and the brain.

With an increasing proportion of foods being consumed away from home, menu planners and cooks are playing an important role as healthy foods and diets are more likely to be adopted if they involve tasty and enjoyable meals. In turn, chefs of *haute-cuisine* restaurants are innovators that continuously incorporate new raw materials and cooking technologies in a quest for unique textures, flavors and eating experiences. As a consequence of the continuous worldwide increase in life expectancy a huge market of 400 million elderly over 80 years of age is expected by 2050. Food scientists and food technologists have the opportunity to contribute to the design of nutritious and enjoyable food products targeting at the needs of the frail elderly. Advances in these two particular alimentations, the gourmet and the elderly, call for interdisciplinary work between food scientists, nutritionists, chefs, neuroscientists, gerontologists and psychologists in order to take into consideration the multiple aspects involved in supplying them with foods and meals that bring health, pleasure and well-being from cradle to grave. In general, particular alimentations present interesting marketing opportunities for the food business as they comprise several thousand of specialized products and ingredients.

ACKNOWLEDGMENTS

Part of this work was supported by the Korea Food Research Center under the project "Microgels as Food Materials for the Elderly." Authors acknowledge the assistance of Alicia Leon in providing some of the entries of Table 2 and reviewing the references.

REFERENCES

Aguilera, J. M. (2013). *Edible structures: The basic science of what we eat.* Boca Raton: CRC Press.
Aguilera, J. M. (2018). Food engineering into the XXI century. *AICHE Journal, 64,* 2–11. https://doi.org/10.1002/aic.16018.
Aguilera, J. M., & Park, D. J. (2016). Texture-modified foods for the elderly: Status, technology and opportunities. *Trends in Food Science and Technology, 57,* 156–164. https://doi.org/10.1016/j.tifs.2016.10.001.
Albors-Garrigos, J., Barreto, V., García-Segovia, P., Martínez-Monzó, J., & Hervás-Oliver, J. L. (2013). Creativity and innovation patterns of haute cuisine chefs. *Journal of Culinary Science and Technology, 11,* 19–35. https://doi.org/10.1080/15428052.2012.728978.
André, B., Canhão, H., Espnes, G. A., Ferreira, A. M., Rodrigues, M., Gregorio, J., et al. (2017). Is there an association between food patterns and life satisfaction among Norway's inhabitants ages 65 years and older? *Appetite, 110,* 108–115. https://doi.org/10.1016/j.appet.2016.12.016.

Annear, M. J., Otani, J., & Sun, J. (2016). Experiences of Japanese aged care: The pursuit of optimal health and cultural engagement. *Age and Ageing*, *45*, 753–756. https://doi.org/10.1093/ageing/afw144.

Anonymous. (2008). The foresight mental capital and wellbeing project. In *Final report—Executive summary*. London: The Government Office for Science.

Anton, S., & Leeuwenburgh, C. (2013). Fasting or caloric restriction for healthy aging. *Experimental Gerontology*, *48*, 1003–1005. https://doi.org/10.1016/j.exger.2013.04.011.

Arboleya, J. C., Lasa, D., Oliva, O., Vergara, J., & Aduriz, A. L. (2012). The pleasure of eating. In C. Vega, J. Ubbink, & E. van der Linden (Eds.), *Kitchen as laboratory* (pp. 254–263). New York: Columbia University Press.

Arganini, C., & Sinesio, F. (2015). Chemosensory impairment does not diminish eating pleasure and appetite in independently living older adults. *Maturitas*, *82*(2), 241–244. https://doi.org/10.1016/j.maturitas.2015.07.015.

Ban, V. (2012). *Analysis of the upscale/fine dining sector in the restaurant industry. In Paper 10MBA student scholarship:* Johnson & Wales University. http://scholarsarchive.jwu.edu/mba_student/10.

Bang, H. O., Dyerberg, J., & Sinclair, H. M. (1980). The composition of the Eskimo food in north western Greenland. *The American Journal of Clinical Nutrition*, *33*, 2657–2661. https://doi.org/10.1093/ajcn/33.12.2657.

Barlow, S. M., Boobis, A. R., Bridges, J., Cockburn, A., Dekant, W., Hepburn, P., et al. (2015). The role of hazard- and risk-based approaches in ensuring food safety. *Trends in Food Science & Technology*, *46*, 176–188. https://doi.org/10.1016/j.tifs.2015.10.007.

Bauer, J., Biolo, G., Cederholm, T., Cesari, M., Cruz-Jentoft, A. J., Morley, J. E., et al. (2013). Evidence-based recommendations for optimal dietary protein intake in older people: A position paper from the PROT-AGE Study Group. *Journal of the American Medical Directors Association*, *14*, 542–559. https://doi.org/10.1016/j.jamda.2013.05.021.

Beaudart, C., Zaaria, M., Pasleau, F., Reginster, J.-Y., & Bruyère, O. (2017). Health outcomes of sarcopenia: A systematic review and meta-analysis. *PLoS One*, *12*(1), e0169548. https://doi.org/10.1371/journal.pone.0169548.

Berenbaum, H. (2002). Varieties of joy-related activities and feelings. *Cognition and Emotion*, *16*, 473–494. https://doi.org/10.1080/0269993014000383.

Berridge, K. C. (2009). "Liking" and "wanting" food rewards: Brain substrates and roles in eating disorders. *Physiology & Behavior*, *97*(5), 537–550. https://doi.org/10.1016/j.physbeh.2009.02.044.

Berridge, K. C., & Kringelbach, M. L. (2013). Neuroscience of affect: Brain mechanisms of pleasure and displeasure. *Current Opinion in Neurobiology*, *23*(3), 294–303. https://doi.org/10.1016/j.conb.2013.01.017.

Blades, M. (2009). Food and happiness. *Nutrition & Food Science*, *39*(4), 449–454. https://doi.org/10.1108/00346650910976310.

Block, L. G., Grier, S. A., Childers, T. L., Davis, B., Ebert, J. E. J., Kumanyika, S., et al. (2011). From nutrients to nurturance: A conceptual introduction to food well-being. *Journal of Public Policy & Marketing*, *30*, 5–13. https://doi.org/10.1509/jppm.30.1.5.

Bollinger, A. (1993). *Así se Alimentaban los Inkas*. Cochabamba, Bolivia: Los Amigos del Libro.

Booth, F. W., Roberts, C. K., & Laye, M. J. (2012). Lack of exercise is a major cause of chronic diseases. *Comprehensive Physiology*, *2*, 1143–1211. https://doi.org/10.1002/cphy.c110025.

Borek, C. (2017). Dietary antioxidants and human cancer. *Journal of Restorative Medicine*, *6*, 53–61. https://doi.org/10.14200/jrm.2017.6.0105.

Boyce, J. M., & Shone, G. R. (2006). Effects of ageing on smell and taste. *Postgraduate Medical Journal, 82*(966), 239–241. https://doi.org/10.1136/pgmj.2005.039453.

Brillat-Savarin, J. A. (1826). *Physiologie du Gout, ou Méditations de Gastronomie Transcendante.* Paris: A. Sautelet et Cie Libraires.

Brown, L., Edwards, J., & Hartwell, H. (2013). Eating and emotion: Focusing on the lunchtime meal. *British Food Journal, 115*, 196–208. https://doi.org/10.1108/00070701311302186.

Bruce, N. (1999). Classification and hierarchy in the discourse of wine: Émile Peynaud's *The Taste of Wine. ASp, 23–26*, 146–164. https://doi.org/10.4000/asp.2376.

Bruder, C. E. G., Piotrowski, A., Gijsbers, A. A. C. J., Andersson, R., Erickson, S., Diaz de Ståhl, T., et al. (2008). Phenotypically concordant and discordant monozygotic twins display different DNA copy-number-variation profiles. *American Journal of Human Genetics, 82*(3), 763–771. https://doi.org/10.1016/j.ajhg.2007.12.011.

Buchler, S., Smith, K., & Lawrence, G. (2010). Food risks, old and new. Demographic characteristics and perceptions of food additives, regulation and contamination in Australia. *Journal of Sociology, 46*, 353–374. https://doi.org/10.1177/1440783310384449.

Cichero, J. A. Y. (2016). Adjustment of food textural properties for elderly patients. *Journal of Texture Studies, 47*(4), 277–283. https://doi.org/10.1111/jtxs.12200.

Cichero, J. A. Y., Lam, P., Steele, C. M., Hanson, B., Chen, J., Dantas, R. O., et al. (2017). Development of international terminology and definitions for texture-modified foods and thickened fluids used in dysphagia management: The IDDSI Framework. *Dysphagia, 32*, 293–314. https://doi.org/10.1007/s00455-016-9758-y.

Cichero, J. A. Y., Steele, C., Duivestein, J., Clavé, P., Chen, J., Kayashita, J., et al. (2013). The need for international terminology and definitions for texture-modified foods and thickened liquids used in dysphagia management: Foundations of a global initiative. *Current Physical Medicine and Rehabilitation Reports, 1*, 280–291. https://doi.org/10.1007/s40141-013-0024-z.

Cohen, J. F. W., Smit, L. A., Parker, E., Bryn Austin, E., Lindsay Frazier, A., Economos, C. D., et al. (2012). Long-term impact of a chef on school lunch consumption: Findings from a 2-year pilot study in Boston middle schools. *Journal of the Academy of Nutrition and Dietetics, 112*, 927–933. https://doi.org/10.1016/j.jand.2012.01.015.

Cordain, L., Eaton, S. B., Sebastian, A., Mann, N., Lindeberg, S., Watkins, B. A., et al. (2005). Origins and evolution of the Western diet: Health implications for the 21st century. *American Journal of Clinical Nutrition, 81*(2), 341–354. https://doi.org/10.1093/ajcn.81.2.341.

Corner, L., Brittain, K., & Bond, J. (2006). Social aspects of ageing. *Women's Health Medicine, 3*, 78–80. https://doi.org/10.1383/wohm.2006.3.2.78.

Costa, A. I. A., & Jongen, W. M. F. (2010). Designing new meals for an ageing population. *Critical Reviews in Food Science & Nutrition, 50*, 489–502. https://doi.org/10.1080/10408390802544553.

Coveney, J., & Santich, B. (1997). A question of balance: Nutrition, health and gastronomy. *Appetite, 28*, 267–277. https://doi.org/10.1006/appe.1996.0083.

Danesi, G. (2012). Pleasures and stress of eating alone and eating together among French and German young adults. *The Journal of Eating and Hospitality, 1*, 77–91.

Dibb, M., Teubner, A., Theis, V., Shaffer, J., & Lal, S. (2013). Review article: The management of long-term parenteral nutrition. *Alimentary Pharmacology & Therapeutics, 37*, 587–603.

Dicksved, J., Halfvarson, J., Rosenquist, M., Järnerot, G., Tysk, C., Apajalahti, J., et al. (2008). Molecular analysis of the gut microbiota of identical twins with Crohn's disease. *The ISME Journal, 2*, 716–727. https://doi.org/10.1038/ismej.2008.37.

Dodge, R., Daly, A., Huyton, J., & Sanders, L. (2012). The challenge of defining wellbeing. *International Journal of Wellbeing, 2*(3), 222–235. https://doi.org/10.5502/ijw.v2i3.4.

Doets, E. L., & Kremer, S. (2016). The silver sensory experience—A review of senior consumers' food perception, liking and intake. *Food Quality and Preference*, *48*, 316–332. https://doi.org/10.1016/j.foodqual.2015.08.010.

Edfors, E., & Westergren, A. (2012). Home-living elderly people's views on food and meals. *Journal of Aging Research*. 2012. Article ID 761291. https://doi.org/10.1155/2012/761291.

Farmer, S., Poirier, C., & Tuan, C. (2017). *Bon Appétit! Enhancing the enjoyment of texture modified food*. Manitoba, Canada: Selkirk Mental Health Centre. Retrieved from: https://www.alzheimer.mb.ca/handouts/2C%20Bon%20Appetit!%20Enhancing%20the%20Enjoyment%20of%20Texture%20Modified%20Foods.pdf.

Figus, C. (2014). *375 Million vegetarians worldwide. All the reasons for a green lifestyle*. http://www.expo2015.org/magazine/en/lifestyle/375-million-vegetarians-worldwide.html.

Fildes, A., van Jaarsveld, C. H. M., Llewellyn, C. H., Fisher, A., Cooke, L., & Wardle, J. (2014). Nature and nurture in children's food preferences. *American Journal of Clinical Nutrition*, *99*, 911–917.

Fischler, C. (1980). Food habits, social change and the nature/culture dilemma. *Social Science Information*, *19*(6), 937–953. https://doi.org/10.1177/053901848001900603.

Fischler, C. (2011). Commensality, society and culture. *Social Science Information*, *50*(3–4), 528–548. https://doi.org/10.1177/0539018411413963.

Fischler, C. (2013). *Les Alimentations Particulières*. Paris: Odile Jacob.

Food Makers. (2017). *Food makers providing easier-to-eat fare for Japan's growing ranks of elderly*. Retrieved from https://www.japantimes.co.jp/news/2017/10/31/national/food-makers-providing-easier-eat-fare-japans-growing-ranks-elderly/#.Wx0-C0iFPIU.

Fox, R. (2014). *Food and eating: An anthropological perspective*. Oxford: Social Issues Research Centre.

Fu, Z. (2011). In *Chinese strategy on anti-aging research trends. Proceedings of the 6th conference on nutrition and aging, September 28–30, 2011* (pp. 165–173), Japan: ILSI.

German, J. B., Zivkovic, A. M., Dallas, D. C., & Smilowitz, J. T. (2011). Nutrigenomics and personalized diets: What will they mean for food? *Annual Review of Food Science and Technology*, *2*, 97–123.

Ghosh, D. (2010). Personalised food: How personal is it? *Genes & Nutrition*, *5*(1), 51–53. https://doi.org/10.1007/s12263-009-0139-0.

Gibbons, A. (2014). The evolution of diets. *National Geographic*, *226*(3), 30–61.

Gibis, M. (2016). Heterocyclic aromatic amines in cooked meat products: Causes, formation, occurrence, and risk assessment. *Comprehensive Reviews in Food in Food Science and Food Safety*, *15*, 269–302. https://doi.org/10.1111/1541-4337.12186.

Gómez-Pinilla, F. (2008). Brain foods: The effect of nutrients on brain function. *Nature Reviews. Neuroscience*, *9*, 568–578. https://doi.org/10.1038/nrn2421.

Greene, C. (2017). In *The outlook for organic agriculture. 94th Annual USDA Agricultural Outlook Forum, Crystal City, VA, February 22, 2017*.

Halfon, T. (2018). Le difficile exercice de la communication des risques liés à l'alimentation. *La Semaine Vétérinaire*, *1750*, 40–45.

Hansen, M. R., & Bentzen, J. (2016). High-risk sun-tanning behaviour: A quantitative study in Denmark, 2008–2011. *Public Health*, *128*, 777–783. https://doi.org/10.1016/j.puhe.2014.07.002.

Hausman, A. (2005). Hedonistic rationality: The duality of food consumption. *Advances in Consumer Research*, *3*, 404–405.

Heilbronn, L. K., & Ravussin, E. (2003). Calorie restriction and aging: Review of the literature and implications for studies in humans. *American Journal of Clinical Nutrition*, *78*, 361–369. https://doi.org/10.1093/ajcn/78.3.361.

Higashiguchi, T. (2015). Development of new home care foods "smile care foods" for the future social nutrition in Japan. *Journal of Japanese Society for Parenteral and Enteral Nutrition, 30*, 1091–1094. https://doi.org/10.11244/jspen.30.1091.

Hill, B. (2009). *Molecular gastronomy: Research and experience.* Melbourne, Australia: International Specialised Skills Institute.

Hill, A. J., Magson, L. D., & Blundell, J. E. (1984). Hunger and palatability: Tracking ratings of subjective experience before, during and after the consumption of preferred and less preferred food. *Appetite, 5*, 361–371. https://doi.org/10.1016/S0195-6663(84)80008-2.

Hill, J. O., & Peters, J. C. (1998). Environmental contributions to the obesity epidemic. *Science, 280*, 1371–1374.

Hirvonen, S. (2014). Ethics and food taste. In P. B. Thompson & D. M. Kaplan (Eds.), *Encyclopedia of food and agricultural ethics* (pp. 630–636). Dordrecht: Springer.

Hoffman, R. (2016). Convenience foods and health in the elderly. *Maturitas, 86*, 1–2. https://doi.org/10.1016/j.maturitas.2015.12.002.

Hong, X., Tsai, H.-J., & Wang, X. (2009). Genetics of food allergy. *Current Opinion in Pediatrics, 21*(6), 770–776. https://doi.org/10.1097/MOP.0b013e32833252dc.

Hyseni, I., Atkinson, M., Bromley, H., Orton, L., Lloyd-Williams, F., McGill, R., et al. (2017). The effects of policy actions to improve population dietary patterns and prevent diet-related non-communicable diseases: Scoping review. *European Journal of Clinical Nutrition, 71*, 694–711. https://doi.org/10.1038/ejcn.2016.234.

IDDSI. (2015). *International dysphagia diet standardization initiative. www.iddsi.org.*

Iisaka, S., Tadaka, E., & Sanada, H. (2008). Comprehensive assessment of nutritional status and associated factors in the healthy, community-dwelling elderly. *Geriatrics & Gerontology International, 8*(1), 24–31. https://doi.org/10.1111/j.1447-0594.2008.00443.x.

Imarat Consultants. (2015). *An overview of the global Halal market.* http://www.halalrc.org/images/Research%20Material/Presentations/overview%20of%20global%20halal%20market.pdf.

Jacobs, D. H., & Tapsell, L. (2013). Food synergy: The key to a healthy diet. *Proceedings of the Nutrition Society, 72*(2), 200–206. https://doi.org/10.1017/S0029665112003011.

Jagger, C. (2011). *The 25 most common causes of death.* https://www.medhelp.org/general-health/articles/The-25-Most-Common-Causes-of-Death/193?page=1.

Jakszyn, P., & González, C. A. (2006). Nitrosamine and related food intake and gastric and oesophageal cancer risk: A systematic review of the epidemiological evidence. *World Journal of Gastroenterology, 12*, 4296–4303. https://doi.org/10.3748/wjg.v12.i27.4296.

JSDR (2013) Japanese Society of Dysphagia Rehabilitation, https://www.jsdr.or.jp/wp-content/uploads/file/doc/classification2013-manual.pdf.

Just, D. R., Wansink, B., & Hanks, A. S. (2014). Chefs move to schools. A pilot examination of how chef-created dishes can increase school lunch participation and fruit and vegetable intake. *Appetite, 83*, 242–247. https://doi.org/10.1016/j.appet.2014.08.033.

Jyväkorpi, S. K., Pitkälä, K. H., Kautiainen, H., Puranen, T. M., Laakkonen, M. L., & Suominen, M. H. (2014). Nutrition education and cooking classes improve diet quality, nutrient intake, and psychological well-being of home-dwelling older people—A pilot study. *Journal of Aging Research & Clinical Practice, 3*, 120–124.

Khan, A., Carmona, R., & Traube, M. (2014). Dysphagia in the elderly. *Clinics in Geriatric Medicine, 30*, 43–53. https://doi.org/10.1016/j.cger.2013.10.009.

KHIDI. (2013). *Health industry statistics annual 2013.* Korea Health Industry Development Institute.

Kim, S., & Joo, N. (2015). The study on development of easily chewable and swallowable foods for elderly. *Nutrition Research and Practice, 9*(4), 420–424. https://doi.org/10.4162/nrp.2015.9.4.420.

Konishi, M., Yasuhara, Y., Nagasaki, T., Hossain, A., Tanimoto, K., & Rohlin, M. (2018). Differences of aspiration between liquid and solid foods in video-fluoroscopic swallowing study: A review of literature. *International Journal of Physical Medicine & Rehabilitation, 6,* 446. https://doi.org/10.4172/2329-9096.1000446.

Kringelbach, M. L. (2015). The pleasure of food: Underlying brain mechanisms of eating and other pleasures. *Flavor, 4,* 20. https://doi.org/10.1186/s13411-014-0029-2.

Kringelbach, M. L., & Berridge, C. K. (2009). Towards a functional neuroanatomy of pleasure and happiness. *Trends in Cognitive Sciences, 13*(11), 479–487. https://doi.org/10.1016/j.tics.2009.08.006.

Krutmann, J., Morita, A., & Chung, J. H. (2012). Sun exposure: What molecular photodermatology tells us about its good and bad sides? *Journal of Investigative Dermatology, 132,* 976–984. https://doi.org/10.1038/jid.2011.394.

Kuhar, M. (2012). *The addicted brain.* New Jersey: FT Press.

Kwon, S. H., Park, H. R., Lee, Y. M., Kwon, S. Y., Kim, O. S., Kim, H. Y., et al. (2017). Difference in food and nutrient intakes in Korean elderly people according to chewing difficulty: Using data from the Korea National Health and Nutrition Examination Survey 2013. *Nutrition Research and Practice, 11*(2), 139–146. https://doi.org/10.4162/nrp.2017.11.2.139.

Lee, M., Yoon, E., & Kropf, N. (2007). Factors affecting burden of South Koreans providing care to disabled older family members. *International Journal of Aging and Human Development, 64*(3), 245–262. https://doi.org/10.2190/C4U5-078N-R83L-P1MN.

Lesso, W. G., & Kenyon, E. (1972). *Astronauts' menu problem.* https://ntrs.nasa.gov/search.jsp?R=19720043776.

Liberman, V. (2014). Molecular gastronomy. In P. B. Thompson & D. M. Kaplan (Eds.), *Encyclopedia of food and agricultural ethics* (pp. 1382–1387). New York. Springer.

Liberman, D., & Cheung, A. (2015). A practical approach to osteoporosis management in the geriatric population. *Canadian Geriatrics Journal, 18*(1), 29–34. https://doi.org/10.5770/cgj.18.129.

Liu, H.-H., & Li, J.-J. (2015). Aging and dyslipidemia: A review of potential mechanisms. *Ageing Research Reviews, 19,* 43–52. https://doi.org/10.1016/j.arr.2014.12.001.

Liu, Z., Zhang, M., Bhandari, B., & Wang, Y. (2017). 3D printing: Printing precision and application in food sector. *Trends in Food Science & Technology, 69,* 83–94. https://doi.org/10.1016/j.tifs.2017.08.018.

López-Otín, C., Blasco, M. A., Partridge, L., Serrano, M., & Kroemer, G. (2013). The hallmark of ageing. *Cell, 153,* 1194–1217. https://doi.org/10.1016/j.cell.2013.05.039.

Macht, M., Meininger, J., & Roth, J. (2005). The pleasures of eating: A qualitative analysis. *Journal of Happiness Studies, 6,* 137–160. https://doi.org/10.1007/s10902-005-0287-x.

MAFF. (2017). *Food industry affairs bureau.* Japan: Ministry of Agriculture, Forestry and Fisheries. http://www.maff.go.jp/e/policies/food_ind/attach/pdf/index-9.pdf.

Marx, W., Moseley, G., Berk, M., & Jacka, F. (2017). Nutritional psychiatry: The present state of the evidence. *Proceedings of the Nutrition Society, 76,* 427–436. https://doi.org/10.1017/S0029665117002026.

Mastication Friendly Foods. (2016). *JAS for "Mastication-friendly food" established.* http://labelbank.com/newsletter/issues/201610.html.

Mattar, R., de Campos Mazo, D. F., & Carrilho, F. J. (2012). Lactose intolerance: Diagnosis, genetic, and clinical factors. *Clinical and Experimental Gastroenterology, 5,* 113–121. https://doi.org/10.2147/CEG.S32368.

McGrath-Hanna, N. K., Greene, D. M., Tavernier, R. J., & Bult-Ito, A. (2003). Diet and mental health in the Arctic: Is diet an important risk factor for mental health in circumpolar peoples? A review. *International Journal of Circumpolar Health, 62*(3), 228–241.

McNamara, D. J. (2015). The fifty year rehabilitation of the egg. *Nutrients, 7,* 8716–8722. https://doi.org/10.3390/nu7105429.

Mealey, L. (2018). *Restaurant fine dining.* https://www.thebalancesmb.com/what-is-fine-dining-2888688.

Meggiato, R. (2016). *Numbers behind chefs and restaurants.* https://www.finedininglovers.com/stories/chef-restaurant-facts-figures/.

Mick, D. G., Burroughs, J. E., Hetzel, P., & Brannen, M. Y. (2004). Pursuing the meaning of meaning in the commercial world: An international review of marketing and consumer research founded on semiotics. *Semiotica, 152*(1/4), 1–74. https://doi.org/10.1515/semi.2004.2004.152-1-4.1.

Milton, K. (2000). Hunter-gatherer diets-a different perspective. *The American Journal of Clinical Nutrition, 71*(3), 665–667. https://doi.org/10.1093/ajcn/71.3.665.

Misselbrook, D. (2014). W is for wellbeing and the WHO definition of health. *The British Journal of General Practice, 64*(628), 582. https://doi.org/10.3399/bjgp14X682381.

Monsivais, P., Aggarwal, A., & Drewnowski, A. (2014). Time spent on home food preparation and indicators of healthy eating. *American Journal of Preventive Medicine, 47*, 796–802. https://doi.org/10.1016/j.amepre.2014.07.033.

Monteiro, C. A., Moubarac, J. C., Cannon, G., Ng, S. W., & Popkin, B. (2013). Ultra-processed products are becoming dominant in the global food system. *Obesity Reviews, 2*, 21–28. https://doi.org/10.1111/obr.12107.

Montero-Fernández, N., & Serra-Rexach, J. A. (2013). Role of exercise on sarcopenia in the elderly. *European Journal of Physical and Rehabilitation Medicine, 49*, 131–143.

Mozaffarian, D., & Ludwig, D. S. (2015). The 2015 US dietary guidelines. Lifting the ban on total dietary fat. *Journal of the American Medical Association, 313*, 2421–2422. https://doi.org/10.1001/jama.2015.5941.

Mujcic, R., & Oswald, A. (2016). Evolution of well-being and happiness after increases in consumption of fruit and vegetables. *American Journal of Public Health, 106*(8), 1504–1510. https://doi.org/10.2105/AJPH.2016.303260.

Mulholland, C. A., & Benford, D. J. (2007). What is known about the safety of multivitamin-multimineral supplements for the generally healthy population? Theoretical basis for harm. *American Journal of Clinical Nutrition, 85*(1), 318S–322S. https://doi.org/10.1093/ajcn/85.1.318S.

Muramatsu, N., & Akiyama, H. (2011). Japan: Super-aging society preparing for the future. *The Gerontologist, 51*, 425–432. https://doi.org/10.1093/geront/gnr067.

National Geographic. (2014). Field rations go gourmet. *National Geographic, 226*(3), 62–65.

Navarro, V., Serrano, G., Lasa, D., Aduriz, A. L., & Ayo, J. (2012). Cooking and nutritional science: Gastronomy goes further. *International Journal of Gastronomy and Food Science, 1*, 37–45. https://doi.org/10.1016/j.ijgfs.2011.11.004.

NHS. (2015). *Appropriate prescribing of thickeners for dysphagia in adults.* https://www.prescqipp.info/thickeners-for-dysphagia/send/169-thickeners-for-dysphagia/1939-bulletin-100-thickeners-for-dysphagia.

Nickerson, R. S. (1998). Confirmation bias: A ubiquitous phenomenon in many guises. *Review of General Psychology, 2*(2), 175–220. https://doi.org/10.1037/1089-2680.2.2.175.

Ogawa, N., Mason, A., Chawla, A., & Matsukura, R. (2010). Japan's unprecedented aging and changing intergenerational transfers. In T. Ito & A. Rose (Eds.), *The economic consequences of demographic change in East Asia* (pp. 131–160). Chicago: University of Chicago Press.

Ohlhorst, S. D., Russell, R., Bier, D., Klurfeld, D. M., Li, Z., Mein, J. R., et al. (2013). Nutrition research to affect food and a healthy life span. *Journal of Nutrition, 143*(8), 1349–1354. https://doi.org/10.3945/jn.113.180638.

Osdoba, K. E., Mann, T., Redden, J. P., & Vickers, Z. (2015). Using food to reduce stress: Effects of choosing meal components and preparing a meal. *Food Quality and Preference, 39*, 241–250. https://doi.org/10.1016/j.foodqual.2014.08.001.

Ozdemir, B., & Caliskan, O. (2015). Menu design: A review of literature. *Journal of Foodservice Business Research*, *18*, 189–206. https://doi.org/10.1080/15378020.2015.1051428.

Painter, V., Le Couteur, D. G., & Waite, L. M. (2017). Texture-modified food and fluids in dementia and residential aged care facilities. *Clinical Interventions in Aging*, *12*, 1193–1203. https://doi.org/10.2147/CIA.S140581.

Parada, J., & Aguilera, J. M. (2007). Food microstructure affects the bioavailability of several nutrients. *Journal of Food Science*, *7*, R21–R32. https://doi.org/10.1111/j.1750-3841.2007.00274.x.

Penso, D. (2017). *Engay food spreading to the world.* http://www.myeyestokyo.com/16519.

Perchonok, M., & Bourland, C. (2002). NASA food systems: Past, present, and future. *Nutrition*, *18*, 913–920. https://doi.org/10.1016/S0899-9007(02)00910-3.

Persistence Market Research. (2017). https://www.persistencemarketresearch.com/market-research/kosher-food-market.asp.

Pilgrim, A., Robison, S., Sayer, A. A., & Roberts, H. (2015). An overview of appetite decline in older people. *Nursing Older People*, *27*(5), 29–35. https://doi.org/10.7748/nop.27.5.29.e697.

Plumer, B. (2015). *Map: Here's how much each country spends on food.* https://www.vox.com/2014/7/6/5874499/map-heres-how-much-every-country-spends-on-food.

Pollan, M. (2008). *In defense of food: An eater's manifesto.* London: Penguin.

Prentice, A. M. (2005). Starvation in humans: Evolutionary background and contemporary implications. *Mechanisms of Ageing and Development*, *126*, 976–981. https://doi.org/10.1016/j.mad.2005.03.018.

Quinn, S. (2016). *Number of vegans in Britain rises by 360% in 10 years.* https://www.telegraph.co.uk/food-and-drink/news/number-of-vegans-in-britain-rises-by-360-in-10-years/.

Redman, L. M., & Ravussin, E. (2011). Caloric restriction in humans: Impact on physiological, psychological, and behavioral outcomes. *Antioxidants & Redox Signaling*, *14*, 275–287. https://doi.org/10.1089/ars.2010.3253.

Regenstein, J. M., Chaudry, M. M., & Regenstein, C. E. (2003). The Kosher and Halal food laws. *Comprehensive Reviews in Food Science and Food Safety*, *2*, 111–127. https://doi.org/10.1111/j.1541-4337.2003.tb00018.x.

Reilly, R., Frankel, F., & Edelstein, S. (2013). Molecular gastronomy: Transforming diets for dysphagia. *Journal of Nutritional Health & Food Science*, *1*(1), 1–6. https://doi.org/10.15226/jnhfs.2013.00101.

Rice, J. M. (2005). The carcinogenicity of acrylamide. *Mutation Research*, *580*, 3–20. https://doi.org/10.1016/j.mrgentox.2004.09.008.

Robine, J.-M., & Cubaynes, S. (2017). Worldwide demography of centenarians. *Mechanisms of Ageing and Development*, *165*(Part B), 59–67. https://doi.org/10.1016/j.mad.2017.03.004.

Rolls, B. J. (1993). Appetite, hunger, and satiety in the elderly. *Critical Reviews in Food Science and Nutrition*, *33*, 39–44. https://doi.org/10.1080/10408399309527610.

Ross Watson, R. (2017). *Nutrition and functional foods for healthy aging.* London: Academic Press.

Rozin, P. (2005). The meaning of food in our lives: A cross-cultural perspective on eating and well-being. *Journal of Nutrition Education and Behavior*, *37*, S107–S112.

Rozin, P., Fischler, C., Imada, S., Sarubin, A., & Wrzesniewski, A. (1999). Attitudes to food and the role of food in life in the U.S.A., Japan, Flemish Belgium and France: Possible implications for the diet–health debate. *Appetite*, *33*, 163–180. https://doi.org/10.1006/appe.1999.0244.

Ryff, C. D. (1989). Happiness is everything, or is it? Explorations on the meaning of psychological well-being. *Journal of Personality and Social Psychology*, *57*(6), 1069–1081. https://doi.org/10.1037/0022-3514.57.6.1069.

Schösler, H., & de Boer, J. (2018). Towards more sustainable diets: Insights from the food philosophies of "gourmets" and their relevance for policy strategies. *Appetite, 127*, 59–68. https://doi.org/10.1016/j.appet.2018.04.022.

Scrinis, G. (2013). *Nutritionism: The science and politics of dietary advice.* New York: Columbia University Press.

Shepherd, G. M. (2012). *Neurogastronomy. How the brain creates flavor and why it matters.* New York: Columbia University Press.

Shlisky, J., Bloom, D. E., Beaudreault, A. R., Tucker, K. L., Keller, H. H., Freund-Levi, Y., et al. (2017). Nutritional considerations for healthy aging and reduction in age-related chronic disease. *Advances in Nutrition, 8*(1), 17–26. https://doi.org/10.3945/an.116.013474.

Sidenvall, B., Nydahl, M., & Fjellström, C. (2000). The meal as a gift—The meaning of cooking among retired women. *Journal of Applied Gerontology, 19*(4), 405–423. https://doi.org/10.1177/073346480001900403.

Spence, C. (2017). Comfort food: A review. *International Journal of Gastronomy and Food Science, 9*, 105–109. https://doi.org/10.1016/j.ijgfs.2017.07.001.

Stahlman, L. B., Garcia, J. G., Hakel, M., & Chambers, E. (2000). Comparison ratings of pureed versus molded fruits: Preliminary results. *Dysphagia, 15*, 2–5. https://doi.org/10.1007/s004559910002.

Strandberg, T. E., Strandberg, A. Y., Pitkälä, K., Salomaa, V. V., Tilvis, R. S., & Miettinen, T. A. (2008). Chocolate, well-being and health among elderly men. *European Journal of Clinical Nutrition, 62*(2), 247–253. https://doi.org/10.1038/sj.ejcn.1602707.

Sugimoto, K. (2013). *Meals for elderly people in Japan.* https://www.rvo.nl/sites/default/files/2013/09/Special%20food%20for%20the%20elderly%20in%20Japan_0.pdf.

Tani, Y., Sasaki, Y., Haseda, M., Kondo, K., & Kondo, N. (2015). Eating alone and depression in older men and women by cohabitation status: The JAGES longitudinal survey. *Age and Ageing, 44*(6), 1019–1026. https://doi.org/10.1093/ageing/afv145.

Teramoto, S. (2014). Clinical significance of aspiration pneumonia and diffuse aspiration bronchiolitis in the elderly. *Journal of Gerontology and Geriatric Research, 3*, 142. https://doi.org/10.4172/2167-7182.1000142.

Tesch-Roemer, C. (2010). *Active ageing and quality of life in old age.* Geneva, Switzerland: United Nations Economic Commission for Europe. Document ECE/WG.1/16.

Testa, G., Biasi, F., Poli, G., & Chiarpotto, E. (2014). Calorie restriction and dietary restriction mimetics: A strategy for improving healthy aging and longevity. *Current Pharmaceutical Design, 20*, 2950–2977. https://doi.org/10.2174/13816128113196660699.

The Cabinet Office of Japan. (2011). *Commission on measuring well-being: A report.* http://www5.cao.go.jp/keizai2/koufukudo/pdf/koufukudosian_english.pdf.

The World 50 Best. (2017). https://www.theworlds50best.com.

Thomas, K. S., Akobundu, U., & Dosa, D. (2016). More than a meal? A randomized control trial comparing the effects of home-delivered meals programs on participants' feelings of loneliness. *Journals of Gerontology Social Science, 71*(6), 1049–1058. https://doi.org/10.1093/geronb/gbv111.

Truman, E., Lane, D., & Elliott, C. (2017). Defining food literacy: A scoping review. *Appetite, 116*, 365–371. https://doi.org/10.1016/j.appet.2017.05.007.

Turnbull, J. L., Adams, H. N., & Gorard, D. A. (2015). Review article: The diagnosis and management of food allergy and food intolerances. *Alimentary Pharmacology & Therapeutics, 41*, 3–25. https://doi.org/10.1111/apt.12984.

Ubeda, N., Achón, M., & Varela-Moreiras, G. (2012). Omega 3 fatty acids in the elderly. *British Journal of Nutrition, 107*(Suppl. 2), S137–S151. https://doi.org/10.1017/S0007114512001535.

Uchida, Y., & Oishi, S. (2016). The happiness of individuals and the collective. *Japanese Psychological Research, 58*, 125–141. https://doi.org/10.1111/jpr.12103.

United Nations. (2015). World Population Prospects: The 2015 revision, key findings and advance tables. In *Working paper no. ESA/P/WP.241*. Department of Economic and Social Affairs, Population Division.

ViaMichelin. (2018). https://www.viamichelin.com/web/Restaurants.

Vidgen, H. A., & Gallegos, D. (2014). Defining food literacy and its components. *Appetite*, 76, 50–59. https://doi.org/10.1016/j.appet.2014.01.010.

Vigarello, G. (2013). Pratiques individualisantes et histoires des régimes. In C. Fischler (Ed.), *Les Alimentations Particulièrs* (pp. 107–116). Paris: Odile Jacob.

Warner, A. (2017). *The angry chef. Bad science and the truth about healthy eating.* London: Oneworld Publications.

Wendin, K., Ekman, S., Bülow, M., Ekberg, O., Johansson, D., Rothenberg, E., et al. (2010). Objective and quantitative definitions of modified food textures based on sensory and rheological methodology. *Food & Nutrition Research*, 54, 5134–5145. https://doi.org/10.3402/fnr.v54i0.5134.

WHO. (1947). Constitution of the World Health Organisation. *Chronicle of the World Health Organization*, 1(1–2), 1–3.

WHO. (2013). *A global brief on hypertension*. Geneva: World Health Organization. www.who.int/cardiovascular_diseases/publications/global_brief_hypertension/en/.

WHO. (2016). *Global report on diabetes*. Geneva: World Health Organization. www.who.int/diabetes/global-report/en/.

WHO. (2017a). *World hunger again on the rise, driven by conflict and climate change, new UN report says*. Geneva: World Health Organization. www.wfp.org/news/news-release/world-hunger-again-rise-driven-conflict-and-climate-change-new-un-report-says.

WHO. (2017b). *Obesity and overweight*. Geneva: World Health Organization. www.who.int/mediacentre/factsheets/fs311/en/.

Wilkins, J. (2015). Good food and bad: Nutritional and pleasurable eating in ancient Greece. *Journal of Ethnopharmacology*, 167, 7–10. https://doi.org/10.1016/j.jep.2014.12.016.

Willet, W. (1987). Nutritional epidemiology: Issues and challenges. *International Journal of Epidemiology*, 16, 312–317. https://doi.org/10.1093/ije/16.2.312.

Yiengprugsawan, V., Banwell, C., Takeda, W., Dixon, J., Seubsman, S., & Sleigh, A. C. (2015). Health, happiness and eating together: What can a large Thai cohort study tell us? *Global Journal of Health Science*, 7(4), 270–277. https://doi.org/10.5539/gjhs.v7n4p270.

Zepf, F. D., Hood, S., & Guillemin, G. J. (2015). Food and your mood: Nutritional psychiatry. *The Lancet Psychiatry*, 2(7), e19. https://doi.org/10.1016/S2215-0366(15)00241-2.

FURTHER READING

Anonymous. (2017). *The truth about fats: The good, the bad, and the in-between.* https://www.health.harvard.edu/staying-healthy/the-truth-about-fats-bad-and-good.

Kimura, Y., Wada, T., Okumiya, K., Ishimoto, Y., Fukutomi, E., Kasahara, Y., et al. (2012). Eating alone among community-dwelling Japanese elderly: Association with depression and food diversity. *Journal of Nutrition Health and Aging*, 16(8), 728–731. https://doi.org/10.1007/s12603-012-0067-3.

Landry, M., Lemieux, S., Lapointe, A., Bédard, A., Bélanger-Gravel, A., Bégin, C., et al. (2018). Is eating pleasure compatible with healthy eating? A qualitative study on Quebecers' perceptions. *Appetite*, 125, 537–554. https://doi.org/10.1016/j.appet.2018.02.033.

McMahon, A., Williams, P., & Tapsell, L. C. (2010). Reviewing the meanings of wellness and well-being and their implications for food choice. *Perspectives in Public Health*, 130, 282–286. https://doi.org/10.1177/1757913910384046.

Thompson, D. (2017). The paradox of American restaurants. *The Atlantic*. https://www. theatlantic.com/business/archive/2017/06/its-the-golden-age-of-restaurants-in-america/ 530955/.

Trubek, A., & Doggett, T. (2014). Gustatory pleasure and food. In P. B. Thompson & D. M. Kaplan (Eds.), *Encyclopedia of food and agricultural ethics* (pp. 1147–1154). Dordrecht: Springer.

Wahl, D. R., Villinger, K., König, L. M., Ziesemer, K., Schupp, H. T., et al. (2017). Healthy food choices are happy food choices: Evidence from a real life sample using smartphone based assessments. *Scientific Reports*, 7, 17069. https://doi.org/10.1038/s41598-017-17262-9.

CHAPTER EIGHT

Meat as a *Pharmakon*: An Exploration of the Biosocial Complexities of Meat Consumption

Frédéric Leroy[1]
Research Group of Industrial Microbiology and Food Biotechnology (IMDO), Faculty of Sciences and Bioengineering Sciences, Vrije Universiteit Brussel, Brussels, Belgium
[1]Corresponding author: e-mail address: frederic.leroy@vub.be

Contents

Advances in Food and Nutrition Research, Volume 87
ISSN 1043-4526
https://doi.org/10.1016/bs.afnr.2018.07.002

Abstract

In contemporary dietary advice, meat is depicted as a *pharmakon*: it is believed to either heal or poison the human body (and mind). Often, it also serves as a scapegoat for a wide range of public health issues and other societal problems. Related attitudes, practices, and beliefs pertain to a demarcated mode of thinking or *episteme* that is characteristic for the so-called post-domestic or industrialized societies. The latter are not only typified by an abundant yet largely concealed production of meat, but increasingly also by moral crisis and confusion about its nutritional meaning. For an improved appreciation of the ambiguous position of meat in human health and disease, as well as the concomitant scattering into different *subject positions* (*e.g.*, the omnivore, flexitarian, vegetarian, vegan, permaculturalist, and carnivore position), an interdisciplinary approach is required. To this end, the current study tentatively combines food research with a selection of (post-structuralist) concepts from the humanities. The aim is to outline a historical and biosocial *need for meat* (as well as its rejection) and to analyze how its transformative effects have contributed to a polarized discourse on diet and well-being in academia and society at large. Excessive categorization (for instance with respect to meat's alleged naturalness, normalness, necessity, and niceness) and Manichean thinking in binary opposites are among the key factors that lead to impassioned yet often sterile debates between the advocates and adversaries of meat eating in a post-truth context.

1. INTRODUCTION

The eating of meat is recurrently depicted in present-day nutrition as having either a very positive or a very negative influence on health (Leroy, Brengman, Ryckbosch, & Scholliers, 2018). Meat, in all its unresolved ambiguity, acts as a *pharmakon* (φάρμακον), of which the etymology can equally denote a medicine and a poison (Derrida, 1981). Additionally, the term relates to φαρμακός or the purifying ritual of scapegoating (Burkert, Girard, & Smith, 1987). Whether or not this is tied to its connotations of slaughter and sacrifice (Leroy & Praet, 2017), meat does indeed serve as a culprit for a series of societal dysfunctions, including a general disquiet about agri-food systems, the poor status of animal welfare, the corrosion of public health, and the destructive impact of humans on the environment (Smil, 2013). Such views are pervasive and culminate in pleas for the application of a "meat tax" (FAIRR, 2017) or enforcement of veganism (Deckers, 2013).

Although Lucretius' claim that "what is food to one man may be fierce poison to others" dates back to the first century BCE, the pharmakon is not part of the usual idiom of food sciences. It could, however, prove a useful corrective for excessive binary thinking and the formulation of unsubstantiated conclusions. Nutritional convictions are all-too often inconsistent, which is largely due to empirical complexity and poor epistemological practice (Young & Karr, 2011). In epidemiology, for instance, confounded observational data generate a Janus effect whereby specific (micro) nutrients can be linked to both a higher *and* lower risk for mortality (Patel, Burford, & Ioannidis, 2015). In addition, ambiguous findings on the healthiness of a given food readily crystallize into culturally constructed opposites (*e.g.*, natural/unnatural; Piazza et al., 2015), whereby one element of the binary is typically privileged over the other (Derrida, 1981). Thus, either the consumption *or* the avoidance of meat can be presented as beneficial (or even as *essential*), depending on the knowledge base, belief system, and agenda behind the affirmation. To assess the contemporary role of knowledge and truth within the food sciences (*cf.* Leroy, Brengman, et al., 2018) and to discredit unjustified claims of authority based on nonfactual assumptions or vested interests (Harcombe, 2017a, 2017b; Leroy, Aymerich, et al., 2018; Noakes & Sboros, 2018; Teicholz, 2015), one may need to grasp the formation of dominant practices, statements, and institutions (*cf.* Foucault, 1977).

To this end, the present study appeals to some key concepts from the domain of continental philosophy, critical theory, and post-structuralism (Belsey, 2002; Bronner, 2011), as to provide the theoretical framework that is needed to challenge traditional beliefs and expose overly rigid systems of thought. Because the position of meat in the human diet is becoming exceedingly controversial indeed—with potential upshots for societal well-being—such a multidisciplinary approach is not only pertinent but also urgent.

2. INTRODUCING THE POST-DOMESTIC SOCIETY

With respect to their meat production and consumption, industrialized societies have been branded as *post-domestic* (Bulliet, 2005). They diverge profoundly from communities of which the sustenance is based on either hunter-gathering or pre-industrial domestication (Leroy & Praet, 2017). This not only refers to the particularities of the meat chain, the prevailing foodways, the disruption of human–animal interactions, and the emergence of animal right movements, but also to a much broader societal context (DeMello, 2012). Indeed, the manners by which a society manages its eating behavior not only reify its technological setups and market conditions, but also serve as an expression of its ideas, desires, and aspirations

(Appadurai, 1981; Visser, 2003). The post-domestic paradigm is shaped by a specific *episteme* (*cf.* Foucault, 1970), which determines the *conditions of possibility* for a mode of thinking and behaving that is contingent on a given place and period, originating in the Western world during the 19th century. Typically, the supply of meat is high and delivered via a mostly hidden production system, whereby the killing of livestock is no longer a visible part of civic life and often gives rise to serious animal welfare issues (Leroy & Degreef, 2015). Meat, therefore, functions as a commodity and no longer has the meaning of a shared good (*cf.* Vivero-Pol, 2017). Following a shift from a *zoophagic* (*i.e.*, the eating of animals) to an alienating *sarcophagic* perspective (*i.e.*, the eating of meat), it is no longer coupled to its animal origins or culturally rooted in sacrificial praxis (Fischler, 2001; Leroy & Praet, 2017).

The episteme corresponding with the post-domestic societal model has resulted in the development of a conceptual minefield, involving troublesome notions such as the so-called *meat paradox* (Loughnan, Haslam, & Bastian, 2010). The latter states that people eagerly consume meat but are at the same time appalled by animal slaughter (Leroy & Praet, 2017). It is claimed that this cognitive dissonance can nonetheless be downplayed by denial, rationalization, or appeal to moral imperatives and beliefs. A rigid mental separation between meat and its animal origins is thereby primordial (Kunst & Hohle, 2016). The belief system by which people uphold the meat paradox has been coined *carnism*, a denomination that is particularly popular among vegan militants (Joy, 2010).

Put together, the central questions related to the post-domestic paradigm are fourfold. First, how does it translate into practices, attitudes, discourses, and beliefs? Second, how are all those driven by a *need for meat*, if any? Third, why and how did this episteme emerge? Fourth, which instabilities does it currently incorporate, and are they sufficiently important to trigger the transition to a new episteme? This study aims at providing some preliminary answers to these interrogations, suggesting options for more dedicated and empirically driven research. To guide the reader, Fig. 1 serves as a schematic representation of the main concepts and jargon involved.

3. PRACTICES, ATTITUDES, DISCOURSES, AND BELIEFS WITHIN THE POST-DOMESTIC ERA

3.1 Applying the Toolbox: Epistemic Archaeology

To infer clues about the internal structure of a prevailing episteme, Foucault (1970) has proposed an *archaeological* investigation that consists in exploring

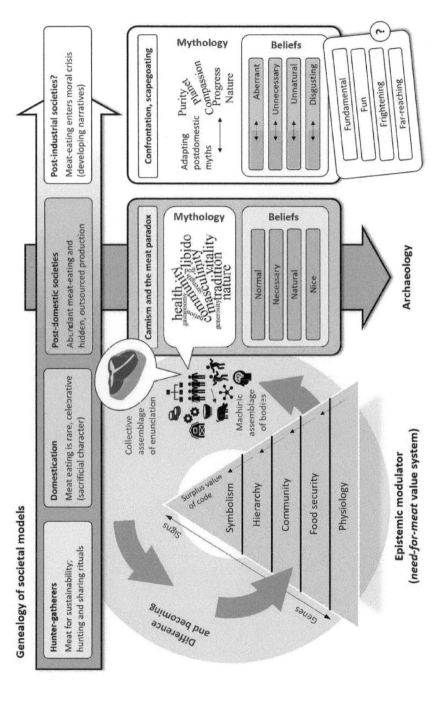

Fig. 1 Schematic overview displaying a genealogical flow of the various societal models that are related to meat eating and an archaeology of discourse and beliefs within the current post-domestic setup (including a proposition for transition). Modulation by a biosocial "need for meat" is depicted, acting as a creative flux of desire.

the surface effects of language and culture, *i.e.*, a mapping of the enunciative field (Fig. 1). The method is therefore also a study of discourse (Jansen, 2008). Although not always robust when generalized too broadly on an empirical basis, it has shown its value for the interpretative analysis of societal paradigms (Gutting, 2005). Evidently, inconsistencies will always be found; even if the cultural macrolevel of enunciation can be used to broadly define a society, the microlevel is suppler and may escape linguistic overcoding (Holland, 2013). Still, an episteme should be able to reflect (and make sense of) a wide range of factual beliefs and practices (Gutting, 2005). Focus of the present section will be on the discursive formation of post-domestic meat eating, as a manner to excavate hidden assumptions and regularities.

3.2 Mapping of the Enunciative Field

Post-domestic meat eaters have been described as systematically rationalizing their behavior by presenting it as *normal, natural, necessary*, and *nice* (Piazza et al., 2015). These elements, which have been empirically explored by focus groups and interviews (*e.g.*, Macdiarmid, Douglas, & Campbell, 2016), act as group beliefs that correlate with the justification of meat eating (Fig. 2). Moreover, they parallel a series of *myths*. The latter must be judged

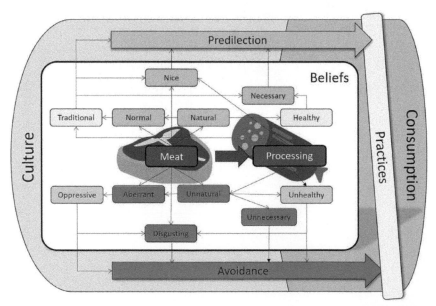

Fig. 2 A symmetric belief system lays at the basis of the current societal practices related to meat consumption (or avoidance).

not so much on their truth value but rather as rich meta-language, in the way this has been framed by Barthes (1957), who for instance exposed public imageries of the French beefsteak as a metaphorical means to confer patriarchal ideas in defense of traditional diets. Other emblematic myths elaborate on how meat is related to vitality, muscularity, the heightening of libido, community spirit, and divine or natural imperatives (Fiddes, 1991; Leroy & Praet, 2015). It vitalizes the body and mind (Leroy, Brengman, et al., 2018), reflects power, virility, and sometimes patriotism (Avieli, 2013; Ruby & Heine, 2011), and affirms belongingness, tradition, and collective identity (Graça, Calheiros, & Oliveira, 2014). According to Barthes (1957), myths feed on symbols, stereotypes, and partial analogies, thereby becoming historical and proverbial amalgams of associative ideas, essences, and values. They naturalize, expand, and impose themselves onto the public. Mass media, for instance, often reflect meat's central myths of vitality and strength, by referring to the bodily effects of its nutrients (Leroy, Brengman, et al., 2018). Heavily medicalized narratives are enunciated by a variety of social actors, including health professionals, scientists, authorities, celebrities, food writers, and industry. This is not trivial, as engagement by specialized members of organizations (*e.g.*, doctors or scientists) in public argumentation creates classifications and norms, so that loose set of beliefs are transformed into epistemic discourse (DeLanda, 2006). On its turn, this opens the way to legitimization and enforcement (Foucault, 1977).

3.3 Beliefs of the Carnist (the 4Ns)

Branded as an ideology (Joy, 2010), carnism is said to incorporate a set of recurrent self-justifications. Be that as it may, a first commonly stated argument situated in post-domestic discourse is that it is *normal* to do so (Piazza et al., 2015). Almost half of the respondents in an Australian study by Lea and Worsley (2003) affirmed that "humans are meant to eat meat." Normalizing judgment is characteristic of Western societies, whereby individual behavior is not valued intrinsically but based on a ranked scale to allow comparison with others (Foucault, 1977). This is also the case for nutritional practice, whereby sets of data are obsessively rank-ordered to qualify modes of behavior (*e.g.*, eating) and situations (*e.g.*, body weight) as socially (un)acceptable (Gutting, 2005; Lecerf, 2015; Rozin, 1999). In so doing, people are under the incessant menace of being marginalized by a multitude of standards and other fragmented power centers, eventually converging into normality

(Foucault, 1977). The idea of normality in foodscapes also involves notions of culinary *tradition*, which functions as an anchoring point in times of hyperpaced and threatening change (Geyzen, Scholliers, & Leroy, 2012). Meat incontestably has a crucial signification within Western traditional diets (Barthes, 1957), serving as the gravitational center of the meal (Belasco, 2008; Leroy & Degreef, 2015). Meat products in specific (*e.g.*, dry-cured hams and fermented sausages) are deeply rooted in gastronomic heritage and thus emblematic of many food traditions (Leroy, Geyzen, Janssens, De Vuyst, & Scholliers, 2013). Within foodscapes, traditions are cultural constellations that encompass elements of geographical origin, historical usage, and craftsmanship, often according to strictly defined codes and demarcations (Leroy, Scholliers, & Amilien, 2015). Whether or not associations with normality and tradition explain part of an often-mentioned penchant of meat eaters toward conservatism (*e.g.*, Hodson & Earle, 2018) needs exploration.

Further breakdown indicates that normative evaluations are construed on social and biological (dys)functions. Related discourse thus also refers to notions of *naturalness*, by setting the standards of physiological and medical normality that define the healthy individual (Foucault, 1979). Part of the argumentation relates to the "natural" position of meat in the evolutionary diet (Piazza et al., 2015), however ill-defined that may be (Turner & Thompson, 2013). The body of evidence pointing toward a key role of meat eating in human evolutionary development and sustenance is indeed substantial (Ben-Dor, Gopher, Hershkovitz, & Barkai, 2011; Cordain et al., 2000; Gupta, 2016). Yet, clear-cut distinctions between nature and culture are always problematic as biology also entails normative social significance, and vice versa (Descola, 2005; Gutting, 2005; Latour, 1993). Depictions of nationalism and bourgeois ideology, for instance, are habitually reinforced by an appeal to both the natural and the normal (Barthes, 1957). Deconstruction of the nature/culture binary, and the often-privileged status of nature therein, further confirms this. The former, with all its connotations of "wildness," is defined from *within* the concept of the latter (Belsey, 2002). It may thus be better to describe relationships between humans and meat as *biosocial* instead (Leroy & Praet, 2015).

The concepts of normality and naturalness generate on their turn a multitude of *necessities*. Individuals internalize the norms whereby they are controlled and thus become creators of needs (Gutting, 2005). The eating of meat is presented as a vital human need (Piazza et al., 2015), *i.e.*, the required dietary path to achieve optimal health and fitness (Leroy, Brengman, et al., 2018). Usually, this refers to biological mechanisms and physiological requirements for protein, iron, zinc, vitamin B12, etc. (Leroy, Brengman, et al., 2018; Pereira & Vicente, 2013). The healthiness of meat is thereby situated within its long-standing role in evolutionary settings, crosslinking naturalness and nutritional need (Frassetto, Schloetter, Mietus-Synder, Morris, & Sebastian, 2009; Leroy & Praet, 2015).

Finally, meat and its derived products are highly valued as *nice* foods in culinary traditions worldwide (Piazza et al., 2015). It is not fully clear whether this is related purely to cultural heritage (*e.g.*, Leroy et al., 2013, 2015) or also to physiological regulation and *meat hunger* (Leroy & Praet, 2015; Williams & Hill, 2017). In this context it may be worthwhile to look into the phenomenon of bushmeat hunger, a psychocultural form of meat craving that occurs in periods that are otherwise nutritionally balanced (Dounias & Ichikawa, 2007). Overall, Darwinian interpretations of food hedonism reflect in their own manner the above-mentioned ideas of normality, naturalness, and necessity.

3.4 Conclusion and Perspectives

Majoritarian attitudes toward meat eating are shaped (but also constrained) by the post-domestic mindset. Further exploration of public narratives is needed to consolidate these findings, to verify their social geography and heterogeneity between and within population segments, and to estimate their reflection of *cultural capital* (*cf.* Bourdieu, 1984). An improved understanding of the discursive interplay will also help to establish if discourse is either the emanation or cause of societal values and beliefs. Source materials should be varied, including data from television, radio, books, journals, magazines, school books, the world wide web, social media, health policies, sales, and advertisements (Jansen, 2008). More fine-grained analysis may help to clarify how the main prevailing myths attempt to *naturalize* their intentions by generating essence and by suggesting causal associations between the signifier and signified (Barthes, 1957). Taken together, post-domestic beliefs (*i.e.*, normalness, naturalness, necessity, and niceness) turn out to be different interrelated facets of the same underlying concept of an engrained human *need for meat*, *i.e.*, a multidimensional *desire* which necessitates further dissection into its major components.

4. A BIOSOCIAL *NEED FOR MEAT*

4.1 Applying the Toolbox: Assemblage Theory

Humans are desiring subjects, their languages often tribal and sloganized. Yet, the link between a society's wants and its expressions is not to be understood too narrowly. The development of a need is never secluded; it is always coupled to other *becomings* (Deleuze & Guattari, 1987). Societies and their phenomena do not, therefore, develop predictably based on linear mechanisms but do so via complex assemblages, emerging as meta-stable *solutions* to *problems* (DeLanda, 2006; Holland, 2013). Organization into social groups, for instance, provides a solution to the problem of "survival" in a given ecosystem (*e.g.*, hunting males). Although assemblages are ever-evolving manifestations of life's creative flow, they can nonetheless sediment into self-organizing structures of meaning, or *strata* (Deleuze & Guattari, 1987). By doing so, they acquire a provisional consistency based on *territorialization* (*i.e.*, constitutive action) and *coding* (*i.e.*, stabilization by genetic or linguistic codes).

It is argued that the *need for meat* can be regarded as such a dynamic assemblage, exposing a stratified flux of desire (Leroy & Praet, 2015). Its layered emergence (Fig. 1) represents a value system (*i.e.*, outlining that what is looked-for and, thus, *valued*) that somewhat mimics the pyramid of human needs proposed by Maslow (1943), albeit in a less hierarchical manner. Five biosocial evolutionary processes can be diagrammed as its constituents, including a (1) physiology, (2) food security, (3) community, (4) hierarchical, and (5) symbolic stratum. The upper strata are mostly situated into what Deleuze and Guattari (1987) have named the *alloplastic* megastratum, dominated by language and signs. This setup suggests a gradient ranging from biologically encoded processes (*e.g.*, meat craving) to cultural manifestations encoded by signs (*e.g.*, symbolism and artistic expression). We advise nonetheless to interpret the various strata as not only biosocial, co-existing, and non-prioritized, but also as interdependent: they *emerge* but do not follow a strictly demarcated progression, even if this may partly correspond with an evolutionary order of events. In all cases, process should be prioritized above structure. It is from and within the strata of this value system that the beliefs, attitudes, and practices related to meat eating arise.

4.2 Analysis of the Biosocial Strata

The two lowest strata of the model situate meat eating as a desire for nutrients and food security. They represent *territorialized* assemblages of nutrients, cofactors, enzymes, organs, bodies, flesh, etc., whereby incorporation of meat in the ancestral diet has been the result of a search for accessible food and biosocial adaptations to ecological pressure (Ben-Dor et al., 2011; Flinn, Geary, & Ward, 2005). During this process, a gradual transition to an omnivore diet took place with repercussions on the anatomy and functional aspects of the human digestive system and brain (Aiello & Wheeler, 1995). Cerebral expansion during evolution has thus been fueled by meat, through the liberation of energy and the provision of building blocks and cofactors (Pfefferle et al., 2011; Previc, 2009; Williams & Hill, 2017). It is noteworthy that early humans have been described with a very Deleuzian concept by paleoanthropologist Domínguez-Rodrigo (quoted in Gupta, 2016) as "meat-eating machines," in contrast to the "plant-processing machines" embodied by other primates. Whether or not this has led to an *innate* biological desire for meat, stratified via genetic coding, contemporary narratives about *naturalness* and *need* often involve references to bodily constitution and vitality (Leroy, Brengman, et al., 2018). Scientied variants may refer, for instance, to a cognitive bonus based on meat's creatine content (*cf.* Allen, 2012; Rae, Digney, McEwan, & Bates, 2003), or to research that has correlated vegetarian deficiencies in vitamin B12 and iron with depression and neurological disorders (Hibbeln, Northstone, Evans, & Golding, 2018; Kapoor, Baig, Tunio, Memon, & Karmani, 2017). Given its evolutionary setting and ancestral role in food security, the fact that meat is highly looked-for by humans may not be surprising. Refraining or defecting from vegetarian diets can usually be ascribed to an irresistible craving for meat (Barr & Chapman, 2002; Lea & Worsley, 2003). Protein homeostasis or a predilection for its umami taste and the aromas that are released upon its grilling may partly explain meat's preconditioned *niceness* (Griffioen-Roose et al., 2012; Morrison, Reed, & Henagan, 2012).

The next two strata explain how the pursuit of a meat-based food security system has catalyzed the structuring of human communities, *i.e.*, as social assemblages that contain meat as a key constituent in their very core. In addition to this point, it needs to be specified that it has not been meat eating as such that has allowed for brain expansion, but a requirement to develop the conceptual and tactical skills for cooperative foraging and sharing of the loot

(Stanford & Bunn, 2001; Tomasello, Melis, Tennie, Wyman, & Herrmann, 2012). Although such territorial assemblages of hunters, animals, spears, earth, etc., are no longer pertinent in the post-domestic era (being replaced by husbandry, slaughterhouses, markets, retail, etc.), some residual effects of their initial biosocial coding cannot be excluded. Because cooperative meat-hunting in groups was a male-dominated activity, it is for instance tempting yet debatable to interpret this as the cornerstone of meat's long-standing masculine aureole (Leroy & Praet, 2015). Exceptions to a common, *normalized* link between meat eating and masculinity do exist, suggesting cultural plasticity and serving as a warning for all-too Eurocentric interpretations (Morris, 1994). Also, the early-human necessity to set up effective meat-sharing schemes has been linked to the intensification of cultural transmission, ritual, and language (Leonetti & Chabot-Hanowell, 2011; Stanford, 1999). Corporeal and non-corporeal assemblages of meals, calendar events, rites, chants, stories, etc., have thus been established. Even in post-domestic societies, these aspects of meat eating and story-telling are still played out regularly in many (religious) celebrative meals to express belonging, tradition, generosity, and friendship (Graça et al., 2014; Jones, 2007; Seleshe, Jo, & Lee, 2014), yielding both *niceness* and *normality*. Adding to this horizontal structuring of communities, meat traditions generate a vertical gradient, encoding notions of power, wealth, masculinity, and strength. In hunter–gatherer societies, this is achieved via costly signaling by the best hunters (Hawkes, 1991), whereas sacrificial rites consolidate aristocratic privileges in sedentary cultures (Bulliet, 2005; deFrance, 2009). Post-domestic societies rely more indirectly on such hierarchical effects, although prioritized access to meat eating still can serve as a manner to express elite status, potentially resulting in nutritional asymmetries (Ross, 1987). Taken together, the primordial nutritional aspects of meat have been partially *re-territorialized* into community-structuring strata, strongly *encoded* by ritual and ceremony. Meat traditions can thus serve to either maintain hierarchal power lines (*cf.* the Turkana; Lokuruka, 2006) or reinforce egalitarianism (*cf.* the Ju/'hoansi people; Suzman, 2017).

   Lastly, the symbolic stratum is the most flexible and therefore problematic to delineate. It harbors what Deleuze and Guattari (1987) have named *semiotic machines* that hyperactively stretch out to all other strata, being governed by flexible linguistic codes and *regimes of signs*. At this level, cultural expressivity emerges as *surplus value* of code (Fig. 1), beyond the above-mentioned coding by genetics, ecological adaptation, and social communication that are needed to satisfy the primary physiological drives

(Young, Genosko, & Watson, 2013). Despite vast heterogeneity and unbridled dynamics, some general cultural patterns can be distinguished. Whereas hunting and meat rites pertaining to the lower strata have the power to gather communities around territorializing marks (*e.g.*, totem animals), further investment via language and signs leads to a symbolic surplus of (proto)religious and nationalistic *refrains* (Holland, 2013). Within this stratum, contemporary consumers desire to reconcile identity, convenience, and responsibility, in a myriad of ways (*cf.* Belasco, 2008). For a more detailed overview of the varied sacred, artistic, cultural, and consumerist emanations involved, we refer to the studies by Fiddes (1991) and Leroy and Praet (2015).

4.3 Conclusion and Perspectives

The human *need for meat* emerges as a value system, *i.e.*, a biosocially stratified desire for (1) nutrient delivery (and the stabilization thereof through food security), (2) a sense of community (and the stabilization thereof through hierarchical structuring), and (3) the cultural manifestations that are needed for the coding of the former two (*i.e.*, depictions of vitality, power, tradition, etc.). More research is required to verify the cross-cultural validity of each stratum, for instance via a study of *anthropological invariants* (*cf.* Descola, 2005), and by comparing with non-meat foods that may exert similar effects (*e.g.*, rice in South Asia; Appadurai, 1981). In other words, it still needs to be investigated if this stratified ensemble is either firmly established in time and space or prone to rewiring and destabilization. Further integration in the more inclusive framework of epistrata and parastrata proposed by Deleuze and Guattari (1987) could be rewarding, especially with respect to its visions on difference and becoming. Doing so will indicate more specifically how the various strata are coded assemblages of *bodies*, both biologically and politically (Colebrook, 2002), with implications that reach far beyond the specific issue of meat. It may shed new light on the (historical) emergence of such biosocial manifestations as brain development, organized hunting, bonding and altruistic behavior, gender, and even language (Leroy & Praet, 2015). This should clarify better how the investment in the various *intensities* of meat (*e.g.*, its redness, blood, smell, flavor, and other bodily effects) and its subsequent institutionalization (*e.g.*, the hunt, ritual, the tribe, power, and wealth) have elevated it as a crucial signifier. As is the case for food in general (Appadurai, 1981), meat is to be approached as a "highly condensed social fact" and a "plastic kind of collective representation." The latter is then to be considered as reliant on a *collective assemblage of enunciation*, whereby desire is invested into the social field within a specific context of place and time, engendering a shared sense of the world and what can be said about it (Holland, 2013). Due to continuous fluctuations in both form and content, the stratified *need-for-meat* assemblage has the intrinsic potential to modulate meaning-generating epistemic assemblies. This bring us to the next interrogation, namely: how do epistemes emerge and evolve?

5. HISTORIOGRAPHICAL INTERPRETATION
5.1 Applying the Toolbox: Epistemic Genealogy

From a historical perspective, the post-domestic model can be designated as a *modern* or even as *post-modern* condition (Leroy & Praet, 2017). Such terms have no real content themselves and only exist with respect to their deviation from other periods (Belsey, 2002). In this case, the definition is based on a contrast with domestic societies and hunter-gatherer lifestyles (Bulliet, 2005). What is of interest here is not so much the moment of transition but rather the mechanisms driving it. To explain change, historiography usually provides associations with underlying socioeconomic forces. All-too often, this leads to general, vague causes (*e.g.*, the rise of bourgeoisie) and grand narratives. Such generalizations can be misleading: besides their linear progression, under the control of a society's power center, historical events are also shaped by more unpredictable becomings (Holland, 2013). As an alternative methodology, we refer to the *genealogical* approach developed by Foucault (1977) in his seminal study of the history of criminal punishment. Accordingly, changes in thought are no products of thought as such but result from a multiplicity of dispersed causes without teleological purpose. These causes are power-driven nevertheless but operate primarily on individual bodies rather than on ideas and institutions, which are mere contractions of underlying diversity. It is within the *machinic* assemblages of bodies, as well as in collective assemblages of *enunciation* (*cf.* Deleuze & Guattari, 1987), that the mechanisms for configuration and transition can be found. As depicted in Fig. 1, epistemes are modulated by everchanging fluxes of *desire* (*in casu*, the need for meat), translating into corporeal as well as non-corporeal constellations and the construction of meaning. The genealogical method thus not only has the ambition to deconstruct discourse but also to expose its material background, as well as the power structures by which it is governed (Gutting, 2005).

5.2 Emergence of the Post-domestic Era

As argued above, the transformation of a society's conceptual basis is a profound, discontinuous, and multi-causal process based on what may often seem trivial facts (Gutting, 2005). The world-shattering transition from hunter-gathering to the domestic model, for instance, is obviously related to the adoption of agriculture and husbandry (Rozin, 1999). However, more subtle effects seem to lay at the basis of some of the deeper transformations. Differences in ownership of large farm animals for plowing, for instance, have been paramount to a disproportionate accumulation of wealth, a major increase in social inequality, and the reinforcement of hierarchies (Kohler et al., 2017). It is within this framework of farming and social disparity that meat became a rare but celebrative food for the populace, ritually institutionalized by the elites as a symbol of wealth, power, and prosperity (Leroy & Praet, 2017).

Likewise, the advent of the post-domestic era needs to be attributed to a variety of factors, of which some appear of lesser importance at first sight. Even if a prior history of destabilizing riots and class struggle has been a significant factor in the progress toward a public access to meat (DeMello, 2012; Smil, 2013), this is not all-explanatory. A dedicated study would be needed to investigate this in full detail, but at least part of the account would have to include the role of specific technological innovations of the 19th century. The latter comprise the development of cooled transportation by railway, improved distribution systems, large-scale intensive farming, the import of cheap feed to land-short countries, increased efficiency of the meat packaging industry, and the industrial adoption of canning and preservation (Leroy & Degreef, 2015; Ogle, 2013; Pilcher, 2005). The amplification of meat supply was rapidly translated into higher demands, paralleling an increase in population, purchase power, and urbanization. Late 19th-century Londoners thus increased their yearly per capita consumption of meat to some 40 kg because almost half of the supply could by then be imported as frozen beef, for instance from Australia (Renton, 2013). Also, purchasing a chicken in the early 1900s still matched the equivalent of 3 h of labor for an average US worker, but only 15 min at the end of the same century (Renton, 2013). The post-domestic transition was further reinforced by the 20th-century invention of chemical fertilizers (Fairlie, 2011) and the creation of a meat-based fast food industry for the convenience of chauffeurs (Leroy & Degreef, 2015).

Besides technological and economical drivers, another catalyst was provided by the bourgeois wish for urban life in clean cities, free of the smells and sounds of butchery and its youth-corrupting "scenes of blood and violence" (Leroy & Praet, 2017; Ogle, 2013). In previous periods, slaughter was still performed within the city walls to offer the reassurance that the meat was freshly obtained from disease-free livestock (Ferrières, 2005). Under the mandate of *consumer demand*, meat production was therefore outsourced to a professional minority and animal killing became confined to well-hidden slaughterhouses (Leroy & Degreef, 2015). Not only meat but food production in general was made more invisible since—at least—the 19th century, thus becoming less central as a burden of concern and according to a process that was embraced by progressive forces in society. This evolution led to a serious erosion of traditional rituals and food-related sensibilities, "mystifying the link between the farm and dinner table" (Belasco, 2008). If one compares such modus operandi to the joyful attitudes of the San people celebrating the convulsions of their preys as a blessed moment of food acquisition (Burkert et al., 1987), the immensurability of the epistemic gap becomes obvious.

5.3 Conclusion and Perspectives

The onset of the post-domestic system has been more than just a swing in attitudes and practices; it entailed an entirely novel way of thinking. Or, to put it otherwise, concepts such as carnism and the meat paradox simply would not have been conceivable in pre-industrial epochs. The present study may focus on the case of meat only, but its findings should ultimately be inserted within a broader historical analysis. For instance, they will have to be related to the more inclusive debates on *signifying semiotics* (Deleuze & Guattari, 1987), modernity (Latour, 1993), the food industry (Noakes & Sboros, 2018; Teicholz, 2015), and capitalism (Colebrook, 2002; Vivero-Pol, 2017). Clearly, humans and their aspects of culture, language, thought, and behavior have been continuously mutating into new forms over the past and will continue to do so (Holland, 2013). Such changes are reflected in (and driven by) the various evolving parts of the need-for-meat assemblage. For one, the urgencies of the nutritional and food security strata have deflated in a foodscape of abundancy, even if meat craving may still apply to some degree and if low-quality diets are undermining public health, leading to major health concerns. Also, the once fundamental aspects of commensality and community-shaping now mostly include more mundane variants of hospitality and celebrative consumption. Finally, it is from and within the symbolic stratum that most of the reconfiguration has been done, whereby ritual and spiritual encodings have been superseded by vulgar imagery (Barthes, 1957). Identifying and acknowledging such dynamics is not only needed to define the present but also to investigate its opening, excess, and instability, as well as to anticipate transition (Colebrook, 2002). The central query thereby is whether or not the need for meat has diminished to such an extent that it can be overtaken by novel developing sets of values, desires, and requirements.

6. INSTABILITIES WITHIN THE CURRENT EPISTEME: MEAT AS A PHARMAKON

6.1 Internal Contradictions and Subject Positions

Based on the above, it can be argued that the *need for meat* is not a fixed characteristic of some "human nature"—being a highly problematic concept in the first place—but, rather, a historic unfolding of a much ampler biosocial investment. During humankind's vast hunter-gatherer past this has been essential for both physiological sustenance and communal functioning (Gupta, 2016), whereas within the post-domestic context so-called carnist attitudes seem to have taken over (Joy, 2010). Within any given discursive formation, however, major inconsistencies can always be found (Jansen, 2008). Extensive signification—as in post-domestic societies—typically parallels binarization of thought; it converts differences into fixed oppositions and consolidates subject positions (Davies & Harré, 1990; Deleuze & Guattari, 1987). Binarization acts in a normalizing manner and operates by the logic of yes and no: one either conforms to the norm or not (Holland, 2013). It is within such profoundly normalizing context that the meat pharmakon is situated. Meat is either healthy *or* damaging; therefore, it should be cherished *or* be avoided. As a result, and in addition to majoritarian omnivore attitudes, the following (emerging) subject positions can be identified: vegetarianism and veganism, flexitarianism, permaculturalism, and carnivorism (Table 1). Each position only serves as a template with its own complex in-group heterogeneity (*cf.* Mariotti, 2017), which will not be covered here for the sake of conciseness.

The first of the minoritarian positions listed includes subjects that systematically refuse to eat meat (vegetarianism) or animal products in general (veganism). This attitude has been ascribed to a diversity of reasons that are usually related to concerns about ethics, health, and/or the environment, or to taste preferences (Mariotti, 2017; Ruby, 2012). Besides these

Table 1 Subject Positions Within the Post-domestic Society (Characteristics Are Indicative and May Mask In-Group Heterogeneity)

Subject Position	Societal Proportion	Meat Consumption	Ethical Concerns	Health Beliefs
Omnivores	Majority	Moderate-high	Low-moderate	Part of normal diet
Flexitarians	Substantial	Low	High	Neutral to harmful
Vegan/ vegetarians	Minority	Absent	Very high	Harmful
Permaculturalists	Minority	Moderate-high	High	Health food
(Zero-carb) carnivores	Minority	Very high	Low-moderate	Health food

self-declared motivations, other less explicit forces may be at play such as a penchant for orthorexia (Barthels, Meyer, & Pietrowsky, 2018; Wilson, 2017), emotional bonding with pet animals and anthropomorphism (Heiss & Hormes, 2017; Niemyjska, Cantarero, Byrka, & Bilewicz, 2018), feminist views on meat and patriarchy (Adams, 1990; Belasco, 2008), religious dogma (Sánchez Sábaté, Gelabert, Badilla, & Del Valle, 2016), as well as a search for purity and a fear of death (Testoni, Ghellar, Rodelli, De Cataldo, & Zamperini, 2017; Wilson, 2017). Even if vegetarian and vegan communities have been growing over the last years, the overall movement remains small (Ruby, 2012). In the UK, for instance, vegetarians and vegans are estimated at about 3% and 1% of the adult population, respectively (Ipsos, 2016). Studies sometimes suggest that adherents have specific demographic profiles, more often being female, empathic, and progressive (Ruby, 2012). Yet, such characteristics are not clear-cut within the group of meat avoiders. A French study, for instance, indicated that vegetarians and vegans are more likely to have, respectively, a higher and lower educational level than omnivores (Allès et al., 2017). Vegans also tend to focus more on animal suffering than vegetarians do and see their diets more as a general lifestyle (Mariotti, 2017).

Whereas historical examples of radical meat avoidance date to Antiquity and somewhat gained momentum during the embryonic stages of the post-domestic era in the 19th century (Leroy & Praet, 2017), flexitarianism is a more recently described phenomenon. The word "flexitarian" was proclaimed in 2003 as the most useful novel word of the year by the American Dialect Society (2003). As a sort of semi-vegetarianism, it consists of the eating of a mostly plant-based diet with the occasional inclusion of meat (for instance, during weekends). Flexitarians are less prone to meat-induced disgust than vegans and vegetarians, probably because of their occasional exposure to animal flesh (Rothgerber, 2014). Despite its recent coining, adherents of flexitarianism are rising quickly. In the UK, rough estimates suggest that about one-third to half of the citizens are identifying with the diet to some degree (Ellson & Mancini, 2017; Hinde, 2016).

As a variant of the latter ecological and health-driven trend, a yet poorly mapped permaculturalist movement seems to be developing (Fairlie, 2011). In contrast to flexitarianism, it approaches husbandry from a fundamentally positive rather than negative premise, advocating a post-industrial setup. It generally favors de-industrialization, community supported agriculture, organic farming, a preference for local foods, a strive for improved environmental and public health objectives, animal welfare measures, and a rejection of ultra-processing. The latter is increasingly being cited as a major cause of chronic disease (Fiolet et al., 2018; Monteiro et al., 2018), not only discharging meat as a culprit but also reinstalling its ancient status as health food within traditional diets. A focus on grass-fed beef is discernible, heavily leaning on research that underlines the beneficial ecological role of integrated livestock management (Fairlie, 2011; Smil, 2013; Stanley, Rowntree, Beede, DeLonge, & Hamm, 2018; Teague et al., 2016).

Finally, carnivorism encompasses a recent lifestyle trend, of which the zero-carb variant commonly relies on a diet of only meat (especially beef) and water, which is for instance popular within the bitcoin community (Mann, 2018). It advocates health optimization based on a meat-eating evolutionary past. The movement is still mostly under the academic and even societal radar, so the number of its followers is unknown. Carnivore populations can be found on Twitter (*e.g.*, using hashtags as #meatheals or #carnivore), where narratives mostly refer to the role of meat in health, muscle building, and fitness (Kendall, 2018). It also serves as a strong identarian statement against the detrimental effects of the Western diet, based on the abundant use of refined carbohydrates and seed oils, and the failure of nutritional guidelines (*cf.* Noakes & Sboros, 2018). Although ecological concern is less outspoken within this community, a clear-cut differentiation from permaculturalism is not always straightforward and discourses sometimes overlap.

6.2 Meat in Health and Disease: The Post-truth Context

Leroy, Brengman, et al. (2018) have compiled an overview of all statements on the relationship between meat eating and aspects of health and disease in

British mass media (*i.e.*, the MailOnline) for the first 15 years of the 21st century. By doing so, they sketched the relationship between the narratives and the non-discursive areas (*e.g.*, institutions, health agencies, economic practices, and political structures). The study indicated that news items were either positive (35%), ambiguous (13%), or negative (52%) about the impact of meat on health. In half of the cases, the statements made were supported by references to health agencies, research institutions, or scientific studies. It needs to be pointed out that opinions are divided even within the academic landscape (Tables 2 and 3). Popular literature reflects this debate, for instance in books as "The meat fix: How a lifetime of healthy eating nearly killed me!" (Nicholson, 2012) versus "Dying for a hamburger: Modern meat processing and the epidemic of Alzheimer's disease" (Waldman & Lamb, 2014).

Even if meat has been evoking strength and vitality to most cultures (Leroy & Praet, 2015), its robust image is now seriously beleaguered. The instability of the health image was, nevertheless, already embryonically present at the onset of the post-domestic era, arguably as an intrinsic feature of the epistemic configuration. Starting from a minoritarian position, institutionalization of meat rejection was slow but has gained ground during the last decades. Analysis of this process should take into account that it paralleled several interlinked societal evolutions such as scare stories on adulterated meat, the foundation of animal right movements, the rise of vegetarian societies (often rooted in religious movements, such as the Seventh-day Adventists), and the practice of bourgeois pet keeping (Leroy & Degreef, 2015; Leroy & Praet, 2017). Food safety issues (*e.g.*, Boyd, Jardine, & Driedger, 2009; Thévenot, Dernburg, & Vernozy-Rozand, 2006) have been picked up by both the public and authorities, formalizing them in legislation (Olmstead & Rhode, 2015), but already trace back to the early 20th century (Sinclair, 1906). The same holds true for nutritional concerns: in the slipstream of recent studies and reports that have linked meat to cardiometabolic diseases and (colon) cancer (*e.g.*, International Agency for Research on Cancer, 2015; see also Table 3), dietary guidelines worldwide are now often depicting the consumption of too much (red) meat and meat products as detrimental to health (*cf.* FAO, 2018), despite serious scientific controversy (*e.g.*, Alexander, Weed, Miller, & Mohamed, 2015; Guyatt & Djulbegovic, 2015; Klurfeld, 2015; Kruger & Zhou, 2018; Leroy, Aymerich, et al., 2018; Lippi, Mattiuzzi, & Sanchis-Gomar, 2015; Turner & Lloyd, 2017). Nevertheless, discrete warnings against too copious

Table 2 Meat as a Pharmakon in Peer-Reviewed Academic Literature: An Indicative, Non-exhaustive Overview of Studies That Explicitly Link the Eating of Meat to Health (or Its Avoidance to Disease)

Title of Study	Reference
Vegetarianism produces subclinical malnutrition, hyperhomocysteinemia and atherogenesis	Ingenbleek and McCully (2012)
Red meat in global nutrition	McNeill and Van Elswyk (2012)
Nutrient-rich meat proteins in offsetting age-related muscle loss	Phillips (2012)
Effect of vegetarian diets on zinc status: A systematic review and meta-analysis of studies in humans	Foster, Chu, Petocz, and Samman (2013)
Proxy indicators for identifying iron deficiency among anemic vegetarians in an area prevalent for thalassemia and hemoglobinopathies	Wongprachum et al. (2012)
Iron status and dietary iron intake of vegetarian children from Poland	Gorczyca, Prescha, Szeremeta, and Jankowski (2013)
Meat nutritional composition and nutritive role in the human diet	Pereira and Vicente (2013)
Novel aspects of health promoting compounds in meat	Young et al. (2013)
Protein-enriched diet, with the use of lean red meat, combined with progressive resistance training enhances lean tissue mass and muscle strength and reduces circulating IL-6 concentrations in elderly women: A cluster randomized controlled trial	Daly et al. (2014)
Avoidance of meat and poultry decreases intakes of omega-3 fatty acids, vitamin B12, selenium and zinc in young women	Fayet, Flood, Petocz, and Samman (2014)
Animal source foods have a positive impact on the primary school test scores of Kenyan schoolchildren in a cluster-randomised, controlled feeding intervention trial	Hulett et al. (2014)
Cerebral atrophy in a vitamin B12-deficient infant of a vegetarian mother	Kocaoglu et al. (2014)
Inclusion of red meat in healthful dietary patterns	McNeill (2014)
The prevalence of cobalamin deficiency among vegetarians assessed by serum vitamin B12: A review of literature	Pawlak, Lester, and Babatunde (2014)

Continued

Table 2 Meat as a Pharmakon in Peer-Reviewed Academic Literature: An Indicative, Non-exhaustive Overview of Studies That Explicitly Link the Eating of Meat to Health (or Its Avoidance to Disease)—cont'd

Title of Study	Reference
High protein intake from meats as complementary food increases growth but not adiposity in breastfed infants: A randomized trial	Tang and Krebs (2014)
Vegan diet, subnormal vitamin B-12 status and cardiovascular health	Woo, Kwok, and Celermajer (2014)
Novel insights on intake of meat and prevention of sarcopenia: All reasons for an adequate consumption	Rondanelli et al. (2015)
Neuropsychiatric and neurological problems among vitamin B12 deficient young vegetarians	Kapoor et al. (2017)
Inadequate iodine intake in population groups defined by age, life stage and vegetarian dietary practice in a Norwegian convenience sample	Brantsæter et al. (2018)
Is vegetarianism healthy for children?	Cofnas (2018)
Vegetarian diets and depressive symptoms among men	Hibbeln et al. (2018)
Identification of vitamin B12 deficiency in vegetarian Indians	Naik, Mahalle, and Bhide (2018)

meat eating were already issued before the First World War (Scholliers, 2013) and increased in amplitude after the Second World War, when theories on the association between (saturated) fat and Western disease were being developed (Keys et al., 1980).

In the traditional sociology of scientific knowledge, biological facts are recognized only if it can be said that the scientific community has established closure of its controversies. Yet, the definition of such community is difficult to pinpoint as it now moves beyond conventional academia into a vast network of different interdependent players, including health practitioners, opinion makers, journalists, politicians, etc. (de Vries, 2016). Generally, actors in the political–moral arena are known to be effective in shaping food attitudes, often overruling traditional beliefs (Rozin, 1999). Meat itself, as any nonhuman entity (de Vries, 2016), is not to be neglected as an active constituent of such network: it not only functions as a "study object" but also continuously (re)defines the position of all other (human and non-human) entities involved. Scientific knowledge is thus created *in* rather than

Table 3 Meat as a Pharmakon in Peer-Reviewed Academic Literature: An Indicative, Non-exhaustive Overview of Studies That Explicitly Link the Eating of Meat to Disease (or Its Avoidance to Health)

Title of Study	Reference
Red and processed meat consumption and risk of incident coronary heart disease, stroke, and diabetes mellitus: A systematic review and meta-analysis	Micha, Wallace, and Mozaffarian (2010)
Red and processed meat consumption and risk of stroke: A meta-analysis of prospective cohort studies	Chen, Lv, Pang, and Liu (2013)
Red and processed meat intake and risk of esophageal adenocarcinoma: A meta-analysis of observational studies	Huang, Han, Xu, Zhu, and Li (2013)
Association between total, processed, red and white meat consumption and all-cause, CVD and IHD mortality: A meta-analysis of cohort studies	Abete, Romaguera, Vieira, Lopez de Munain, and Norat (2014)
The health advantage of a vegan diet: Exploring the gut microbiota connection	Glick-Bauer and Yeh (2014)
Red meat and processed meat consumption and all-cause mortality: A meta-analysis	Larsson and Orsini (2014)
Meat consumption and colorectal cancer risk: An evaluation based on a systematic review of epidemiologic evidence among the Japanese population	Pham et al. (2014)
Carcinogenicity of consumption of red and processed meat	Bouvard et al. (2015)
Meat subtypes and their association with colorectal cancer: Systematic review and meta-analysis	Carr, Walter, Brenner, and Hoffmeister (2016)
Red meat consumption and the risk of stroke: A dose-response meta-analysis of prospective cohort studies	Yang et al. (2016)
Red and processed meat consumption and mortality: Dose-response meta-analysis of prospective cohort studies	Wang et al. (2016)
The Strong Heart Study: Adding biological plausibility to the red meat-cardiovascular disease association	Banegas and Rodríguez-Artalejo (2017)
Meat consumption reduction in Italian regions: Health co-benefits and decreases in GHG emissions	Farchi, De Sario, Lapucci, Davoli, and Michelozzi (2017)
Red meat consumption and cardiovascular target organ damage (from the Strong Heart Study)	Haring et al. (2017)

Continued

Table 3 Meat as a Pharmakon in Peer-Reviewed Academic Literature: An Indicative, Non-exhaustive Overview of Studies That Explicitly Link the Eating of Meat to Disease (or Its Avoidance to Health)—cont'd

Title of Study	Reference
Red meat intake is positively associated with non-fatal acute myocardial infarction in the Costa Rica Heart Study	Wang, Campos, and Baylin (2017)
Potential health hazards of eating red meat	Wolk (2017)
Is meat consumption associated with depression? A meta-analysis of observational studies	Zhang et al. (2017)
Meat intake and incidence of cardiovascular disease in Japanese patients with type 2 diabetes: Analysis of the Japan Diabetes Complications Study (JDCS)	Horikawa et al. (2018)
Vegans report less bothersome vasomotor and physical menopausal symptoms than omnivores	Beezhold, Radnitz, McGrath, and Feldman (2018)
Meat intake and risk of diverticulitis among men	Cao et al. (2018)
Meat consumption and risk of metabolic syndrome: Results from the Korean population and a meta-analysis of observational studies	Kim and Je (2018)
High red and processed meat consumption is associated with non-alcoholic fatty liver disease and insulin resistance	Zelber-Sagi et al. (2018)

by society. In the current societal fabric, convergence of opinion on nutritional facts seems to be seriously hampered by multipolarity within the overall epistemological network. Rather than stabilizing into a single "truth," knowledge has dispersed over several conflicting centers of truths, even within academia. Reasons for this are multiple, including the fact that nutritional sciences rely extensively on the unsharp observational data of cross-sectional studies and epidemiological associations. In fact, the problem is systemic: observational claims usually fail to replicate when assessed in clinical trials and lead to oversimplifications by the media and the public (Boffetta et al., 2008; Patel et al., 2015; Young & Karr, 2011). Methodological pitfalls, heterogeneity, and confounders of all sorts make it extremely difficult to generate valid and clear conclusions regarding the effect of either meat eating or vegetarianism on health (*cf.* Abete et al., 2014; Alexander et al., 2015; Carr et al., 2016; Guyatt & Djulbegovic, 2015;

Mariotti, 2017). Low-quality evidence allows for conflicting theories, whereby everything ranging from entire diets over specific foods to single nutrients can be shown to both promote and deteriorate health.

Also, the arrival of what has been named the *post-truth era* and the *attention economy* has been instrumental (Leroy, Brengman, et al., 2018). The latter has been identified as one of the main reasons why multimedia has evolved to sensationalism (Marwick & Lewis, 2017). Due to internet-driven information overload and the link between page views and advertisement income, news items are designed as *click baits*. As a result, fact-checking and qualitative journalism are declining, leading to content that is novel, unsubstantiated, emotional, and thus misleading and disinformative (Fengler & Ruß-Mohl, 2008; Marwick & Lewis, 2017; Starr, 2012). Moreover, this is hardly accompanied by education on true risks and benefits, probabilities, and the relative value of single nutritional studies versus larger bodies of evidence (Rozin, 1999). A generalized "fake news" culture negatively affects public trust in media, as well as in expert opinion (Thorson, 2016). Nutritional authority has been substantially eroded and confidence in institutionalized expertise is historically low (Freidberg, 2016; Harcombe, 2017a, 2017b; Noakes & Sboros, 2018), whereas dietic individualism is on the rise (Fischler, 2015). The public deals with the information overload by either ignoring it, overacting, or developing simplifications, usually by categorizing food as either good *or* bad (Rozin, 1999; Wilson, 2017). It is within such overly complex foodscapes that the *pharmakon* serves as a simplified heuristic.

6.3 Inverting the Carnist Beliefs

Common vegetarian or vegan discourse states that meat is cruel, barbaric, murderous, inhumane, offensive, disgusting, fattening, unhealthy, ecologically destructive, and expensive (Mariotti, 2017). It often starts from values that are based on the rejection of traditionalism, natural order, and hierarchy, which are precisely the societal elements that are heavily symbolized by meat. This is not always the case as, for instance, Seventh-day Adventist vegetarianism pertains to a conservative rather than progressive worldview (Sánchez Sábaté et al., 2016). Novel metanarratives do not only claim that meat eating is harmful for health and the planet (Macdiarmid et al., 2016) but also that it perpetuates hegemonic masculinity, and thus reifies other power hierarchies (Adams, 1990; DeLessio-Parson, 2017). The overall value system for meat avoidance, from which beliefs and attitudes emerge, seems to be a deeply reterritorialized and recoded evolution of the historical need-for-meat model (Fig. 2). Its strata refer to a desire for health (meat is said to damage physiology), ecological protection (meat is said to threaten sustainability), belonging (meat avoidance as a lifestyle-based bonding mechanism), rejection of authority (meat as a perceived cause of oppression), and corresponding symbolism (meat avoidance as a search for identity and *purity*) (Mariotti, 2017). A dedicated study would have to clarify the genealogical background of this new set of beliefs, but it can be tentatively related to a vanishing of monolithic traditional norms and an increasing feeling of anxiety within a globalizing and changing world. The latter induces a status of exile and a search for identity (Belsey, 2002), during which food often serves as a nexus for both belonging and distinctiveness (Geyzen et al., 2012).

From a minoritarian position, story-telling that uses meat avoidance as an identifier establishes itself around inversions of the majoritarian mythologies (Leroy, Brengman, et al., 2018). Contemporary vegan militants fully understand the strategic need for the debunking of meat's aura of prestige, strength, health, and virility, as well for the endorsement of meat avoidance by celebrities (Zaraska, 2016). They aim at both dejecting the positive symbolism of meat and emphasizing its potential harmfulness. Rather than being the *natural* state-of-affairs, meat eating is depicted as a temporary aberration of our original makeup, which is said to be one of compassion (Joy, 2017). In a search for origins, the omnivore history of early humans is thereby superseded by a more remote and pre-ancestral past as fructivores. In this context of origins, fruits and vegetables are glorified as "direct products of the earth" (Testoni et al., 2017). With respect to *necessity*, claims that promote meat for the achievement of nutritional completeness, improved muscle building, and optimal vitality are being countered by anectodical references to vegan (celebrity) athletes (Frazier & Romine, 2017). At times this can be very explicit, as in the book title "Meat is for pussies: A how-to guide for dudes who want to get fit, kick ass, and take names" (Joseph, 2014). More restrained approaches argue that vegan diets *can* be used in sports and exercise, but only if they are well-designed and supplemented with meat-associated compounds such as creatine (Rogerson, 2017). The need for supplementation, in particular of vitamin B12 (Schüpbach, Wegmüller, Berguerand, Bui, & Herter-Aeberli, 2017), remains an element of lingering instability when developing narratives on the naturalness of vegan diets. Finally, a lot of investment in the exaltation of meat-free meals (*niceness*) is discernible, as can be witnessed from the plethora of vegan recipe books, often with voluptuous titles such as "Chloe flavor: Saucy, crispy, spicy, vegan" (Coscarelli, 2018). In contrast, the flavor of meat is explicitly linked to one of blood and gore (Mariotti, 2017).

Within the symbolic realm, meat thus also represents *death* (Hamilton, 2006), whereby the development of disgust relates to the exposure to specific sensorial inputs, such as butchering scenes or explicit body parts, textural and olfactorial perceptions of meat and blood, etc. (Testoni et al., 2017). Subtracting the processes of husbandry and animal slaughter from everyday life has paralleled the emergence of a general feeling of unease and disgust when meat is (graphically) linked to its animal origins, despite the general public's fondness for meat consumption. A cross-cultural comparison of consumers from Ecuador and the United States has shown that a decreased exposure to unprocessed meat (*e.g.*, presence of the head on the carcass) indeed enhances the penchant for vegetarianism based on feelings of disgust (Kunst & Palacios Haugestad, 2018). Somewhat linked is the belief that meat represents arrogance, violence, and *oppression* (Adams, 1990; Hamilton, 2006), as well as bodily and spiritual *impurity* (Testoni et al., 2017). A recurrent question is to which degree veganism is also part of a more general ascetic trend within contemporary societies. A "beginning of the moralization of food indulgence" was already identified by Rozin (1999), whereby preservation of corporeal purity is seen as a moral duty, especially in puritan circles (Sánchez Sábaté et al., 2016). A link between pleasure of eating and health is said to differentiate Americans (food–poison attitude) from French culture (food–pleasure attitude) (Rozin, 1999). In this context, veganism has also been criticized as a prude belief system and a form of body fascism (Wilson, 2017).

6.4 Conclusion and Perspectives

Self-scrutiny and opposition to majoritarian belief can feel as an act of liberation when societal norms have become dysfunctional. Yet, rerouting toward a "truer nature" leads on its turn to the internalization of a novel set of external norms, ethical codes, and power relations (Gutting, 2005). As such, the depreciation of meat is depicted as the new *normal* by rising vegetarian, vegan, and—especially—flexitarian population segments. Nevertheless, it seems premature at this point to speak of a true epistemic shift. Despite the availability of a large body of empirical data, there are still a lot of uncertainties on the various categories of meat avoiders, especially with respect to their numbers and the consistency and heterogeneity of their dietary choices, motivations, beliefs, attitudes, and behaviors (Mariotti, 2017).

Also, defection rates are still important, usually due to a craving for meat, inconvenience, health concerns related to nutrient deficiencies, stigma, and social awkwardness, especially so for those adherents typified by a conservative worldview (Hodson & Earle, 2018). Current societies, therefore, may not be all that willing to robustly absorb radical interventions such as the enforcement of extensive "meat taxes" or veganism (*cf.* Deckers, 2013; FAIRR, 2017), although pleas to move toward a halving of meat consumption are gaining ground (Greenpeace, 2018). More problematic even is the current incapability of imagining a functional framework that would allow for sustainable human–animal interactions, meeting both the ethical concerns related to animal killing and, at the same time, preventing catastrophe in the realms of public health and the environment. With respect to the issue of animal suffering, for instance, drastic interference in nature at large has been proposed by pursuing the genetic engineering of carnivores into Garden-of-Eden-like herbivores or by phasing out wildlife populations altogether, using large-scale sterilization and confinement of the remaining animals to wildlife parks (Moen, 2016).

Whether or not veganism and vegetarianism will eventually reach a dominant, paradigm-changing influence will depend on the possibility to reach a *tipping point*. The latter corresponds with the level of societal representation whereby a rare practice upheld by an inflexible minority rapidly becomes mainstream. As for existing historical precedents (*e.g.*, the suffragette movement or the American civil-rights movement), this is said to happen when the 10% mark of representation is crossed (Xie et al., 2011). Additionally, it is yet unclear how the competing meat-embracing alternatives (permaculturalism and carnivorism) will evolve within the shifting post-domestic model, relative to the anti-meat movements. Williams and Hill (2017), for instance, suggest that the current global crisis should not so much be tackled by a dismantling of meat production than by a sustainable optimization of its production and consumption, as well as by a better redistribution of animal products toward the poor, which they see as an important way to control population size and improve human capital. This line of thought meets the empirical data on cognitive improvement of children in developing countries based on the administration of meat (Hulett et al., 2014). A last point of attention, which was not addressed in the present study but for which we refer to earlier works by Leroy and Praet (2017) and Bryant and Barnett (2018), is the developing research on lab-cultured meat. Several bottlenecks are still preventing its societal implementation but, if successfully implemented, this technology may shape beliefs and attitudes in important ways.

In the meanwhile, a war is raging between the different subject positions to gain sufficient ground and to defend territory; whereby both the consumption and avoidance of meat have been designed as political acts (DeLessio-Parson, 2017). Contradictory views on the necessity of meat eating are at the very heart of the issue (*cf.* Lestel, 2011 versus McWilliams, 2015). The problem, of course, is that authority is always referential and that an incontestable and overarching ruling system that could lead to reconciliation is lacking. In the post-truth era, argumentation is instead driven by contradiction and stereotyping, and the scientific method has become subordinated to the cherry-picking and invention of data (Baggini, 2017). Such incommensurability has been defined by Lyotard (1988) as a *differend*, whereby both parties aim at winning the dispute but rely on different idioms and myths and cannot agree on common criteria for judgment. Convincing the other may not even be the envisaged aim, especially in the case of "bad faith." A quest for ideological "justice" may be more important than the search for "truth." Therefore, the good and the bad are in constant need of actualization (Deleuze & Guattari, 1987). Since human societies rely heavily on ideology, policy makers should endeavor to work toward a set of more "open" beliefs. The designation of food by Rozin (1999) a *fundamental, fun, frightening*, and *far-reaching* may be a good start (Fig. 1).

7. CONCLUSIONS

Meat is a pharmakon: it is cutting across a range of associated categories that are generally seen as mutually exclusive. Ways forward may have to explore the in-between (which is not necessarily the same as working toward a compromise). Meat, as a concept, needs to be profoundly re-configurated. An interesting approach, however theoretical and (arguably) unlikely at this moment in time, would be the re-introduction of animal products within the commons, which can but hearten social engagement and lead to a more broadly shared accountability.

The present study was built largely on concepts developed previously by Derrida (1981), Foucault (1970, 1977, 1979), and Deleuze and Guattari (1987), yet with a substantial amount of liberty concerning jargon and content. We wish to assume that these thinkers would not have objected, as they prioritized creativity above orthodoxy. It is our hope that the present writing may serve as a stimulus for further exploration of an urgent problem: the place of animal husbandry and meat eating in healthy societies, *sensu lato*.

ACKNOWLEDGMENTS

F.L. acknowledges financial support of the Research Council of the Vrije Universiteit Brussel, including the SRP7 and IOF342 projects, and in particular the Interdisciplinary Research Programs "Food quality, safety, and trust since 1950: Societal controversy and biotechnological challenges" (IRP2) and "Tradition and naturalness of animal products within a societal context of change" (IRP11). The author wishes to thank Dr. Istvan Praet (Roehampton University) for his constructive suggestions.

REFERENCES

Abete, I., Romaguera, D., Vieira, A. R., Lopez de Munain, A., & Norat, T. (2014). Association between total, processed, red and white meat consumption and all-cause, CVD and IHD mortality: A meta-analysis of cohort studies. *British Journal of Nutrition*, *112*, 762–775.

Adams, C. (1990). *The sexual politics of meat. A feminist-vegetarian critical theory*. New York: Continuum.

Aiello, L. C., & Wheeler, P. (1995). The expensive-tissue hypothesis: The brain and the digestive system in human and primate evolution. *Current Anthropology*, *36*, 199–221.

Alexander, D. D., Weed, D. L., Miller, P. E., & Mohamed, M. A. (2015). Red meat and colorectal cancer: A quantitative update on the state of the epidemiologic science. *Journal of the American College of Nutrition*, *34*, 521–543.

Allen, P. J. (2012). Creatine metabolism and psychiatric disorders: Does creatine supplementation have therapeutic value? *Neuroscience and Biobehavioral Reviews*, *36*, 1442–1462.

Allès, B., Baudry, J., Méjean, C., Touvier, M., Péneau, S., Hercberg, S., et al. (2017). Comparison of sociodemographic and nutritional characteristics between self-reported vegetarians, vegans, and meat-eaters from the NutriNet-Santé study. *Nutrients*, *9*, 1023.

American Dialect Society. (2003). *Words of the year*. https://www.americandialect.org/2003_words_of_the_year.

Appadurai, A. (1981). Gastro-politics in Hindu South Asia. *American Ethnologist*, *8*, 494–511.

Avieli, N. (2013). Grilled nationalism. Power, masculinity and space in Israeli barbeques. *Food, Culture and Society*, *16*, 301–320.

Baggini, J. (2017). *A short history of truth: Consolations for a post-truth world*. London: Quercus.

Banegas, J. R., & Rodríguez-Artalejo, F. (2017). The Strong Heart Study: Adding biological plausibility to the red meat-cardiovascular disease association. *Journal of Hypertension*, *35*, 1782–1784.

Barr, S. I., & Chapman, G. E. (2002). Perceptions and practices of self-defined current vegetarian, former vegetarian, and non-vegetarian women. *Journal of the American Dietetic Association*, *102*, 354–360.

Barthels, F., Meyer, F., & Pietrowsky, R. (2018). Orthorexic and restrained eating behaviour in vegans, vegetarians, and individuals on a diet. *Eating and Weight Disorders*, *23*, 159–166.

Barthes, R. (1957). *Mythologies*. Paris: Éditions du Seuil.

Beezhold, B., Radnitz, C., McGrath, R. E., & Feldman, A. (2018). Vegans report less bothersome vasomotor and physical menopausal symptoms than omnivores. *Maturitas*, *112*, 12–17.

Belasco, W. (2008). *Food. The key concepts*. Oxford: Berg.

Belsey, C. (2002). *Poststructuralism*. Oxford: Oxford University Press.

Ben-Dor, M., Gopher, A., Hershkovitz, I., & Barkai, R. (2011). Man the fat hunter: The demise of *Homo erectus* and the emergence of a new hominin lineage in the Middle Pleistocene (ca. 400 kyr) Levant. *PLoS One*, *6*, e28689.

Boffetta, P., McLaughlin, J. K., La Vecchia, C., Tarone, R. E., Lipworth, L., & Blot, W. J. (2008). False-positive results in cancer epidemiology: A plea for epistemological modesty. *Journal of the National Cancer Institute, 100*, 988–995.

Bourdieu, P. (1984). *Distinction*. Abingdon: Routledge.

Bouvard, V., Loomis, D., Guyton, K. Z., Grosse, Y., El Ghissassi, F., Benbrahim-Tallaa, L., et al. (2015). Carcinogenicity of consumption of red and processed meat. *Lancet Oncology, 16*, 1599–1600.

Boyd, A. D., Jardine, C. G., & Driedger, S. M. (2009). Canadian media representations of mad cow disease. *Journal of Toxicology and Environmental Health. Part A, 72*, 1096–1105.

Brantsæter, A. L., Knutsen, H. K., Johansen, N. C., Nyheim, K. A., Erlund, I., Meltzer, H. M., et al. (2018). Inadequate iodine intake in population groups defined by age, life stage and vegetarian dietary practice in a Norwegian convenience sample. *Nutrients, 10*, E230.

Bronner, S. E. (2011). *Critical theory*. Oxford: Oxford University Press.

Bryant, C., & Barnett, J. (2018). Consumer acceptance of cultured meat: A systematic review. *Meat Science, 143*, 8–17.

Bulliet, R. W. (2005). *Hunters, herders, and hamburgers. The past and future of human-animal relationships*. New York: Columbia University Press.

Burkert, W., Girard, R., & Smith, J. Z. (1987). *Violent origins: Ritual killing and cultural formation*. Stanford: Stanford University Press.

Cao, Y., Strate, L. L., Keeley, B. R., Tam, I., Wu, K., Giovannucci, E. L., et al. (2018). Meat intake and risk of diverticulitis among men. *Gut, 67*, 466–472.

Carr, P. R., Walter, V., Brenner, H., & Hoffmeister, M. (2016). Meat subtypes and their association with colorectal cancer: Systematic review and meta-analysis. *International Journal of Cancer, 138*, 293–302.

Chen, G. C., Lv, D. B., Pang, Z., & Liu, Q. F. (2013). Red and processed meat consumption and risk of stroke: A meta-analysis of prospective cohort studies. *European Journal of Clinical Nutrition, 67*, 91–95.

Cofnas, N. (2018). Is vegetarianism healthy for children? *Critical Reviews in Food Science and Nutrition*. https://doi.org/10.1080/10408398.2018.1437024. (in press).

Colebrook, C. (2002). *Gilles Deleuze*. London: Routledge.

Cordain, L., Brand Miller, J., Boyd Eaton, S., Mann, N., Holt, S. H. A., & Speth, J. D. (2000). Plant-animal subsistence ratios and macronutrient energy estimations in worldwide hunter-gatherer diets. *American Journal of Clinical Nutrition, 71*, 682–692.

Coscarelli, C. (2018). *Chloe flavor: Saucy, crispy, spicy, vegan*. New York: Clarkson Potter.

Daly, R. M., O'Connell, S. L., Mundell, N. L., Grimes, C. A., Dunstan, D. W., & Nowson, C. A. (2014). Protein-enriched diet, with the use of lean red meat, combined with progressive resistance training enhances lean tissue mass and muscle strength and reduces circulating IL-6 concentrations in elderly women: A cluster randomized controlled trial. *American Journal of Clinical Nutrition, 99*, 899–910.

Davies, B., & Harré, R. (1990). Positioning: The discursive production of selves. *Journal for the Theory of Social Behaviour, 20*, 43–63.

Deckers, J. (2013). In defence of the vegan project. *Bioethical Inquiry, 10*, 187–195.

deFrance, S. D. (2009). Zooarcheology in complex societies: Political economy, status, and ideology. *Journal of Archaeological Research, 17*, 105–168.

DeLanda, M. (2006). *A new philosophy of society: Assemblage theory and social complexity*. London & New York: Continuum.

DeLessio-Parson, A. (2017). Doing vegetarianism to destabilize the meat-masculinity nexus in La Plata, Argentina. *Gender, Place and Culture, 50*, 46Z.

Deleuze, G., & Guattari, F. (1987). *A thousand plateaus*. Minneapolis: University of Minnesota Press.

DeMello, M. (2012). *Animals and society. An introduction to human-animal studies*. New York: Columbia University Press.

Derrida, J. (1981). *Dissemination*. Chicago: University of Chicago Press.

Descola, P. (2005). *Par-delà nature et culture*. Paris: Gallimard.

de Vries, F. (2016). *Bruno Latour*. Cambridge: Polity Press.

Dounias, E., & Ichikawa, M. (2007). Seasonal bushmeat hunger in the Congo Basin. *Nutrition and Dietetics, 64*, S102–S107.

Ellson, A., & Mancini, D. P. (2017). Age of the flexitarian: Millions now only eat meat at weekends. *The Times*. https://www.thetimes.co.uk/article/age-of-the-flexitarian-millions-now-only-eat-meat-at-weekends-k3sbmfrlr.

Fairlie, S. (2011). *Meat: A benign extravagance*. East Meon: Permanent Publications.

FAIRR. (2017). *Climate tax on meat becoming 'increasingly probable'*. http://www.fairr.org/news-item/climate-tax-meat-becoming-increasingly-probable.

FAO. (2018). *Food-based dietary guidelines*. http://www.fao.org/nutrition/nutrition-education/food-dietary-guidelines/en.

Farchi, S., De Sario, M., Lapucci, E., Davoli, M., & Michelozzi, P. (2017). Meat consumption reduction in Italian regions: Health co-benefits and decreases in GHG emissions. *PLoS One, 12*, e0182960.

Fayet, F., Flood, V., Petocz, P., & Samman, S. (2014). Avoidance of meat and poultry decreases intakes of omega-3 fatty acids, vitamin B12, selenium and zinc in young women. *Journal of Human Nutrition and Dietetics, 27*, 135–142.

Fengler, S., & Ruß-Mohl, S. (2008). Journalists and the information-attention markets: Towards an economic theory of journalism. *Journalism, 9*, 667–690.

Ferrières, M. (2005). *Sacred cow, mad cow: A history of food fears*. New York: Columbia University Press.

Fiddes, N. (1991). *Meat. A natural symbol*. London: Routledge.

Fiolet, T., Srour, B., Sellem, L., Kesse-Guyot, E., Allès, B., Méjean, C., et al. (2018). Consumption of ultra-processed foods and cancer risk: Results from NutriNet-Santé prospective cohort. *British Medical Journal, 360*, k322.

Fischler, C. (2001). *L'homnivore. Le gout, la cuisine et le corps*. Paris: Odile Jacob.

Fischler, C. (2015). Is sharing meals a thing of the past? In C. Fischler (Ed.), *Selective eating. The rise, meaning and sense of personal dietary requirements*. Paris: Odile Jacob.

Flinn, M. V., Geary, D. C., & Ward, C. V. (2005). Ecological dominance, social competition, and coalitionary arms races: Why humans evolved extraordinary intelligence. *Evolution and Human Behavior, 26*, 10–46.

Foster, M., Chu, A., Petocz, P., & Samman, S. (2013). Effect of vegetarian diets on zinc status: A systematic review and meta-analysis of studies in humans. *Journal of Science and Food Agriculture, 93*, 2362–2371.

Foucault, M. (1970). *The order of things, an archaeology of the human sciences*. New York: Pantheon Books.

Foucault, M. (1977). *Discipline and punish, the birth of the prison*. New York: Pantheon Books.

Foucault, M. (1979). The history of sexuality: An introduction. *Vol. 1*. London: Allen Lane.

Frassetto, L. A., Schloetter, M., Mietus-Synder, M., Morris, R. C., & Sebastian, A. (2009). Metabolic and physiologic improvements from consuming a paleolithic, hunter-gatherer type diet. *European Journal of Clinical Nutrition, 63*, 947–955.

Frazier, M., & Romine, S. (2017). *The no meat athlete cookbook: Whole food, plant-based recipes to fuel your workouts—And the rest of your life*. New York: The Experiment.

Freidberg, S. (2016). Wicked nutrition: The controversial greening of official dietary guidance. *Gastronomica, 16*, 69–80.

Geyzen, A., Scholliers, P., & Leroy, F. (2012). Innovative traditions in swiftly transforming foodscapes: An exploratory essay. *Trends in Food Science and Technology, 25*, 47–52.

Glick-Bauer, M., & Yeh, M. C. (2014). The health advantage of a vegan diet: Exploring the gut microbiota connection. *Nutrients, 6,* 4822–4838.

Gorczyca, D., Prescha, A., Szeremeta, K., & Jankowski, A. (2013). Iron status and dietary iron intake of vegetarian children from Poland. *Annuals of Nutrition and Metabolism, 62,* 197–291.

Graça, J., Calheiros, M. M., & Oliveira, A. (2014). Moral disengagement in harmful but cherished food practices? An exploration into the case of meat. *Journal of Agricultural and Environmental Ethics, 27,* 749–765.

Greenpeace. (2018). *Halve meat and dairy production to protect climate, nature and health.* http://www.greenpeace.org/eu-unit/en/News/2018/halve-meat-protect-climate-nature-health.

Griffioen-Roose, S., Mars, M., Siebelink, E., Finlayson, G., Tomé, D., & de Graaf, C. (2012). Protein status elicits compensatory changes in food intake and food preferences. *American Journal of Clinical Nutrition, 95,* 32–38.

Gupta, S. (2016). Brain food: Clever eating. *Nature, 531,* S12–S13.

Gutting, G. (2005). *Foucault.* Oxford: Oxford University Press.

Guyatt, G., & Djulbegovic, B. (2015). *Mistaken advice on red meat and cancer.* http://junkscience.com/wp-content/uploads/2015/11/Microsoft-Word-Red-meat-and-cancer_Final.docx-file1.pdf.

Hamilton, M. (2006). Eating death: Vegetarians, meat and violence. *Food, Culture, and Society, 9,* 155–177.

Harcombe, Z. (2017a). Dietary fat guidelines have no evidence base: Where next for public health advice? *British Journal of Sports Medicine, 51,* 769–774.

Harcombe, Z. (2017b). Designed by the food industry for wealth, not health: The "Eatwell Guide". *British Journal of Sports Medicine, 51,* 1730–1731.

Haring, B., Wang, W., Fretts, A., Shimbo, D., Lee, E. T., Howard, B. V., et al. (2017). Red meat consumption and cardiovascular target organ damage (from the Strong Heart Study) *Journal of Hypertension, 35,* 1794–1800.

Hawkes, K. (1991). Showing off tests of a hypothesis about men's foraging goals. *Ethology and Sociobiology, 12,* 29–54.

Heiss, S., & Hormes, J. M. (2017). Ethical concerns regarding animal use mediate the relationship between variety of pets owned in childhood and vegetarianism in adulthood. *Appetite, 123,* 43–48.

Hibbeln, J. R., Northstone, K., Evans, J., & Golding, J. (2018). Vegetarian diets and depressive symptoms among men. *Journal of Affective Disorders, 225,* 13–17.

Hinde, N. (2016). Flexitarians are on rise in UK, with one in three people identifying as 'semi-vegetarian'. *Huffington Post.* http://www.huffingtonpost.co.uk/entry/flexitarians-on-rise-in-uk_uk_57fcbeace4b01fa2b9052e9b.

Hodson, G., & Earle, M. (2018). Conservatism predicts lapses from vegetarian/vegan diets to meat consumption (through lower social justice concerns and social support). *Appetite, 120,* 75–81.

Holland, E. W. (2013). *Deleuze and Guattari's 'a thousand plateaus': A reader's guide.* London: Bloomsbury Academic.

Horikawa, C., Kamada, C., Tanaka, S., Tanaka, S., Araki, A., Ito, H., et al. (2018). Meat intake and incidence of cardiovascular disease in Japanese patients with type 2 diabetes: Analysis of the Japan Diabetes Complications Study (JDCS). *European Journal of Nutrition,* 1–10. https://doi.org/10.1007/s00394-017-1592-y. (in press).

Huang, W., Han, Y., Xu, J., Zhu, W., & Li, Z. (2013). Red and processed meat intake and risk of esophageal adenocarcinoma: A meta-analysis of observational studies. *Cancer Causes and Control, 24,* 193–201.

Hulett, J. L., Weiss, R. E., Bwibo, N. O., Galal, O. M., Drorbaugh, N., & Neumann, C. G. (2014). Animal source foods have a positive impact on the primary school test scores of Kenyan schoolchildren in a cluster-randomised, controlled feeding intervention trial. *British Journal of Nutrition, 111,* 875–886.

Ingenbleek, Y., & McCully, K. S. (2012). Vegetarianism produces subclinical malnutrition, hyperhomocysteinemia and atherogenesis. *Nutrition, 28*, 148–153.

International Agency for Research on Cancer. (2015). *Press release: IARC Monographs evaluate consumption of red meat and processed meat.* https://www.iarc.fr/en/media-centre/pr/2015/pdfs/pr240_E.pdf.

Ipsos. (2016). *Vegan society poll.* https://www.ipsos.com/ipsos-mori/en-uk/vegan-society-poll.

Jansen, I. (2008). Discourse analysis and Foucault's "archaeology of knowledge". *International Journal of Caring Sciences, 1*, 107–111.

Jones, M. (2007). *Feast: Why humans share food.* Oxford: Oxford University Press.

Joseph, J. (2014). *Meat is for pussies: A how-to guide for dudes who want to get fit, kick ass, and take names.* New York: Harper Wave.

Joy, M. (2010). *Why we love dogs, eat pigs, and wear cows. An introduction to carnism.* San Francisco: Conari Press.

Joy, M. (2017). *Beyond beliefs: A guide to improving relationships and communication for vegans, vegetarians, and meat eaters.* Petaluma: Roundtree Press.

Kapoor, A., Baig, M., Tunio, S. A., Memon, A. S., & Karmani, H. (2017). Neuropsychiatric and neurological problems among vitamin B12 deficient young vegetarians. *Neurosciences, 22*, 228–232.

Kendall, M. (2018). *Dr. Shawn Baker's carnivore diet: A review—Optimising nutrition.* https://optimisingnutrition.com/2018/03/14/dr-shawn-bakers-carnivore-diet-a-review.

Keys, A., Aravanis, C., Blackburn, H., Buzina, R., Djordjevic, B. S., Dontas, A. S., et al. (1980). *Seven countries. A multivariate analysis of death and coronary heart disease.* Cambridge: Harvard University Press.

Kim, Y., & Je, Y. (2018). Meat consumption and risk of metabolic syndrome: Results from the Korean population and a meta-analysis of observational studies. *Nutrients, 10*, E390.

Klurfeld, D. M. (2015). Research gaps in evaluating the relationship of meat and health. *Meat Science, 109*, 86–95.

Kocaoglu, C., Akin, F., Caksen, H., Böke, S. B., Arslan, S., & Aygün, S. (2014). Cerebral atrophy in a vitamin B12-deficient infant of a vegetarian mother. *Journal of Health, Population, and Nutrition, 32*, 367–371.

Kohler, T. A., Smith, M. E., Bogaard, A., Feinman, G. M., Peterson, C. E., Betzenhauser, A., et al. (2017). Greater post-Neolithic wealth disparities in Eurasia than in North America and Mesoamerica. *Nature, 551*, 619–622.

Kruger, C., & Zhou, Y. (2018). Red meat and colon cancer: A review of mechanistic evidence for heme in the context of risk assessment methodology. *Food Chemistry and Toxicology, 118*, 131–153. https://doi.org/10.1016/j.fct.2018.04.048.

Kunst, J. R., & Hohle, S. M. (2016). Meat eaters by dissociation: How we present, prepare and talk about meat increases willingness to eat meat by reducing empathy and disgust. *Appetite, 105*, 758–774.

Kunst, J. R., & Palacios Haugestad, C. A. (2018). The effects of dissociation on willingness to eat meat are moderated by exposure to unprocessed meat: A cross-cultural demonstration. *Appetite, 120*, 356–366.

Larsson, S. C., & Orsini, N. (2014). Red meat and processed meat consumption and all-cause mortality: A meta-analysis. *American Journal of Epidemiology, 179*, 282–289.

Latour, B. (1993). *We have never been modern.* Cambridge: Harvard University Press.

Lea, E., & Worsley, A. (2003). Benefits and barriers to the consumption of a vegetarian diet in Australia. *Public Health Nutrition, 6*, 505–511.

Lecerf, J. M. (2015). The dark side of diets: Reason and folly. In C. Fischler (Ed.), *Selective eating. The rise, meaning and sense of personal dietary requirements.* Paris: Odile Jacob.

Leonetti, D. L., & Chabot-Hanowell, B. (2011). The foundation of kinship. *Human Nature, 22*, 16–40.

Leroy, F., Aymerich, T., Champomier-Vergès, M.-C., Cocolin, L., De Vuyst, L., Flores, M., et al. (2018). Fermented meats (and the symptomatic case of the Flemish food pyramid): Are we heading towards the vilification of a valuable food group? *International Journal of Food Microbiology, 274*, 67–70.

Leroy, F., Brengman, M., Ryckbosch, W., & Scholliers, P. (2018). Meat in the post-truth era: Mass media discourses on health and disease in the attention economy. *Appetite, 125*, 345–355.

Leroy, F., & Degreef, F. (2015). Convenient meat and meat products: Societal and technological issues. *Appetite, 94*, 40–46.

Leroy, F., Geyzen, A., Janssens, M., De Vuyst, L., & Scholliers, P. (2013). Meat fermentation at the crossroads of innovation and tradition: A historical outlook. *Trends in Food Science and Technology, 31*, 130–137.

Leroy, F., & Praet, I. (2015). Meat traditions: The co-evolution of humans and meat. *Appetite, 90*, 200–211.

Leroy, F., & Praet, I. (2017). Animal killing and postdomestic meat production. *Journal of Agricultural and Environmental Ethics, 30*, 67–86.

Leroy, F., Scholliers, P., & Amilien, V. (2015). Elements of innovation and tradition in meat fermentation: Conflicts and synergies. *International Journal of Food Microbiology, 212*, 2–8.

Lestel, D. (2011). *Apologie du carnivore*. Paris: Fayard.

Lippi, G., Mattiuzzi, C., & Sanchis-Gomar, F. (2015). Red meat consumption and ischemic heart disease. A systematic literature review. *Meat Science, 108*, 32–36.

Lokuruka, M. N. I. (2006). Meat is the meal and status is by meat: Recognition of rank, wealth, and respect through meat in Turkana culture. *Food and Foodways, 14*, 201–229.

Loughnan, S., Haslam, N., & Bastian, B. (2010). The role of meat consumption in the denial of moral status and mind to meat animals. *Appetite, 55*, 156–159.

Lyotard, J.-F. (1988). *The differend: Phrases in dispute*. Minneapolis: University of Minnesota Press.

Macdiarmid, J. I., Douglas, F., & Campbell, J. (2016). Eating like there's no tomorrow: Public awareness of the environmental impact of food and reluctance to eat less meat as part of a sustainable diet. *Appetite, 96*, 487–493.

Mann, S. (2018). I ate nothing but meat for 2 weeks. Here's what it was like. *Inc.* https://www.inc.com/sonya-mann/carnivory-zero-carb-experiment.html.

Mariotti, F. (2017). *Vegetarian and plant-based diets in health and disease prevention*. London: Academic Press.

Marwick, A., & Lewis, R. (2017). *Media manipulation and disinformation online*. New York: Data and Society Research Institute.

Maslow, A. H. (1943). A theory of human motivation. *Psychological Review, 50*, 370–396.

McNeill, S. H. (2014). Inclusion of red meat in healthful dietary patterns. *Meat Science, 98*, 452–460.

McNeill, S., & Van Elswyk, M. E. (2012). Red meat in global nutrition. *Meat Science, 92*, 166–173.

McWilliams, J. E. (2015). *The modern savage. Our unthinking decision to eat animals*. New York: Thomas Dunne Books.

Micha, R., Wallace, S. K., & Mozaffarian, D. (2010). Red and processed meat consumption and risk of incident coronary heart disease, stroke, and diabetes mellitus: A systematic review and meta-analysis. *Circulation, 121*, 2271–2283.

Moen, O. M. (2016). The ethics of wild animal suffering. *Nordic Journal of Applied Ethics, 10*, 91–104.

Monteiro, C. A., Cannon, G., Moubarac, J. C., Levy, R. B., Louzada, M. L. C., & Jaime, P. C. (2018). The UN decade of nutrition, the NOVA food classification and the trouble with ultra-processing. *Public Health Nutrition, 21*, 5–17.

Morris, B. (1994). Animals as meat and meat as food: Reflections on meat eating in southern Malawi. *Food and Foodways, 6,* 19–41.

Morrison, C. D., Reed, S. D., & Henagan, T. M. (2012). Homeostatic regulation of protein intake: In search of a mechanism. *The American Journal of Physiology—Regulatory, Integrative and Comparative Physiology, 302,* 917–928.

Naik, S., Mahalle, N., & Bhide, V. (2018). Identification of vitamin B12 deficiency in vegetarian Indians. *British Journal of Nutrition, 119,* 629–635.

Nicholson, J. (2012). *The meat fix: How a lifetime of healthy eating nearly killed me!.* London: Biteback Publishing.

Niemyjska, A., Cantarero, K., Byrka, K., & Bilewicz, M. (2018). Too humanlike to increase my appetite: Disposition to anthropomorphize animals relates to decreased meat consumption through empathic concern. *Appetite, 127,* 21–27.

Noakes, T., & Sboros, M. (2018). *Lore of nutrition: Challenging conventional dietary beliefs.* Cape Town: Penguin Random House South Africa.

Ogle, M. (2013). *In meat we trust: An unexpected history of carnivore America.* New York: Houghton-Mifflin-Harcourt.

Olmstead, A. L., & Rhode, P. (2015). *Arresting contagion: Science, policy, and conflicts over animal disease control.* Cambridge: Harvard University Press.

Patel, C. J., Burford, B., & Ioannidis, J. P. A. (2015). Assessment of vibration of effects due to model specification can demonstrate the instability of observational associations. *Journal of Clinical Epidemiology, 68,* 1046–1058.

Pawlak, R., Lester, S. E., & Babatunde, T. (2014). The prevalence of cobalamin deficiency among vegetarians assessed by serum vitamin B12: A review of literature. *European Journal of Clinical Nutrition, 68,* 541–548.

Pereira, P. M., & Vicente, A. F. (2013). Meat nutritional composition and nutritive role in the human diet. *Meat Science, 93,* 586–592.

Pfefferle, A. D., Warner, L. R., Wang, C. W., Nielsen, W. J., Babbitt, C. C., Fedrigo, O., et al. (2011). Comparative expression analysis of the phosphocreatine circuit in extant primates: Implications for human brain evolution. *Journal of Human Evolution, 60,* 205–212.

Pham, N. M., Mizoue, T., Tanaka, K., Tsuji, I., Tamakoshi, A., Matsuo, K., et al. (2014). Meat consumption and colorectal cancer risk: An evaluation based on a systematic review of epidemiologic evidence among the Japanese population. *Japanese Journal of Clinical Oncology, 44,* 641–650.

Phillips, S. M. (2012). Nutrient-rich meat proteins in offsetting age-related muscle loss. *Meat Science, 92,* 174–178.

Piazza, J., Ruby, M. B., Loughnan, S., Luong, M., Kulik, J., Watkins, H. M., et al. (2015). Rationalizing meat consumption. The 4Ns. *Appetite, 91,* 114–128.

Pilcher, J. (2005). *Food in world history.* London, UK: Routledge.

Previc, F. H. (2009). *The dopaminergic mind in human evolution and history.* Cambridge: Cambridge University Press.

Rae, C., Digney, A. L., McEwan, S. R., & Bates, T. C. (2003). Oral creatine monohydrate supplementation improves brain performance: A double-blind, placebo-controlled, cross-over trial. *Proceedings of the Royal Society B: Biological Sciences, 270,* 2147–2215.

Renton, A. (2013). *Planet carnivore.* London: Guardian Books.

Rogerson, D. (2017). Vegan diets: Practical advice for athletes and exercisers. *Journal of the International Society of Sports Nutrition, 14,* 36.

Rondanelli, M., Perna, S., Faliva, M. A., Peroni, G., Infantino, V., & Pozzi, R. (2015). Novel insights on intake of meat and prevention of sarcopenia: All reasons for an adequate consumption. *Nutrición Hospitalaria, 32,* 2136–2143.

Ross, E. B. (1987). An overview of trends in dietary variation from hunter-gatherer to modern capitalist societies. In M. Harris & E. B. Ross (Eds.), *Food and evolution: Toward a theory of human food habits* (pp. 19–23). Philadelphia: Temple University Press.

Rothgerber, H. (2014). A comparison of attitudes toward meat and animals among strict and semi-vegetarians. *Appetite, 72,* 98–105.

Rozin, P. (1999). Food is fundamental, fun, frightening, and far-reaching. *Social Research, 66,* 9–30.

Ruby, M. B. (2012). Vegetarianism. A blossoming field of study. *Appetite, 58,* 141–150.

Ruby, M. B., & Heine, S. J. (2011). Meat, morals, and masculinity. *Appetite, 56,* 447–450.

Sánchez Sábaté, R., Gelabert, R., Badilla, Y., & Del Valle, C. (2016). Feeding holy bodies: A study on the social meanings of a vegetarian diet to Seventh-day Adventist church pioneers. *HTS Theological Studies, 72,* a3080.

Scholliers, P. (2013). Food recommendations in domestic education, Belgium 1890–1940. *Paedagogica Historica, 49,* 645–663.

Schüpbach, R., Wegmüller, R., Berguerand, C., Bui, M., & Herter-Aeberli, I. (2017). Micronutrient status and intake in omnivores, vegetarians and vegans in Switzerland. *European Journal of Nutrition, 56,* 283–293.

Seleshe, S., Jo, C., & Lee, M. (2014). Meat consumption culture in Ethiopia. *Korean Journal for Food Science of Animal Resources, 34,* 7–13.

Sinclair, U. (1906). *The jungle.* New York: Doubleday, Jabber & Company.

Smil, V. (2013). *Should we eat meat? Evolution and consequences of modern carnivory.* Chichester, UK: Wiley-Blackwell.

Stanford, C. B. (1999). *The hunting apes: Meat eating and the origin of human behavior.* Princeton: Princeton University Press.

Stanford, C. B., & Bunn, H. (2001). *Meat-eating and human evolution.* Oxford: Oxford University Press.

Stanley, P. L., Rowntree, J. E., Beede, D. K., DeLonge, M. S., & Hamm, M. W. (2018). Impacts of soil carbon sequestration on life cycle greenhouse gas emissions in Midwestern USA beef finishing systems. *Agricultural Systems, 162,* 249–258.

Starr, P. (2012). An unexpected crisis: The news media in postindustrial democracies. *International Journal of Press/Politics, 17,* 234–242.

Suzman, J. (2017). Why 'Bushman banter' was crucial to hunter-gatherers' evolutionary success. *The Guardian online.* https://www.theguardian.com/inequality/2017/oct/29/why-bushman-banter-was-crucial-to-hunter-gatherers-evolutionary-success?.

Tang, M., & Krebs, N. F. (2014). High protein intake from meats as complementary food increases growth but not adiposity in breastfed infants: A randomized trial. *The American Journal of Clinical Nutrition, 100,* 1322–1328.

Teague, W. R., Apfelbaum, S., Lal, R., Kreuter, U. P., Rowntree, J., Davies, C. A., et al. (2016). The role of ruminants in reducing agriculture's carbon footprint in North America. *Journal of Soil and Water Conservation, 71,* 156–164.

Teicholz, N. (2015). How dietary guidelines are out of step with science. *British Medical Journal, 351,* h4962.

Testoni, I., Ghellar, T., Rodelli, M., De Cataldo, L., & Zamperini, A. (2017). Representations of death among Italian vegetarians: An ethnographic research on environment, disgust and transcendence. *European Journal of Psychology, 13,* 378–395.

Thévenot, D., Dernburg, A., & Vernozy-Rozand, C. (2006). An updated review of *Listeria monocytogenes* in the pork meat industry and its products. *Journal of Applied Microbiology, 101,* 7–17.

Thorson, E. (2016). Belief echoes: The persistent effects of corrected misinformation. *Political Communication, 33,* 460–480.

Tomasello, M., Melis, A. P., Tennie, C., Wyman, E., & Herrmann, E. (2012). Two key steps in the evolution of human cooperation: The interdependence hypothesis. *Current Anthropology, 53,* 673–692.

Turner, N. D., & Lloyd, S. K. (2017). Association between red meat consumption and colon cancer: A systematic review of experimental results. *Experimental Biology and Medicine, 242,* 813–839.

Turner, B. L., & Thompson, A. L. (2013). Beyond the Paleolithic prescription: Incorporating diversity and flexibility in the study of human diet evolution. *Nutrition Reviews, 71*, 501–510.

Visser, M. (2003). Etiquette and eating habits. In S. Katz (Ed.), *Encyclopaedia of food and culture* (pp. 586–592). New York: Scribner's.

Vivero-Pol, J. L. (2017). The idea of food as commons or commodity in academia. A systematic review of English scholarly texts. *Journal of Rural Studies, 53*, 182–201.

Waldman, M., & Lamb, M. (2014). *Dying for a hamburger: Modern meat processing and the epidemic of Alzheimer's disease.* New York: Thomas Dunne Books.

Wang, D., Campos, H., & Baylin, A. (2017). Red meat intake is positively associated with non-fatal acute myocardial infarction in the Costa Rica Heart Study. *British Journal of Nutrition, 118*, 303–311.

Wang, X., Lin, X., Ouyang, Y. Y., Liu, J., Zhao, G., Pan, A., et al. (2016). Red and processed meat consumption and mortality: Dose-response meta-analysis of prospective cohort studies. *Public Health Nutrition, 19*, 893–905.

Williams, A. C., & Hill, L. J. (2017). Meat and nicotinamide: A causal role in human evolution, history, and demographics. *International Journal of Tryptophan Research, 10.* 1178646917704661.

Wilson, B. (2017). Why we fell for clean eating. *The Guardian online.* https://www.theguardian.com/lifeandstyle/2017/aug/11/why-we-fell-for-clean-eating.

Wolk, A. (2017). Potential health hazards of eating red meat. *Journal of Internal Medicine, 281*, 106–122.

Wongprachum, K., Sanchaisuriya, K., Sanchaisuriya, P., Siridamrongvattana, S., Manpeun, S., & Schlep, F. P. (2012). Proxy indicators for identifying iron deficiency among anemic vegetarians in an area prevalent for thalassemia and hemoglobinopathies. *Acta Haematologica, 127*, 250–255.

Woo, K. S., Kwok, T. C., & Celermajer, D. S. (2014). Vegan diet, subnormal vitamin B-12 status and cardiovascular health. *Nutrients, 6*, 3259–3273.

Xie, J., Sreenivasan, S., Korniss, G., Zhang, W., Lim, C., & Szymanski, B. K. (2011). Social consensus through the influence of committed minorities. *Physical Review E, 84*, 011130.

Yang, C., Pan, L., Sun, C., Xi, Y., Wang, L., & Li, D. (2016). Red meat consumption and the risk of stroke: A dose-response meta-analysis of prospective cohort studies. *Journal of Stroke and Cerebrovascular Diseases, 25*, 1177–1186.

Young, E. B., Genosko, G., & Watson, J. (2013). *The Deleuze and Guattari dictionary.* London & New York: Bloomsbury.

Young, S. S., & Karr, A. (2011). Deming, data and observational studies. *Significance, 8*, 116–120.

Young, J. F., Therkildsen, M., Ekstrand, B., Che, B. N., Larsen, M. K., Oksbjerg, N., et al. (2013). Novel aspects of health promoting compounds in meat. *Meat Science, 95*, 904–911.

Zaraska, M. (2016). *Meathooked: The history and science of our 2.5-million-year obsession with meat.* New York: Basic Books.

Zelber-Sagi, S., Ivancovsky-Wajcman, D., Fliss Isakov, N., Webb, M., Orenstein, D., Shibolet, O., et al. (2018). High red and processed meat consumption is associated with non-alcoholic fatty liver disease and insulin resistance. *Journal of Hepatology, 68*, 1239–1246. https://doi.org/10.1016/j.jhep.2018.01.015.

Zhang, Y., Yang, Y., Xie, M. S., Ding, X., Li, H., Liu, Z. C., et al. (2017). Is meat consumption associated with depression? A meta-analysis of observational studies. *BMC Psychiatry, 17*, 409.